PERSPECTIVES IN ETHOLOGY

Volume 12

Communication

CONTRIBUTORS

Jeffrey R. Alberts
Department of Psychology
Indiana University
Bloomington, Indiana 47405

Michael D. Beecher
Animal Behavior Program
Departments of Psychology and Zoology
University of Washington
Seattle, Washington 98195

Mark S. Blumberg
Department of Psychology
University of Iowa
Iowa City, Iowa 52242

John M. Burt
Animal Behavior Program
Departments of Psychology and Zoology
University of Washington
Seattle, Washington 98195

S. Elizabeth Campbell
Animal Behavior Program
Departments of Psychology and Zoology
University of Washington
Seattle, Washington 98195

Marian Stamp Dawkins
Department of Zoology
University of Oxford
Oxford
England

Christopher S. Evans
Department of Psychology
Macquarie University
Sydney, NSW 2109
Australia

Michael D. Greenfield
Department of Entomology
University of Kansas
Lawrence, Kansas 66045

Tim Guilford
Department of Zoology
University of Oxford
Oxford
England

William J. Hamilton III
Division of Environmental Studies
University of California, Davis
Davis, California 95616

Christopher E. Hill
Animal Behavior Program
Departments of Psychology and Zoology
University of Washington
Seattle, Washington 98195

Andrew G. Horn
Department of Biology
Dalhousie University
Halifax, Nova Scotia B3H 4J1
Canada

John W. NcNutt
Wild Dog Research Project
Private Bag 13
Maun
Botswana

Eugene S. Morton
National Zoological Park
Smithsonian Institution
Washington, D.C. 20008

J. Cully Nordby
Animal Behavior Program
Departments of Psychology and Zoology
University of Washington
Seattle, Washington 98195

Adrian L. O'Loghlen
Animal Behavior Program
Departments of Psychology and Zoology
University of Washington
Seattle, Washington 98195

Donald H. Owings
Department of Psychology
University of California, Davis
Davis, California 95616

Michael J. Owren
Department of Psychology
Reed College
3202 SE Woodstock Boulevard
Portland, Oregon 97202

Drew Randall
Department of Psychology
University of Pennsylvania
Philadelphia, Pennsylvania 19104

W. John Smith
Department of Biology
University of Pennsylvania
Philadelphia, Pennsylvania 19104-6018

Nicholas S. Thompson
Departments of Biology and Psychology
Clark University
950 Main Street
Worcester, Massachusetts 01610-1477

Peter L. Tyack
Biology Department
Woods Hole Oceanographic Institution
Woods Hole, Massachusetts 02543-1049

A Continuation Order Plan is available for this series. A continuation order will bring delivery of each new volume immediately upon publication. Volumes are billed only upon actual shipment. For further information please contact the publisher.

PERSPECTIVES IN ETHOLOGY

Volume 12

Communication

Edited by

Donald H. Owings
University of California, Davis
Davis, California

Michael D. Beecher
University of Washington
Seattle, Washington

and

Nicholas S. Thompson
Clark University
Worcester, Massachusetts

Springer Science+Business Media, LLC

The Library of Congress has cataloged this title as follows:

Perspectives in ethology.-Vol. 1–New York: Plenum Press, 1973–
v.: ill.; 24 cm.

Irregular.

Includes bibliographies and indexes.
Editors: v. 1– P.P.G. Bateson and P.H. Klopfer.
ISSN 0738-4394 = Perspectives in ethology.

1. Animal behavior—Collected works. I. Bateson. P.P.G. (Paul Patrick Gordon), 1938– II. Klopfer, Peter H. [DNLM: W1 PE871AN]

QL750.P47 591.5'.1—dc19 86-649219

AACR 2 MARC-S

Library of Congress [8610]

ISBN 978-1-4899-1747-8 ISBN 978-1-4899-1745-4 (eBook)
DOI 10.1007/978-1-4899-1745-4

Originally published by Plenum Publishing Corporation in 1997
Softcover reprint of the hardcover 1st edition 1997

10 9 8 7 6 5 4 3 2 1

Preface

ON THE FUTURE OF *PERSPECTIVES*

When Patrick Bateson and Peter Klopfer offered me the editorship of *Perspectives* in 1992, the world of academic publishing was in one of its periodic upheavals. Subscriptions to series—even distinguished series such as *Perspectives*—had been declining and individual volume prices had been rising, a trend that if continued could only result in the series pricing itself out of the market. In the course of the negotiations around the change of editors, the publishers offered a cost-cutting solution: change the production pattern to "camera ready" and eliminate the costs of indexing and proofreading. While I could see the sense in this proposal, I was reluctant to accept it. Part of what I had always liked about the volumes in this series was that they were real books, intelligently proofread, nicely laid out, and provided with proper indexes. Thus, I in return offered a "Devil's bargain": the publisher should maintain the present quality of the series for two more volumes and make a renewed effort to advertise the series to our ethological and sociobiological colleagues, while I as the new series editor committed myself to a renewed effort to make *Perspectives* the publication of choice for writers who are trying to get their message out to the world intact and readers who are seeking clear, coherent, comprehensive and untrammeled presentations of authors' ideas and research programs. If these efforts did not raise readership substantially, we would agree to bring the series to a graceful end after its 12th volume.

The volume you now hold in your hand is that 12th volume. Certainly, the publisher has held to its side of the bargain and I hope you will think that the series editor has held to his. For this volume, on communication, I have recruited two highly qualified volume editors, Donald Owings and Michael Beecher, and they in turn have recruited some of the most interesting and imaginative authorities in this strange and intricate field.

As you peruse these chapters, please remember the odd contract that exists between a *Perspectives* editor and a *Perspectives* author. Once the author has been offered and has accepted a place in the volume, the editor's role is to assist that author in making the writing as clear, internally consistent, and persuasive as possible. Never is an author required to alter the text to accord with the beliefs of the editor or any other reviewer. If an author chooses to hang himself in the pages of *Perspectives*, the editor's sole commitment is to make sure that he hangs himself good! Thus, what has made the contributions to *Perspectives* so unique over its 11 volumes is that authors have not had to chart a course between hostile reviews by Dr. Scylla on the one hand and Professor Charybdis on the other but have been able to sail directly toward their own, personally chosen, destination.

What of the future of this experiment? That depends. This place is normally where I would announce the topic and the volume editors for the next volume. I have many ideas about future topics and about future editors to recruit authors and organize volumes on these topics. I particularly think the time has come to evaluate the relationship between the new and burgeoning field of evolutionary psychology and traditional concepts of ethology. But this and other plans must await the response of the readership to volumes 11 and 12. I hope you will let me know what you think.

Nicholas S. Thompson
West Road
New Braintree, Massachusetts

CONTENTS

Chapter 4

REFERENTIAL SIGNALS

Christopher S. Evans

Chapter 5

SEXUAL SELECTION AND THE EVOLUTION OF ADVERTISEMENT SIGNALS

Michael D. Greenfield

Chapter 7

INCIDENTAL EMISSIONS, FORTUITOUS EFFECTS, AND THE ORIGINS OF COMMUNICATION

Mark S. Blumberg and Jeffrey R. Alberts

Chapter 8

STUDYING HOW CETACEANS USE SOUND TO EXPLORE THEIR ENVIRONMENT

Peter L. Tyack

Chapter 12

COMMUNICATION AND NATURAL DESIGN

Nicholas S. Thompson

Introduction

PERSPECTIVES ON COMMUNICATION

Donald H. Owings

Department of Psychology
University of California
Davis, California 95616

Michael D. Beecher

Animal Behavior Program
Departments of Psychology and Zoology
University of Washington
Seattle, Washington 98195

Nicholas S. Thompson

Departments of Biology and Psychology
Clark University
950 Main Street
Worcester, Massachusetts 01610-1477

In the 1950s-60s, an information-exchange conception of animal communication arose out of ethological interests in the causation and evolution of signalling behavior (Marler, 1961; Smith, 1963). This traditional approach was challenged in the 1970s by evolutionary approaches founded more explicitly on the logic of natural selection (Maynard Smith & Price, 1973; Dawkins & Krebs, 1978; Caryl, 1979) . This set of adaptive approaches is now 20+ years old, and so has been around long enough to have achieved its own status as a traditional perspective on the study of animal communication.

After a period of ambiguous confrontation between these approaches, some writers came to the conclusion that the two are not true antagonists. In fact, given that the former is proximate and the latter ultimate, the two may be seen as natural complements (Markl, 1985; Smith, 1986). The fusion of these two approaches formed a modern synthesis that combined proximate questions about how mechanisms of information exchange serve the day-to-day goals of individual animals (Cheney & Seyfarth, 1985; Marler, Dufty & Pickert, 1986; Smith, 1991; Macedonia & Evans, 1993) with ultimate questions about how these mechanisms have been favored by natural and sexual selection during the history of the species (Zahavi, 1991; Guilford & Dawkins, 1991; Wiley, 1994; Hauser, 1996).

The authors of this volume take a variety of perspectives on these two approaches and on the desirability of attempting a synthesis between them. Chapters 1 and 2 take complementary looks at the modern synthesis, the first expanding an information-exchange approach partly in light of the logic of natural selection, and the second enriching our understanding of the evolution of signals by looking in more detail at the proximate mechanisms through which they must work. In chapter 1, "The Behavior of Communicating, after Twenty Years," Smith discusses the informational/interactional approach to communication developed in his 1977 book, *The Behavior of Communicating*, in light of the subsequent literature. His focus remains on information exchange and the complex social context of communication, but he also discusses the implications for communication of evolutionary, cognitive and social-role considerations. In chapter 2, "Conspicuousness and Diversity in Animal Signals," Dawkins and Guilford join others in a call for a synthesis of the approaches of behavioral ecologists and of students of the proximate mechanisms of communication. Their particular emphasis is on "extravagant" signals that have traditionally been viewed as conveying information about quality. They propose that this extravagance may have evolved in many cases to enhance the impact of signals on the neural and psychological mechanisms of targets, rather than to emphasize the messages of signals. According to their approach, it is the properties of these proximate mechanisms that are the sources of selection for signal efficacy in managing target behavior.

Chapters 3–6 provide novel insights into the functional significance of communicative behavior in part by taking a closer look at the proximate details of interactions among individuals. In chapter 3, "What Is the Function of Song Learning in Songbirds?," Beecher et al discuss a communication system, birdsong, which is unusual in two important respects: signals are learned and are idiosyncratic to individual singers. Their field studies of song sparrows have identified several design features of the song development system which serve to maximize the number of songs the young bird will share with his neighbors, especially his near neighbors, in his first breeding season. Although not their central point, they suggest that one function of a bird's singing may be to teach young birds his particular songs. They suggest that song tutoring is an alternative to song learning, since either route serves the same function of equipping both sender and receiver with the same songs. In chapter 4, "Referential Signals," Evans explores the extent and limits of our knowledge about this category of signals. Signals are said to be referential if they transfer information to targets about specific external objects and events, rather than about properties of the signaller itself. This chapter calls for more integration of these considerations of the mechanisms of communication with the broader proximate and ultimate literature on communication. For example, Evans notes, as Dawkins and Guilford did, that limited insights into signal structure are provided by studying the referents of signals. Other properties of the signal must be considered, including the possibility that conspecifics are not the targets of all alarm calls (predators may be instead), and that a signal may function tonically, rather than phasically. In chapter 5, "Sexual Selection and the Evolution Of Advertisement Signals," Greenfield reviews the various hypotheses about the evolution of sexually-selected signals, and discusses how one might go about acquiring data that distinguish among those hypotheses. He finishes with case studies that illustrate possible research directions, and their limitations. This chapter includes discussion of Greenfield's work on the role of proximate signal-timing mechanisms in the evolution of both competitive and cooperative patterns of coordinated advertisement signalling behavior by males. In chapter 6, "Determinants of Conflict Outcomes," Hamilton and McNutt discuss the limits of purely evolutionary game-theoretic models for making sense of the outcomes of conflicts between individual animals. They argue that an animal's behavior in situations of conflict is always founded on its estimates of its probability of winning, which are based heavily on its past experience, and that, game-theoretic predictions notwithstanding, settlement on the basis of arbitrary conventions is unlikely. Their model of conflict behavior is derived from the empirical literature on animal conflict, and includes a host of proximate influences on fighting ability and motivation. The dynamic nature of fighting ability and motivation in this model leaves room for communication as one means of updating estimates of the probability of winning.

Chapters 7 and 8 explore noncommunicative processes and the insights they provide into the origins and functioning of communication systems. In chapter 7, "Incidental Emissions, Fortuitous Effects, And The Origins Of Communication," Blumberg and Alberts call for more attention to proximate mechanisms in functional studies of communication. They describe how vocalizations can be produced as by-products of respiration, noting that actively assessing individuals may respond predictably to such cues, even though the sound itself has not been specialized for communicative function. By assuming function and ignoring mechanism, they argue, researchers have arrived at a distorted view of the communicative relations between signallers and targets, and denied themselves a path to understanding the evolutionary origins of communication. In chapter 8, "Studying How Cetaceans Use Sound to Explore Their Environment," Tyack explores how the opportunism of animals can generate new functions for old activities, resulting in a sharing of components by echolocation and communication systems. Cetaceans may, for example, detect concentrations of prey by attending to the distortions such prey groups induce in the structure of acoustic signals reaching them from conspecifics. Similarly, the vocal plasticity evolved for echolocation may have preadapted cetaceans for plasticity in their communicative vocalizations.

Chapters 9–12 each challenge the proximate/ultimate synthesis in some way. In chapter 9, "An Affect-conditioning Model of Nonhuman Primate Vocal Signaling," Owren and Rendall call for expansion of the modern synthesis to include consideration of affective effects on targets of primate vocal signals. According to this model, vocalizations influence the behavior of others through both unconditioned and conditioned affective effects. Individuality in vocal structure does not function to make information available about caller identity; instead, individuality allows for conditioned effects of signals that are specific to the relationship between two individuals. For example, "threat" calls from a dominant individual, A, have in the past been followed by an actual attack, and so through classical conditioning, now elicit conditioned reactions of fear and withdrawal by the subordinate target, B. Because of the individual distinctiveness of threat calls, the same type of threat call from another individual, C, will not have the same effect on B if it has not been paired with attack. In chapter 10, "Speech Acts and Animal Signals," Horn notes that the speech-act literature in linguistics provides a better model for the study of animal communication than traditional linguistic models founded on the view that communication consists of the exchange of true or false propositions. Like speech acts, animal signals are emitted most fundamentally to produce some proximate social effect. This chapter turns typical thinking about the roles of truth, fact and information in communication on its head. Truth, fact and information are ultimate validators of signals rather than proximate mediators of their impact on others. In chapter 11, "The Role of Information in Communication: Assessment/Management Approach," Owings and Morton offer an alternative to the prevailing information-exchange proximate framework. According to this approach, animals are self-interested regulatory systems, meeting their needs in part by regulating, or managing, the behavior of others. Proximately, signals work by exploiting the couplings maintained by the active assessment systems of others, not by making information available. Owings and Morton argue that this pragmatic proximate framework provides maximum compatibility with an ultimate approach founded on the logic of natural selection. In chapter 12, "Communication and Natural Design," Thompson too raises doubts about the informational approach, but from a more philosophical perspective. He argues that it is inappropriate to suppose that the sender is transmitting information to the receiver—or misinformation, for that matter—because "there is no particular reason to believe that the sender has information about what he is disposed to do that is in principle better than information possessed by his antagonist." But he also takes on the manipulation and management perspectives, contending that they rely on the notion that information is used to manipulate and to manage conspecifics. Thompson's position is in essence an argument against mentalism, which he believes is implicit to varying degrees in most contemporary writing on animal communication. The goal of the natural design perspective is to capture the richness of the structure of animal interactions without resort to implicit mentalism.

The perspectives on communication provided by the chapters in this volume imply some prescriptions for future research in animal communication. If our contributors have it right, the following predictions will be borne out. (1) Proximate and ultimate questions about communication will more frequently be addressed in the context of each other, rather than independently. (2) Further exploration of the causal bases of communication will enhance our insights into the developmental and evolutionary origins of communicative systems. (3) Studies of the fitness consequences of communicative behavior should continue and expand. In addition, however, we will pay more attention to the developmental and real-time effects of communicative behavior that mediate those fitness consequences, and to the resulting proximate and ultimate changes in behavior. (4) Proximate researchers will be weaned from their dependence on information exchange as the mediator of communication. (5) The content of proximate studies will expand from the current emphasis on physiological and cognitive mechanisms of individuals to more completely encompass questions about the motivational and emotional bases of communication. This is far from an exhaustive listing of the trends likely to be evident in future studies of animal communication. Nevertheless, these predictions provide stimulating points of reference for those who will be following and contributing to that literature.

BIBLIOGRAPHY

Caryl, P. G. (1979). Communication by agonistic displays: what can games theory contribute to ethology? *Behaviour, 68,* 136–69.

Cheney, D. L. & Seyfarth, R. M. (1985). Vervet monkey alarm calls: manipulation through shared information? *Behaviour, 94,* 150–66.

Dawkins, R. & Krebs, J. R. (1978). Animal signals: Information or manipulation? In J. R. Krebs & N. B. Davies (Eds.) *Behavioural Ecology: An Evolutionary Approach*, (pp. 282–309). Sunderland: Sinauer Associates.

Guilford, T. & Dawkins, M. S. (1991). Receiver psychology and the evolution of animal signals. *Animal Behaviour, 42,* 1–14.

Hauser, M. D. (1996). *The Evolution of Communication.* Cambridge: MIT Press.

Macedonia, J. M. & Evans, C. S. (1993). Variation among mammalian alarm call systems and the problem of meaning in animal signals. *Ethology, 93,* 177–97.

Markl, H. (1985). Manipulation, modulation, information, cognition: Some of the riddles of communication. In B. Holldobler & M. Lindauer (Eds.), *Experimental Behavioral Ecology and Sociobiology* (pp. 163–94). Sunderland: Sinauer Associates.

Marler, P. (1961). The logical analysis of animal communication. *Journal of Theoretical Biology, 1,* 295–317.

Marler, P., Dufty, A. & Pickert, R. (1986). Vocal communication in the domestic chicken: I. Does a sender communicate information about the quality of a food referent to a receiver? *Animal Behaviour, 34,* 188–93.

Maynard Smith, J. & Price, G. R. (1973). The logic of animal conflict. *Nature, 246,* 15–8.

Smith, W. J. (1963). Vocal communication of information in birds. *American Naturalist, 97,* 117–26.

Smith, W. J. (1986). An “informational” perspective on manipulation. In R. W. Mitchell & N. S. Thompson (Eds.) *Deception: Perspectives on Human and Nonhuman Deceit* (pp. 71–86). Albany: State University of New York Press.

Smith, W. J. (1991). Animal communication and the study of cognition. In C. A. Ristau (Ed.), *Cognitive Ethology: The Minds of Other Animals* (pp. 209–30). Hillsdale: Lawrence Erlbaum Associates.

Wiley, R. H. (1994). Errors, exaggeration, and deception in animal communication. In L. A. Real (Ed.), *Behavioral Mechanisms in Evolutionary Ecology* (pp. 157–89). Chicago: University of Chicago Press.

Zahavi, A. (1991). On the definition of sexual selection, Fisher's model, and the evolution of waste and of signals in general. *Animal Behaviour, 42,* 501–3.

Chapter 1

THE BEHAVIOR OF COMMUNICATING, AFTER TWENTY YEARS

W. John Smith

Department of Biology
University of Pennsylvania
Philadelphia, Pennsylvania 19104-6018

ABSTRACT

Communication is a social activity. Its properties coevolve within Darwinian constraints. A central theme of my 1977 book on communicating was that individuals benefit from social opportunities in part by making selected information available to one another. Their signals facilitate and channel the development of orderly interactions that are, on average (and yet often to very different extents), mutually advantageous. I suggested that some essential tasks for research on communication were to determine how this information can be characterized, how signaling specializations themselves can be categorized, and how individuals responding to signals obtain and integrate information from multiple sources. In this chapter I clarify the concept of information that underlies my approach and emphasize the conditional status of all predictions based on information; briefly review and revise descriptions of commonly found 'messages' and message categories; explore the great scope of communication and the centrality of negotiation in social interactions; and clarify the case for arguing that most signals facilitate prediction that is largely reliable. I also address trends in ethologists' interests and goals that I had not anticipated, and that have considerably altered the ways many ethologists view communication. With the development of cognitive ethology, much interest in signals has returned to an organism-centered focus (as opposed to an interactional focus). With

Perspectives in Ethology, Volume 12
edited by Owings *et al.*, Plenum Press, New York, 1997

the advent of sociobiology/behavioral ecology focus has also shifted from proximate social mechanisms to population characteristics and intergenerational time frames, often overstressing individual independence and underestimating the coevolved interdependencies of social life. Yet organismic and evolutionary foci usefully broaden the field, and another goal of this chapter is to bring them and the interactional focus into a mutually enriching perspective.

INTRODUCTION

Singing birds, toads, and crickets, flashing fireflies, scent-marking canids and purring cats fascinate us. What are such animals signaling about? Why? How does the process of communicating work? Although Darwin's (1872) attempts to explain behavior provided the impetus that led to the field of ethology, Darwin did not regard what we would call signaling (e.g., the facial "expressions" of dogs and humans) as functional adaptations but as "vestiges or accidents" (Fridlund, 1994:15). Today, signaling is often seen as a window on the workings of animals' minds, or a subject for cost-benefit analyses. Whatever the perspective, both Darwin's and most current conceptual frameworks for studying communication derive more from prevalent interests and fashions than from attempts to involve basic properties of the communication process. Certainly, more than one framework can be useful if each reveals different aspects of this intricate phenomenon. Yet a framework that I have advocated, and which I believe offers a necessary perspective, is underrepresented. I here revisit the perspective and themes of *The Behavior of Communicating* (hereafter: TBOC), my previous attempt to explore the study of animal communication (Smith, 1977), and attempt to relate that effort to subsequent trends in the field. While not a detailed review of the past two decades of development in ethology, this is my response to those developments, a response that offers clarifications and revisions designed to yield a new, broader, synthesis.

The framework for analyzing communication in TBOC is, briefly: animals are socially interdependent in many ways, particularly within species. Each individual has its own needs, agenda, and capabilities, yet profits from interacting with other individuals. Benefits need not be equally distributed among participants for each to profit from interacting, and considerable costs may be borne to obtain benefits over the long term. To interact in orderly ways, each individual must to some extent be able to predict how others will behave. Prediction is facilitated when participants in an interaction share selected and largely reliable information with one another, which they do at least partly by signaling. We discover the information that their signals make available primarily by determining the correlates (of all kinds) of signaling actions. Much of the information is about the signalers' behavior. This behavioral information fosters predictions that are conditional and probabilistic. Signals also provide information about who and where signalers are.[1] Signaling is complex and based on a number of markedly different kinds of specializations with great scope and subtlety. Yet signaling is not the only important source of information. Animals usually respond to information from multiple sources. In communication, many of these sources are contextual to signals. Responses, quick or delayed, provide an observer with clues to what signals mean to their recipients, in particular situations.

The synthesis in TBOC was attempted explicitly from an interactional perspective. (An 'interaction' is a behavioral event in which participants influence each other's moves.) I still consider the interactional perspective to be the most appropriate way to look at communication. Other approaches have their uses, but it is fundamentally necessary to understand communication in the realm in which it operates. It is social behavior, primarily with social consequences. And in social life of any measurable intricacy and persistence the participating individuals are interdependent and, to some extent, accommodating—characteristics that greatly influence what is required of communication.

My interactional framework, however, attracted less interest than did my emphasis on analyzing the contributions of information to interactional behavior. Throughout TBOC, communication is conceived of in terms of information that is made available by one individual and acted upon by another. To my surprise, many biologists balked, rejecting the centrality of information. Their writings reveal diverse concepts of information, all quite different from the one I espouse. My first goal here is to outline and explain this concept more fully than before, discussing objections that have been raised (implicitly or explicitly) in the biological literature. I continue to believe that an understanding of information, as defined in the next section, is fundamental to understanding communication.

My second goal is to elaborate upon my conviction that there is great scope for communication because the social lives of many animals are rich in opportunities for individuals to profit from providing and responding to selected kinds of information. Negotiation is often more profitable than is unalloyed self-assertion. Over evolutionary time the intricacies of social interactions, relationships, and groups have both shaped and been shaped by animals' capacities for communication.

Further goals are to outline the sources from which animals get the information they need in social life, and the extent to which kinds of information that were outlined in TBOC appear to need discussion or revision after twenty years. Finally, I address major trends not anticipated in TBOC, which appeared in an unusual period for development of concepts in ethology.

[1] When writing TBOC I viewed information about external stimuli to which signalers are responding only as a possibility consistent with the framework; see section "Kinds of Information" and TBOC:69, 74, 181.

Writing of TBOC was completed by summer 1975, before Dawkins and Krebs (1978) got overly enthusiastic about deception, before Don Griffin (1976) inaugurated cognitive ethology, and before Ed Wilson's prediction (1975) that sociobiology would supplant ethology began to seem true. Students of animal signaling behavior switched, many from modeling emotional and motivational mechanisms underlying behavior to focus on proximate cognitive mechanisms. Behavioral ecologists emphasized evolutionary (ultimate) causes, and asked what happens when one individual's interests are pitted against those of another. At least initially, many investigators accepted the misguided notion that for an individual to communicate information about itself "does not accord with the theory of natural selection" (Krebs & Davies, 1978:157). The pace of research surged, but the directions taken were channeled. Some key issues slipped almost out of sight, as workers began to view communication and other social behavior as primarily manipulative, and as they focused exclusively on evolutionary adaptiveness and mental functioning. Yet most social behavior cannot be so simply controlled for one participant's advantage, and social behavior is integrated at more than one level of complexity. Focusing on the individual ignored the means by which contributions of participants are integrated into coherent social events. Communication's role in interactions cannot be predicted from an understanding just of the internal and evolutionary causes of an individual's behavior.

Cognitive and evolutionary research do provide insights into social phenomena. Some crucial questions about communication are cognitive. For example: what information is important when interacting, where does an individual get this information, and how is it processed? And understanding of individual and social mechanics is enhanced, even guided, by considering ultimate causation (and vice versa: behavior can be "the pacemaker of evolutionary change," Mayr, 1982:612). Yet the proximate workings of everyday social interactions remain largely uncharted. Social interactions, and the social relationships, bonds, and group structures that arise from them, are the arena for communication.

Much research on animal communication in the past two decades thus took paths I did not foresee when devising the conceptual framework presented in TBOC. Yet much of what has been learned is relevant to my goals. With the maturation of different perspectives it is appropriate to examine how they relate to mine, reframe my goals in contemporary terms, and take another shot at explaining the centrality of a broad concept of information. One of ethology's holy grails is still an understanding of how and why animals communicate (among recent reviews: Sebeok, 1977; Kroodsma & Miller, 1982, 1996; Payne, 1983; Moynihan, 1985; Estes, 1991a). The pursuit is no less exciting now than it was when Lorenz, von Frisch and Tinbergen began to dissect the striking signaling behavior that had long enthralled naturalists. The task becomes ever more accessible as analytic software, hardware and experimental procedures become more powerful.

INFORMATION AND COMMUNICATION

I consider communication to be any sharing of information between entities—in *social* communication, between individual animals. So defined, the process of communicating cannot be interpreted without a concept of information. The concept is simple but powerful: the acquisition and use of information reduces uncertainty. It thereby helps animals (or plants, cells, even machines) to make choices—to select among alternatives. As an entity acquires information it becomes more able to anticipate and respond appropriately to events, and to use feedback to select changes in its state. Nothing is more basic to coping with an inconstant world than is the acquisition and use of information.

Engineers may have been the first to grasp the significance of the relationship between information and the process of selecting (Hartley, 1928; Shannon & Weaver, 1949; Cherry, 1966; Krippendorff, 1989). The well-known statistical information theory of Shannon and Weaver applies only to the selection of signals from within a signal repertoire, and measures only the amounts of information that different selections provide. The theory's focus on selective power embodies the core of the information concept, but is relevant to only a small part of the process of communicating (i.e., syntactics, the simplest of semiotic's three levels; see TBOC, chapter 1) and is indifferent to the *kinds* of information involved (information "about what").

Most of our common language conceptions equate information with news, knowledge, or fact, and aren't much help. But one, the way we envisage the descriptive information we use in sorting among stimuli, images, perceptions, memories, and ways to respond, does correspond to the engineering sense of information. Sorting underlies making choices, and making choices underlies action.

Ethologists, however, use many different definitions of "information." Some have never been made explicit. For instance, Dawkins and Krebs (1978:286) suggested that "the idea of an exchange of information is a carry-over from human language." Their focus was on the notion of exchange, of giving one thing in return for another. They left information undefined, but implicitly endowed it with spurious properties such as indivisibility. That is, they assumed that if information is involved when an animal signals, all the information that animal has must be made available—a task both impossible and inadvisable. Further, they explicitly lumped together as "informational" perspectives such as mine with perspectives that focus on the motivational causes of behavior. But separate levels of integration are addressed by these perspectives: social and intraindividual, respectively. In traditional ethology, analyses of communication focused on its motivational causes and its apparent functions. The 'releaser' concept (and the alternative of 'inhibiting') largely obscured the need for a concept of information, and none was made explicit. The term 'information' was used rarely, and simply to reflect underlying motivation (see, e.g., Tinbergen's summation of his group's research on the signals of gulls, 1959:36). No explicit concept of information was developed and no central role was envisaged for such a concept.

There have been many sorts of misunderstandings, some quite curious. Dawkins and Guilford (1991:868), for instance, implied that I made "colloquial use" of the term 'information' and they proposed, as a message I might attribute to an animal's signal: "a strong male capable of fighting hard." The relation between such a message and anything proposed in TBOC is remote. While suggesting that information about physiological status can be part of the identifying information of signals, I have stressed that this is only part of the overall information package. Information predictive of behavior (not just capabilities) is always made available by a signal. (Further, the definition in TBOC *is* effectively the one Dawkins and Guilford offered.)[2] The shorthand statement that information increases predictability is equivalent to saying that it reduces uncertainty, and hardly "colloquial" (shorthand because information itself is not a causal agent). Dawkins and Guilford went on to attack the notion (but whose notion?) that 'honesty' might be used to mean "perfect information" (op. cit.:870). Another straw man. I know of no such published definition of honesty, and have always maintained that signaled information permits only probabilistic and conditional predictions about some part of a signaler's activities. A curious case indeed, especially as Dawkins and Guilford were offering an interesting analysis of the limits of signal reliability (see section "Reliability").

Many ethologists appear to conceive of information as if it were a material thing—it is not. It is a property of things and events that, unlike matter, is not diminished by being shared. You cannot give information away, but you can share it with others. When we talk or write to one another we do not lose the information we impart. Neither do we share all of the information we have. Our sharing is selective, as in all social communication. And, although acts of sharing information can have effects, information itself is not causal. In explanations, information is an intervening concept between predictions and their causes.

Even selective sharing of information implies nothing about conscious choice. But it does raise a basic question: just what information *is* shared by specialized signaling? Reviewing what we know about that information was a major focus of TBOC (chapters 3–8).

Most alternatives to models based on information are too mechanical, too simple. They imbue signals with direct causative power, like that of the 'releasers' of early ethology. The concept that signals could release reflex-like, prede-

[2] Some misunderstandings have arisen over the term 'message,' for which I now usually say "information made available." But 'message' is shorter and convenient, if only some would not interpret it to mean 'signal.' Signals are the physical vehicles that make messages available.

termined responses was based on some very interesting cases, but does not generalize well. Cognitive processing of most signals involves more complex integration of information from multiple sources than a lock-and-key releaser model suggests. Krebs and Dawkins (1984) used a releaser-like concept when they offered 'persuasion' as a means by which signals operate. Yet natural selection must be expected to favor skeptical responders that seek additional information, and are not automatically persuaded (Smith, 1986a).

One alternative model is called management/assessment. It views communication as a process emerging from the interplay of two interactional roles, manager and assessor (loosely corresponding to signaler and receiver, see Owings, 1994). Managing individuals attempt to influence assessing individuals, who in turn adjust their behavior to serve their own interests, thereby influencing the initial managers (Owings & Hennessy, 1984). This 'transactional' model differs from interactional/informational models in having no central role for information (information is 'assessed,' but not 'provided') and, insofar as I can determine, in postulating no mechanisms by which a managing individual attempts to influence an assessor. Communication seems to be viewed as a regulatory system. My difficulty with the model is that I cannot see how such a system operates, if not on information. And if information is essential to the mechanics, then this model converges in many ways, from independent origins, with the interactional/informational model I have presented. Certainly, it yields some similar findings such as multiple repertoires of signaling devices, active roles for individuals receiving signals (the 'assessing individuals'), and patterns of communicating that fit different ontogenetic stages. Owings' model also, however, shares some features (e.g., entrainment of assessor states) with the model of Dawkins and Krebs (1978).

In the social communication studied in ethology, links are established between individuals when one shares information with another. Many ethologists (e.g., Wiley, 1983) require that the recipient's behavior be changed, but I suggest that any effect, including a continuation of the behavior, is sufficient. (This does raise an operational problem: an observer seeing no change in behavior can determine whether this was because of, or independent of, the signal only by experimenting.) Wiley (loc. cit.) would also say that information has not been communicated (or 'transmitted,' his equivalent term) but only 'broadcast' when that information is not received. In an event without a recipient I would simply say that information was made available, but communication did not occur. Our conceptions are the same.

Ethologists study primarily communication that is based on specialized signals—on actions and things that have been modified ('formalized,' TBOC:327) to function as sources of information. Formalized acts achieve their effects indirectly. That is, signals are effective only if recipient individuals themselves take action. Their performers (signalers) exert no physical force on other individuals. In contrast, by acts such as pulling or pushing, one individual can move another by direct physical force. The use or absence of physical force in achieving effects is an important distinction that has long been recognized by ethologists (e.g., in the concept of ritualization, Tinbergen, 1952; Moynihan, 1955). Yet pushing and cajoling are both informative. Every thing and every action carries information, and individuals communicate with acts that are directly as well as indirectly functional. There are ethologists who would withhold the term 'communication' from events involving only nonspecialized acts (e.g., Philips & Austad, 1990). In sciences other than ethology, 'communication' is used to refer to any sharing of information, by any means. Good broad terms are scarce. We should not waste one by arbitrarily restricting it.

INFORMATION IN SOCIAL INTERACTIONS

The perspective underlying TBOC involves studying how communicating animals inform one another and facilitate social interactions—an 'interactional' perspective. During social interactions, participants coordinate their activities and influence each other's behavior. To interact efficiently, animals must have some notion of what to expect of each other. Each individual's main contribution is to make some, though by no means all, of its behavior conditionally predictable. Its signals are tools specialized for that job, and must serve in a wide range of social events.

Interactions are the fundamental stuff of social behavior, the ephemeral events from which enduring social relationships and groups arise. Because individuals differ in their current needs and agendas, they must work to coordinate their actions sufficiently for interactions to be orderly and profitable. Some of the work is cooperative: the accommodating to one another's actions, and the signaling which enables others to accommodate. At the same time, each individual may compete to control or manage an interaction, maneuvering for advantage. Currently preoccupied with the competitive possibilities, many ethologists tend to underestimate the cooperative work—the coordination, compromise, and tolerance—and the extent to which orderliness depends on predictability.

A considerable diversity of social encounters must be elicited, facilitated, navigated, or avoided by most vertebrate animals in the course of their lives. They must help order to prevail in these events. They must also negotiate useable outcomes to competitive issues and, often, try to control at least some aspects of the development of interactions. Further, the social arena is a dynamic environment. Its members change through ontogeny, health, social status, emigration, and immigration (see West & King, 1996, for a systems approach to the ontogenetic changes; and Kroodsma, 1996, for divergent effects of site fidelity vs. nomadic habits). All changes add to the demands on communication. It was a major shortcoming of TBOC that I gave no organized account of social events.

Readers saw little basis for my contention that the arena of social interaction affords great opportunities for subtle and complex communication.

A Simplified Example of the Intricacies of Social Activities

Territoriality is but one of many classes of social behavior, yet consider the number of kinds of individuals with whom a territory-holding songbird must interact. By 'kinds' of individuals I mean something close to the concept of 'social identities' formulated for human interaction by Goodenough (1965). I assume that animals must form expectations of events as an essential procedure for coping actively with change, but not that the expectations must correspond in detail to many of those consciously formed by humans.[3]

Even a generalized synopsis (Fig. 1) shows a dozen categories of interactants with whom a songbird must deal in its social identity of "territory-holder." For each category it needs appropriate social procedures and communicative devices. It must, for instance, negotiate with its neighbors over the division of space, arriving at mutually accepted boundaries. It must periodically check its neighbors' continued acceptance of the boundaries, and reaffirm its commitment to defense. Yet it may sometimes allow neighbors access to its territory, shifting to the role of locally dominant individual and requiring them to demonstrate the deference of subordinates. Other territorial intruders will be strangers with whom relationships must be established, familiar individuals such as floaters, migrants in passage, or individuals prospecting widely for space or mates.

As our territory owner interacts with individuals of different social identities, it plays various 'parts'. It can play established neighbor, local dominant, challenger, defender, tolerant associate, mate, parent or other parts in different events. It can play more than one part with the same individual in different circumstances, or with different individuals in the same event. In short, our focal individual will engage in a wide range of different interactions.

The social picture becomes further complicated. An individual can have more than one kind of territory and relate differently to various individuals on that basis. It will usually make a space immediately around it inaccessible to others: its 'individual distance' (for humans, 'personal space'). It may defend a larger mobile territory around its mate or family. The individual may hold disjunct territories for tasks such as mating, nesting, and foraging, with different neighbors and spatial needs at each site. Some territories may be ephemeral (e.g., for foraging on tidal flats) with flexible boundaries and no established conventions with neighbors. Some territories are too large to control as a whole, but

[3] A social identity is a characteristic of an individual that, relative to characteristics of another individual, determines what expectations they usually have of each other.

their sections are defended when in use ('spatiotemporal territories,' Wilson, 1975).

The Diversity of Social Environments

There are many kinds of social experiences. Some pelagic fish live in enormous schools, every member behaving just as the others do, and order emerging largely from their conformity (Williams, 1964). Other animals, foxes, for instance, live largely alone, but with knowledge of various neighbors and sometime associates (Barash, 1975) gleaned from recurrent encounters, signs that have been left, or distant vocalizations. At the other extreme, various primates, cetaceans, rodents, and birds live continuously in complexly structured groups comprised of many individuals of successive generations. Group members differ in gender, age, and kinship, and are organized into intricate webs of relationships. Group structure differs among species and within species as details of the ecological environment and social parameters vary spatially and temporally. Ontogenetic trajectories and contemporary circumstances differ widely among members of a single social group. Individuality blossoms, and each member in many ways follows its own agenda. Yet all members have a stake in sustaining their groups and finding orderly ways to interact. Each must both foster and recognize predictability—they must share crucial information. The importance of maintaining more or less peaceful social relationships is shown by procedures with which individuals reconcile with each other after a dispute (de Waal, 1989).

Fig. 1. A 'social identity' is a characteristic of an individual that (relative to characteristics of another individual) determines their usual expectations of each other (Goodenough, 1965). (1) Territorial neighbors may drop boundaries and forage in a group when conditions are not appropriate for breeding, resuming territorial relations when conditions improve. (2) A territorial individual may be alone or have various associates within its territory. These range from codefenders to individuals who are largely or entirely dependent on it for spatial defense. They are differentiated by various criteria (such as age, familial relationship, dominance rank) that determine (a) competitive and (b) accommodative aspects of their relationships. (3) Such deferent individuals may be passing through. (4) Familiar intruders can include members of a neighborhood who periodically assemble at special foraging or roosting sites, independent offspring who at times return home, or "flock mates" of an individual who has recently established its territory or is on an ephemeral territory from which it will rejoin the flock. Consider the various 'parts' a territory-holding individual might play as it interacts with individuals of the more than 20 different categories shown in the diagram. It can play neighbor, dominant, challenger, attacker, tolerant associate, mate, parent, or other parts as it interacts with different individuals—and more than one of these parts with the same individual in different circumstances. Further, it will not always find 'territory-holder' to be a relevant social identity. Its use of territorial behavior will be limited to some events (such as interacting with neighbors) and irrelevant in other events (such as feeding its offspring).

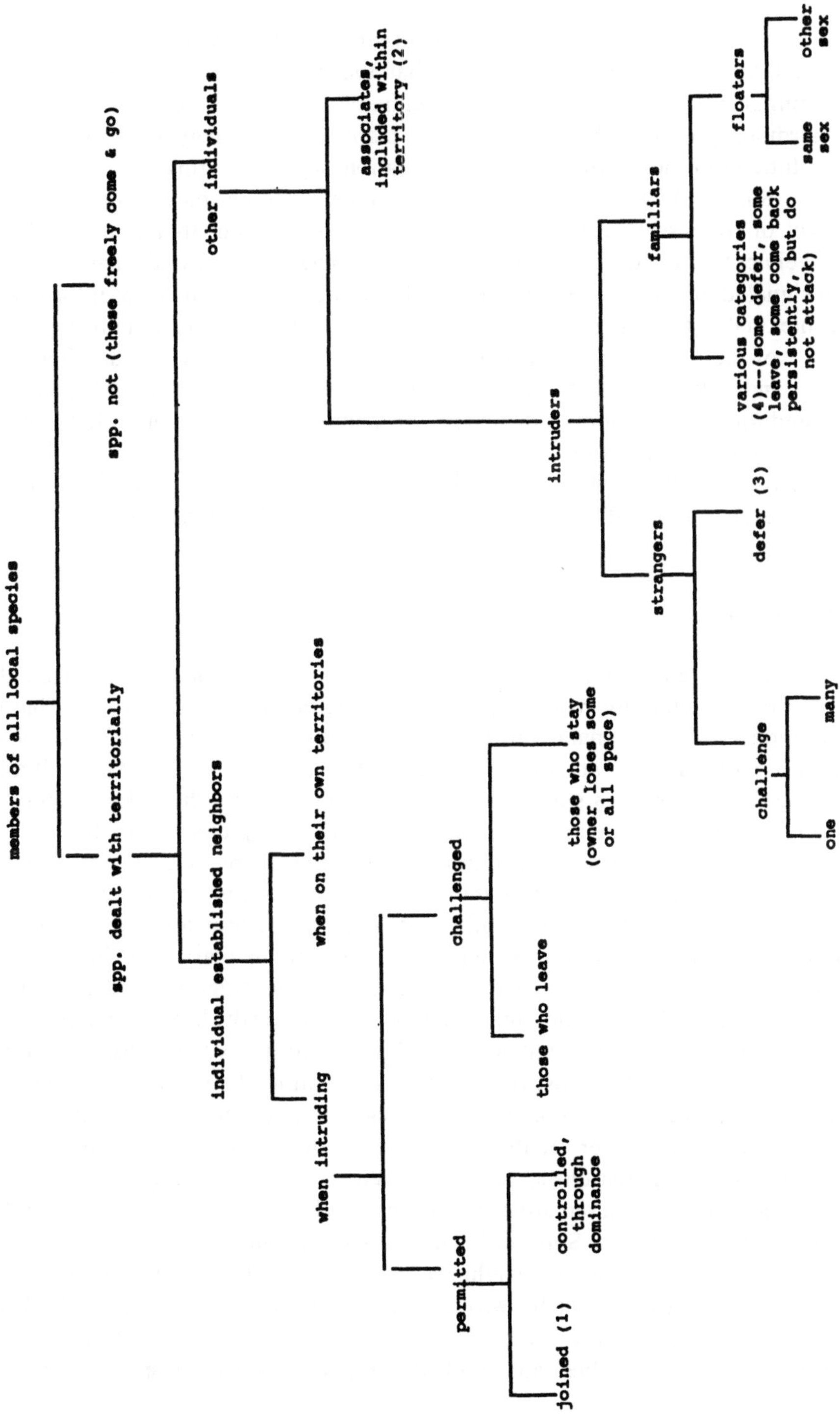

members of all local species
spp. dealt with territorially
spp. not (these freely come & go)
individual established neighbors
other individuals
when intruding
when on their own territories
associates, included within territory (2)
intruders
permitted
challenged
strangers
familiars
joined (1)
controlled, through dominance
those who leave
those who stay (owner loses some or all space)
challenge
defer (3)
various categories (4)--(some defer, some leave, some come back persistently, but do not attack)
floaters
one
many
same sex
other sex

We are just beginning to understand some of the simpler aspects of animals' social lives. It is already apparent that the scope for extensive and subtle communication is great. Although nonhuman species lack devices with the enormous power of human language, many of them may have pushed the evolution of other formalized communication to great lengths.

So, should we expect animal communication to be stuck at the reflex-like level of 'releasers' in which individuals respond to signals automatically, in largely preordained ways? No. The evidence increasingly reveals that responses to signaled information are fundamentally influenced by sources of information contextual to the signals. Should we expect animals to bluster and bluff, putting most of their energies into deceitful attempts to manipulate one another? No. Each individual will find advantage where it can but, especially for individuals in enduring relationships, advantage will come more often from reliable communication in interactions that are at least partly cooperative. Natural selection works on both those who signal and those who respond to signals. The intraspecific result is more often a form of mutualistic coevolution than an arms race.

Negotiating

We should expect negotiation, rather than threat or bluster, to be the primary procedure for resolving most intraspecific competitive events. That is, individuals should signal to one another, continuing to exchange information as each assesses how the other may behave (TBOC:12, 450; Hinde, 1985). Initially, neither participant has sufficient information to evaluate the risks inherent in decisive action. Yet neither need make its moves completely predictable. Usually, neither can. Only as a negotiation proceeds can the negotiators decide when and how to end it. Negotiators thus signal continuing involvement, revealing only that the momentary probabilities of decisive actions are low, relative to indecisive behavior and further signaling. Those momentary probabilities usually change, but change can be temporarily forestalled. Negotiators delay important decisions while eliciting information. Delay is costly, but less costly than a premature commitment might be. Delay is also informative, because an individual's acceptance of the cost (revealed by its continued signaling) is a reliable indicator of its continuing pursuit of the negotiation (Kennan & Wilson, 1990).

Negotiations over territorial borders or dominance ranks may be protracted. Some negotiations appear nearly static, with symmetrical moves and at most only intermittent tentative escalations in which decisive moves are briefly tried and parried (see Simpson's accounts of disputing fighting fish [*Betta splendens*], 1968; Estes' of neighboring wildebeest [*Connochaetes taurinus*], 1969; and Norman et al. on brown-headed cowbirds [*Moluthrus ater*], 1977). To develop a mutually accepted territorial border, two individuals meet, signal, probe, respond to probes, and gradually adjust their use of space with only limited physical contact. Change is initially slow. At different sites either contestant may give ground up to a point at which its resistance stiffens. Boundaries gradually emerge.

Prolonged encounters are not the norm, however. The brevity of most ordinary, everyday encounters suggests that animals usually have enough information to keep negotiations simple. There is also an indirect but significant clue that negotiating is at least briefly involved in many encounters: most signals make available information about the probability of a signaler engaging in one of two or more incompatible activities, and the probability of instead behaving indecisively (see section "Kinds of Information"). This prevalence of information about indecisive behavior strongly implies that most signals are used as individuals negotiate, influencing each other's choices before committing themselves.

Negotiation is not restricted to events in which competition looms large. Maxim (1982:81) described (and nicely diagrammed) the steps in what he called a subtle and "progressive negotiation" of a consort relationship between a male and female rhesus macaque (*Macaca mulatta*) when they were first placed together in a pen. Moreover, the basic features of negotiating can be seen in all orderly interactions. Such events are managed jointly and mutually sustained and shaped, each participant to some extent accommodating to the others' actions (TBOC:12). For one individual to gain full control is difficult, because each participant yields no more than it must, and withholds considerable information. The benefits from interacting are rarely equally distributed, and are often paid in different coin to different participants. But each participant must make it worthwhile for the others to join in, thereby fostering tolerance and interdependence that offer something to everyone. Negotiating is so central to interactional management that it is virtually ubiquitous. There are, of course, many other interactional procedures, from animals simply behaving nearly identically with one another (as in large fish schools), to reconciling after a dispute, giving tactical support, and so on. Yet, to the extent that no individual achieves complete control, compromise, the typical product of negotiating, is always involved.

Each participant in an interaction works toward its own goals or, at least, a compromise it can use. Nothing requires that a participant refrain from sharing selected information about itself with other participants, Dawkins and Krebs (1978) to the contrary. They instead see a participant as persuading those with whom it interacts. But persuasion, in their sense, implies that the signaler gets its own way by overriding the others' needs. Overriding happens, even within the framework of negotiations, when some participants have severely limited options. The bulk of social interaction, however, requires mutual accommodation.

Dawkins and Krebs also proposed that rhythmic stimuli "pulsing away at the brain's own sensitive frequencies...can have dramatic effects on behavior" (op. cit.:307). They thereupon suggested that the singing of birds might operate like "hypnotic persuasion" (loc. cit.). But it is not necessary to propose the commandeering of recipients' brains to account for the quasi-rhythmicity of singing. Singing is primarily a procedure for long-range communicating. In the unpredictable sound environment produced by the momentary utterances of individuals of many species, rhythmic repetition makes orderly sequences of vocalizations stand out. That is, the procedure attracts attention. The rhythm also enables distant listeners to anticipate and prepare perceptually for each successive signal unit (each 'song' in a singing performance, Smith, 1991a). Rhythmic singing enables individuals to conduct considerable social business without making the effort to come into each other's immediate presence. Most of the information separated individuals have about one another may come from singing (Smith, 1991a, 1996; Smith & Smith, 1996), but there is little or no evidence to indicate that, say, two countersinging neighbors are each somehow subverting or persuading the other.

SOURCES OF INFORMATION

Information has been defined by the way its acquisition can facilitate prediction. As animals interact, they often use actions and structures (e.g., crests, colored badges, molecules released as scents) that have been specialized to be informative (i.e., "formalized," TBOC:327). I refer to all such specializations as 'signals' herein. What are signals, in this sense, like? What are the distinctive characteristics and contributions of signals of different kinds? How important are other sources of information, sources that are not formalized and occur both contextually to signals and in their absence?

Signals and Signaling Rules

The formalization of behavior has yielded sets of signal units, and those units have provided the focus, or initial focus, of most ethological studies of communication. However, formalization has also yielded sets of rules for modifying the basic signal units, and for incorporating them into signaling performances. These rules have received much less study.

We can grasp the full scope of animal communication only if we recognize that each individual can signal with a diversity of repertoires, each with distinctive properties, and each recipient can evaluate information from many sources in every event. Vertebrate animals of every species have, first, a repertoire of up to 40–50 behavioral signals ('displays,' TBOC:7). Each individual also has up to five other repertoires of behavioral specializations for signaling (Smith, 1986b, 1991a). Each repertoire has distinctive formal characteristics that enable its components to provide a distinctive class of information (Table 1). At least

Table 1. Kinds of Signaling Repertoires (Because There Are Two Repertoires of Superimposed Formalizations for Both Displays and Formalized Interactions, the Total Number of Repertoires Is Six)

Repertoire	Component units	Contributions
Primary signal units:		
Displays	Audible, visible, tactile, and other formalized acts that can be done by single individuals.	Make available information about a signaler and its probable behavior, rendering activities of social importance to some extent predictable, and facilitating social interactions; make available information about external stimuli to which the signaler is responding, thus alerting recipients to dangers, resources, etc.
Formalized interactions	Programmed routines that coordinate and organize activities of 2 or more signalers, allotting each participant a predictable part to play in the interaction.	Both provide information and enable participants to elicit information. Participating (using special, coordinated signaling) indicates acceptance of predictable frameworks and constraints. Formalized interactions clarify and facilitate negotiations, and allot each participant a share of the control in events that involve change or marked inequalities among participants.
Superimposed formalizations:		
Procedures for varying form	Rules for changing duration, frequency, and other characteristics of primary units.	Enable signalers to modify or augment the information made available by primary units, scaling information continuously as form changes. Reveal directions in which behavioral predictions are (or may be) changing. Most effective when signals are performed in rapid succession, because detection of minor variations requires accessible standards of comparison.

two and sometimes all of these repertoires are brought into play in a signaling event, allowing for subtle communication of considerable richness.

In TBOC I described only two classes of repertoires: displays and formalized interactions. Displays include specialized utterances, movements, postures and the like—behavioral signals that a single individual can perform. Formalized interactions coordinate the actions of two or more performers, permitting shared control and negotiation in events that might otherwise become overly one-sided or unstable. Formalized interactions include the morning greetings of seahorses (Vincent, 1995), duetting of birds, turn-taking by individual pygmy marmosets (*Cebuella pygmaea*), when members of a group utter trills in a largely fixed order (each awaiting its slot; Snowdon, 1983), reconciliations by chimpanzees and the joint overtures preceding them (de Waal, 1982), the human handshake and various interactions that incorporate it, among many other brief or extended routines. Since TBOC, I have recognized four additional repertoires made up of rule-governed procedures for altering the forms of, or combining, primary signal units (Table 1). The effects of those procedures are superimposed on primary signals. For example, singing (briefly described at the end of section "Negotiating") is based on rule-governed sequencing. A singer utters a continuing succession of primary displays at approximately regular intervals and, in many species, in accordance with grammatical rules for sequencing. Those units are mostly 'songs' (displays that occur most often in singing performances) but can include nonsong signals (Smith, 1988). The rate at which song units are uttered is informative. The sequencing and rate of singing performances add information to that contributed by each song.

Rules for varying signal form and for combining signals were both recognized in TBOC, along with some of the kinds of information made available when the rules are applied. Preoccupied with trying to clarify the nature of displays, however, I treated these rule-bound procedures more as obstacles to a simple display concept than as major sources of information.

The sizes of each species' repertoires of displays and formalized interactions appear to be limited. This may result at least partly from inherent limits on the processing capacity of working memory (TBOC:339), limits that may reside in basic properties of neuronal activity in the brain (Lisman & Idiart, 1995). Numbers of rules for varying signal forms and for combining signals may be similarly limited. The complexity that signaling contributes to animal communication comes both from the number of signals within each repertoire and from the diversity and combined effects of the repertoires. Each repertoire contributes a qualitatively distinctive kind of information.

The overall complexity of a species' signaling repertoires can be vast. Our ability to record and analyze signaling behavior has been greatly extended since 1977 by the development of small videorecorders, personal computers, digitizing real-time sound spectrographs, and other tools. In my own work, the Kay Elemetrics DSP 5500 sound station makes possible analyses of complex sounds and their sequencing that were impractical when TBOC was written. Yet it remains challenging to describe complicated actions in ways that facilitate objective comparisons. Among important contributions: Miller (1988) reviewed the description of motor patterns of behavior, indicating issues in distinguishing units, categories, and combinations, and in the use of cluster analysis, time-series analysis, auto- and cross-correlation functions, the coefficient of variation and other measures of variability. McCowan (1995) used cluster analysis to group the correlation coefficients of frequency measurements of dolphins' whistle contours. Her technique categorizes contours into appropriate groups even when durations and frequencies are different. The method should be applicable to narrow-band, frequency-modulated signals of any species. Thompson et al. (1994) tackled the inconsistencies that have arisen in attempts to describe birds' songs. Recognizing that only three methods (one based on pauses, and two on shapes revealed in sound spectrograms) have been used to divide a stream of singing into chunks, Thompson et al. proposed a formula for analyzing units at different levels. Given neutral names, these levels permit better, more detailed comparisons of songs among species than have been achieved previously.

Digitizing procedures have also enabled the synthesis of previously nonexistent sounds, which are then used to test animals' responsiveness. Components of songs have been rearranged, even inserted into recordings of other species (Marler & Peters, 1977) to show that naive young songbirds can extract and then learn to reproduce those characteristic of their species. These youngsters impose appropriate sequencing, even if they have not heard it! The calls of common ancestors of sets of species of frogs (*Physalaemus pustulosis* and relatives) have been estimated from averaged values of spectral and temporal parameters of calls of the extant species (Ryan & Rand, 1995). By playing back synthesized ancestral vocalizations, Ryan and Rand showed that females did not distinguish between calls of males of their species and calls at their most recent ancestral node.

Morton (1977, 1982) suggested yet another way in which the form of vocal signals may be rule-bound. The basic form of some vocalizations used in close encounters may be determined by the motivational state of the signaler. The idea traces back to at least Darwin (1872). It has considerable support from observational (but mostly unquantified) studies. Are relatively low-frequency sounds usually uttered by aggressive or approaching animals, and higher-frequency sounds by fearful or withdrawing ones? Perhaps, but many alternative possibilities must be examined. For instance, low-frequency sounds may be typical of relatively relaxed (i.e., confident) individuals whatever their behavior, and high-frequency sounds of tense individuals. High-frequency sounds may be harder for inappropriate listeners to localize than are low-frequency sounds (Marler, 1955), and may be better whenever long-range effects are disadvantageous (because high-frequency sounds attenuate faster with distance than do low). High-frequency sounds may also be useful because they can be directionally beamed, even by birds with small heads (Witkin, 1977), or because they contrast with most of the vocal repertoire and so elicit immediate attention in emergencies (Shalter, 1978). Surely the physical characteristics of vocalizations are subject to natural selection, but many different selective forces may operate, in concert with or in opposition to each other.

Morton made his interpretations in terms of four motivational states rather than behavior. In addition to aggression and 'friendly or appeasing' (op. cit.:856) he suggested that rising then falling (or vice versa) frequency correlates with an individual's "interest" (op. cit.:861) in a stimulus when undecided how to proceed. (In fact, *most* signals are given during indecisive behavior.) All vocal communication cannot be based on just four internal states, of course, and Morton's scheme may relate to only a small fraction of signaling. Whether it holds remains to be seen, and evidence is accumulating both pro and con. Some supporting evidence comes from observations of captive animals (e.g., Sieber, 1984), but the constraints of captivity limit behavior and usually preclude detailed analyses of signals' messages. Other detailed and thorough studies have yielded correlations other than those Morton predicts. For example, Nelson (1984, 1985) found that high-frequency, brief guillemot (*Cepphus columba*) vocalizations correlated not with avoidance but with local movements toward or away. Dabelsteen (1985) found that, in European blackbird (*Turdus merula*) songs (long-distance signals), the correlates of high frequency and harshness were the opposite of those predicted by Morton. That the structural relations of short-distance vocalizations to motivational states can be *reversed* for long-distance vocalizations makes it seem unlikely that signal form is determined by motivation. Reversing the information provided by the same structural features in different circumstances also risks leading responding individuals into serious errors.

Morton has addressed a real phenomenon, but it may be much more complex (and probably more variable) than he has foreseen (Smith, 1986b). Certainly, it is too early to refer to putative correlations as motivation-structural 'rules,' or to claim that all information is provided by form, making the display concept unnecessary for short-range signals. I would also prefer a behavioral rather than, or in addition to, a motivational hypothesis. Behavior can be predicted from signals, and most animals may not need to base their predictions on abstract concepts such as motivation.

The Importance of Information from Sources Not Formalized

Everything can be informative. As a result, the most abundant sources of information are those not formalized, and much of the information animals share is made available incidentally. All of an animal's activities, eating, sleeping, the direction in which it turns its head—every act can be informative to other animals. Menzel and Halperin (1975) found that a chimpanzee's rate of locomotion provided other chimpanzees with sufficient information to discriminate between more and less highly preferred objects when only the leading individual had knowledge of hidden prizes. Formal signals such as gestures and vocalizations were seldom used. The presence or absence of known associates, their locations relative to habitat features or territorial boundaries, the time of day, season, and weather can all be significant sources of information. The innumerable sources are too many to contemplate, but perception is selective. Some sources are accepted as pertinent or possibly pertinent. Others form a less closely attended background. We need to learn more about the kinds of sources of information that animals select to be focal and background in different circumstances.

Formalized information sources ('signals'), having been specialized to make information available, are often pertinent when presented. Other sources have value contextually to signals, and also in the absence of signals. With sufficient information from ordinary sources, animals can interact in orderly (usually simple or routinely practiced) ways without signaling. Patas monkeys (*Erythrocebus patas*) for instance, can sustain conservative features of their group's organization largely by continually monitoring each other and adjusting (e.g., to being approached) with only infrequent signaling (Rowell & Olson, 1983).

Research on 'information centers' (ICs) has revealed nice cases of the power of ordinary sources of information, and of conditions in which formalized sources can come into play. As originally proposed by Ward and Zahavi (1973), the main transmission of information at ICs involved no formalized behavior. Animals simply gathered at roosts or colonies, and later departed in one of two ways. Animals having information about a patch of still existing food flew directly back to it. Those who had not found food recently followed them. The information might be simply in the confident-appearing flight behavior, as suggested for quelea (*Quelea quelea*) by Ward and Zahavi, and shown experimentally for common ravens (*Corvus corax*) by Marzluff et al. (1996). Or it might be in the sight of food that successful animals brought back, as shown for cliff swallows (*Hirundo pyrrhonota*) (Brown, 1986) and suggested earlier for herons (*Ardea herodias*) by Krebs (1974). Important information is, in such cases, available from entirely ordinary, nonformalized sources. However, in infrequent cases involving relatively desperate situations and ephemeral opportunities, at least cliff swallows appear to have vocal signals (and formalized ways of flying) that successful foragers can use in summoning unsuccessful companions (Stoddard, 1988; Brown et al., 1991).

Even when formalized signaling is used, information from sources contextual to it is often decisive for recipients. Much of the information provided by signaling is only broadly predictive (see section "Kinds of Information"). Such information reduces uncertainty, but not to the extent of enabling precise predictions in different circumstances. Still, such broadly predictive messages are probably economical. Nonhuman animals must get through many kinds of interactions with relatively small numbers of signals. Even with multiple signaling repertoires per species, the number of formalized units available to nonhuman animals is much smaller than the number of words available in any human language. Yet the contrast is a matter of degree. All communication, our speech included, is context-dependent. The less we say, the more we depend on sources contextual to our words. Thus "help me," "what's that?," and "start now" each elicit different responses in different events. And, when we elaborate with words we use them to provide context for each other—as, indeed, a nonhuman animal may use its different song types in a sequence of singing.

Nearly all ethologists agree that animals' responses to signals are context-dependent. Yet the concept of 'context' continues to be diluted by weak use of the term and by failure to trace each kind of information that is important in an event to its source (see also West & King, 1996). The weak sense of context equates it with the circumstances in which a signal is performed. Identifying these circumstances is a first step. However, it does not, in itself, reveal which sources of information are accepted as relevant by individuals responding to the signal. More significant, there is a tendency to credit signals with more information than they actually provide. This inflation, a failure to tease apart the sources of information used by a respondent to a signal, is evident, for instance, when investigators lump what an animal has learned about a signal's occurrences in with the signal's messages (e.g., Seyfarth et al., 1980). But what has been learned about events becomes information available to each respondent from its memory, a source contextual to the signal as it occurs in later events. The messages of a signal do not change as an individual becomes more experienced. Rather, the information available from memory changes. Further, a signal's messages are not different for every responding individual, although memories are.

The significance of the distinction between 'immediate' and 'historical' sources of information (TBOC; Smith, 1965) has now begun to be recognized (Evans et al., 1993). Individuals respond to the information made available by unfolding events, within the context of implications generated by their memories of similar events. Memories remain even when the sources of information that are available concurrently with a signal are restricted. And memories may lead subjects to treat the absence of expected, concurrent sources as an informational problem (see section "Focus on the Individual Turns to Its Mind").

KINDS OF INFORMATION

Knowing the many kinds of information in which animals traffic as they communicate would tell us potential uses and limitations of their formalized signaling. It would provide clues to the uncertainties that can be reduced as communication smooths the course of animals' encounters, to the workings of animals' minds, and to the evolutionary pressures shaping the process of communicating.

How, then, can we describe the information that signals make available? Stated otherwise, what are the referents of signals? By 'referent' I mean the things and events about which signals provide information (TBOC:69). Ethologists do not agree on the definition of this term (see Macedonia & Evans, 1993, for a clear outline of different uses). Some ethologists use it only for things and events external to a signaler, and do not include the signaler's behavior. Any narrowing of the term seems inadvisable, however. If a signal provides information about (refers to) a signaler's behavior, yet behavior cannot be called a referent, how do we express the relationship? On the other hand, one recently invoked compound, 'functional referentiality' (op. cit.) is not a narrowing. It instead requires that context independent responding be demonstrated in addition to appropriate use of the signal by signalers. A nice idea. But contextual information cannot be eliminated (see section "Sources of Information"). The test might better be that responding accord with predictions made about the significance of a signal's messages in different circumstances. For now, keep in mind that 'referent' implies different concepts as used by different authors, and comparisons of their inferences must be done very carefully.

Opinions about referents are influenced by the ways in which we study communication. I argued (TBOC:78) that the best way to discover referents is to seek correlations between the performance of signals and such things as the signaler's concurrent or subsequent behavior, attributes of the signaler such as its species and gender, and even stimuli to which the signaler appears to be responding. Some events pose problems for this procedure. In particular, changing conditions can affect a signaler's behavior, rendering obsolete whatever predictions might have been based on its signaling. The problem has led to serious misunderstandings. For instance, Caryl (1979, 1982) reviewed several studies and concluded that signals performed in disputes give good indications of the probability that the signaler will flee, but not of the probability that it will attack. Paton and Caryl (1986) later tried to test the conclusion using new data from great skuas (*Catharacta skua*), but chose the least appropriate subjects imaginable: inexperienced, immature, noncommitted birds showing no site attachment, subjects that avoided squabbles simply by moving away whenever closely approached. Paton (1986) went on to redefine threat signals by imposing criteria for testing responses, criteria that will often be rendered impractical by poor choices of circumstances and subjects.

Outright fighting is uncommon in nature. When an animal signals that it may attack, its opponent usually backs down or threatens in return. Either response usually obviates or forestalls attack behavior (TBOC:78–79, 128–130, 132–133; Stout, 1975; Nelson, 1984; Popp, 1987; Piersma & Veen, 1988). The prediction of attack, like all signaled predictions, is conditional. In these events responses to signals commonly, and very quickly, alter the conditions. Ethologists must then use clues other than the signaler's next actions to infer a signal's referents, clues such as responses made by appropriate classes of recipients

(TBOC:78–79). Much recent research has used playback of recorded vocal signals to evoke responses from which messages are then inferred (e.g., Cheney & Seyfarth, 1990). This useful technique can mislead. It requires meticulous and often difficult separation of the different sources of information being used by responding subjects. Responses reveal a signal's implications to knowledgeable recipients, but there remains the task of dissecting down to the information inherent in just the signal itself.

I assume that the information provided by a signal is fixed for each form in which that signal is presented (TBOC:73–76, 287–288). Why? Because without consistent relations to their referents, signals would be confusing sources of information (Smith, 1985:55, 67–73). Yet each signal does elicit different responses in different events. As long as the form of the signal is unchanged, flexibility in responding is due entirely to the information available from sources contextual to the signal. Additional flexibility can be obtained through varying the form of a signal, however. This flexibility has at times been mistaken for inconsistent referents (e.g., by Beer, 1977, 1980). Yet the referents remain fixed. What matters is that two different signaling repertoires are involved: a repertoire of basic signal units, and a repertoire of ways of altering the form of these units (see section "Signals and Signaling Rules"). In different events the signal may take different forms, but members of each repertoire have consistent relation to their referents.

Consistency, however, does not imply precision. Many signals provide information that enables only broad predictions. I suggested that "low ambiguity...may often be of secondary importance if the necessary ends are achieved in due time without severe penalties for delays" (TBOC:173). Hinde (1981) argued further that signaling will evolve optimal ambiguity, making no more of an event predictable than is appropriate for the signaler. And ambiguity can be exploited in special ways. For instance, Estes (1991b) has argued that females of many bovid species have evolved horns as a way of rendering ambiguous a formerly male-specific feature. The females can thus buffer their maturing male offspring against attacks by dominant males.

A major source of ambiguity in animal communication is that each signal has more than one behavioral selection message. That is, a signal usually makes two or more kinds of behavior predictable (e.g., attack, escape, or indecisive actions), providing only relative probabilities for each. Indecisive behavior is usually the most probable selection at the moment of signaling, a very ambiguous prediction. Ignoring this kind of ambiguity has led to spurious claims. For instance, Seyfarth et al. (1980) and Marler (1984) assumed that if vervet monkey (*Cercopithecus aethiops*) alarm calls had behavioral referents then a caller would flee. Since calling vervets did not always flee, and fleeing vervets often did not call, the authors concluded that the link between alarm calling and escape behavior could be severed. But the ethological literature gives no basis for expecting a 1:1 'link' between escape and a vocal signal. Tinbergen (e.g., 1959) and Hinde (1981) emphasized the internal 'conflict' underlying most signaling.

I argued that most signals do not make available information about a single kind of behavior (TBOC:87, 127–128). Animals signal primarily when they are faced with choices. They report on their alternatives, preparing perceivers of their signals for various possibilities. They use signals consistently, but without resolving all uncertainty.

In TBOC, information made available by signals, the signals' 'messages,' was categorized as being about the signalers' behavior as well as nonbehavioral issues such as the signalers' identity, certain physiological, ontogenetic, and social attributes, and location. For recent, perceptive reviews of the experimental study of ontogenetic issues see West and King (1996), and of information identifying signalers as individuals see Stoddard (1996). Considerable attention has been given to the ability of vocal signals to reveal or conceal information about one aspect of a signaler's location: distance from a recipient of a signal (Richards, 1981; McGregor et al., 1983; Morton, 1986).

The provisional list of behavioral messages in TBOC still appears to be largely appropriate, albeit with alterations. Recent research has addressed how accurately distance of source from self can be reckoned by recipients of far-carrying vocal signals (Wiley & Richards, 1982; Shy & Morton, 1986; Naguib, 1995). The most significant change, however, is that a list of messages whose referents are external to the signaler must now be added.

Most signals apparently have behavioral messages that predict functionally broad classes of activities, such as locomotory or interactional behavior. Such broadly predictive information should support useful predictions, although not very precise ones, and these broad messages appear to be similar even when found in the signals of markedly different species. Some behavioral classes (e.g., attack, which is a subset of interactional behavior) are more narrowly specified. These messages characteristically refer to acts with immediately or continuously pertinent consequences. All behavioral predictions are conditional. Signaling does not entail a commitment by the signaler to follow through, come what may, with a predicted course of action. Rather, each signal provides information about the relative probabilities with which a signaler will select one of two or more courses of action, given that the situation remains within certain limits or develops in certain ways. The situation itself provides information sources that are contextual to signals and help put their information, however broadly or narrowly predictive, into perspective.

Some ethologists unfortunately continue to impute information to a signal on the basis of an apparent function of that signal. For instance, signals performed in territorial encounters are often presumed to inform primarily about the signaler's probability of attacking. Most such signals are not used as or just before the signaler attacks, however, but while it remains in an encounter and continues to exchange signals with its opponent. There is thus little observational evidence for an interpretation of imminent attack. The signaler acts more like a negotiator than a berserker. A more appropriate interpretation is that the signal is primarily about the indecisive activities that will probably continue to forestall outright attack or capitulation, at least in the short term. (The longer term depends on the course the negotiations take, making it inherently less predictable as negotiations begin.) Signals facilitate prediction of the behavioral steps in such an encounter more readily than they can indicate the final resolution (Smith, 1986a).

Should negotiating be added to TBOC's list of behavioral selection messages? Probably not. A signaling individual can provide information about the probability that it will behave indecisively (TBOC:106–108): that it will pause, may make incomplete movements, will probably not commit immediately to any directly effective action, and will probably continue to signal. It takes two to negotiate, however. Some negotiating routines are patterned as formalized interactions (e.g., those of the fighting fish and wildebeest, cited above), but many others may be largely spontaneous. We can recognize negotiation when two parties forego full resolution of an issue and keep options open, engaging at least briefly in an exchange of signals.

Commonplace negotiating can be seen in the minutiae of any orderly interaction, as participants adjust to each other's moves. Each seeks as much control as it can get (see section "Information in Social Interactions"), yet makes participating worthwhile for the others and fosters tolerance and interdependence. Tolerance is dramatically revealed in the behavior of reconciling (an interaction which must itself be negotiated), now recognized as widespread among species of primates (de Waal, 1982, 1986, 1989; Cords, 1992, 1993; Kappeler, 1993; Kappeler & van Shaik, 1992) and perhaps awaiting discovery in other kinds of animals.

I now repeat a caveat given repeatedly in TBOC (pp. 81, 84, 114, 461): the interpretation, categorization, and naming of messages must remain provisional. Categorization organizes our data and gives us hypotheses, but must be continuously scrutinized and discarded or modified as new evidence comes in. For instance, of the behavioral selection messages suggested in TBOC, I no longer consider 'seeking' to be an appropriate description of a behavioral category. The examples of signals with that message given in TBOC (pp. 118–121) are, except for those involved in seeking to interact, more simply interpreted as conditional predictions of associating, copulating, attacking, or escaping. The case for information about seeking to interact is more complex, and involves an unexpected kind of information. Recent research on birdsongs (Smith, 1988, 1996; Smith & Smith, 1992, 1996a, b, c) reveals that many species of birds (including oscines, suboscines, and even nonpasserines) use songs in making available information about different extents to which a singer may take initiative to interact. In some circumstances, for instance when it has well-established boundaries with its neighbors, a singer can indicate that it will take initiative only to the extent of singing. The singer is usually preoccupied with self-maintenance activities, but will respond if another individual initiates a closer interaction. In effect, the singer defers most social initiative to other individuals. In other circumstances, the same signaler can use songs differently (e.g., by combining two or more song types in different proportions) and indicate that it is actively searching for other individuals and will approach them. Even while distant, neighbors, strangers, and the singer's mate get information with which to decide whether to engage the singer in a closer interaction or to avoid interaction. In fact, information about the extent to which a signaler will take initiative is especially useful when the individuals linked by communication are not close to one another. Searching, approaching, and even singing itself can all foster interactions. However, it strains the category of 'seeking' behavior to include singing while deferring further initiative to others. The message should perhaps simply be interactional behavior, with supplemental information about how actively the signaler will foster encounters.

A serious and vexing problem I tried to address in TBOC remains stubbornly alive: many ethologists still equate information about behavior with information about internal states (e.g., Dawkins & Krebs, 1978; Paton & Caryl, 1986; Seyfarth et al., 1980). Some even describe TBOC and my other work as interpreting communication in terms of internal states (variously: arousal, affect, motivation and emotion), despite my explicit statements to the contrary (e.g., Smith, 1981) and my account of how analyses of inferred internal states and of observed behavior are complementary (TBOC:195–196). The two kinds of analyses belong to entirely different levels of integration (Fig. 2). The irreducibility of biology into physical phenomena (Mayr, 1982) is nowhere more strikingly obvious: most of communication's key features make sense only when understood as somehow forged in the linking of different individuals. Yes, physiological states are part of the proximal causation of behavior, but no, the product is not the same thing as the processes that yield it. Further, casual claims made about animals' 'intentions' are merely restatements of behavioral predictions and incorporate no discernable concept of intending. To describe a signal as revealing behavioral intentions rather than probable behavior unproductively broaches the question of how to know when an animal's actions are intentional. Much research at the individual level of integration now focuses on the cognitive

ANALYSIS	LEVEL OF INTEGRATION	
	Organismic	Interactional
behavioral unit analyzed	an act by a single individual	an orderly interaction involving two or more individuals
causes analyzed	internal states of the individual	actions of individual participants

Fig. 2. Motivational, emotional, and cognitive internal states organize an individual's response to external and internal stimuli. On the other hand, it is actions of individual participants that initiate, guide, sustain, and terminate an interaction.

processing of information (see section "Focus on the Individual Turns to Its Mind: Cognitive Ethology"). Motivational analyses have become uncommon (but for a significant and stimulating case, see Moynihan, in press).

I saw no obvious reason why animals' signals could not provide information about external referents (TBOC: 69, 73–74; i.e., things and events that exist apart from, externally to, a signaler). No clear cases had been established for vertebrate animals, however, and many suggested instances were based on simplistic interpretation. Further, in describing the dances of honey bees (*Apis mellifera*), which clearly have an external referent, I focused on information about behavior. Because behavioral messages are widespread or universal in animal communication, I reasoned that the evolution of honey bee dancing had initially involved natural selection to encode behavioral information. By implication, information about a resource (usually nectar, sometimes water or a site for a hive) might be epiphenomenal. That is, I assumed preexistence of cognitive mechanisms for acting on signaled information about behavior, and that new mechanisms would have to be developed to process information about external referents. Be that as it may, it is an issue in the evolutionary history of cognition. We know too little about the cognitive abilities of bees to assign secondary status to the external referent of their dancing. (Macedonia and Evans, 1993, suggested that the need to signal about "fundamentally incompatible escape responses" drove the evolution of predator-specific alarm calls in primates. They correctly saw that both behavior and predators are referents of such signals.)

Behavioral messages involving external foci presented me with an obvious problem. The behavioral selection message about acts of staying put at a site or (say) within a territorial boundary (TBOC:115–118), and the message about monitoring an event or thing (TBOC:126), both involve actions defined with respect to externalities. Moreover, the behavior described by the majority of known behavioral selection messages requires an external focus. Attack behavior requires an object of attack, escape something to escape from, associating a companion, and so forth. Items external to a signaler are involved in all but four of the dozen behavioral selection messages I described. (The exceptions can be defined wholly in terms of a signaler's behavior.) As in the honey bee case, I resolved the problem by assigning priority to the behavioral information and considering the other information to be implicit. Implicit or otherwise, however, the nonbehavioral information is present. Don Griffin pointed out to me (in conservation, 1979) that there is no reason why a signal should not evolve to provide information about both behavioral and external referents. Owings and Leger (1980) explicitly recognized information about both predators and the signalers' behavior in variants of the chatter vocalization of California ground squirrels (*Spermophilus beecheyi*). More recently, Macedonia and Evans (1993) have made a lucid review of this and related issues. Griffin's point works both ways: signals for which external referents have been described also provide information about signalers' behavior, and most signals with behavioral messages also make available information about external referents.

Research on external referents of vertebrate signaling at first focused on what are known as alarm calls. Cases were found in signals of ground squirrels (initially by Owings & Leger, 1980) and primates (Seyfarth et al., 1980, experimentally testing observations by Struhsaker, 1967) in which vocalizations or their variants make available the information that the signaler has detected a predator—even a designated kind of predator. The primates were more narrowly precise than the ground squirrels, who uttered some of the same variants in intraspecific disputes, making the signals' relations with predators context-dependent. There was initially some tendency to contrast these findings with inferences derived incorrectly from TBOC (see rejoinder by Smith, 1981). The misunderstandings, for instance over the distinction between an animal's behavior and its internal states, over the multiplicity of each signal's referents, and over distinctions among sources of information, have been largely clarified. Subsequent work on primates has revealed signals that provide information about the class of opponent with which the signaler is contending (Gouzoules et al., 1984; Gouzoules & Gouzoules, 1989), the detection of a neighboring troop (Cheney & Seyfarth, 1982), and the discovery of a valuable resource (usually, but not always, food—as in honey bees, Dittus, 1984).

An especially promising suggestion of Gouzoules et al. (1995) may provide a way through an impasse in interpreting the referents of signals. The idea is that we should not categorize the objects and events correlated with a signal more narrowly than the signalers themselves do—that where we have proposed distinct classes, to which a thing or event can either belong or not, animals may instead classify into fuzzy sets. Items can belong to a fuzzy set (say, aerial predators) "to some degree" (op. cit.:248). Although I had recently recognized the utility of fuzzy categorization (Smith, 1991b:221), I had not begun to grasp its full potential significance. Now, in cases where we find only 'imperfect' correlations between things and performances of a signal, the heuristic of fuzzy sets can guide us in devising tests for less precisely constrained conceptual categories. There will be a problem, however, in determining when animals' signals have fuzzy categories of referents and when we have instead simply failed to specify appropriate categories that are not fuzzy. Use of fuzzy sets must not become an excuse for superficial research that is satisfied with rough descriptions of a signal's correlates.

The signals for which external referents have now been explicitly described have interesting commonalities. Most are vocalizations (excepting the honey bee's dance performance). All are performed when either the stimulus (e.g., a predator, or a site with food) is sufficiently distant from signal recipients that they may not have detected it, or a crucial individual is absent (e.g., a young macaque's mother may not be close and able to see the signaler encounter an opponent).

Perhaps these commonalities are an artifact, to disappear as more external referents are studied. Alternatively, they may imply that the evolution of signals with prominent external referents occurs primarily when some degree of distance may cause stimuli important to the signaler to be missed by other individuals. Perhaps, when individuals are close and accessible to one another, they can usually interact without signaling primarily about obvious local externalities. But these individuals do need to know how each other will act, and must make available some information about their probable behavior. If most signals do provide information about both behavioral and external referents, the issue of which is more important in communication becomes a matter for research on cognition.

The precision with which external things are referenced may have to do with how seriously different actions of a species may conflict. This was suggested by Macedonia & Evans (1993). They made a nice case that only when members of a species have had to choose between markedly different escape tactics to evade different kinds of predators have their alarm calls tended to evolve referential specificity, such that different calls are well correlated with different types of predators.

Referents have also been described in terms of situational properties. For instance, when a species' different 'alarm calls' or their variants do not correlate with different kinds of predators, their referents have sometimes been interpreted as different degrees of danger, or risk (e.g., Marler, 1961; Leger et al., 1980; Macedonia & Evans, 1993; Blumstein, 1995). Perhaps degree of danger varies directly with the probability that the signaler will flee, freeze, take cover, or monitor the source of danger. A description of this behavior offers more detail than a description of the level of danger, and so may be better for some analytic and comparative purposes. Also, a signaling difference correlated with different classes of predators (e.g., aerial vs. terrestrial) provides information that need not covary with the degree of urgency in an event (Evans et al., 1993). Otherwise, situational and behavioral or eliciting stimulus formulations are equivalent. When formulations are fully equivalent, who is to say which is better? Again, the question becomes cognitive: how do the animals represent the referents to themselves?

However, if a signal is said to be produced, say, during territorial encounters we can say that closer correlates have usually been overlooked. The signal might come with attack, withdrawal, or (most likely) with indecisive actions perhaps biased toward attacking or fleeing. Or it might be used during standoff behavior in static phases of negotiations. The simple situational interpretation in such a case is inadequate, simply a step toward finding more precise correlates.

Perhaps, as suggested above, when a signaler is not closely interacting with recipients of its signal, information about the things or situation eliciting the signal may be more useful than information about the signaler's behavior. During close interactions much of the information participants need about the situation is available from sources other than signaling.

RELIABILITY

To a great extent, signaling leads to reliable predictions. It has to. Why else would recipients so often base responses on it? Misleading signaling, in which information is provided that can lead responding individuals to make incorrect predictions, is not entirely precluded. However, it is relegated (at least for intraspecific communication, and especially among familiar individuals) to limited, and usually subtle, ancillary uses. And, because misleading signaling can develop, there is evolutionary selection favoring both skeptical responders (Moynihan, 1982; Smith, 1986a) and self-certifying signalers (Zahavi, 1975, 1991; Smith, 1986a).

Some ethologists appear to believe that, in my research on the messages of signals, I *assume* reliable signaling. No. I find correlates of signal performance (whether in behavior, identities, or elsewhere) empirically, by observation and experiments. Those correlates reveal at least some of the information made available to an individual that notices the performance. That information does provide a basis for reliable communication. Perhaps some signalers (undetected by the research) occasionally violate the correlations. But in general, the reported correlations hold.

In TBOC I gave little attention to the issue of reliability, and Krebs (1977) rightly criticized this. Only interspecific examples of misleading signaling were available in 1975, and I mentioned only fleetingly the possibility of intraspecific cases. Many attempts have since been made to detect intraspecific misleading signaling. Few attempts have succeeded. Most have revealed only reliable signaling.

It is essential to understand that there are real limits to the precision with which signals enable behavior to be predicted. The principal limit is that signaling behavior provides information about the relative probabilities of the signaler choosing one of two or more incompatible activities, not definitive probabilities of each possible choice (TBOC; Moynihan, 1982; Hinde, 1985; Smith, 1986a). Further, the predictions that can be made are conditional, and thus maximally viable only at the moment of signaling. Recipients of signals thus cannot know exactly how likely attack (say) is, or precisely how the signaler may respond to (say) counterattacking. (Ethological observers too can make only relative predictions, and these are presumably less precise than are the programmed expectations of the animals themselves.) Limits to achievable precision should give signalers some opportunity to profit from biasing responders' expectations (see, e.g., Dawkins & Guilford, 1991). Yet all current evidence suggests that opportunity to be highly constrained in intraspecific communication.

Why so constrained? In many cases it must be partly because vertebrate animals come to know one another as individuals. Before individuals are familiar with one another (e.g., when autumnal flocks of birds are forming) they may estimate outcomes of potential encounters on the basis of gross cues such as plumage markings. Once familiar, they switch to reliance on their experience (Parsons & Baptista, 1980; Lemel & Wallin, 1993). With familiarity comes accumulated knowledge of the ways in which each individual tends to behave (van Rhijn & Vodegel, 1980; Smith, 1986a). Even with less familiarity, recipients of signals can test predictions, for example by attacking a threatening opponent (e.g., in flocks of house sparrows, Moller, 1987). When faced with the likelihood of eventual change (e.g., seasonal), individuals can repeatedly test one another (van Rhijn & Vodegel, 1980). Neighboring territorial male wildebeest do this daily with formalized sparring and a vomeronasal sampling of urine (apparently assaying androgen levels that reveal aggressiveness, Estes, 1969). Zahavi (1977) proposed that one individual can test another by imposing on it, and will get reliable information because the cost of responding is either borne or not. Some tests can be subtle. A pair bond, for instance, may be tested repeatedly as mates demand food or attention (Smith, 1994), share space, interfere with each other's activities, and periodically squabble.

The evolution of skepticism provides impetus for the evolution of 'handicaps' (Zahavi, 1975) or self-certifying signaling (Smith, 1986a). That is, for signaling that cannot be faked. Examples range from the frequent roaring of red deer (*Cervus elaphus*) stags contesting for possession of harems (Clutton-Brock & Albon, 1979), and the state of musth in male elephants (*Loxodonta africana*) (Poole, 1989a,b), to the aerial displays of male harriers (*Circus cyaneus*) (Simmons, 1988) and bobolinks (*Dolichonyx oryzivorus*) (Mather & Robertson, 1992), and even the use of what were called 'tokens' in TBOC (Moreno et al., 1994). But, again, there are limits to the perfection of reliable signals. Signalers differ from one another in many ways, and even handicaps cannot expose all of this variation (Johnstone & Grafen, 1993).

It was Dawkins and Krebs (1978) who called attention to the evolutionary pressures that could lead to the evolution of misleading communication and other 'manipulative' signaling that elicit responses beneficial only to the signaler. This especially important contribution was, unfortunately, sensationalized. Emphasizing "the struggle between individuals" (op. cit.:309), Dawkins and Krebs treated us to the remarkable assertions that "if information is shared at all it is likely to be false information" (op. cit.:289) and "cooperation, if it occurs, should be regarded as something surprising" (loc. cit.). Although they later repudiated both claims (Krebs & Dawkins, 1984:388), their initial account continues to be widely cited and its spirit upheld (e.g., Gross, 1994, categorized cooperation as an "enigma" in his review of behavioral ecology). Both accounts of Dawkins and Krebs reveal serious misunderstanding of the ethological literature on communication. Both versions are rife with theatrical metaphors. ("Mind reading," for instance, is no more—in fact less—than the capacities described in TBOC:9.) Although evocative and entertaining, such terms circumvent reasoned discussion.

Dawkins and Krebs made the surprising, unrealistic assumptions that reliable signaling requires a signaler to make available all the information it has, and to follow through with the behavior its signals predict. Yet natural selection cannot favor signals specialized to make public information that works to a signaler's net disadvantage. That sort of 'leakage' (term from Ekman & Friesen, 1969) is instead selected against. All signaling is selective, some information is made available and much is withheld (Smith, 1986a). And all predictions are conditional. To reiterate, signaling does not commit an animal to take action that becomes inappropriate.

'Informational' and 'motivational' analyses were contrasted in TBOC. Nonetheless, Dawkins and Krebs (1978) equated them, misrepresenting the informational as simply a rephrasing of the motivational. Yet the informational approach I described requires an ethologist to find information in visible, audible, or otherwise accessible correlates of signaling. No inferences are made about motivational, emotional, intentional, or other internal states, including 'arousal.' In contrast, ethology's more traditional motivational analyses do make inferences about the internal workings of individuals. In effect, such analyses seek the information available within an individual that is integrated to yield a coherent behavioral output. Such information is used at the level of the individual organism, not at the social level of integration. Disparaged by Griffin (1991) as the "groans of pain" approach, traditional analyses must assume hypothetical CNS states about which we know little. It is not clear to what extent information about internal states is useful social currency.

Dawkins and Krebs themselves hewed close to the traditional (and largely outmoded) 'releaser' model of communication. This model requires an animal to respond automatically, in preordained ways, on receipt of a signal. Individuals in immediate peril and infants sometimes do respond almost this simply, and invertebrate animals may in many circumstances. But infants initially lack information stored as memories, individuals in peril often lack the time to gather and assess information from multiple sources, and many invertebrates may lack sufficient processing capacity to be flexible. Other responding to signaled information by vertebrate animals (and in some cases honey bees, fire ants (*Solenopsis invicta*), and other invertebrates) is always context-dependent.

Stimulated to look for deceptive use of intraspecific signaling, ethologists found a few nice examples in interactions among invertebrates (e.g., Thornhill, 1979; Caldwell, 1986; Adams & Caldwell, 1990; Adams & Mesterton-Gibbons, 1995). Much more often, however, signaling even in highly competitive situations was found to be reliable (e.g., in butterflies: Rutowski, 1979; fish: Nicoletto, 1993; lizards: Thompson & Moore, 1991; newts: Halliday & Houston, 1978; Green, 1991; frogs: Wagner, 1992; birds: Petrie, 1988; Simmons, 1988; Hill, 1990; Evans & Thomas, 1992; Mather & Robertson, 1992; Moreno et al., 1994; Andersson, 1994; ungulates: Clutton-Brock & Albon, 1979; Poole,

1989a,b; and primates: Boinski et al., 1994). Misleading intraspecific signaling seems hard to find. Perhaps, if not rare, it is usually subtle.

Big lies are usually impractical for individuals that encounter each other frequently. Perhaps slight, low-cost deceptions are often overlooked, and are more common (Otte, 1974). But minor cases of misleading may be as hard for ethologists to detect as they are for deceived animals. So, too, may be cases in which an animal manipulates another's responses by failing to signal when signaling would be expected—when it has detected a predator, for instance. I have read no reports of definitive cases, although Cheney and Seyfarth (1990) summarize some suggestive reports involving nonhuman primates.

Various observations may imply some level of 'tactical deception' (Byrne & Whiten, 1985, 1992) in the use of signals. That is, use of an otherwise reliable signal in an event in which it is not appropriate and elicits responses useful only to the signaler. Alternative explanations are hard to exclude, however. Munn (1986), for example, describes utterances of 'hawk alarm calls' that distract competitors from the pursuit of food, to the signaler's advantage (see also Matsuoka, 1980). But whether these vocalizations are truly 'hawk alarm calls' is unknown, and Munn found them uttered to hawks *less* often than they occurred in competition over food. Close study might reveal them in any events in which the signaler may suddenly flee. (Many 'alarm calls,' e.g., those of California ground squirrels (Owings & Leger, 1980), are commonly uttered by losing participants in intraspecific fights. These vocalizations thus are not simply 'alarm calls' in the usual sense of responses to predators.) In an "aerial tumble" (Munn, 1986:170) among birds as they chase a flushed prey item, the calls might be uttered as a signaler faces a decision to avoid its flockmates—a decision that is obviated if the call causes them to flee. Is this deception, or serendipity? (Competitors might flee on hearing the call if, while concentrating on the contest, they were not alert for predators and so used a 'worst-case scenario'; see section "Focus on the Individual Turns to Its Mind.")

We cannot safely recognize misleading use of a signal until we know the full range of its reliable uses. Only then do we know all the kinds of information the signal makes available (its messages; for a detailed example of the need for complete evidence see TBOC:74–76). That is, to know that a performance in circumstance y is a simulation (Thompson's term, 1986) of performances in circumstances a-n, we must know all that is included in a-n. Too often a signal is interpreted (even named, e.g., as a 'food call,' see criticism by Byrne, 1989) on the basis of limited knowledge of its use, and then subsequent cases of wider use are assumed to show deception.

Evolution could not produce exclusively signals that mislead, and produce none that provide reliable information. Signals function for signalers only by eliciting responses. Readiness to respond can evolve only if such behavior is, in balance, adaptive for responding individuals. Predispositions can evolve to accept and act upon misleading information only if (a) it is provided by signals that resemble some preexisting sources of information that *do* require the target response, and (b) if the existence of misleading signals does not overly diminish the advantage of responding to their reliable models. That is, misleading signals evolve through mimicry (TBOC:395–396; Smith, 1986a; or simulation, Thompson, 1986). They buy into already adaptive responses, subverting them.

The obvious examples are interspecific and life-threatening. Many poisonous animals have warning colors (structural 'badges') which predators avoid. Populations of edible, Batesian mimics copy the warning badges. Mimics that do not become too numerous relative to their models are avoided. Camouflaged animals mimic background environments and are overlooked to the extent that predators cannot afford excessive scrutinizing of the background. The predators go hungry. Natural selection thus favors predators that can distinguish models from mimics. This leads to selection for more accurate mimicry and the evolutionary unstable progression that Dawkins and Krebs (1978) liken to an "arms race."

It may be easier to mislead about identity than about behavior. Forthcoming behavior can be betrayed by many slight movements and postures with which an animal prepares for action, and which can contradict misleading signals (Moynihan, 1982; see related concept of 'leakage,' Ekman & Friesen, 1969).

Life-threatening events are not the norm of social interaction, most of which is intraspecific. Instead, encounters tend to be low-risk and orderly. Each participant profits in some way, although individuals need not profit equally, in the same ways, or in every event. In contesting over a resource, a dominant may gain and a subordinate forego it, but the latter individual avoids injury and remains a member of the dominant's group. Group membership pays off over time. Only with each other's tolerance and accommodation can individuals get social food-finding or antipredator benefits, share territorial borders without endless testing, or raise offspring together.

Social behavior is rooted in coordination and cooperation. Although individuals are necessarily selfish and do compete, they also need one another (Smith, 1986a). They compete in addition to cooperating, not "rather than" (Dawkins & Krebs, 1978:289). We must allow for disparities among individuals, and focus not simply on short-term, ephemeral issues but also on long-term needs and game plans. Some level of misleading signaling (deception, or manipulation) must occur (Dawkins & Guilford, 1991; Johnstone & Grafen, 1993). Deceitful withholding of information in circumstances in which its provision is expected may also occur. The task is to detect such phenomena, which are specialized to avoid detection. At this stage, we cannot assess their overall significance, beyond saying that reliable signaling is the norm—complex social behavior is not built on deception.

ADAPTIVE SIGNIFICANCE OF COMMUNICATIVE BEHAVIOR

The cost/benefit perspective is central to sociobiology and behavioral ecology. These related schools began to blossom in the late 1970s. Their focus on the causes of evolutionary fitness is narrower than ethology's. Ethologists are interested in both ultimate (evolutionary) and proximate (current) causes of behavior. Proximate causes include, at different levels of integration, the internal states of individual animals and the moves participants make in setting the direction or maintaining the momentum of an interaction.

The search for fitness implications of behavior patterns rapidly dominated in behavioral ecology. As Wilson (1975) predicted, sociobiology (and behavioral ecology) replaced ethology. For a while. Even though Wilson himself (1976:718) graciously accepted Barlow's 1976 prediction that his newly named field would instead become part of ethology, realization of that relationship has been slow to emerge.

Much has been gained, most basically: well-focused questions, great strides in experimental testing procedures, and mathematical modeling (albeit not always related to the real world). Kin selection, altruism, evolutionarily stable strategies, parent-offspring conflicts, resource defense, and other issues have been identified and subjected to cost/benefit analyses. We are increasingly aware of ecological constraints and opportunities that affect such issues as the local structures of populations and social groups, the circumstances of cooperative breeding, the ways trade-offs shape patterns of life histories, different patterns of mating, and optimal decision making.

But much has been lost. Interest in proximate mechanisms declined, to be revived primarily in just one area, the study of cognition. To many adherents of the new evolutionary school, the only significant questions are about fitness. Yet fundamental biological questions do exist outside the intergenerational time-frame and population level. The ontogenetic and moment-by-moment frames have been obscured and made unpopular. So has the interindividual scale at which communicating is done. Simplified assumptions about the functions of behavior (e.g., a vocalization uttered on sighting a predator is termed an 'alarm call' without further observations to determine if it has wider uses and provides other information; a bird's songs are labeled 'territorial' without recognizing their many other functions) have led detailed observation (ethology's hallmark) to be neglected. Social events have been modeled largely as battles for individual advantage. Communication has been called manipulation. Cooperation has been modeled less than have ways in which individuals might control and manage social events.

With sociobiology/behavioral ecology ascendent, there has been too little recent interest in how animals come together with disparate agendas and yet conduct orderly social interactions. The proximate issue of how social behavior is conducted, however, is no less central to a unified biology than is evolutionary fitness.

Note: my goal is not to disparage sociobiology and behavioral ecology, their ultimate questions, and their accomplishments. It is, rather, to point to the loss of interest in proximate phenomena, and to say that biology needs to understand both the proximate and the ultimate. Surely this is beyond debate. Just as surely, though, proximate questions have been out of fashion in the rush to merge behavioral and population biology.

FOCUS ON THE INDIVIDUAL TURNS TO ITS MIND: COGNITIVE ETHOLOGY

A mind is a subjective state, not necessarily conscious, that arises from the activity of a brain working with information. A mind formulates representations of things and events in an individual's world, organizes these into coherent relationships, and exploits those relationships—for instance, by making predictions and devising responses. To understand how animals' minds work, we need to know how animals make decisions. What information do they use? Where do they get it? How do they categorize and organize it? What predictions do they make, and what is their grasp of conditionality? What evidence do they show of information stored in memory? Does channeling of their flexibility reveal cognitive constraints? Concerning communication: knowing input (the information available from a signal and at least some contextual sources) and output (the behavior of a recipient of that information) can we begin to infer aspects of the recipient's information processing and decision rules?

Consistent with my interactional perspective, I view cognitive research from an "information processing approach" (Yoerg & Kamil, 1991:279). Such an approach is characteristic of cognitive psychology. Unlike many cognitive ethologists (e.g., Griffin, 1984; Ristau, 1991) I do not focus on the issues of consciousness and intentionality, of mental experience, but instead on the mental processing of information. To say that animals use information and select responses does not imply conscious cognitive process. We unconsciously process large amounts of information selectively gleaned from our sensory organs and memories just to walk about in any complex setting. We constantly process information from exteroceptors, proprioceptors, CNS analyses of blood chemistry, memories of skills, and so forth. We, in all of our subsystems, live on information. Most of it is beyond conscious access. And most does not get communicated, but remains 'private,' shared only among cells and systems within our bodies.

TBOC was intended to focus attention on interactional implications of communication, yet a crucial issue was cognitive: responses made to signals are context-dependent. Signals alone provide insufficient information to enable recipients to choose appropriate responses. A male moth detecting a female's sex pheromone must also detect the wind direction. A territorial bird responding to a song must determine whether the singer is within or beyond his borders. A ground squirrel hearing an alarm call must know the direction of the nearest burrow entrance, and is helped if it also knows the direction of the predator. Careful observation of such simple cases, plus experiments, can show us what sources of information are sought as animals respond to a signal, and how information from different sources is weighted in different circumstances.

Context-dependency is basic, underlying all perception. Evaluation of signals is influenced by concurrent sources of information, experience, and genetics ("cognition has not evolved as a wholly neutral filter," Wilson, 1994:352). Context-dependent evaluation of information implies appreciably complex cognitive processes. Many of these processes may be skills that become highly automatic, such as binocular depth perception. But cognitive processes must also categorize and organize unique information from many sources in events that are far from fully predictable.

Categorization of information involves focal:contextual distinctions. As in the figure:ground distinctions of visual perception, minds select, rank, and evaluate information sources. Patterns change from moment to moment, and many kinds of changes fall into familiar, quasi-predictable sequences. At the simplest, a receding object is sensed as smaller by the retina but is perceived as constant in size if appropriate depth cues are available. More complex changes occur in the patterns of conduct of social interactions, patterns that are usually probabilistic and conditional. Information from many sources is integrated, potential trajectories envisaged, and predictions continually examined, compared with alternative sequences, and adjusted or rejected.

Animals' responses to signals enable us to explore the cognitive processes used in evaluating significance in the context of information from multiple sources. We can ask how animals fit such information into patterns and how they test their predictions as an event unfolds. We can explore their grasp of conditionality, and the alternative paths that are revealed.

Animals must need, as we do, heuristic formulas to guide cognitive processing. For instance, the first step in responding to signals with only broadly predictive referents (such as 'interactional' behavior or 'food resource') might be to seek additional evidence of its most probable expectation (Smith, 1985). Expectations would be based in part on experience of common sequences of events. For each unique instance, an individual might tailor a 'predictive scenario' (Smith, 1990, 1991b), a flexible guide to finding or imposing a pattern. When having only limited information, the individual might choose among options such as delaying response while it seeks further information, or, if the event appears nonthreatening, responding in terms of an open 'typical case scenario' that can be made more precise as the episode develops. Conversely, if limited information suggests immediate peril, an individual might adopt a 'worst case scenario' and behave cautiously or preemptively.

Implications for the design and interpretation of experiments arise from the possibility that animals will respond to signals on the basis of simplified scenarios when information from other sources is limited. For instance, when an alarm call is played from a hidden speaker, a responding subject cannot look toward the caller and see the direction of its gaze. Even if members of the species utter the call in a range of situations (e.g., when losing an intraspecific fight as well as when fleeing from a predator), a subject responding to playback may 'assume the worst' and rush to cover. To conclude from this response that the call provides information about a predator is premature. The problem can be addressed by making more observational studies of naturally occurring uses of the call, a time-consuming but traditional ethological procedure.

The problem with the design of the above experiment was that useful sources of information were kept from the subject. They are not always available in natural events either, and animals may be prepared to deal with a paucity of information by selecting fail-safe responses. The experimental limitation ('control') of sources of information is thus not an unalloyed blessing, for it creates a special class of situations. There can be no context-free stimuli: paucity of concurrent contextual sources of information is itself a circumstance that affects selection of responses. And there are always stored ('historical') sources of information, many knowable only through developmental studies (Evans et al., 1993).

Despite such problems, experimental presentations of stimuli are valuable. For instance, working with vervets, Cheney and Seyfarth (1980, 1990) have used playback of 2-year-old juveniles' screams to test for knowledge of social associations. In these experiments, groups of three females each all recognized the identity of the 'calling' juvenile. Its mother looked toward the playback, but other females looked toward the mother, often even before she made any overt response. Cheney and Seyfarth (1990) have also demonstrated vervets' ability to calibrate one another (see section "Reliability"). Repeated playback of one individual's 'leopard alarm' vocalization got subjects to habituate, but only to the vocalization as uttered by that individual. Playback of the same vocalization recorded from other individuals continued to elicit responses. So did playback of 'eagle alarm' vocalizations from the first individual—calibration was very specific! For animals that live in enduring social groups, the cognitive ability to calibrate each other's signaling is important. It must considerably affect the ubiquitous process of negotiating, and it permits social complexity that comes from a group having individualistic members.

Many animals do not rely simply on signals when assessing the capabilities of foes or potential mates. Assessors calibrate other individuals in play (Simpson, 1976; Mitchell & Thompson, 1986) and in competition. Smaller-horned rams (*Ovis canadensis*) watch contests between larger-horned individuals, and may challenge the loser of such a contest (Geist, 1971). Female elephant seals

(*Mirounga angustirostris*) incite fights among males, accepting winners as mates (Cox & LeBoeuf, 1977). Most defenses against unreliable signaling are means of obtaining additional information. This implies that assessing is often skeptical, a complex cognitive process of checking signals and other sources of information against each other.

Be alert to the cognitive implications of unstated assumptions. For instance, the idea that signals evolve to be misleading rests on several assumptions about cognitive processes. Largely unskeptical gullibility is one. Another is in the suggestion by Dawkins and Krebs (1978:304–305) that "advertisements...have little to do with the conveying of information...they are there to *persuade*." As a persuasive technique, Dawkins and Krebs offered redundancy: "repetition to the point of what seems like inanity" (loc. cit.). But is inane repetition really characteristic of signaling? Dawkins and Krebs borrowed their imputation of apparent inanity from Wilson (1975), but neglected both Wilson's caveat that redundancy can be largely in the mind of the observer, and his list of reasons why seeming redundancy provides highly useful information. For instance, advertising animals (e.g., singing birds, frogs, or crickets) often cannot know if any suitable audience is even present. Continual singing makes information available for that unpredictable moment when an audience, still unseen, comes within auditory range. For an audience within range, changes in the rate and rhythmicity of singing, and in the songs being sung, can provide information about changes in activities. That is, singing patterns are not usually static. Even when singing does remain uniform, its repetition provides the information that the singer is continuing as indicated. Information about its persistence is 'new' with every song that is uttered.

The study of communication raises many other cognitive issues. How do animals sort their signals into different categories (Horn & Falls, 1996)? Is a recipient of a signal always attentive to all of the information that is provided? Or is it differently selective in different circumstances, perhaps focusing on information about behavior in close interactions, but sometimes focusing on information about external referents in more distant events? How is the information from a signal represented in an animal's mind? Not with words. We ourselves often do not represent referents with words when responding to our own nonspeech signals. For example, we found that a facial signal called tongue showing indicated a low probability of the signaler either initiating or responding readily to social interaction (Smith et al., 1974). Dolgin and Sabini (1982) then predicted that human subjects should be slower to make interactional overtures to individuals using the signal than to those with similar facial appearance but not tongue showing. The effect emerged strongly in experiments. Yet subjects questioned after their trials were unaware that they had seen the signal. They had responded to it without conscious awareness.

Are our descriptions of the information in animals' signals adequate for probing cognitive processes that cannot be based on words? They may be. At least our descriptions of many of our own unconscious cognitive processes such as depth perception are useful.

Animal communication offers a unique window, revealing many kinds of information that animals process. As we learn what kinds of information they make available with signals, we glimpse matters salient to them in many circumstances. As animals respond to signaling we glimpse how they supplement the information from signals by selecting from contextual sources of information. We can check our interpretations by comparing observations and through experimental alteration and presentation of information sources. We can try to bias animals' choices from among multiple sources. Our understanding of animal communication will require more than a grasp just of the interactional level. Cognitive studies will contribute to our understanding of communication just as studies of that social process contribute to cognitive ethology.

CODA

Were I to write *The Behavior of Communicating* today, there is much I would change. Yet most changes would be tactical rather than substantive. The material in the book retains its value. Certainly much would have to be added about advances in our knowledge of animal communication over the past two decades. The growing fields of sociobiology, behavioral ecology, and cognitive ethology have enormous influence on what we know and how we know it. But I continue to believe that the interactional, informational framework and the synthesis I espoused in 1977 remains fundamental to research on animal communication. Important as it is to understand how communication evolves, how signals function, and how animals' minds work, a central issue will always be to investigate the information in which animals traffic with their formalized signaling, and the information they factor in from sources contextual to this signaling as they select their responses and interact. Communication is a process. It is as indispensable to understand the mechanics of the process and the social arena within which it operates, as it is to understand the significance of its results for evolutionary fitness.

At the social level, our essential task is still "to understand...how the behavior of communicating contributes to the management of interactions and the orderliness of relationships among individuals" (TBOC:465). This goal is no less exciting now than it was 20 years ago. And we need now add the goal of finding ways in which our knowledge of animal communication can contribute to the urgent agenda of conservation biology and other needs of human society.

ACKNOWLEDGMENTS

For the past decade my experimental research program investigating the kinds of information made available by birds' songs has been done with Anne Marie Smith, who has contributed immeasurably to it. The editors of this volume, Nick Thompson and Don Owings, have challenged me on many issues, and greatly increased the clarity of my presentation. They and I hold views of the process of communication that are often opposed but, in the best traditions of editing, they have sought only to see that I made my views clear. Many others, including Don Kroodsma, Ted Miller, Meredith West, and Jane Brockman have worked similarly with my prose in recent years. And many, including Irene Pepperberg, Philip Stoddard, and Benson Smith have helped me refine concepts. To all of them, and to those who have helped me finish the manuscript, especially Anne Marie Smith and Lorraine Palita, my thanks.

REFERENCES

Adams, E., & Caldwell, L. (1990). Deceptive communication in asymmetric fights of the stomatopod crustacean *Gonodactylus bredini*. *Anim. Behav., 39,* 706–716.

Adams, E. S., & Mesterton-Gibbons, M. (1995). The cost of threat displays and the stability of deceptive communication. *J. Theor. Biol., 175,* 405–421.

Andersson, S. (1994). Costs of sexual advertising in the lekking Jackson's widowbird. *Condor, 96,* 1–10.

Barash, D. P. (1975). Neighbor recognition in two "solitary" carnivores: the raccoon (*Procyon lotor*) and the red fox (*Vulpes fulva*). *Science, 185,* 794–796.

Barlow, G. W. (1976). (untitled review of *Sociobiology*.) *Anim. Behav., 24,* 700–701.

Beer, C. G. (1977). What is a display? *Amer. Zool., 17,* 155–165.

Beer, C. G. (1980). The communication behavior of gulls and other seabirds. In J. Burger, B. L. Olla, & H. E. Winn (eds.), *Behavior of Marine Animals, 4: Marine Birds* (pp.169–205). New York: Plenum Press.

Blumstein, D. T. (1995). Golden-marmot alarm calls. I. The production of situationally specific vocalizations. *Ethology, 100,* 113–125.

Boinski, S., Moraes, E., Kleiman, D. G., Dietz, J. M., & Baker, A. J. (1994). Intra-group vocal behaviour in wild golden lion tamarins, *Leontopithecus rosalia*: honest communication of individual activity. *Behaviour, 130,* 53–75.

Brown, C. R. (1986). Cliff swallow colonies as information centers. *Science, 234,* 83–85.

Brown, C. R., Brown, M. B., & Shaffer, M. L. (1991). Food-sharing signals among socially foraging cliff swallows. *Anim. Behav., 42,* 551–564.

Byrne, R. W. (1989). (review of *Primate Vocal Communication*). *Anim. Behav., 38,* 729–730.

Byrne, R. W., & Whiten, A. (1985). Tactical deception of familiar individuals in baboons (*Papio ursinus*). *Anim. Behav., 33,* 669–673.

Byrne, R. W., & Whiten, A. (1992). Cognitive evolution in primates: evidence from tactical deception. *Man (N.S.), 27,* 609–627.

Caldwell, R. L. (1986). The deceptive use of reputation by stomatopods. In R. W. Mitchell & N. S. Thompson (eds.), *Deception. Perspectives on Human and Nonhuman Deceit* (pp.129–145). Albany, NY: State University of New York Press.

Caryl, P. G. (1979). Communication by agonistic displays: what can games theory contribute to ethology? *Behaviour, 68,* 136–169.

Caryl, P. G. (1982). Telling the truth about intentions. *J. Theor. Biol., 97,* 679–689.

Cheney, D. L., & Seyfarth, R. M. (1980). Vocal recognition in free ranging vervet monkeys. *Anim. Behav., 28,* 362–367.

Cheney, D. L., & Seyfarth, R. M. (1982). How vervet monkeys perceive their grunts: field playback experiments. *Anim. Behav., 30,* 739–751.

Cheney, D. L., & Seyfarth, R. M. (1990). *How Monkeys See the World.* Chicago, IL: University of Chicago Press.

Cherry, C. (1966). *On Human Communication.* Cambridge, MA: MIT Press.

Clutton-Brock, T. H., & Albon, S. D. (1979). The roaring of red deer and the evolution of honest advertisement. *Behaviour, 69,* 145–169.

Cords, M. (1992). Post-conflict reunions and reconciliation in long-tailed macaques. *Anim. Behav., 44,* 57–61.

Cords, M. (1993). On operationally defining reconciliation. *Amer. J. Primatol., 29,* 255–267.

Cox, C. R., & LeBoeuf, B. J. (1977). Female incitation of male competition: a mechanism in sexual selection. *Amer. Nat., 111,* 317–335.

Dabelsteen, T. (1985). Messages and meanings of bird song with special reference to the blackbird (*Turdus merula*) and some methodology problems. *Biol. Skr. Dan. Vid. Selsk., 25,* 173–208.

Darwin, C. (1872). *The Expression of the Emotions in Man and Animals.* London: Appleton.

Dawkins, M. S., & Guilford, T. (1991).The corruption of honest signaling. *Anim. Behav., 41,* 865–873.

Dawkins, R., & Krebs, J. R. (1978). Animal signals: information or manipulation? In J. R. Krebs & N. B. Davies (eds.), *Behavioral Ecology. An Evolutionary Approach* (pp.282–309). Sunderland, MA: Sinauer.

de Waal, F. (1982). *Chimpanzee Politics.* New York: Harper and Row.

de Waal, F. (1986). The integration of dominance and social bonding in primates. *Quart. Rev. Biol., 61,* 459–479.

de Waal, F. (1989). *Peacemaking Among Primates.* Cambridge, MA: Harvard University Press.

Dittus, W. P. J. (1984). Toque macaque food calls: semantic communication concerning food distribution in the environment. *Anim. Behav., 32,* 470–477.

Dolgin, K. G., & Sabini, J. (1982). Experimental manipulation of a human non-verbal display: the tongue-show affects an observer's willingness to interact. *Anim. Behav., 30,* 935–936.

Ekman, P., & Friesen, W. V. (1969). Nonverbal leakage and clues to deception. *Psychiatry, 32,* 88–105.

Estes, R. D. (1969). Territorial behavior of the wildebeest (*Connochaetes taurinus* Burchell, 1823). *Zeits. Tierpsychol., 26,* 284–370.

Estes, R. D. (1991a). *The Behavior Guide to African Mammals.* Berkeley, CA: University of California Press.

Estes, R. D. (1991b). The significance of horns and other male secondary sexual characteristics in female bovids. *Applied Animal Behavior Science, 29,* 403–451.

Evans, C. S., Evans, L. & Marler, P. (1993). On the meaning of alarm calls: functional reference in an avian vocal system. *Anim. Behav., 46,* 23–38.

Evans, M. R., & Thomas, A. L. R. (1992). The aerodynamic and mechanical effects of elongated tails in the scarlet-tufted malachite sunbird: measuring the cost of a handicap. *Anim. Behav., 43,* 337–347.

Fridlund, A. J. (1994). *Human Facial Expression.* New York: Academic Press.

Geist, V. (1971). *Mountain Sheep. A Study in Behavior and Evolution.* Chicago, IL: University of Chicago Press.

Goodenough, W. H. (1965). Rethinking 'status' and 'role' toward a general model of the cultural organization of social relationships. *Assoc. Social Anthropologists, Monogr., 1,* The relevance of models for social anthropology: 1–24.

Gouzoules, H., & Gouzoules, S. (1989). Design features and developmental modification of pigtail macaque, *Macaca nemestrina*, agonistic screams. *Anim. Behav., 37,* 383–401.

Gouzoules, H., Gouzoules, S., & Ashley, J. (1995). Representational signaling in nonhuman primate vocal communication. In E. Zimmerman, J. D. Newman & U. Jurgens (eds.), *Current topics in primate vocal communication* (pp.235–252). New York: Plenum Press.

Gouzoules, S., Gouzoules, H., & Marler, P. (1984). Rhesus monkey (*Macaca mulatta*) screams: representational signaling in the recruitment of agonistic aid. *Anim. Behav., 32,* 182–193.

Green, A. J. (1991). Large male crests, an honest indicator of condition, are preferred by female smooth newts, *Triturus vulgaris* (Salamandridae) at the spermatophore transfer stage. *Anim. Behav., 41,* 367–369.

Griffin, D. R. (1976). *The Question of Animal Awareness.* New York: Rockefeller University Press.

Griffin, D. R. (1984). *Animal Thinking.* Cambridge, MA: Harvard University Press.

Griffin, D. R. (1991). Progress toward a cognitive ethology. In C. A. Ristau (ed.), *Cognitive Ethology. The Minds of Other Animals* (pp.3–17). Hillsdale, NJ: Lawrence Erlbaum Associates.

Gross, M. R. (1994). The evolution of behavioural ecology. *Trends Ecol. Evol., 9,* 358–360.

Halliday, T., & Houston, A. (1978). The newt as an honest salesman. *Anim. Behav., 26,* 1273–1274.

Hartley, R. V. L. (1928). Transmission of information. *Bell System Technology Journal, 7,* 535–563.

Hill, G. E. (1990). Female house finches prefer colourful males: sexual selection for a condition-dependent trait. *Anim. Behav., 40,* 563–572.

Hinde, R. A. (1981). Animal signals: ethological and games-theory approaches are not incompatible. *Anim. Behav., 29,* 535–542.

Hinde, R. A. (1985). Expression and negotiation. In G. Zivin (ed.), *The Development of Expressive Behavior* (pp.103–116). New York: Academic Press.

Horn, A. G., & Falls, J. B. (1996). Categorization and the design of signals: the case of song repertoires. In D. E. Kroodsma & E. H. Miller (eds.), *Ecology and Evolution of Acoustic Communication in Birds* (pp.121–135). Ithaca, NY: Cornell University Press.

Johnstone, R. A., & Grafen, A. (1993). Dishonesty and the handicap principle. *Anim. Behav., 46,* 759–764.

Kappeler, P. M. (1993). Reconciliation and post-conflict behaviour in ringtailed lemurs, *Lemur catta* and redfronted lemurs, *Eulemur fulvus rufus*. *Anim. Behav., 45,* 901–915.

Kappeler, P. M., & van Shaik, C. P. (1992). Methodological and evolutionary aspects of reconciliation among primates. *Ethology, 92,* 51–69.

Kennan, J., & Wilson, R. (1990). Theories of bargaining delays. *Science, 249,* 1124–1128.

Krebs, J. R. (1974). Colonial nesting and social feeding as strategies for exploiting food resources in the great blue heron (*Ardea herodias*). *Behaviour, 51,* 99–134.

Krebs, J. R. (1977). Animal communication in ethological research. *Nature, 270,* 120.

Krebs, J. R., & Davies, N. B. (1978). Introduction (to Part 2). In J. R. Krebs & N. B. Davies (eds.), *Behavioural Ecology. An Evolutionary Approach* (pp.155–158). Sunderland, MA: Sinauer.

Krebs, J. R., & Dawkins, R. (1984). Animal signals: mind-reading and manipulation. In J. R. Krebs & N. B. Davies (eds.), *Behavioural Ecology* (2nd ed.) (pp.380–402). Oxford: Blackwell.

Krippendorf, K. (1989). Information theory. *International Encyclopedia of Communications, 2,* 314–320. New York: Oxford University Press.

Kroodsma, D. E. (1996). Ecology of passerine song development. In D. E. Kroodsma and E. H. Miller (eds.), *Ecology and Evolution of Acoustic Communication in Birds* (pp.3–19). Ithaca, NY: Cornell University Press.

Kroodsma, D. E., & Miller, E. H. (eds.). (1982). *Acoustic Communication in Birds, 1 & 2.* New York: Academic Press.

Kroodsma, D. E., & Miller, E. H. (eds.). (1996). *Ecology and Evolution of Acoustic Communication in Birds*. Ithaca, NY: Cornell University Press.

Leger, D. W., Owings, D. H., & Gelfand, D. (1980). Single-note vocalizations of California ground squirrels: graded signals and situation-specificity of predator- and socially-evoked calls. *Z. Tierpsychol., 52,* 227–246.

Lemel, J., & Wallin, K. (1993). Status signaling, motivational condition and dominance: an experimental study in the great tit, *Parus major* L. *Anim. Behav., 45,* 549–558.

Lisman, J. E., & Idiart, M. A. P. (1995). Storage of 7 +/- 2 short-term memories in oscillatory subcycles. *Science, 267,* 1512–1515.

Macedonia, J. M., & Evans, C. S. (1993). Variation among mammalian alarm call systems and the problem of meaning in animal signals. *Ethology, 93,* 177–197.

Marler, P. (1955). Characteristics of some animal calls. *Nature, 176,* 6–8.

Marler, P. (1961). The logical analysis of animal communication. *J. Theor. Biol., 1,* 295–317.

Marler, P. (1984). Animal communication: affect or cognition? In K. R. Scherer & P. Ekman (eds.), *Approaches to Emotion* (pp.345–365). Hillsdale, NJ: Erlbaum Associates.

Marler, P., & Peters, S. (1977). Selective vocal learning in a sparrow. *Science, 198,* 519–521.

Marzluff, J. M., Heinrich, B., & Marzluff, C. S. (1996). Raven roosts are mobile information centres. *Anim. Behav., 51,* 89–103.

Mather, M. H., & Robertson, R. J. (1992). Honest advertising in flight displays of bobolinks (*Dolichonyx oryzivorus*). *Auk, 109,* 869–873.

Matsuoka, S. (1980). Pseudo warning call in titmice. *Tori: Bull. Ornithol. Soc. Japan, 29,* 87–90.

Maxim, P. E. (1982). Contexts and messages in macaque social communication. *Amer. J. Primatol., 2,* 63–85.

Mayr, E. (1982). *The Growth of Biological Thought*. Cambridge, MA: Harvard University Press.

McCowan, B. (1995). A new quantitative technique for categorizing whistles using simulated signals and whistles for captive bottlenose dolphins (Delphinidae, *Tursiops truncatus*). *Ethology, 100,* 177–193.

McGregor, P. K., Krebs, J. R., & Ratcliffe, L. M. (1983). The reaction of great tits (*Parus major*) to playback of degraded and undegraded songs: the effect of familiarity with the stimulus song type. *Auk, 100,* 898–906.

Menzel, E. W., & Halperin, S. (1975). Purposive behavior as a basis for objective communication between chimpanzees. *Science, 189,* 652–654.

Miller, E. H. (1988). Description of bird behavior for comparative purposes. In R. F. Johnston (ed.), *Current Ornithology, 5,* 347–394.

Mitchell, R. W., & Thompson, N. S. (1986). Deception in play between dogs and people. In R. W. Mitchell & N. S. Thompson (eds.), *Deception. Perspectives on Human and Nonhuman Deceit* (pp.193–204). Albany, NY: State University of New York Press.

Moller, A. P. (1987). Social control of deception among status signaling house sparrows *Passer domesticus*. *Behav. Ecol. Sociobiol., 20,* 307–311.

Moreno, J., Solder, M., Moller, A. P., & Linden, M. (1994). The function of stone carrying in the black wheatear, *Oenanthe leucura*. *Anim Behav., 47,* 1297–1309.

Morton, E. S. (1977). On the occurrence and significance of motivation structural rules in some bird and mammal sounds. *Amer. Nat., 111,* 855–869.

Morton, E. S. (1982). Grading, discreteness, redundancy, and motivation structural rules. In D. E. Kroodsma & E. H. Miller (eds.), *Acoustic Communication in Birds, 1. Production, Perception, and Design Features of Sounds* (pp.183–212). New York: Academic Press.

Morton, E. S. (1986). Predictions from the ranging hypothesis for the evolution of long distance signals in birds. *Behaviour, 99,* 65–86.

Moynihan, M. H. (1955). Remarks on the original sources of display. *Auk, 72,* 240–246.

Moynihan, M. (1982). Why is lying about intentions rare during some kinds of contests? *J. Theor. Biol., 97,* 7–12.

Moynihan, M. (1985). Communication and noncommunication in cephalopods. Bloomington, IN: Indiana University Press.

Moynihan, M. (in press). *The Social Regulation of Competition and Aggression, or The Importance of Tactics*. Washington, D.C.: Smithsonian Institution Press.

Munn, C. A. (1986). The deceptive use of alarm calls by sentinel species in mixed-species flocks of neotropical birds. In R. W. Mitchell & N. S. Thompson (eds.), *Deception. Perspectives on Human and Nonhuman Deceit* (pp.169–175). Albany, NY: State University of New York Press.

Naguib, M. (1995). Auditory distance assessment of singing conspecifics in Carolina wrens: the role of reverberation and frequency-dependent attenuation. *Anim. Behav., 50,* 1297–1307.

Nelson, D. A. (1984). Communication of intentions in agonistic contexts by the pigeon guillemot, *Cepphus columba. Behaviour, 88,* 145–189.

Nelson, D. A. (1985). The syntactic and semantic organization of pigeon guillemot (*Cepphus columba*) vocal behavior. *Zeits. Tierpsychol., 67,* 97–130.

Nicoletto, P. F. (1993). Female sexual response to condition-dependent ornaments in the guppie, *Poecilia reticulata. Anim. Behav., 46,* 441–450.

Norman, R. F., Taylor, P. D., & Robertson, R. J. (1977). Stable equilibrium strategies and penalty functions in a game of attrition. *J. Theor. Biol., 65,* 571–578.

Otte, D. (1974). Effects and functions in the evolution of signaling systems. *Ann. Rev. Ecol. Syst., 5,* 385–417.

Owings, D. H. (1994). How monkeys feel about the world: a review of *How Monkeys See the World. Language and Communication, 14,* 15–30.

Owings, D. H., & Hennessy, D. F. (1984). The importance of variation in sciurid visual and vocal communication. In J. O. Murie & G. R. Michener (eds.), *The Biology of Ground-dwelling Squirrels* (pp.169–200). Lincoln, NE: University of Nebraska Press.

Owings, D. H., & Leger, D. W. (1980). Chatter vocalizations of California ground squirrels: predator-and social-role specificity. *Zeits. Tierpsychol., 54,* 163–184.

Parsons, J., & Baptista, L. F. (1980). Crown color and dominance in the white-crowned sparrow. *Auk, 97,* 807–815.

Paton, D. (1986). Communication by agonistic displays: II. Perceived information and the definition of agonistic displays. *Behaviour, 99,* 157–175.

Paton, D., & Caryl, P. G. (1986). Communication by agonistic displays: I. Variation in information content between samples. *Behaviour, 98,* 213–239.

Payne, R. (ed.). (1983). Communication and behavior of whales. *AAAS Selected Symposium, 76,* 1–643. Boulder, CO: Westview Press.

Petrie, M. (1988). Intraspecific variation in structures that display competitive ability: large animals invest relatively more. *Anim. Behav., 36,* 1174–1179.

Philips, M., & Austad, S. N. (1990). Animal communication and social evolution. In M. Bekoff & D. Jamieson (eds.), *Interpretation and Explanation in the Study of Animal Behavior, 1,* 254–268. Boulder, CO: Westview Press. (Reprinted in M. Bekoff & D. Jamieson (eds.)., *Readings in Animal Cognition* (1996). Cambridge, MA: MIT Press.

Piersma, T., & Veen, J. (1988). An analysis of the communication function of attack calls in little gulls. *Anim. Behav., 36,* 773–779.

Poole, J. H. (1989a). Announcing intent: the aggressive state of musth in African elephants. *Anim. Behav., 37,* 140–152.

Poole, J. H. (1989b). Mate guarding, reproductive success and female choice in African elephants. *Anim. Behav., 37,* 842–849.

Popp, J. W. (1987). Agonistic communication among wintering purple finches. *Wilson Bull., 99,* 97–100.

Richards, D. G. (1981). Estimation of distance of singing conspecifics by the Carolina wren. *Auk, 98,* 127–133.

Ristau, C. A. (1991). *Cognitive Ethology. The Minds of Other Animals*. Hillsdale, NJ: Erlbaum Associates.

Rowell, T. E., & Olson, D. K. (1983). Alternative mechanisms of social organization in monkeys. *Behaviour, 86,* 31–54.

Rutowski, R. L. (1979). The butterfly as an honest salesman. *Anim. Behav., 27,* 1269–1270.

Ryan, M. J., & Rand, A. S. (1995). Female responses to ancestral advertisement calls in tungara frogs. *Science, 269,* 390–392.

Sebeok, T. A. (1977). *How Animals Communicate*. Bloomington, IN: University of Indiana Press.

Seyfarth, R. M., Cheney, D. L., & Marler, P. (1980). Vervet monkey alarm calls: semantic communication in a free-ranging primate. *Anim. Behav., 28,* 1070–1094.

Shalter, M. D. (1978). Location of passerine seet and mobbing calls by goshawks and pygmy owls. *Zeits. Tierpsychol., 46,* 260–267.

Shannon, C. E., & Weaver, W. (1949). *The Mathematical Theory of Communication*. Urbana, IL: University of Illinois Press.

Shy, E., & Morton, E. S. (1986). The role of distance, familiarity, and time of day in Carolina wren responses to conspecific songs. *Behav. Ecol. Sociobiol., 19,* 393–400.

Sieber, O. J. (1984). Vocal communication in raccoons (*Procyon lotor*). *Behaviour, 90,* 80–113.

Simmons, R. (1988). Honest advertising, sexual selection, courtship displays, and body condition of polygynous male harriers. *Auk, 105,* 303–307.

Simpson, M. J. A. (1968). The display of the Siamese fighting fish, *Betta splendens*. *Anim. Behav. Monogr., 1,* 1–73.

Simpson, M. J. A. (1976). The study of animal play. In P. P. G. Bateson & R. A. Hinde (eds.), *Growing Points in Ethology* (pp.385–400). New York: Cambridge University Press.

Smith, W. J. (1965). Message, meaning, and context in ethology. *Amer. Nat., 99,* 405–409.

Smith, W. J. (1977). *The Behavior of Communicating. An Ethological Approach*. Cambridge, MA: Harvard University Press.

Smith, W. J. (1981). Referents of animal communication. *Anim. Behav., 29,* 1273–1275.

Smith, W. J. (1985). Consistency and change in communication. In G. Zivin (ed.), *The Development of Expressive Behavior* (pp.51–76). New York: Academic Press.

Smith, W. J. (1986a). An "informational" perspective on manipulation. In R. W. Mitchell & N. S. Thompson (eds.), *Deception: Perspectives on Human and Non-human Deceit* (pp.71–86). New York: State University of New York Press.

Smith, W. J. (1986b). Signaling behavior: contributions of different repertoires. In R. J. Schusterman, J. A. Thomas & F. G. Wood (eds.), *Dolphin Cognition and Behavior: A Comparative Approach* (pp.315–330). Hillsdale, NJ: Lawrence Erlbaum Associates.

Smith, W. J. (1988). Patterned daytime singing of the eastern wood-pewee, *Contopus virens*. *Anim. Behav., 36,* 1111–1123.

Smith, W. J. (1990). Communication and expectations: a social process and the cognitive operations it depends upon and influences. In M. Bekoff & D. Jamieson (eds.), *Interpretation and Explanation in the Study of Animal Behavior, 1,* 234–253, Boulder, CO: Westview. Reprinted in M. Bekoff & D. Jamieson (eds.). (1996). *Readings in Animal Cognition* (pp.243–255). Cambridge, MA: MIT Press.

Smith W. J. (1991a). Singing is based on two markedly different kinds of signaling. *J. Theor. Biol., 152,* 241–253.

Smith, W. J. (1991b). Animal communication and the study of cognition. In C. Ristau (ed.), *Cognitive Ethology. The Minds of Other Animals* (pp.209–230). Hillsdale, NJ: Erlbaum Associates.

Smith, W. J. (1994). Animal duets: forcing a mate to be attentive. *J. Theor. Biol., 166,* 221–223.

Smith, W. J. (1996). Using interactive playback to study how songs and singing contribute to communication about behavior. In D. E. Kroodsma & E. H. Miller (eds.), *Ecology and Evolution of Acoustic Communication in Birds* (pp.375–397). Ithaca, NY: Cornell University Press.

Smith, W. J., Chase, J., & Lieblich, A. K. (1974). Tongue showing: a facial display of humans and other primate species. *Semiotica, 11,* 201–246.

Smith, W. J., & Smith, A. M. (1992). Behavioral information provided by two song forms of the eastern kingbird, *T. tyrannus*. *Behaviour, 120,* 90–102.

Smith, W. J., & Smith, A. M. (1996a). Information about behaviour provided by Louisiana waterthrush, *Seiurus motacilla* (Parulinae), songs. *Anim. Behav., 51,* 785–799.

Smith, W. J., & Smith, A. M. (1996b). Vocal signaling of the great crested flycatcher, *Myiarchus crinitus* (Aves, Tyrannidae). *Ethology, 102,* 725–735.

Smith, W. J., & Smith, A. M. (1996c). Playback interactions with great crested flycatchers, *Myiarchus crinitus* (Aves, Tyrannidae). *Ethology, 102,* 725–735.

Snowdon, C. T. (1983). Language parallels in the vocal communication of callitrichids. *A Primatologia no Brasil* (pp.221–232). (An. first Brazilian Congress of Primatology, Belo Horizonte).

Stoddard, P. K. (1988). A rare avian food signal: the cliff swallow bugs call. *Condor, 90,* 714–715.

Stoddard, P. K. (1996). Vocal recognition of neighbors by territorial passerines In D. E. Kroodsma & E. H. Miller (eds.), *Ecology and Evolution of Acoustic Communication in Birds* (pp.356–374). Ithaca, NY: Cornell University Press.

Stout, J. F. (1975). Aggressive communication in *Larus glaucescens*. III. Description of the displays related to territorial protection. *Behaviour, 19,* 208–218.

Struhsaker, T. T. (1967). Auditory communication among vervet monkeys (*Cercopithecus aethiops*). In S. A. Altman (ed.), *Social Communication Among Primates* (pp.281–324). Chicago, IL: University of Chicago Press.

Thompson, C. W., & Moore, M. C. (1991). Throat colour reliably signals status in male tree lizards, *Urosaurus ornatus*. *Anim. Behav., 42,* 745–753.

Thompson, N. S. (1986). Deception and the concept of behavioral design. In R. W. Mitchell & N. S. Thompson (eds.), *Deception. Perspectives on Human and Nonhuman Deceit* (pp.53–65). Albany, NY: University of New York Press.

Thompson, N. S., LeDoux, K., & Moody, K. (1994). A system for describing bird song units. *Bioacoustics, 5,* 267–279.

Thornhill, R. (1979). Adaptive female-mimicking behavior in a scorpionfly. *Science, 205,* 412–414.

Tinbergen, N. (1952). 'Derived' activities: their causation, biological significance, origin and emancipation during evolution. *Quarterly Rev. Biol., 27,* 1–32.

Tinbergen, N. (1959). Comparative studies of the behaviour of gulls (Laridae): a progress report. *Behaviour, 15,* 1–70.

van Rhijn, J. G., & Vodegel, R. (1980). Being honest about one's intentions: an evolutionary stable strategy for animal conflicts. *J. Theor. Biol., 85,* 623–641.

Vincent, A. C. J. (1995). A role for daily greetings in maintaining sea horse pair bonds. *Anim. Behav., 49,* 258–260.

Wagner, W. E., Jr. (1992). Deceptive or honest signaling of fighting ability? A test of alternative hypotheses for the function of changes in call dominant frequency by male cricket frogs. *Anim. Behav., 44,* 449–462.

Ward, P., & Zahavi, A. (1973). The importance of certain assemblages of birds as "information centres" for food-finding. *Ibis, 115,* 517–534.

West, M., & King, A. (1996). Eco-gen-actics: a systems approach to avian communication. In D. E. Kroodsma & E. H. Miller (eds.), *Ecology and Evolution of Acoustic Communication in Birds* (pp.20–38). Ithaca, NY: Cornell University Press.

Wiley, R. H. (1983). The evolution of communication: information and manipulation. In T. R. Halliday & P. J. B. Slater (eds.), *Animal Behavior, Vol. 2: Communication* (pp.156–189). San Francisco, CA: W. H. Freeman.

Wiley, R. H., & Richards, D. G. (1982). Adaptations for acoustic communication in birds: Sound transmission and signal detection. In D. E. Kroodsma & E. H. Miller (eds.), *Acoustic Communication in Birds, 1,* (pp. 131–181). New York: Academic Press.

Williams, G. C. (1964). Measurement of consociation among fishes. *Mich. State Univ. Mus. Publ., 2,* 349–384.

Wilson, E. O. (1975). *Sociobiology: The New Synthesis.* Cambridge: Belknap Press.

Wilson, E. O. (1976). Author's reply. *Anim. Behav., 24,* 716–718.

Wilson, E. O. (1994). *Naturalist.* Washington, D.C.: Island Press.

Witkin, S. R. (1977). The importance of directional sound radiation in avian vocalization. *Condor, 79,* 490–493.

Yoerg, S. I., & Kamil, A. C. (1991). Integrating cognitive ethology with cognitive psychology. In C. A. Ristau (ed.), *Cognitive Ethology. The Minds of Other Animals* (pp.271–289). Hillsdale, NJ: Erlbaum Associates.

Zahavi, A. (1975). Mate selection—a selection for a handicap. *J. Theor. Biol., 53,* 205–214.

Zahavi, A. (1977). The testing of a bond. *Anim. Behav., 25,* 246–247.

Zahavi, A. (1991). On the definition of sexual selection, Fisher's model, and the evolution of waste and of signals in general. *Anim. Behav., 42,* 501–503.

Chapter 2

CONSPICUOUSNESS AND DIVERSITY IN ANIMAL SIGNALS

Marian Stamp Dawkins
Tim Guilford

Department Zoology
University of Oxford, United Kingdom

ABSTRACT

Animal signals that are conspicuous or extravagant are commonly explained as costly advertisements of 'quality.' There are, however, many other evolutionary explanations of conspicuousness, most notably those to do with 'efficacy' or factors making signals effective at getting their message across to receivers. Efficacy can be a major evolutionary force driving signals toward conspicuousness even if there is no selection pressure to advertise quality. Its effects have been widely underestimated through a failure to understand the variety of different ways in which it can act. Efficacy offers a variety of alternative explanations for the evolution of signals, including extravagant sexual signals and can also explain a further feature of animal signals, namely the diversity between species in the signals they use.

INTRODUCTION

There are two very striking features of current evolutionary theories of animal signalling. The first is the heavy emphasis on quality advertisement and

Perspectives in Ethology, Volume 12
edited by Owings *et al.*, Plenum Press, New York, 1997

honesty-through-cost as the main explanation of why animal signals—especially sexually selected ones—become conspicuous. The other is the relative lack of attention to the question of why signals take the particular form that they do. Thus, conspicuous ornaments such as the peacock's train are interpreted as indicators of male 'quality' (Petrie, 1994) but we do not have any very convincing explanations of why quality advertisement should take the form of a great fan with iridescent eyes in the peacock but of feathers in the shape of a lyre in the lyrebird or crests or throat pouches or patches of color in other birds.

We want to argue that in order to explain both the conspicuousness of animal signals and the diversity of signal design among animals, we have to take into account a much wider range of selection pressures than is usually done. Quality advertisement is one, but only one of the many selection pressures leading to the evolution of conspicuous signals and by itself is unlikely to explain their striking diversity.

We will argue that animal signals carry many different sorts of message (Smith, 1977), other than just the 'quality' of the signaller, and that many of these other messages can lead to the evolution of both conspicuousness and diversity quite independently of quality advertisement. A particularly rich source of both signal conspicuousness and signal diversity comes from what we have called 'efficacy' (Guilford & Dawkins, 1991)—a whole range of factors affecting all signals and responsible for making them effective carriers of information by, for example, enabling them to travel over long distances, to carry through distorting media or to 'tap into' the psychology of the receiver. Many of the ideas we discuss are modern versions of concepts developed by 'classical' ethologists such as Tinbergen (1951) and Huxley (1966) but are often overlooked in contemporary discussions of animal signals. As a result, the power of 'efficacy' to lead to conspicuousness even in the absence of quality advertisement has been greatly underestimated and has led to the mistaken but widespread view that if a signal is conspicuous, then this must represent some sort of quality advertisement. A comprehensive view of animal communication has to take account of the fact that conspicuousness in signalling can arise under the influence of many different selection pressures.

THE EVOLUTION OF CONSPICUOUSNESS

Animal signals are, by definition, specially evolved or designed to modify the behavior of other animals (Krebs & Davies, 1993; Hasson, 1994). What has long puzzled biologists is that this 'special evolution' can sometimes lead to very extreme results, even to the point of endangering the life of the signaller. Nestling birds, for example, give conspicuous vocal as well as visual begging signals, potentially drawing the attention of predators to the nest, even though their parents are only a few centimeters away from them and could presumably see or hear much smaller and less conspicuous signals. Coral reef fish use colors designed to be conspicuous over considerable distances (Lythgoe, 1979), prey animals may be brightly colored rather than cryptic or give energy-consuming displays (Cresswell, 1994). Perhaps the most spectacular examples come from the sexual displays of many male animals such as birds and fish where the signals are attractive to predators as well as females (e.g. Endler, 1988).

So far, two kinds of answers have been given to the question of why such signals can be so conspicuous. The first derives from so-called 'classical ethology' under the influence of Konrad Lorenz and Niko Tinbergen. Classical ethologists had some very clear and perceptive ideas about what made animal signals conspicuous, most of which centered on the idea of effective information transfer. Animal signals were described as being 'ritualized'—a word originally coined by Julian Huxley and used by Cullen (1966) to mean "changes in the signalling system during the course of evolution [which] have come about to make the signals more effective in evoking appropriate behaviour."

Signals were, in other words, designed to alter the behavior of other animals by transferring information from one to another and 'ritualization' encompassed all the evolutionary changes that made them better at doing it: changes in form, exaggeration of amplitude, increased regularity of rhythm, acquisition of color or sound and so on. All these possible effects of ritualization provided a vast and what seemed to be quite adequate source of raw material for the evolution of both conspicuousness and diversity in animal signals.

For example, signals would be under selection pressure to travel well through the medium separating sender and receiver—simply to arrive at the destination in a form that could be picked up and interpreted by the receiver. If sender and receiver were separated by a long distance and particularly if there were dense vegetation or turbid water between them, the sender might have to shout or do the equivalent in olfactory or visual terms simply to get its message across (e.g. Alexander, 1962; Konishi, 1970; Marler, 1959, 1967, 1968; Morton, 1975; Hailman, 1977; Lythgoe, 1979). This could lead to conspicuousness and could also open the way to diversity between species, since different animals would be operating in different environments, have different sorts of obstructions between them, be subject to varying amounts of counter selection from predators and so on.

A further way in which ritualization was thought to lead to both conspicuousness and diversity was seen to be the selection pressure to stimulate the receiver. There is no point in a signal travelling through a medium only to reach a receiver that is blind or deaf to it or even just not particularly interested.

Effective signals are ones that stimulate the sense organs and brains of receivers particularly strongly. Tinbergen (1951) describes a number of cases where animals respond to 'sign stimuli'—certain key features of their environment having a particularly powerful effect, such as the red belly of the male stickleback stimulating aggression in a rival or the gape patterns of baby birds stimulating their parents to feed them. He also describes cases where animals respond more to super-normal stimuli or exaggerated versions of the real thing, such as oystercatchers preferring to incubate a giant ostrich egg to their own.

The fact that a wide variety of animals (butterflies (Magnus, 1958), fish (Rowland, 1989), birds (Tinbergen, 1951) and even humans) respond more strongly to supernormal versions of their environment means that we have a ready-made explanation for why signals become exaggerated. Here is conspicuousness waiting to evolve. If a response to exaggerated versions of natural stimuli is built into the reception systems of many animals, then effective signals would be ones that were particularly stimulating supernormal versions of stimuli to which receivers were already sensitive. They would, in other words, be 'conspicuous' versions of normal stimuli.

Throughout the 1950s and 1960s, there was a widespread acceptance of the idea that the way in which sensory and perceptual systems were organized for nonsignal functions was the key to understanding why signals evolved in the way they did. Textbooks such as those by Marler and Hamilton (1966) and Manning (1967) emphasized the similarity between the way animals responded to the external world in general and the way they responded to that subset of it made up of other animals. Signals were a part—albeit a particularly interesting part—of the ways animals responded to the world around them.

These ways were acknowledged to be both specific and general. For example, the highly specialized system of the silk moth in which the scent of the female (bombykol) is detected by the antennae of the male, which appear to be almost insensitive to anything else (Schneider, 1969) was an example of a highly specific signal-receiver system. But equally some ritualized signals were thought to work, at least partly, through their general attention-catching properties rather than as the result of specifically evolved responsiveness (Marler, 1961; Cullen, 1966).

What was striking at this time was the general acceptance of the idea that to understand signal evolution, it was essential to understand how sense organs and brains received and processed incoming information. Signaller and receiver were seen to be intimately connected in a coevolutionary process in which signals evolved to 'exploit' the perceptual mechanisms of receivers (Barlow, 1977). Ethologists had to know about neurophysiological mechanisms and neurophysiologists needed to know what use these mechanisms were being put to in communication between animals (Marler, 1961). The unfortunate intellectual split that occurred in the mid-1970s between 'behavioral ecology' on the one hand and neuroethology on the other had not yet happened (Dawkins, 1989).

This split was the result of a revolution—the sociobiological or selfish gene revolution—that swept through biology and changed radically the way people interpreted interactions between animals including the signals they used in communication (Wilson, 1975; R. Dawkins, 1976). It also indirectly gave rise to a second kind of explanation as to why signals might become conspicuous.

Classical ethology, along with much of the rest of biology had often confused the interests of the individual with the interests of the species as a whole and had not even considered gene selection. Mutual cooperation between animals, particularly a common interest in transferring information, was widely assumed. The time was certainly right for a critical look at the different selection pressures operating on the different parties to an interaction. But in the revolution that followed, it was not just the 'good of the species' way of thinking that was thrown out, but many ideas about the evolution of animal signalling that could be valuable even now. 'Classical ethology' became synonymous with being old-fashioned and group-selectionist. Its insights into the selection pressures on signals were lost sight of in the enthusiasm to reinterpret animal behavior in the light of gene selfishness.

R. Dawkins and Krebs (1978) argued that 'transfer of information' was the wrong way to interpret animal signals: on the contrary, animal communication should be seen as one animal attempting to manipulate the other. To support their view, they drew on the new sociobiological interpretations of interactions between animals in terms of genetic conflict. Even parents and their offspring or male and female parents were acknowledged to have potential conflicts of interest and consequently the signals that passed between them were not necessarily transferring information to mutual advantage. The signals could potentially be carriers of false information, information that it would be detrimental for receivers to respond to. They might 'exploit' or 'manipulate' receivers. The evolution of signals was indeed an arms race but one with deception and counterdeception as its theme. Dawkins and Krebs argued that the two views of animal communication—the 'classical ethological' view of information transfer and the newer more 'cynical' view of conflict and manipulation — were alternatives in opposition to one another, which in many ways they were not (Hinde, 1981; Owings & Hennessy, 1984; Smith, 1986; Thompson, 1986). It was perhaps unfortunate for our understanding of signal evolution that the new genetic insights into ethology were presented as an alternative, not a complementary view of animal signalling. It was doubly unfortunate that the view of signalling as information transfer was saddled with the label 'classical.' Who wants to be called 'classical' when there is a revolution afoot?

One of the most seductive things about the manipulation view of animal signalling was that it appeared to offer an entirely new explanation of the evolution of conspicuousness in animal signals. Since a major feature of signalling was now seen to be signallers evolving to manipulate receivers, this would in turn put pressure on receivers not to be manipulated against their own interests. For example, a weak fighter might attempt to manipulate a stronger animal into retreating by giving out false information about his fighting ability. Zahavi (1975, 1977) put forward a theory of how signal evolution might proceed under these circumstances. He argued that receivers would be selected not to be manipulated into responding to false information and that they should respond only to 'honest' signals, for example, signals that contained reliable information about fighting ability. His great insight was to propose a way in which signals might become 'honest.' He suggested that honest signals would be ones that were costly for a signaller to produce. Senders would 'handicap' themselves by giving a costly signal and the cost of the handicap would be the receivers' guarantee that the signaller genuinely possessed the quality of interest to the receiver (fighting ability, for example) and so demonstrate to the receiver that it possessed this quality by being able to fight despite its handicap.

The handicap theory has gone through several changes of interpretation since then (West-Eberhard, 1979; Andersson, 1982; Grafen, 1990; Maynard Smith, 1987, 1991; Johnstone, 1995) and there are now at least two distinct versions (Maynard Smith, 1991). With conditional or strategic choice handicaps (Grafen, 1990a,b), the signal is itself costly to produce (for example a long tail is costly to grow or giving a loud signal at a high rate is exhausting). The signal reduces the fitness of the signaller, so much so that it 'uses up' some of the quality being signalled about (Zahavi, 1991). However, because a given size of signal is costlier for low-quality signallers than for higher-quality ones, only genuinely high-quality signallers will be selected to give exaggerated signals (Grafen, 1990 a,b). Receivers will be selected to only respond to costly signals and these will tend to be ones that are energetic or time-consuming—in other words conspicuous.

The other version of the handicap theory explains conspicuous signals, not as 'using up' the quality being signalled about but simply reliably showing it through 'revealing signals' (termed 'revealing handicaps' by Maynard Smith, 1991). For example, female toads are attracted to males with low-frequency calls (Davies & Halliday, 1978). They benefit from mating with large males because large males have more sperm. Larger males have lower-pitched voices not because it is costlier to have a low-pitched voice but because low-pitched sounds can only be made by a male with long vocal cords, which are only possible in a large body. The pitch of a male's croak therefore reveals his body size in a reliable way, not as a costly strategic choice handicap but as a 'revealing handicap' (Maynard Smith, 1991). The qualities that receivers, such as females or rivals are most interested in having revealed are likely to be size, strength, physical stamina and so on and signallers are likely to signal that they are bigger or stronger than their rivals, which will once again result in 'conspicuous' signals. The gradual acceptance of one of these two forms of the handicap theory came to be seen as a major cause of the evolution of conspicuousness in animal signals.

There is, however, a potential source of confusion in applying the handicap theory—which lays emphasis on the costs of signals—to the evolution of conspicuousness. This is that signals that are reliable through being costly are not *necessarily* conspicuous (Wiley, 1994), although they often turn out to be in practice. The handicap theory demands that signals are costly and many of the ways that signals can be costly also involve them being conspicuous, for example, making a sound where the loudness is correlated with physical strength or giving a display where the amplitude or duration is correlated with stamina. Loud sounds and high-amplitude displays also happen to be conspicuous. But it is entirely possible for a signal to be costly to produce—for example, an isotonic tensing of the muscles—that is not at all conspicuous. Of course, isotonic tensing that is difficult for the receiver to detect is unlikely to be used as a signal because it would be a very ineffective conveyor of information, even though it might be a highly reliable indicator of fighting strength. The signal has to be 'amplified' (Hasson, 1989, 1991) in order to be read. The point is that the two aspects of signal design—the reliability of the signal's message and how effective it is at getting its message across—can both independently lead to the evolution of conspicuousness. Sometimes, selection for reliability can itself produce a conspicuous signal but sometimes reliable signals that are inherently inconspicuous can become conspicuous under further selection for efficiency of information transfer.

In an attempt to distinguish between these two processes, we proposed (1991, 1993) that the selection pressures on all signals should be seen as arising from two sources: efficacy—effective signal transmission—and content—the information or message transmitted (the original term used was 'strategy' rather than 'content' but subsequent criticism of this term has led us to change it). Thus 'efficacy' includes selection pressures such as those to make a sound signal travel well through a forest or to make a visual signal particularly detectable that have been well studied over a number of years (e.g. Wiley, 1983; Gerhardt, 1994). It also includes more recent discoveries about the psychology of receivers and the 'hidden preferences' revealed by neural net models of perception (Arak & Enquist, 1993, 1995). 'Content' covers the selection pressures such as those involved in the design of a signal to contain reliable information about fighting ability or escape potential. 'Content' is about messages.

It is important to point out, however, that 'content' selection does not always lead to a signal being large, costly or conspicuous. Sometimes it will—specifically where there is conflict of interest and selection for 'honest' advertisement of some underlying quality (Markl, 1985; Arak & Enquist, 1995). But content also includes design for messages about species, sex, identity, readiness to mate and so on, which may not necessarily involve any potential for deception at all (Krebs & Dawkins, 1984; Markl, 1985). If it is in the interests of both animals that one should signal and the other should respond, the receiver will be selected to become more and more sensitive to the signal and so under many circumstances the sender can make its signal smaller and still achieve the desired result of altering the behavior of the receiver while reducing costs such as being spotted by a predator. Krebs and Dawkins called these small inconspicuous signals 'conspiratorial whispers.' An example would be the contact calls heard in groups of birds or primates—often almost imperceptible to an observer standing only a short distance away, but clearly of importance to the animals themselves.

The point of the distinction between efficacy and content is to make it clear that quality advertisement through costly handicaps is only one of several different explanations for the evolution of conspicuousness. We will argue that although much recent emphasis has been on the evolution of conspicuousness through selection on reliability of content, consideration of efficacy also leads to conspicuousness in its own right in ways that are much more numerous than commonly realized. Maynard Smith and Harper (1995) distinguish between what they call 'minimal signals' whose cost is no greater than that needed to transmit information effectively and 'cost-added' signals where the cost is greater than that required to transmit information. (Maynard Smith (1994) also refers to minimal signals as 'road signs.') The very use of the term 'minimal' signal, however, implies that selection for information transfer (efficacy) will never lead to a signal being conspicuous without some other factor acting on it.

Taking up this implied challenge, we will now look at various possible ways in which signals that Maynard Smith and Harper would call 'minimal' can become very conspicuous indeed. In so doing, we draw on two features of ethology that should never have become lost. The first we have already alluded to: the concept of efficacy includes early ideas on signal transmission and reception as well as more recent applications of 'receiver psychology.' The second is the emphasis that ethology has always given to asking a broad range of questions about animal behavior, in particular to asking about both function and mechanism. Behavioral ecology was, for a long time, concerned largely with functional or adaptive questions, leaving neuroethology to investigate mechanism. Only recently has the importance of a broad-ranging—essentially ethological—approach been recognized. Behavioral ecology has rediscovered the importance of studying mechanism (e.g. Real, 1994), phylogeny (e.g.Ryan & Rand, 1993; Johnstone, 1995) and even development. The essential features of a broad-based ethological approach have reemerged as important as ever.

EFFICACY AND CONSPICUOUSNESS

All signals, both cooperative conspiratorial whispers and the more costly signals given in conflict situations, will be subject to both efficacy and content selection pressures. It can therefore sometimes be difficult to decide in any given case whether conspicuousness is the result of 'design' for efficacy or 'design' for honest content or a mixture of both. The task becomes easiest, however, when we look at signals where there is no conflict and where the content or strategic selection pressures are toward giving small and inconspicuous signals. If, despite this, such signals are large and conspicuous, we can reasonably safely conclude that we are looking at conspicuousness driven by 'pure' efficacy. To use Maynard Smith and Harper's (1995) terminology, we are looking at 'minimal signals' and asking 'how minimal is minimal when there is strong selection pressure to transfer information effectively?' To use Krebs and Dawkins' distinction, we are asking whether they were correct in suggesting that cooperative signals should mainly take the form of 'conspiratorial whispers' or whether we should also expect to find examples of 'conspiratorial shouting.'

We suggest the following situations where efficacy might be particularly important:

1. Where *catching the attention* of the receiver is particularly important, for example, when the first signaller to be noticed among a crowd of others is the one chosen. Here the emphasis is on the 'attention-grabbing' properties of the signal, not the quality it signifies. An analogy might be, as Dawkins and Krebs (1978) pointed out, with the advertising industry. Advertising may be very intense when the difference in quality between different products is very small. Thus a firm may be prepared to spend a great deal of money pushing its product over a rival's when there is little to choose between them on quality and all the effort goes on attracting consumers' attention and impressing one particular brand name on their minds. The product that 'wins' may be little different from any others. It may just be better known.

   In animals, signallers may compete to give 'alerting signals' (Wiley, 1983) that are particularly good at gaining the attention of a receiver. Examples of such signals among animals might include begging signals by nestling birds (Godfray, 1991; Guilford & Dawkins, submitted) or choruses of calling males (Wiley, 1991; Kirkpatrick & Ryan, 1991; Greenfield, 1994). Of course, in both these cases, alternative explanations based on the receiver assessing the relative quality or need of the various signallers have been put forward. Our purpose is not to dismiss such explanations, merely to point out that they are just one possibility and that alternatives based on efficacy should at least be considered.

2. Where *synchrony* is important. In athletics, where it is in the mutual interest of both runners and race organizers for everyone to start at the same time, the starting signal is not a mumbled 'Go now.' It is likely to be a loud shot from a starting pistol preceded by a loud and rhythmical warning of when the shot will appear in the form of a shouted "On your marks...get set...." The synchrony at the start of a race is achieved by a conspicuous set of signals designed for effective transfer of information about the exact moment when the race is to start.

   A possible example is found during the courtship of the bluehead wrasse (*Thalassoma bifasciatum*). These fish are found on patch reefs throughout the Caribbean. They have external fertilization in which the male and female suddenly dash upwards together from coral reef heads to the surface of the water where eggs and sperm are discharged together (Warner, 1987). The fertilized eggs then float away in the plankton. The spawning dash itself may take as little as 0.5 sec and the bulk of the time between spawnings is occupied by the males aggressively chasing other males away from their spawning territories while the females wait just below the coral head.

It is in the mutual interests of both sexes to synchronize their spawning dash. The female leads and the male follows closely behind, so it is clearly important for the female when she starts the spawning dash to be sure that the male is ready and not, say, about to start defending his territory against intruders. Equally, it is in the interests of the male not to 'waste' the limited number of females in his territory by having them spawn when he is not ready for them. At the same time, territorial defense is an important priority for the male owing to the presence of large numbers of nonterritorial males that constantly try (and sometimes succeed) to obtain matings by intruding into territories.

The male bluehead wrasse has a series of signals that indicate his readiness to spawn on increasingly fine time scales. First, the color of his body changes depending on whether he is about to chase other males or to court females (Dawkins & Guilford, 1993). When he is defending his territory, the back part of his body is a bright yellow-green but when he starts to court females using a circling movement in which he 'flutters' his pectoral fins, his body loses this bright coloration and goes an opalescent neutral gray. The aggressive yellow-green color is predominantly 569 nm (70.5%) and contrasts maximally with the background blue of the seawater so that it is visible over a long distance. The neutral grey of the courtship colours, on the other hand, contrasts much less with the background and is visible only over short distances (Dawkins & Guilford, 1994).

At the same time as these color changes take place, the pattern on the male's fins changes. When he is chasing other males, his pectoral fins are quite colourless. But as he courts a female, a black 8-mm diameter oval spot appears at the end of each pectoral fin. The male moves his fins rapidly backwards and forwards as he circles above the female, the spots accentuating the visibility of the fin movements. In the few seconds before the spawning dash, the rate of fin movement increases dramatically so that the precise speed of the fins with their conspicuous moving spots gives a clear indication of exactly when spawning will occur on a time scale of seconds.

The male bluehead therefore provides an example of a conspicuous display involving body color changes, appearance of spot patterns and movement. The most obvious interpretation is that the male is demonstrating his 'quality' by moving his fins in this energetic way, perhaps particularly important for wrasses since this group of fish is characterized by using pectorals to swim with. Ability to move the pectorals rapidly in a courtship display might therefore correlate well with swimming ability. However, it seems unlikely that the females are assessing male quality at all, since Warner (1987) has shown that the quality of a territory as a spawning site is more important to female choice than is the male himself.

We suggest an alternative explanation, based on selection for efficacy, specifically synchrony. This is that the message of the signal is to transmit information about when spawning will take place, functioning to synchronize the act of spawning. It is therefore a cooperative signal in which neither side would gain from cheating. The conspicuousness of the signal may therefore not arise from selection for honesty-through-cost but through selection for effective information transfer about when spawning is to occur. It is an example of a conspiratorial shout, not a conspiratorial whisper.

3. Where *decisions* have to be made rapidly. If a receiver has difficulty in detecting the signaller or in discriminating one signaller from another, then there will be a premium on signallers that, through being conspicuous, make themselves easy to detect or easy to discriminate. It will be to the advantage of the receiver to respond to such signallers *even though they may not be of higher quality than any of the others*, if by doing so, the receiver saves time or the costs of searching or deciding. Quality advertisement may also be involved. The point is that conspicuousness could evolve even without it.
4. Where *learning and memory* are important. Conspicuous warning colors, for example, seem to aid predator learning (Gittleman & Harvey, 1980) or enhance predator recall (reviewed in Guilford, 1992; Schuler & Roper, 1992; Guilford and Dawkins, submitted).

All of these explanations are open to experimental test. They open up a whole range of possibilities for thinking about the evolution of animal signalling. As we will now see, they are even relevant to that most controversial of all topics in animal signalling, sexual selection.

Efficacy and Sexual Selection

We have already seen that in at least one example of sexual signalling—the courtship of the bluehead wrasse—it is possible to explain the evolution of a display's conspicuousness without having to invoke quality advertisement through cost. This appears, however, to have a large element of cooperation where both sexes benefit from synchronizing their spawning activities. In most cases of sexual interaction, it might be argued, the interactions between males and females are those of conflict. The male benefits if he mates with many females, the female benefits if she is choosy and rejects most mates (Parker, 1983). Nowhere has the theory of honest advertisement had so great a hold on the way people think about conspicuous signalling as in this type of conflict. Males are held to advertise their 'quality' by using costly displays. Females use these displays to infer male quality, such as whether he has 'good genes' (e.g. Moore, 1994; Petrie, 1994; Nicoletto, 1995).

There are, however, other adaptive ways besides quality advertisement in which conspicuous sexual signals can evolve. Selection for signal efficacy again provides a wide range of alternatives for the evolution of conspicuousness but the problem is to decide whether the observed conspicuousness of a given signal evolved primarily under the influence of selection to transfer information efficiently or the selection for signal content to be honest and reliable. Many sexual signals may have been pushed toward conspicuousness by both kinds of selection pressure, but it still makes sense to ask which has been the most important in the evolution of any given signal. Obviously, if we have no clear idea about how much conspicuousness is needed for effective information transfer in sexual encounters, we may be tempted to inflate the contribution of selection for reliability; hence the importance of studying the ways in which efficacy alone can produce conspicuousness, even when 'content' selection would not necessarily predict the evolution of exaggerated signals.

There are many important kinds of information that are transferred between males and females, not just about mate quality, but also about species, location, readiness to mate and so on. There is undoubtedly conflict of interest between males and female over some of these but not necessarily all. 'Quality' of a male is a clear case where conflict does potentially exist in the sense that a low-quality male might gain from falsely signalling that he was of high quality and so females have been selected to respond only to signals that reliably convey information about male quality. Signals whose content was male quality would consequently be expected to become honest indicators of quality through being costly. However, other kinds of information do not involve conflict, or at least do so to a very much reduced extent. A male rarely gains from falsely signalling about what species he was or that he was ready to mate when he wasn't or by giving misleading information about where he is. Signals whose content is about species, readiness to mate or location would not be expected to incur costs associated with honesty. For this reason, many authors have argued that signals conveying such information will not evolve to be particularly conspicuous and so cannot be the explanation for the exaggerated sexual signals that have caused so much controversy. For example, Andersson (1994) argues that "Even if species recognition is involved, it appears unlikely to explain fully the most conspicuous sex traits such as the peacock's tail, which are much more extreme than necessary for species recognition' (p. 27).

But how do we know what is 'necessary' for species recognition? Or for location or synchrony? It has become very apparent from recent studies of sexual selection that an important component is the cost of female choice (Pomiankowski et al., 1991). If females have to pay costs in searching for and choosing mates, then the benefits accruing to choosy as opposed to randomly mating females begin to diminish. Indeed, costs of female choice have been raised as a serious objection to Fisher's runaway theory of female choice in that, at equilibrium, when all the males are genetically similar and there will be no genetic advantage to a female in mating with one male rather than another, then any costs to female choice will result in unchoosy females being favored over choosy ones (Parker, 1982).

A consequence of this is that females that mate with minimal costs will be at an advantage and that males might increase the number of females that mate with them by producing signals that reduce the costs of female choice. Such signals might well be expected to be conspicuous. Conspicuous signals would effectively increase a male's 'direct detection distance' (Smith, 1974) and enable a female to find him with lower search costs than she would have to expend on looking for a male that conveyed information about his location less effectively. Conspicuous distinctive signals would also enable females to rapidly and without error find a male of the right species and once again cut down on her search costs because she would have to spend less time making certain he was the right species.

Females that responded to such signals would gain (their choice would be adaptive) even though the males they mated with might be of no better quality than any others. The female gain would be the reduction in search costs (Dawkins & Guilford, in press). The result would be male signal conspicuousness evolved through effective information transfer enhanced by selection on females to reduce their search costs.

There are many different ways in which sexual signals can be designed for effective information transfer, but one particularly important idea that has been recently introduced into discussions of sexual selection is through 'sensory exploitation' or 'sensory bias' (Basolo, 1990, 1995; Ryan & Rand, 1993; Endler, 1993). 'Sensory bias' or 'sensory traps' (West-Eberhard, 1984; Christy, 1995) refer to phenomena which have been discovered in a wide range of animals where females respond strongly in a sexual context to certain stimuli rather than others even though these stimuli appear to have nothing to do with mate choice. Female jumping spiders initially respond to males as if they were food and the male appears to 'tap in' to the female's preexisting tendency to turn toward small moving potential food objects in order to get her attention in a sexual context (Clark & Uetz, 1992). A similar use of a female's 'bias' to respond to food occurs in water mites where the first part of the male's courtship appears to mimic the movement of prey (Proctor, 1991). In the Oriental fruit moth, the male exudes ethyl transcinnamate, a component of fermented fruit juice, which is the female's favorite food (Löfstedt, 1993).

A male jumping spider that attracts the attention of females by eliciting her prey-catching response might seem to be playing with fire. But males will be selected to attract the attention of females—their signals will be selected for efficacy—and one way of doing this is to mimic food. Those that do so will attract the attention of more females than males that do not. Provided they can subsequently convince the female that they are not, after all, prey, such males will be at an advantage.

Some authors have implied that it is only the males that gain from such signals and that females responding to sensory biases are behaving in a nonadaptive or arbitrary way (e.g. Hill, 1994). However, it is important to stress that females also gain by 'being exploited,' for example, finding a male of the right species to mate with with minimal effort on her part. We would not expect 'sensory biases' to persist unless females did gain. So the theory of 'sensory bias,' taken by itself, is essentially a theory of the evolutionary origin of signals (Tinbergen, 1952). Signals, as effective means of altering the behavior of other animals, do not come out of an evolutionary nowhere. They evolve because some preexisting nonsignal movement or structure in one animal alters the behavior of another animal (Tinbergen, 1952; Krebs & Davies, 1993). Such movements would not have become signals at all if they had not been perceivable by receivers. To that extent, all signals evolved from preexisting biases of various sorts, in just the same way that all other aspects of living organisms evolved out of the raw material of what was available at the time (Foster, 1995). The subsequent evolutionary history of the preexisting structure or behavior that 'just happened' to have an effect on another animal will depend on factors such as the advantages and disadvantages to the two animals of altering the behavior of others and having their own behavior altered. It is important not to confuse the evolutionary origins of a signal with the selective factors leading to their subsequent evolution (Reeve & Sherman, 1993).

Sensory biases provide dramatic examples of where signals may have come from, but signals can persist only if it is on average adaptive for a signaller to give them and for a receiver to respond. Their subsequent adaptive advantage could come from a variety of different sources, including improving the efficacy of a quality advertisement, providing the basis of a Fisherian runaway process or even, as we have stressed here, by simply making the male that makes use of them in his signal able to attract females by reducing their search costs. The further point we wish to emphasize is that the efficacy function of a signal does not have to be confined to the specific 'mimicry' implied by sensory bias or sensory traps. It could also be served by 'tapping in' to more general properties of receivers' sense organs and brains that we discussed in an earlier section, such as a general sensitivity to movement, color or contrast. Since these are common to a wide range of organisms, it will often be impossible to test an 'efficacy' theory of signal evolution simply by looking at the phylogenetic distribution of signal and response as has been attempted for more specific sensory bias theories (Ryan & Rand, 1993), because ancestral forms are likely to possess these sensitivities anyway. In some cases, it may be difficult to be sure how specific the sensory bias is. For example, the response of female platyfish to the 'swords' of male swordtails is described as a sensory bias on the grounds that male platyfish do not have swords so that there is a preexisting bias in the whole *Xiphophorus* genus of poecilid fishes that preceded the evolution of the sword itself (Basolo, 1990, 1995). In fact, in preferring males with long swords, both female platyfish and female swordtails may be following an even more widespread 'bigger is better' rule (Rosenthal et al., in press) that normally leads females of many different species to choose large males of their own species (e.g. Rowland, 1989).

Particularly where females follow 'bigger is better' rules for mate choice, the possibilities for signal exaggeration—and conspicuousness—are considerable, as has long been recognized in the concept of the supernormal stimulus (Tinbergen, 1951). Recent simulations of signal evolution using artificial neural networks have also shown that where female biases are relatively open-ended, substantial exaggeration of the male signal easily occurs (Arak & Enquist, 1993). Where similar female biases are common to whole groups of species, they may lead to exaggerated signals through reduction of female search costs but will be useless for species identification. Where the biases are confined to one species, however, they could become exaggerated signals through the coevolution of increased female certainty for species identification and male competition to provide the signal that reliably gives females information about which species they belong to.

Signal Conspicuousness and Signal Diversity

The final issue we want to address is whether the broad framework of dividing the selective pressures on signals into 'content' and 'efficacy' is up to the job of explaining the great diversity we observe in animal signals, particularly those that have evolved to be conspicuous. The stunning array of courtship signals in different species of birds, fish and lizards, for example, demands explanation. Why are they all so different? We have discussed some of the reasons why signals are often conspicuous and 'exaggerated' but can we also explain why they show such variety in the way they are designed?

Some of the diversity in animal signals is readily explicable in terms of gross differences in the ways of life pursued by different species. Electric fish, for example, use electric location to find objects in the murky river waters in which they live and it is therefore not particularly surprising to find that they use their electric sense for communication as well. On the other hand, birds use mainly vision and hearing for communication as these are the main senses they use for other functions. But over and above these major differences we find diversity of a sort that at times seems astonishing. Butterfly fish (Chaetodontidae) are an example of a large group of animals exhibiting a high degree of variation of signals between closely related species. The general yellowish color and use of bold black stripes by this group can be understood as adaptation to signalling underwater (Lythgoe, 1979) but different species living within the same area are markedly different in both color and pattern so that 'adaptations to signalling underwater' do not explain this diversity.

The framework we have proposed does at least provide the potential for considering a variety of selection pressures that would lead to diversity. Quality advertisement is one, in that different signals will display different qualities (e.g. escape ability, health etc.) and so can be expected to be different depending on the message the signaller is trying to convey. Quality advertisement is, however, somewhat limited in its scope for explaining variation in signals with similar function in closely related species. Unless closely related species are thought to be advertising different qualities, for example, displaying freedom from different diseases, it is difficult to see why they should not be convergent rather than divergent in the form that their signals take.

Broadening the range of explanations for conspicuous signalling to include both other messages (e.g. species identification) and efficacy considerations, however, gives a much greater potential for explaining signal diversity. Species recognition, for example, predicts diversity between closely related species, particularly those inhabiting the same area (Sanderson, 1989; Butlin, 1989; Paterson, 1993). Add to that a selection pressure for, say, quick and cost-free detection in water that absorbs light and we have potentially an explanation for signals that are both conspicuous and different in different species.

Or, consider the example we discussed earlier, the conspicuous set of signals that apparently serves to synchronize spawning in bluehead wrasse—color change, donning of fin spots and fin-fluttering performed by the male. Two other species of wrasse, the saddle wrasse (*Thalassoma duperrey*) and the bird wrasse (*Gomphosus varius*) also show fin-fluttering before a synchronized spawning dash by male and female (personal observations). Like the bluehead, fin-fluttering and circling are accentuated by pectoral fins that are not only patterned but contrast strongly with the body against which they are moved. But there are differences. The bird wrasse has a bright green body with a brilliant lime green slash just behind the pectoral fins, which are themselves ornamented with black and bright turquoise stripes. The saddle wrasse has fins with black edges, not spots on the ends like the bluehead and, although it superficially resembles the bluehead in having a white stripe against which the fins flutter conspicuously, the white is in fact on a different part of the body. Also neither the saddle wrasse nor the bird wrasse is like the bluehead in turning the fin patterns on and off rapidly depending on whether the male is courting females or chasing other males.

The similarities between the three species (circling display, fin-fluttering with fins moved against a contrasting background) appear to be due to the common problem of synchronizing male and female spawning dash. But why the differences? It could be that species identification is more of a problem for saddle and bird wrasse as they spawn in the same areas at the same time. Or, subtle differences in habitat such as the depth at which spawning takes place may equally affect what is the most effective signal for each species. Both different messages and different needs for effectively getting those messages across to the intended receiver could potentially contribute to observed differences in design.

CONCLUSIONS

We have argued that in order to understand the evolution of conspicuous signals in animal communication as well as the diversity of form these can take, we need to take account of a broad range of selective forces that potentially operate on them. While 'quality advertisement' and the selection on receivers not to be deceived by signals containing false information has been an important factor in the evolution of many animal signals and a powerful cause of their conspicuousness, it is not the only one. A whole range of factors relating to the effective transfer of information between signallers and receivers may also lead to conspicuousness. To make sense of what is otherwise a long list of 'causes of conspicuousness,' the selective forces on all signals can be divided into those concerned with 'content' (message or strategy) and those concerned with efficacy (effective information transfer). This makes it clear that effective information transfer is not a mechanism in itself (although it clearly invites an explanation of how signals are transduced and processed) but a selective force in its own right, to be seen alongside selection for 'content' components of design which itself may sometimes lead to larger and sometimes to smaller signals, depending on the degree of conflict between signaller and receiver.

Conspicuous cooperative signals, where there is little or no conflict between signaller and receiver, are particularly interesting as here receivers have not been subject to the dangers of being manipulated by false information and so have not been selected to demand conspicuous, costly cheat-proof signals. Their conspicuousness shows the effects of efficacy particularly clearly. Our plea, therefore, is for a broadening of current discussions of animal communication. We should see quality advertisement as one 'cause of conspicuousness' in a list that also includes a large number of other factors to do with efficacy now expanded through recent work on 'receiver psychology.'

Any complete theory of animal communication should be able to explain not only conspicuousness but also diversity. We have hardly begun to understand the factors that lead animal signals to differ so much from each other but we hope the framework we have proposed will help people to think about this phenomenon in constructive new ways.

REFERENCES

Alexander, R. D. (1962). Evolutionary change in cricket acoustical communication. *Evolution, 16*, 443–367.

Andersson, M. (1982). Sexual selection, natural selection and quality advertisement. *Biol. J. Linn. Soc., 1*, 375–393.

Arak, A. (1988). Female mate selection in the natterjack toad: active choice or passive attraction? *Behav. Ecol. Sociobiol., 22*, 317–327.

Arak, A. & Enquist, M. (1993). Hidden preferences and the evolution of signals. *Phil. Trans. Roy. Soc. Lond. B, 340*, 207–213.

Arak, A. & Enquist, M. (1995). Conflict, receiver, bias and the evolution of signal form. *Phil. Trans. Roy. Soc. Lond. B, 349*, 337–344.

Barlow, G. W. (1977). Modal action patterns. In T. A.Sebeok (Ed.), *How Animals Communicate* (pp.98–134). Bloomington, IN: Indiana University Press.

Basolo, A. (1990). Female preference predates the evolution of the sword in swordtail fish. *Science, 250*, 808–810.

Basolo, A. (1995). Phylogenetic evidence for the role of a pre-existing bias in sexual selection. *Proc. Roy. Soc. Lond. B, 259*, 307–311.

Butlin, R. (1989). Reinforcement or premating isolation. In D. Otte & J. A. Endler (Eds.), *Speciation and its Consequences* (pp. 158–179). Sunderland, MA: Sinauer.

Christy, J. (1995). Mimicry, mate choice and the sensory trap hypothesis. *Amer. Nat., 146*, 171–181.

Clark, D. L. & Uetz, G. W. (1992). Morph-independent mate selection in a dimorphic jumping spider: demonstration of movement bias in female bias using video-controlled courtship behaviour. *Anim. Behav., 43*, 247–254.

Cresswell, W. (1994). Song as a pursuit-deterrence signal and its occurrence relative to other anti-predator behaviour of the skylark on attack by merlins. *Behav. Ecol. Sociobiol., 34*, 217–223.

Cullen, J. M. (1966). Reduction of ambiguity through ritualisation. *Phil. Trans. Roy. Soc. Lond. B, 215*, 363–374.

Davies, N. B. & Halliday, T. R. (1978). Deep croaks and fighting assessment in toads *Bufo bufo*. *Nature, 274*, 683–685.

Dawkins, M. S. (1989). The future of ethology: how many legs are we standing on? In P. P. G. Bateson & P. H. Klopfer (Eds.), *Perspective in Ethology, 8*, 47–83.

Dawkins, M. S. & Guilford, T. (1993). Colour and pattern in relation to sexual and aggressive behaviour in the bluehead wrasse *Thalassoma bifasciatum*. *Behav. Proc., 30*, 245–252.

Dawkins, M. S. & Guilford, T. (1994). Design of an intention signal in the bluehead wrasse (*Thalassoma bifasciatum*). *Proc. Roy. Soc. Lond. B, 257*, 123–128.

Dawkins, M. S. & Guilford, T. Sensory bias and the adaptiveness of female choice. *Amer. Nat.*, in press.

Dawkins, R. (1976). *The Selfish Gene*. Oxford: Oxford University Press.

Dawkins, R. & Krebs, J. R. (1978). Animal signals: information or manipulation? In J. R. Krebs & N. B. Davies (Eds.). *Behavioural Ecology: an Evolutionary Approach* (pp. 282–309). Oxford: Blackwell Scientific Publications.

Endler, J. A. (1988). Sexual selection and predation risk in guppies. *Nature, 232*, 593–594.

Endler, J. A. (1993). Some general comments on the evolution and design of animal communication systems. *Phil. Trans. Roy. Soc. Lond. B, 340*, 215–225.

Fitzgibbon, C. D. & Fanshaw, J. H. (1988). Stotting in Thomson's gazelles: an honest signal of condiion. *Behav. Ecol. Sociobiol., 23*, 69–74.

Foster, S. A. (1995). Constraint, adaptation, and opportunism in the design of behavioral phenotypes. In N. S.Thompson (Ed.), *Perspectives in Ethology, 11*, 61–81. New York: Plenum.

Gerhardt, H. C. (1994). The evolution of vocalisation in frogs and toads. *Annu. Rev. Ecol. Syst., 25,* 293–324.

Gittleman, J. L. & Harvey, P. H. (1980). Why are distasteful prey not cryptic? *Nature, 286,* 149–150.

Godfray, H. C. (1991). Signalling of need by offspring to their parents. *Nature, 352,* 328–330.

Grafen, A. (1990a). Sexual selection unhandicapped by the Fisher process. *J. Theoret. Biol., 144,* 475–516.

Grafen, A. (1990b). Biological signals as handicaps. *J. Theoret. Biol., 144,* 517–546.

Greenfield, M. D. (1994). Cooperation and conflict in the evolution of signal interactions. *Annu. Rev. Ecol. Syst., 25,* 97–126.

Guilford, T. (1992). Predator psychology and the evolution of prey coloration. In M. J. Crawley (Ed.), *Natural Enemies: The Population Biology of Predators, Parasites and Diseases* (pp. 377–394). Oxford: Blackwell Scientific Publications.

Guilford, T. & Dawkins, M. S. (1991). Receiver psychology and the evolution of animal signals. *Anim. Behav., 42,* 1–14.

Guilford, T. & Dawkins, M. S. (1993). Receiver psychology and the design of animal signals. *Trends in NeuroScience, 16,* 430–436.

Guilford, T. & Dawkins, M. S. The efficiency of exaggerated signals. *Animal Behaviour*. (submitted).

Hailman, J. P. (1977). Optical Signals. *Animal Communication and Light.* Bloomington, IN: Indiana University Press.

Hasson, O. (1989). Amplifiers and the handicap principle: different emphasis. *Proc. Roy. Soc. Lond. B, 235,* 383–406.

Hasson, O. (1991). Sexual displays as amplifiers: practical examples with an emphasis on feather decorations. *Behav. Ecol., 2,* 189–197.

Hasson, O. (1994). Cheating signals. *J. Theoret. Biol., 167,* 223–238.

Hill, G. E. (1994). Geographic variation in male ornamentation and female mate preference in the house finch: a comparative test of models of sexual selection. *Behav. Ecol., 5,* 64–73.

Hinde, R. A. (1981). Animal signals: ethological and games theory approaches are not incompatible. *Anim. Behav., 29,* 535–542.

Huxley, J. (1966). A discussion of ritualisation of behaviour in animals and man: introduction. *Phil. Trans. Roy. Soc. Lond. B, 251,* 247–271.

Johnstone, R. A. (1995). Sexual selection, honest advertisement and the handicap principle: reviewing the evidence. *Biol. Rev., 70,* 1–65.

Kirkpatrick, M. & Ryan, M. J. (1991). The evolution of mating preferences and the paradox of the lek. *Nature, 350,* 33–38.

Konishi, M. (1970). Evolution of design features in the coding of species-specificity. *Amer. Zool., 10,* 67–72.

Krebs, J. R. & Davies, N. B. (1993). *An Introduction to Behavioural Ecology.* (3rd ed.). Oxford: Blackwell Scientific Publications.

Krebs, J. R. & Dawkins, R. (1984). Animal signals: mind-reading and manipulation. In J. R. Krebs & N. B. Davies (Eds.), *Behavioural Ecology: An Evolutionary Approach.* (2nd ed., pp. 380–402). Oxford: Blackwell Scientific Publications.

Löfstedt, C. (1993). Moth pheromone genetics and evolution. *Phil. Trans. Roy. Soc. Lond. B, 340,* 167–177.

Lythgoe, J. N. (1979). *The Ecology of Vision.* Oxford: Clarendon Press.

Magnus, D. B. E. (1958). Experimental analysis of some overoptimal sign stimuli in the mating behaviour of the fritillary butterfly *Argynnis paphia L. Proc. 10th Int. Cong. Ent. Montreal, 2,* 405–418.

Manning, A. (1967). *An Introduction to Animal Behaviour.* London: Edward Arnold.

Markl, H. (1985). Manipulation, modulation, information, cognition: some riddles of communication. In B. Holldobler & M. Lindauer (Eds.), *Experimental Behavioral Ecology and Sociobiology* (pp. 163–194). Stuttgart: Gustav Fisher Verlag.

Marler, P. (1959). Developments in the study of animal communication. In P. R. Bell (Ed.), *Darwin's Biological Work* (pp.150–206). Cambridge: Cambridge University Press.

Marler, P. (1961). The filtering of external stimuli during instinctive behaviour. In W. H. Thorpe & O. L. Zangwill (Eds.), *Current Problems in Animal Behaviour* (pp. 150–166). Cambridge: Cambridge University Press.

Marler, P. (1967). Animal communication signals. *Science, 157,* 769–774.

Marler, P. (1968). Visual systems. In T. A. Sebeok (Ed.), *Animal Communication* (pp. 103–126). Bloomington, IN: Indiana University Press.

Marler, P. & Hamilton, W. J. (1966). *Mechanisms of Animal Behavior*. New York: Wiley.

Maynard Smith, J. (1987). Sexual selection—a classification of models. In J. W. Bradbury & M. Andersson (Eds.), *Sexual Selection: Testing the Alternatives* (pp. 9–20). New York: Wiley.

Maynard Smith, J. (1991). Theories of sexual selection. *Trends Ecol. Evol., 6,* 146–151.

Maynard Smith, J. (1994). Must reliable signals always be costly? *Anim. Behav., 47,* 1115–1120.

Maynard Smith, J. & Harper, D. C. G. (1995). Animal signals: models and terminology. *J. Theoret. Biol., 177,* 305–311.

Moore, A. J. (1994). Genetic evidence for the 'good genes' process in sexual selection. *Behav. Ecol. Sociobiol., 35,* 235–241.

Morton, E. S. (1975). Ecological sources of selection on avian sounds. *Amer. Nat., 111,* 855–869.

Nicoletto, P. F. (1995). Offspring quality and female choice in the guppy. *Anim. Behav., 49,* 377–387.

Owings, D. H. & Hennessy, D. F. (1984). The importance of variation in scuirid visual and vocal communication. In J. O. Murie & G. R. Michener (Eds.), *The Biology of Ground-Dwelling Sciurids: Annual Cycles, Behavioral Ecology and Sociality* (pp. 169–226). Lincoln, NE: University of Nebraska Press.

Parker, G. A. (1982). Phenotype-limited evolutionarily stable strategies. In: King's College Sociobiology Group (Eds.), *Current Problems in Sociobiology* (pp. 173–202).

Parker, G. A. (1983). Mate quality and mating decisions. In P. Bateson (Ed.), *Mate Choice* (pp. 141–166). Cambridge: Cambridge University Press.

Paterson, H. E. H. (1993). The specific-mate recognition system and variation in mobile animals. In P. P. G. Bateson, P. H. Klopfer & N. S. Thompson (Eds.), *Perspectives in Ethology, 10* (pp. 209–227). New York: Plenum.

Petrie, M. (1994). Improved growth and survival of offspring of peacocks with more elaborate trains. *Nature, 371,* 598–599.

Pomiankowski, A., Iwasa, Y. & Nee, S. (1991). The evolution of costly mate preferences 1. Fisher and biased mutation. *Evolution, 45,* 1422–1430.

Proctor, H. (1991). Courtship in the water mite *Neumania papillator*: males capitalise on female adaptations for predation. *Anim. Behav., 42,* 589–598.

Real, L. A. (Ed.) (1994). *Behavioral Mechanisms in Evolutionary Ecology*. Chicago, IL: University of Chicago Press.

Reeve, H. K. & Sherman, P. W. (1993). Adaptation and the goals of evolutionary research. *Q. Rev. Biol., 68,* 1–31.

Rosenthal, G. et al (in press). *Animal Behaviour.*

Rowland, W. J. (1989). Mate choice and the supernormality effect in female sticklebacks (*Gasterosteus aculeatus*). *Behav. Ecol. Sociobiol., 24,* 433–438.

Ryan, M. J. & Rand, A. S. (1993). Sexual selection and signal evolution: the ghosts of biases past. *Phil. Trans. Roy. Soc. Lond. B, 340,* 187–195.

Sanderson, N. (1989). Can gene flow prevent reinforcement? *Evolution, 43,* 1223–1235.

Schneider, D. (1969). Insect olfaction: deciphering system for chemical messages. *Science, 163,* 1031–1037.

Schuler, W. & Roper, T. J. (1992). Responses to warning coloration in avian predators. In P. J. B. Slater, J. S. Rosenblatt, C. Beer & M. Milinski (Eds.), *Advances in the Study of Behavior, 21* (pp.111–146). New York: Academic Press.

Smith, J. N. M. (1974). The food searching behaviour of two European thrushes. II The adaptiveness of the search patterns. *Behaviour, 49,* 1–61.

Smith, W. J. (1977). *The Behavior of Communicating.* Cambridge, MA: Harvard University Press.

Smith, W. J. (1986). An "informational" perspective on manipulation. In R. W. Mitchell & N. S. Thompson (Eds.), *Deception: Perspectives on Human and Nonhuman Deceit* (pp. 71–86). Albany, NY: SUNY Press.

Thompson, N. S. (1986). Deception and the concept of behavioural design. In R. W. Mitchell & N. S. Thompson (Eds.), *Deception: Perspectives on Human and Nonhuman Deceit* (pp. 53–65). Albany, NY: SUNY Press.

Tinbergen, N. (1951). *The Study of Instinct.* Oxford: Clarendon Press.

Tinbergen, N. (1952). Derived activities; their causation, biological significance, origin and emancipation during evolution. *Q. Rev. Biol., 27,* 1–32.

Warner, R. R. (1987). Female choice of sites versus males in a coral reef fish, *Thalassoma bifasciatum. Anim. Behav., 35,* 1470–1478.

West-Eberhard, M. J. (1984). Sexual selection, competitive communication and species-specific signals in insects. In T. Lewis (Ed.), *Insect Communication* (pp. 281–324). London: Academic Press.

Wiley, R. H. (1983). The evolution of communication: information and manipulation. In T. R. Halliday & P. J. B. Slater (Eds.), *Animal Behaviour, Vol: 2. Communication* (pp. 156–189). Oxford: Blackwell Scientific Publications.

Wiley, R. H. (1994). Errors, exaggeration, and deception in animal communication. In L. A. Real (Ed.), *Behavioral Mechanisms in Evolutionary Ecology* (pp. 157–189). Chicago, IL: University of Chicago Press.

Wilson, E. O. (1975). *Sociobiology.* Cambridge, MA: Harvard University Press.

Zahavi, A. (1975). Mate selection – a selection for a handicap. *J. Theoret. Biol., 53,* 205–214.

Zahavi, A. (1977). Reliability in communication systems and the evolution of altruism. In B. Stonehouse & C. M. Perrins (Eds.), *Evolutionary Ecology* (pp. 253–259). London: Macmillan.

Zahavi, A. (1991). On the definition of sexual selection, Fisher's model and the evolution of waste and of signals in general. *Anim. Behav., 42,* 501–503.

Chapter 3

WHAT IS THE FUNCTION OF SONG LEARNING IN SONGBIRDS?

Michael D. Beecher
J. Cully Nordby
S. Elizabeth Campbell
John M. Burt
Christopher E. Hill
Adrian L. O'Loghlen

Animal Behavior Program
Departments of Psychology and Zoology
University of Washington
Seattle, Washington 98195

ABSTRACT

In this paper we approach the question of the function of song learning in songbirds by addressing the more particular question of why a bird chooses the particular songs he does from among the many songs he hears during his song learning period. Our generalizations about song learning are derived from observations of a sedentary (nonmigratory) population of song sparrows (*Melospiza melodia*). We examine this question in the field, rather than in the laboratory, because we believe that the variables controlling song selection are social and ecological. We proceed from the hypothesis that song learning is an adaptive strategy, and attempt to identify the specific design features and overall function of this strategy in our particular study species. From our studies, we have identified several design features which, taken together, serve to maximize the number of songs the young bird will share with his neighbors, especially his near neighbors, in his first breeding

Perspectives in Ethology, Volume 12
edited by Owings *et al.*, Plenum Press, New York, 1997

season. Why should it be advantageous for the bird to have songs he shares with his neighbors? We suggest four possible, non-mutually exclusive advantages, and discuss the evidence in support of the fourth one. First, shared songs may be attractive to females. Second, shared songs may provide a mechanism by which two neighbors might effectively codefend their territories against other birds: each bird would effectively be mimicking the other while repelling prospective intruders. Third, shared songs may function as a "badge" of familiarity among territorial males; shared songs are a reliable signal of familiarity since they must be learned in the local neighborhood. Fourth, at least in song sparrows, shared songs appear to facilitate communication among neighboring birds. In particular, a bird uses the songs he shares with a neighbor to direct his song to that bird. We conclude by noting the paradox that songbirds have the ability to improvise new songs (demonstrated in lab experiments), yet in the field birds of most species faithfully copy the songs of their older neighbors. We suggest that song researchers need to ponder this paradox, and figure out why it is so important for a songbird to have the same songs as his neighbors.

INTRODUCTION

Loud, complex, species-distinctive vocal signals, or "song," occurs in many animal groups, including insects, frogs, primates, whales and, most prominently, the songbirds (or oscine passerines). Song in the songbirds is unusual in two important ways: songbirds learn their songs and, in many species, individuals learn multiple song types.

In most animals that have song, individuals raised in isolation from conspecifics still produce normal song as adults. In songbirds, on the other hand, an individual's song is strongly influenced by early experiences. Perhaps the most interesting comparison with the songbirds is the other suborder of the Passeriformes, the suboscine passerines. In both the oscines and the suboscines, birds appear to use song in essentially the same way, to attract mates and/or to repel conspecific rivals, especially in the context of territoriality. Yet song learning has been found to occur in all of the oscine species and none of the suboscine species studied to date, suggesting that song learning evolved in the original oscine passerine (Kroodsma 1996).

In most animals that sing, all individuals of a species sing a well-defined species-specific song. This pattern is found in some songbird species, but in the majority of songbirds, individuals sing multiple song types that, while species-typical, differ among themselves and from the songs of other individuals of the species. Individuals have multiple song types, or "song repertoires" in about three-quarters of the songbirds (Kroodsma 1996). These song repertoires can range from a few to more than 100 song types.

The variety of singing styles, and especially the variation in repertoire size in songbird species has thwarted attempts to develop comprehensive theories of song function. For example, theories of song repertoires abound (review in Catchpole & Slater 1995) yet none of them can adequately explain why one-quarter of the songbirds do not have repertoires, nor why repertoire size varies from modest (a few song types) to very large. As a consequence, most theories are designed around particular species, or particular cases (e.g., species with large repertoires).

In this chapter we develop a theory of the function of song learning in one particular species. We then explore whether this theory can be generalized to other species with different patterns of singing and song learning. The keys to our approach are (1) reframing the question of what is learned in song learning, and (2) examining song learning in its natural context, in the field.

The question of what is learned in song learning can be reframed in terms of Marler and Peters' (1982) demonstration that a bird's final repertoire represents only a fraction of the songs the bird memorized. Although a bird may have only 1 song (white-crowned sparrow) or 3–4 songs (swamp sparrow) in its final repertoire, during the early plastic stage of singing the bird sings versions of many other tutor songs it heard earlier. According to Marler and colleagues, the young bird memorizes many songs during his early sensitive period, many of which can be identified later in plastic song. Many or most of these "overproduced" songs will disappear from the bird's repertoire by a process of "selective attrition," perhaps strongly influenced by social interactions with his neighbors just prior to his first breeding season. Although "overproduction" has been described in detail only for the swamp sparrow (Marler & Peters 1982), and in lesser detail for the white-crowned sparrow (Nelson & Marler 1994), early production of songs later dropped has also been demonstrated in the field for several species (song sparrows, Nordby et al unpub.; field sparrows, Nelson 1992; white-crowned sparrows, Baptista & Morton 1988, DeWolfe et al 1989; indigo buntings, Payne 1983).

The fact the bird's final repertoire is only a fraction of the songs he heard in his sensitive period, and only a fraction of the songs he memorized, suggests that we should pay as much attention to the songs the bird does not retain for his final song repertoire as to those he does. From this perspective the difference between species with and without song repertoires seems much less important, since in both cases we are left with the same question: why do birds learn or retain the particular songs of their final repertoire (be it one, ten or a hundred songs) rather than the many other songs to which they were exposed? Why do they choose the songs they do? We view this question of *song selection* as the fundamental question of bird song learning.

The question of song selection must be addressed in the field, rather than the laboratory, if, as we believe, the variables controlling song selection are social and ecological. Laboratory studies permit control and manipulation of the relevant variables, but social and ecological variables are not easily simulated in the lab, and in any case, one cannot attempt to simulate these variables before knowing what they are, which can be discovered only in the field (Beecher 1996).

THE SONG LEARNING STRATEGY OF THE SONG SPARROW

Our generalizations about song learning in this species are derived from observations of a sedentary (nonmigratory) population of song sparrows (*Melospiza melodia*) in an undeveloped 200-ha park bordering Puget Sound in Seattle, Washington. We attempt to color band all (or most) of the males in this population, about 150 birds in an average year.

A male song sparrow typically has 6–10 distinct song types (mode = 8) in his song repertoire. The song types in his repertoire are as distinct from one another as are the song types of different birds. Nevertheless, neighboring birds will often share some of their song types, that is, have very similar song types. Song sharing is very local, with birds more than 4 or 5 territories away rarely sharing song types. This pattern of neighbor song sharing—which has been observed in many different songbird species (see Catchpole & Slate 1995)—suggests that young birds learn the songs of the neighborhood to which they immigrate following natal dispersal. Briefly, the key details of song sparrow natural history in our population are as follows. After fledging, young birds leave the natal area at about a month of age, and, following dispersal, remain in their new area for the remainder of their lives (Beecher et al 1994; Nordby et al submitted; see also Arcese 1987, 1989). A young bird usually begins singing subsong and plastic song in the late summer or early fall, but does not sing adult-like song until the following spring. He usually crystallizes his repertoire by early or mid March, shortly before the breeding season begins in earnest. Song sparrows do not modify their song repertoire after their first breeding season.

We have found it useful to proceed from the hypothesis that song learning is an adaptive strategy (Kroodsma 1983, 1988, 1996; Beecher 1996). Our approach is to identify the specific design features and overall function of this strategy in our particular study species. We refer to these design features as "rules" of song learning, which collectively define the overall strategy of song learning (Beecher et al 1994; Nordby et al submitted).

Rule 1: Copy song types completely and precisely

In the field, young song sparrows usually develop near-perfect copies of the songs of their older neighbors (Beecher et al 1994; example in Fig. 1). It is this very fact that made us realize that we could trace song learning in the field. The song similarities are striking, with the differences between tutor and student

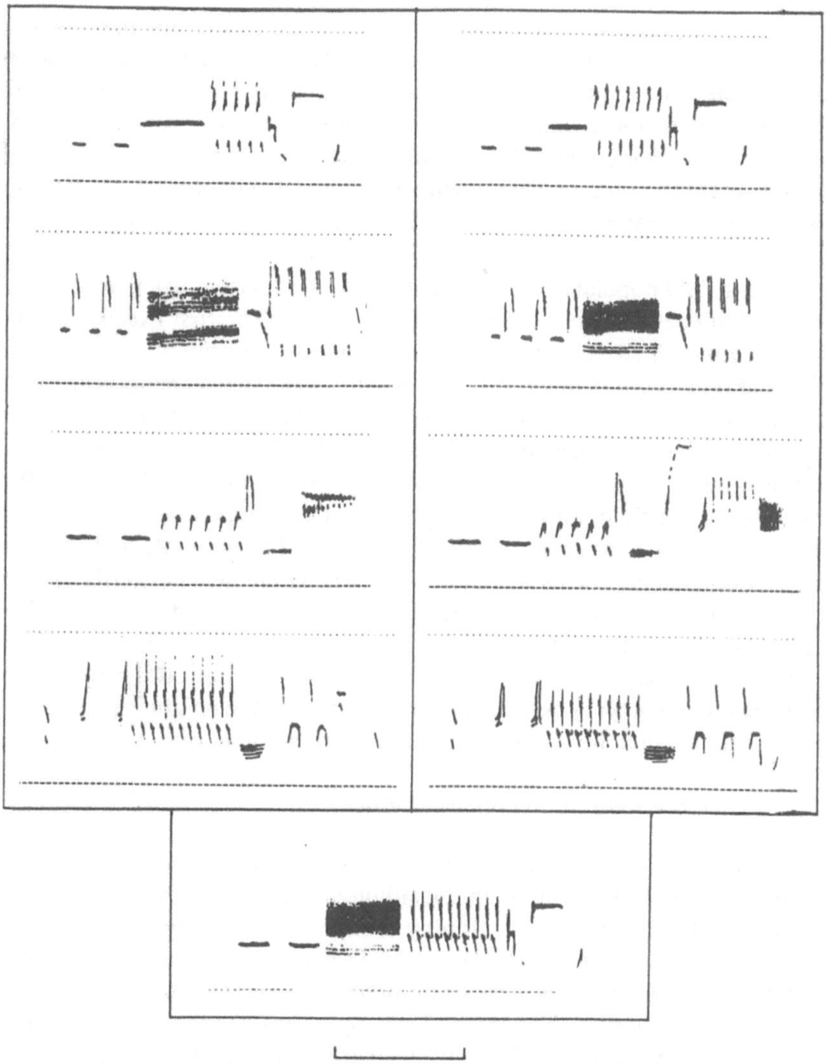

Fig. 1. In the field, young song sparrows learn nearly perfect copies of their tutor songs. Left panel: four of the 9 song types of a young bird. Right panel: corresponding songs of one of his tutors. Bottom panel: the bird might develop a song like this if, instead of learning entire song types, he learned elements and rearranged them into hybrid songs. Note that this hypothetical hybrid song contains one element each from the four tutor songs on the right. Such hybrid songs are common in lab experiments, but we do not find them in our field population. Frequency markers at the bottom and top of each sonogram are 0 and 10 kHz, time marker 1 sec, bandwidth 117 Hz. [From Beecher 1996.]

often being no greater than one normally sees in repetitions of the same song sung by one bird (Stoddard et al 1988).

These field results differ remarkably from the laboratory findings using tape tutors (Marler & Peters 1987, 1988). In the laboratory setting, song sparrows copy song elements quite precisely, but they frequently combine elements from different songs to form what we will call "hybrid" songs—songs made up of parts of different song types. That is, songs are often improvised from learned elements. Figure 1 contains a hypothetical example of a hybrid song that the bird illustrated might have improvised had he copied his tutors as birds in the lab often do.

In our field population, two exceptions to the "perfect copy" rule actually clarify the rule, and suggest the meaning of the rule. The first exception occurs when the young bird "blends" two tutors' somewhat different versions of the same song type (vs. copying one or the other). These songs are not true hybrids because although elements are selected from two different tutors, they are selected from the same, or a very similar, song type. The second exception occurs when the young bird combines elements from two dissimilar song types of the *same* tutor. In the field, birds rarely hybridize a song type of one singer with a distinctly dissimilar song type of a different singer. The principle underlying the two exceptions, then, seems to be to combine song elements of different songs only if they are different tutors' versions of the same type, or if they are different song types sung by the same tutor. We summarize this principle by saying the student "preserves tutor and/or type" in his songs. We believe this rule functions to give the bird song types he shares with other birds in the neighborhood. Hybrid songs, being personal inventions, are shared with no other birds.

Rule 2: Learn the songs of neighboring birds

On average, it takes 3 or 4 "tutors" to account for the young bird's entire repertoire of 8 or 9 song types (Fig. 2), although some birds learn from fewer or more tutors (Beecher et al 1994; Nordby et al sub.). Invariably, these tutors turn out to have been neighbors in the young bird's hatching summer. Usually by the following spring (the young bird's first breeding season) some of these tutor-neighbors will have died. The young bird appears to commence song learning shortly after he has dispersed from his natal area, and it appears that most or all of song memorization probably occurs in the traditional lab-determined sensitive period, which is roughly the second and third months of life (Marler & Peters 1987).

Rule 3: Attempt to establish a territory near your tutor-neighbors

The young bird usually establishes his territory within the territorial range of his song tutors, often replacing a dead tutor. In the exceptional cases where

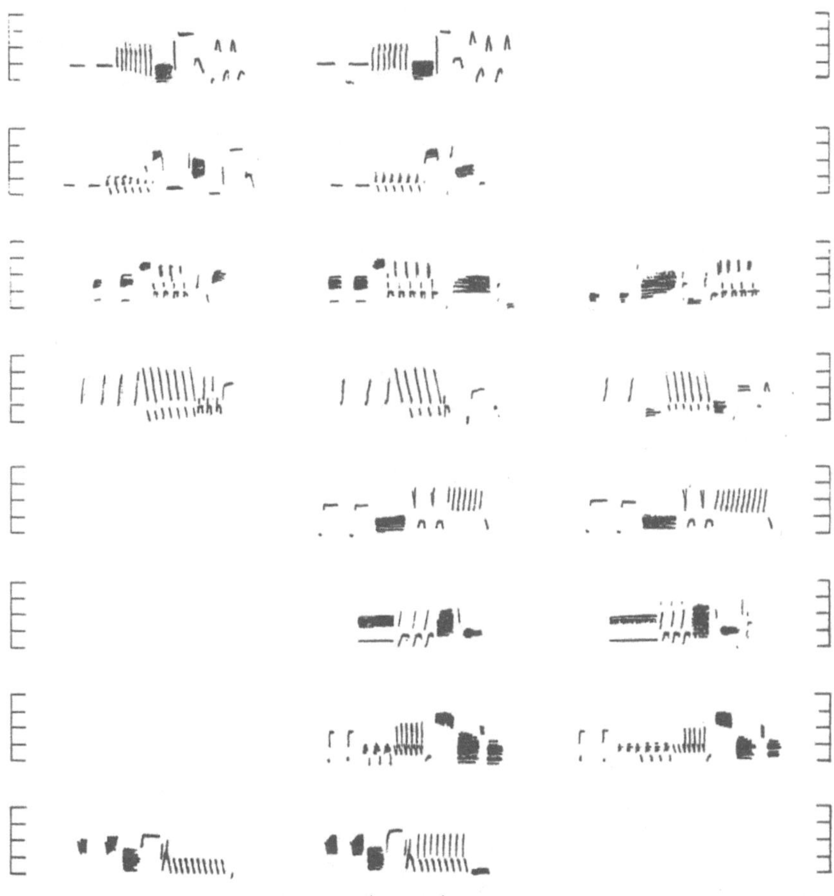

Fig. 2. In the field, young male song sparrows learn their songs from 3 or more older males. Eight of the 11 song types of young bird center, and of three tutors (one upper left, one lower left, and one right). For the two cases where two tutor songs are shown, the one on the left is clearly the better match. Frequency scale 2 to 10 kHz in 2-kHz steps, time marker 1 sec, bandwidth 117 Hz. [From Beecher 1996.]

the young bird does not establish his territory among his tutor-neighbors, it invariably turns out that he didn't because he couldn't—because none of his tutors had died and/or because other young birds moved into this area (Nordby et al submitted). It should be emphasized that birds spending their postdispersal natal summer in area A and moving to area B that autumn or later, wind up with the songs of birds in area A, not of their neighbors in area B, pointing to the importance of the first few months following dispersal (months 2 and 3 of life, the "classical" sensitive period) for song memorization. This, however, is not to imply that a bird's final repertoire is unaffected by events subsequent to this period, i.e., in autumn, winter and spring. At least for tutors present both early and late, this later period is quite important in shaping the young bird's final repertoire (see Rule 5).

Rule 4: Preferentially learn tutor-shared songs

As we noted earlier, in our field population neighbors typically share a portion of their song repertoires, on average about 2–4 of their 8–9 song types. We have found that the young bird preferentially learns (or retains) song types shared by two or more of his tutors (Beecher et al 1994). For example, the young bird diagrammed in Fig. 3 (along with his tutors) retained 7 types that were shared by two or more of his tutors, and only 2 that were unique to one of these

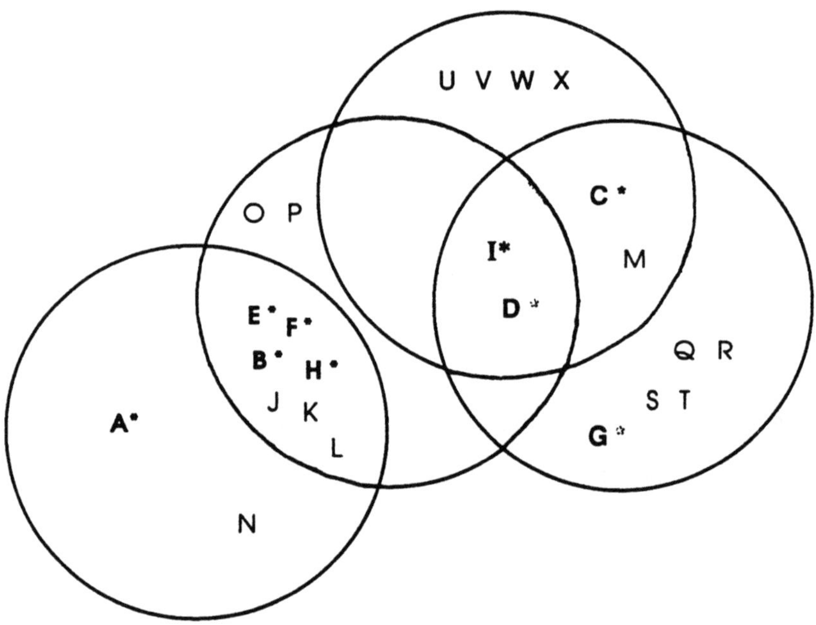

Fig. 3. Young males preferentially learn songs shared by 2 or more tutors. A Venn diagram of the 4 tutors (4 circles) of subject mxpx. Each song type is indicated by a letter, and shared songs of 2 or more birds are given the same letter. Shared types are thus in the intersections of the tutor circles (e.g., types I and D were shared by 3 tutors). The 9 song types learned by mxpx are indicated in boldface and starred. He learned 7 of the 11 tutor-shared types, but only 2 of the 13 tutor-unique types. [From Beecher 1996.]

tutors, despite the fact that in his tutor group there were 11 shared types and 13 unshared types.

Our interpretation of this rule is that it represents a "bet-hedging" strategy to guarantee that the young bird has song types he will share with his neighbors in his first breeding season. If instead the bird learned tutor-unique songs, he would have songs "individualized" for these particular tutor-neighbors (i.e., shared with that neighbor only). But these individualized songs would be good only until the tutor dies or moves, whereas a shared song is good until all the birds having it in the neighborhood die or move, and even beyond then because other young birds moving into the area also preferentially learn shared types.

Rule 5: Preferentially retain song types of tutors surviving to your first breeding season

Birds often have song types that can be traced to tutors that were alive in the young bird's natal summer but died before the next breeding season. Nevertheless, they generally retain more songs of tutors that survive into the next breeding season than of tutors that do not (Nordby et al submitted). We describe this late learning as "late influence" because it does not appear to be de novo learning. We have yet to find a case where a young bird learned a new song in the early spring (January–March) that he could not have encountered the previous summer. As mentioned earlier, in those cases where a bird moves away from his natal summer tutors (because none of them die), he does not learn the songs of his new neighbors. He may well retain those song types he memorized earlier that best match those of his new neighbors (we are examining this hypothesis) but his songs still match those of his old tutor-neighbors better than those of his new neighbors.

These results are quite consistent with the two-phase model of song learning of Nelson and Marler (1994), according to which songs are memorized during a sensitive period in the natal summer and songs are kept for the final repertoire or dropped from it on the basis of later social interactions with neighbors of the next breeding season ("selective attrition").

Rule 6: Preferentially retain song types of your primary tutor if you establish a territory next to him

A bird's primary tutor—the bird that had the greatest influence on the young bird's song repertoire—can vary from having total influence to only slightly more influence than other tutors. We have discovered a correlate for this range of influence: whether or not the young bird and the primary tutor are contiguous neighbors in the bird's first breeding season (Nordby et al submitted). About half the young birds do not wind up next to their primary tutor, and when a bird does not, the influence of his primary tutor is less. That birds do not always wind up strongly influenced by a contiguous neighbor suggests that the commitment to the primary tutor occurs early, probably in the natal summer. According to this hypothesis, if the young bird manages to set up next to that primary tutor, he retains more of that tutor's songs for his final repertoire. If the young bird fails to set up next to the primary tutor, winding up instead next to one of his lesser tutors, or farther away yet, he will not show a strong influence from any one tutor.

SONG SHARING AND COUNTERSINGING WITH NEIGHBORS

What Is the Function of Song Learning in Song Sparrows?

We suggested in the Introduction that the fundamental question of song learning is: why do birds learn the songs they do? We believe we have provided an answer to that question for at least one population of song sparrows. On the basis of our examination of song learning in a natural population of this species, we have identified six design features of an adaptive learning strategy. These six rules of song learning, taken together, work to maximize the number of songs the young bird will share with his neighbors, especially his near neighbors, in his first breeding season. We propose that equipping the young bird with songs he shares with his neighbors, perhaps tinged with some individuality, is the function of this song learning strategy.

This song learning strategy does have its imperfections. A major imperfection is the failure of a bird to develop good copies of his territorial neighbors' songs in those rare cases where he is unable to establish his territory in the area where he memorized his songs in his natal summer. Although "song choice" among earlier-memorized songs continues into the bird's first breeding season—he retains more songs from tutors that survive until then, and especially from those that are on adjacent territories—it does seem that selection would have favored the ability to memorize new songs into that first breeding season. In fact some songbirds have the ability to learn songs de novo in the first breeding season and sometimes in subsequent seasons as well (see further discussion below).

It is as if in song sparrows the early sensitive period is a constraint which the rest of the song learning strategy must accommodate. Perhaps the cleverest example of such accommodation is Rule 4. If the young bird preferentially memorizes tutor-shared songs in his natal summer, he protects against losing shared types should a particular tutor not survive to the next breeding season. Another example is Rule 5: by keeping a large pool of memorized songs open for selective attrition the following spring, the bird protects against being stuck with too many unshared song types should some of his tutors die over the winter.

However, as just discussed, the song learning strategy does not seem to protect against the eventuality of the bird's being forced out of the area in which he memorized songs in his natal summer.

Previous accounts of song sparrows have tended to focus on repertoire size as the key variable (Searcy 1984; Hiebert et al 1989). Our examination of song learning suggests that song sharing may be the major target of selection. If sharing songs with his neighbors does afford a bird some advantage, this should be reflected in measures of fitness. Although the test for fitness correlates is the most direct test of the hypothesis that song sharing is advantageous, very little evidence on this point has been gathered to date. Song sharing has been shown to correlate with some measures of fitness in indigo buntings, but the effect has not been consistent (Payne 1982; Payne et al 1988). Preliminary studies of our song sparrow population suggest that song sharing is in fact a better predictor of male territorial success than is repertoire size (Beecher et al unpublished study). More research is needed, however, on this most direct test of the song sharing hypothesis.

What Are the Advantages of Song Sharing?

Why should it be advantageous for the bird to have songs he shares with his neighbors? We will suggest four possible advantages, which are not mutually exclusive, and focus on the fourth candidate. First, shared songs may be advantageous in male-female interactions. For example, females might prefer the common song types of a neighborhood. Although the role of song in the mate choice context has received much attention, research and theory have focused almost exclusively on song repertoires (e.g., Catchpole 1986, 1987; Searcy 1984, 1992; Searcy & Andersson 1986; Searcy & Yasukawa 1990; Hasselquist et al 1996). The only attention paid to song types per se has been in studies indicating female preferences for local vs. alien dialects (e.g., Baker 1983). We have only just recently begun examining female song sparrow preferences for different types of songs (O'Loghlen & Beecher in press), and so will leave discussion of this important selection pressure on male song for a later time.

Second, Craig and Jenkins (1982) and Payne (1983) have suggested that shared songs may be a form of mimicry, an attempt by young birds to deceive older, established birds as to their identity. We suggest here an alternative mimicry hypothesis. An important function of song is "posting" the territory: birds appear to avoid the area so long as they hear appropriate song from that area. This function of song has been well demonstrated by experiments in which the bird is removed and replaced with broadcast recorded song (the song keeps his territory vacant longer), or in which the bird is left on territory but his voice removed via devocalization (he fails to hold his territory; these studies are reviewed in Catchpole & Slater 1996). Our hypothesis is that shared songs may be mutually beneficial to two adjacent neighbors in posting their territory against third parties. The special effectiveness of shared song with respect to this posting function is that if your neighbor is singing a shared song near your mutual boundary, you may be relieved of that responsibility, as he may be when you are singing in that area.

Third, shared songs may function as a "badge" of familiarity among territorial males. Getty (1987) suggested that territorial birds should prefer familiar to unfamiliar neighbors. Beletsky and Orians (1989) showed that male red-winged blackbirds with familiar neighbors have greater breeding success than do males with unfamiliar neighbors. Neither preferring nor cooperating with familiar neighbors requires shared songs, of course, but shared songs are a reliable signal of familiarity since they must be learned in the local neighborhood.

The fourth advantage, and the one we will discuss in most detail, is that a bird may use songs he shares with a neighbor to direct his song to that bird. Many studies have shown that birds will, at least under some circumstances, reply to a shared song with the same song type: "song matching," or more precisely, "type matching" (reviews in Krebs et al 1981; Smith 1991; McGregor 1991). Most workers have viewed type matching as a way of directing one's response at the bird that has just sung. A puzzling finding has been that higher rates of type matching occur when the stimulus song is the bird's own song or a song of a stranger that is similar to one of the bird's own songs, than when it is neighbor song (song sparrows, McArthur 1986, Stoddard et al 1992; western meadowlarks, Falls 1985; great tits, Falls et al 1982). We were able to shed some light on this unusual finding in a study of established song sparrow neighbors, tested mid to late in the breeding season. While these birds, as expected, did not type match the broadcast neighbor song, they did consistently reply with some other song they shared with that neighbor (Beecher et al 1996). We have called this pattern of song selection "repertoire matching." That is, suppose the bird and his neighbor share song types A, B and C, but not the rest of the songs in their repertoires, and the neighbor sings A. Then the subject can "type match" by replying with A, or "repertoire match" by replying with B or C (or not match at all by singing any of his other songs, or not sing at all). Note that repertoire matching implies knowledge of the singer's repertoire; the bird must know which songs he shares with the singer in order to consistently reply in this way. These observations extend our earlier demonstrations of individual neighbor song recognition in song sparrows (Stoddard et al 1990, 1991) and show that this recognition is more sophisticated that we might have imagined.

As indicated, a song sparrow will respond to the song of a stranger quite differently. If he has a similar song type, he usually will respond with that type. We cannot describe how he selects his reply song in the case where he does not have a similar type (the usual case), except for the following observation. If the stranger song is played from a neighbor's territory, the subject will reply with a song type he does *not* share with the neighbor there. It is as if he reserves his shared songs for that neighbor. In one study we presented the same nonmatching stranger playback song to the subject on two different days; birds tended to respond with different songs on the two days, which suggests that they may not have been selecting a particular reply song (Stoddard et al 1992).

Finally, we have begun to identify some of the conditions under which song sparrows will in fact type match neighbor song. In the field we had noticed more type matching early in the breeding season, and suspected it was related to the territorial instability and higher levels of aggression that occur at that time, especially between new neighbors. We thus tested response to neighbor song by new neighbors twice during the breeding season: early, in April, and again a month and a half later (Beecher et al unpub. ms.). Early in the breeding season, new neighbors will have only recently established their territorial boundary, which may still be in dispute, and territorial skirmishes will have occurred recently or may still be occurring. A new neighbor singing at the boundary early in the season represents a more serious challenge than a well-established neighbor singing at that same boundary and we predicted higher levels of type matching at that time. As predicted, early in the season birds usually replied to a shared neighbor song with a type match, whereas a month and a half later they usually replied with a repertoire match (they did not respond with unshared song either early or late). These results are consistent with the hypothesis that type matching is a more aggressive or escalated response than repertoire matching. Repertoire matching in turn may be an escalated response compared to non-matching or not singing at all. We are presently investigating this hypothesis.

Our results suggest that when neighbors share some song types and not others, two neighbors may use their shared songs to modulate their social interactions. We suggest here one possible hypothesis for how the birds might do this. The hypothesis is consistent with our data so far, but so are numerous others, and we offer it here only to illustrate a way in which birds might use their shared songs in a communication context. According to the hypothesis, the bird uses the songs he shares with a particular neighbor to engage that neighbor, who may acknowledge by repertoire matching, escalate by type matching, or break off the interaction by singing an unshared song type. Consider again the two hypothetical neighbors sharing songs A, B and C but no others. Bird 1 can engage bird 2 by singing any one of A, B or C (toward bird 2, since other neighbors may also share some types). Let us say bird 1 sings A. Bird 2 then can acknowledge the signal by replying with B or C (repertoire match), can escalate by replying with A, or can ignore by singing one of his other types or not singing at all. We are presently examining natural countersinging interactions to see whether we can predict patterns of increasing or decreasing behavioral engagement from patterns of switching into or out of repertoire matching and type matching. We also plan interactive playback experiments in which we simulate a neighbor replying to the subject either by type matching or repertoire matching.

What happens when the interactants have only shared songs? We occasionally find pairs of song sparrows that share all their songs. Complete song sharing occurs both in some large-repertoire species (e.g., marsh wrens, Verner 1976) and some small-repertoire species (e.g., corn buntings, McGregor et al 1988; bobolinks, Woods 1995) as well as, of course, many single-song species. In the species with repertoires, birds with all shared songs still have the opportunity to escalate by switching from repertoire matching to type matching, or deescalate by switching from type matching to repertoire matching and to deescalate further by stopping singing. The birds also have timing options, such as following vs. leading vs. asynchronous singing, as well as options such as overlapping (e.g., Kramer et al 1985; Horn & Falls 1988; Dabelsteen & McGregor 1996).

Song Sharing in Other Songbirds

We suggested in the Introduction that an approach that focused on the species' song learning strategy in its natural context might yield generalizable theories. The hypothesis generated by our study of song learning in song sparrows is that song sharing with neighbors is the function of the song learning strategy of song sparrows. Is song sharing with neighbors likely to be the function of the song learning strategy with other songbird species? We believe considerable evidence supports this hypothesis.

To begin with, two pioneer studies demonstrated postdispersal learning of songs from neighbors (Bewick's wren, Kroodsma 1974; saddlebacks, Jenkins 1978). Other studies have showed that in several species, birds modify their song repertoires in the first or subsequent breeding seasons so as to increase song sharing with neighbors (indigo buntings, Payne 1983, 1996; white-crowned sparrows, Baptista & Morton 1988; great tits, McGregor & Krebs 1989; field sparrows, Nelson 1992; American redstarts, Lemon et al 1994; cowbirds, O'Loghlen & Rothstein 1993; European starlings, Mountjoy & Lemon 1995). Finally, populations in which birds share songs more closely with neighbors than with distant birds provide at least circumstantial evidence of learning from neighbors or neighbors-to-be (see Catchpole & Slater 1995). Although a full-scale study of song learning is required to provide strong evidence, the pattern of neighbor song sharing is at least consistent with the hypothesis that birds are settling next to their song tutors and other birds with which they share songs learned from a "recent common ancestor."

Turning to species with only a single song type, the notable fact is that no species has yet been discovered in which birds do not share song types with their neighbors. The most common case is the "dialect" species in which birds over a rather large area share the same song type (e.g., white-crowned sparrows). A somewhat different case occurs when many distinct song types occur in the population, but particular song types are not randomly distributed in the population but rather clustered in neighborhoods, i.e., neighbors share song types (e.g., indigo buntings, Payne & Payne 1993; Payne 1996). When song is examined at the micro level, these two cases may not be that different, for even in the "dialect" species, neighbors have more similar versions of the dialect type than do nonneighbors (e.g., Bell et al in press). So far as we know, in no nonrepertoire species do birds have songs distinctly different from their neighbors. The ubiquity of song sharing, despite tremendous variation in other song features, suggests, once again, an advantage to being able to reply to your neighbor with the *same* song type.

Populations without Neighbor Sharing?

In some populations, however, birds are reported not to share more songs with neighbors than with more distant birds. Most reports of neighbor sharing or lack thereof do not provide enough information to be evaluated, so we will confine ourselves to one reasonably well-documented comparison, between different populations of song sparrows. Three different western, sedentary populations have been observed to have high levels of song sharing (Washington state, our work; British Columbia, Cassidy 1993; California, Nielsen & Vehrencamp 1995). In contrast, three different eastern, migratory populations have been described as having minimal neighbor song sharing (Ontario: Kramer & Lemon 1983; Pennsylvania: W. A. Searcy et al unpub. obs.; New York state: J. L. Bower unpub. obs.). This contrast suggests the migratory/sedentary difference or a racial difference is the key variable determining the extent of neighbor song sharing in the population.

It is easy to imagine that in migratory populations, a bird that completes his song learning in his natal summer might learn his songs in one area in the natal summer but settle away from his tutors on returning from migration, leading to a pattern of low song sharing among territorial neighbors. No data have been gathered directly on this point, however, and migration does not necessarily preclude young birds returning from migration to settle near their song tutors of their natal summer. In most passerine species that have been studied, adults return quite precisely to their breeding sites of the previous year. Yearling birds are often described as less philopatric than adults, but for the most part all that has been shown is that adults return to their breeding area but yearlings do not return to their natal neighborhoods. But young birds disperse from their natal neighborhoods, and would go through the song memorization phase of song learning in their postdispersal area in their natal summer. Thus the question is whether they return to this postdispersal area and breed there the next spring. If they did, they and their tutors would be returning to the same area, and perhaps to neighboring territories. In one of the few studies to actually track young birds after their natal summer dispersal (usually young birds are banded only predispersal, in the nest), Morton (1992) showed that a substantial fraction of yearling white-crowned sparrow males return faithfully to the postdispersal area of their natal year. Birds that spent enough time in the postdispersal area (4 or more weeks) return as accurately to this area as adults return to their breeding area of the previous summer. Thus, the song learning pattern in this population of migratory white-crowned sparrows could be very similar to that found in sedentary populations of the species, with birds learning songs from birds that will eventually be their territorial neighbors (Baptista & Morton 1988; DeWolfe et al 1989).

We have recently studied a migratory population of song sparrows in Washington (nesting at 3000 feet in the Cascade Mountains) and find a pattern of neighbor song sharing quite similar to that of our sedentary population (Hill et al unpub. data). At this point we cannot say why this western migratory population is more like the western sedentary populations than it is like the eastern migratory populations. Clearly more data are needed, but we can make the following points. First, we need more complete descriptive studies of song sharing in natural populations. The tendency to simply describe a population as having or not having song sharing among neighbors is no longer sufficient. Second, if song sharing is less in some populations than others, it could still be the case that neighbors show more similarities in their songs than nonneighbors. We need good quantitative data on this point. Third, and most important, even where song similarity is not as marked as in our population, the birds themselves may classify their songs as "shared" (similar to the neighbor's) or "unshared" (dissimilar). A tendency to classify some songs as similar even when detailed physical similarities are absent should be detectable by playback experiments. For example, over the course of a number of playback trials using all the songs of a particular neighbor, the subject should reply only (or mostly) with the types he regards as "shared" songs.

Diversity of Song Patterns

We suggested in the Introduction that the diversity of song patterns in songbirds has thwarted development of general theories of song. Some of the parameters of this diversity are sketched out in Table 1. Nevertheless, neighbor song sharing is as close to a songbird universal as has been found to date (with the exception, of course, of song learning) As suggested above, if close song sharing with your neighbors is advantageous, one might expect selection for later learning, in the spring of the first breeding season. Indigo buntings are an example of a species in which males evidently learn songs de novo at the beginning of their first breeding season (Payne 1996). And in some species, as mentioned earlier, individuals continue to modify their repertoires each breeding season (or at least have the potential to do so). From this perspective, a species

Table 1. Some Dimensions of Song Variation in Songbirds

1. When song is learned	From very early in life (zebra finch) to throughout the lifetime (canary, thrush nightingale).
2. Song repertoire size	From a single species-specific song (white-crowned sparrow) to just a few song types (chaffinch, great tit, < 5), to many (marsh wren, nightingale, >100) to very many (brown thrasher, > 1000).
3. Individual variation (over some defined geographic area)	From all birds with the same song type (white-crowned sparrows) to birds with markedly different song types (indigo buntings).
4. Mimicry	From species that copy only within tight species-specific parameters to mimics that copy virtually anything they hear (mockingbirds, thrashers, catbirds).
5. Copying fidelity	From precise copying (marsh wren) to improvisation (sedge wren).

like the song sparrow seems puzzling at first: a bird gets new neighbors each breeding season, yet never changes his repertoire after the first season. The bird has an alternative route to maintaining song sharing, however, which is to teach young birds his songs. If sharing is mutually advantageous (an untested hypothesis), then just as young birds are selected to learn the songs of their neighbors, so are the older birds favored to teach these songs to them. We have found that song sparrows maintain song sharing over their lifetimes despite neighbor turnover each year, evidently because young birds learn their songs; the tendency to learn neighbor-shared songs helps as well. Song tutors have generally been thought of as passively providing songs which young birds actively seek out. Perhaps we have been blinded to the possibility of "active tutoring," however, by our heritage of laboratory song learning studies in which the tutor is a tape recording. It could be that we have overestimated the importance of the difference between song learning programs that close in the bird's first year and those that remain open longer, for if the goal is to maintain song sharing with his neighbors, the bird has two means to this end: learn their songs or teach them his.

CONCLUSIONS

We have come full circle and arrived at a paradox, or at least an irony. Songbirds have the ability to learn songs, and the ability to improvise, to sing what no bird has sung before. In the laboratory, they often make full use of their ability to improvise (see Beecher 1996; Kroodsma 1996). But in the field most birds instead slavishly copy the songs of their neighbors or neighbors-to-be. In the extreme case of some of the single-song species—the white-crowned sparrow comes to mind—the songs of neighboring birds are as similar as those of their hard-wired suboscine cousins. One wonders why the oscines have retained the ability to learn song! At the other extreme are the birds that use their abilities of invention and improvisation to the max, such as the catbird (Kroodsma et al in press). And somewhere in the middle we have birds such as the song sparrow whose repertoire, although certainly constructed with creativity, is essentially the common denominator of what the bird heard his neighbors singing during his first year of life.

The task of song researchers will be to face this paradoxical result, and figure out why it is so important for a songbird to have the same songs as his neighbors.

ACKNOWLEDGMENTS

Special thanks to Philip Stoddard for getting this all started. We thank Discovery Park for hosting our fieldwork and the National Science Foundation for supporting this research. Finally, we thank Nick Thompson and Don Owings for their valuable comments and considerable patience.

REFERENCES

Arcese, P. (1987). Age, intrusion pressure and defence against floater by territorial male song sparrows. *Animal Behavior, 35*, 773–784.

Arcese, P. (1989). Intrasexual competition, mating system and natal dispersal in song sparrows. *Animal Behavior, 37*, 45–55.

Baker, M. C. (1983). The behavioral response of female Nuttall's white-crowned sparrows to male song of natal and alien dialects. *Behav Ecol Sociobiol 12*, 309–315.

Baptista, L. F. & Morton, M. L. (1988). Song learning in montane white-crowned sparrows: from whom and when. *Animal Behavior, 36*, 1753–1764.

Beecher, M. D. (1996). Bird song learning in the laboratory and the field. In D. E. Kroodsma & E. L. Miller (Eds.), *Ecology and evolution of acoustic communication in birds* (pp. 61–78). Cornell, Ithaca, NY.

Beecher, M. D., Campbell, S. E. & Stoddard, P. K. (1994). Correlation of song learning and territory establishment strategies in the song sparrow. *Proc Natl Acad Sci 91*, 1450–1454.

Beecher, M. D., Stoddard, P. K., Campbell, S. E. & Horning, C. L. (1996). Repertoire matching between neighbouring songbirds. *Anim Behav 51*, 917–923.

Beletsky, L. D. & Orians, G. H. (1989). Familiar neighbors enhance breeding success in birds. *Proc Natl Acad Sci 86*, 7933–7936.

Bell, D. A., Trail, P. W. & Baptista, L. F. (In press). Song learning and vocal tradition in Nuttall's white-crowned sparrows. *Anim Behav.*

Cassidy, A. L. (1993). *Song variation and learning in island populations of song sparrows.* Unpublished doctor dissertation, University of British Columbia, Canada.

Catchpole, C. K. (1986). Song repertoires and reproductive success in the great reed warbler *Acrocephalus arundinaceus. Behav Ecol Sociobiol 19,* 439–445.

Catchpole, C. K. (1987). Bird song, sexual selection and female choice. *Trends Ecol Evol 2,* 94–97.

Catchpole, C. K. & Slater, P. J. B. (1995). *Bird song: Biological themes and variations.* Cambridge.

Craig, J. L. & Jenkins, P. F. (1982). The evolution of complexity in broadcast song in passerines. *J Theor Biol 95,* 415–422.

Dabelsteen, T. & McGregor, P. K. (1996). Dynamic acoustic communication and interactive playback. In D. E. Kroodsma & E. L. Miller (Eds.), *Ecology and evolution of acoustic communication in birds* (pp. 377–397). Cornell, Ithaca, NY.

DeWolfe, B. B., Baptista, L. F. & Petrinovich, L. (1989). Song development and territory establishment in Nuttall's white-crowned sparrows. *Condor 91,* 397–407.

Falls, J. B. (1985). Song matching in western meadowlarks. *Can J Zool 63,* 2520–2524.

Falls, J. B., Krebs, J. R. & McGregor, P. K. (1982). Song matching in the great tit (*Parus major*) : The effect of similarity and familiarity. *Anim Behav 30,* 997–1009.

Getty, T. (1987). Dear enemies and the prisoner's dilemma: why should territorial neighbors form defensive coalitions? *Am Zool 27,* 327–336.

Hasselquist, D., Bensch, S. & Schantz, T. (1996). Low frequency of cuckoldry in the polygynous great reed warbler: extra-pair young are sired by attractive neighbors. *Nature.*

Hiebert, S. M., Stoddard, P. K. & Arcese, P. (1989). Repertoire size, territory acquisition and reproductive success in the song sparrow. *Anim Behav 37,* 266–273.

Horn, A. G. & Falls, J. B. (1988). Repertoires and countersinging in western meadowlarks. *Ethology 77,* 337–343.

Jenkins, P. F. (1978). Cultural transmission of song patterns and dialect development in a free-living bird population. *Anim Behav 26,* 50–78.

Kramer, H. G. & Lemon, R. E. (1983). Dynamics of territorial singing between neighbouring song sparrows (*Melospiza melodia*). *Behaviour,* 85, 198–223.

Kramer, H. G., Lemon, R. E. & Morris, M. J. (1985). Song switching and agonistic stimulation in the song sparrow: five tests. *Anim Behav 33,* 135–149.

Krebs, J. R., Ashcroft, R. & van Orsdol, K. (1981). Song matching in the great tit (*Parus major* L). *Anim Behav, 29,* 918–923.

Kroodsma, D. E. (1974). Song learning, dialects, and dispersal in the Bewick's wren. *Z Tierpsychol 35,* 352–380.

Kroodsma, D. E. (1983). The ecology of avian vocal learning. *33,* 165–171.

Kroodsma, D. E. (1988). Contrasting styles of song development and their consequences. In R. B. Bolles & M. D. Beecher (Eds.), *Evolution and learning.* Erlbaum, New Jersey.

Kroodsma, D. E. (1996). Ecology of passerine song development. In D. E. Kroodsma & E. L. Miller (Eds.), *Ecology and evolution of acoustic communication in birds* (pp. 3–19). Cornell, Ithaca, NY.

Kroodsma, D. E., Houlihan, P.W., Fallon, P.A. & Wells, J. A. (In press). Song development by grey catbirds. *Anim Behav.*

Lemon, R. E., Perrault, S. & Weary, D. M. (1994). Dual strategies of song development in American redstarts, *Setophaga ruticilla. Anim Behav 47,* 317–329.

Marler, P. & Peters, S. (1982). Developmental overproduction and selective attrition; new processes in the epigenesis of birdsong. *Dev Psychobiol 15,* 369–378.

Marler, P. & Peters, S. (1987). A sensitive period for song acquisition in the song sparrow, *Melospiza melodia*: a case of age-limited learning. *Ethology 76,* 89–100.

Marler, P. & Peters, S. (1988). The role of song phonology and syntax in vocal learning preferences in the song sparrow, *Melospiza melodia*. *Ethology 77,* 125–149.

McArthur, P. D. (1986). Similarity of playback songs to self song as a determinant of response strength in song sparrows. *Anim Behav 34,* 199–204.

McGregor, P. K. (1991). The singer and the song: on the receiving end of bird song. *Biol Rev 66,* 57–79.

McGregor, P. K. & Krebs, J. R. (1989). Song learning in adult great tits (*Parus major*): effects of neighbours. *Behaviour 108,* 139–159.

McGregor, P. K., Walford, V. R. & Harper, D. G. C. (1988). Song inheritance and mating in a songbird with local dialects. *Bioacoustics 1,* 107–129.

Morton, M. L. (1992). Effects of sex and birth date on premigration biology, migration schedules, return rates and natal dispersal in the mountain white-crowned sparrow. *Condor 94,* 117–133.

Mountjoy, D. J. & Lemon, R. E. (1995). Extended song learning in wild European starlings. *Anim Behav 49,* 357–366.

Nelson, D. A. (1992). Song overproduction and selective attrition lead to song sharing in the field sparrow (*Spizella pusilla*). *Behav Ecol Sociobiol 30,* 415–424.

Nelson, D. A. & Marler, P. (1994). Selection-based learning in bird song development. *Proc Natl Acad Sci 91,* 10498–10501.

Nielsen, B. M. B. & Vehrencamp, S. L. (In press). Responses of song sparrows to song type matching via interactive playback. *Behav Ecol Sociobiol.*

Nordby, J. C., Campbell, S. E. & Beecher, M. D. (submitted). Ecological correlates of song learning in song sparrows.

O'Loghlen, A. L. & Beecher, M. D. (In press). Sexual preferences for mate song types in female song types. *Anim Behav.*

O'Loghlen, A. L. & Rothstein, S. I. (1993). An extreme example of delayed vocal development: song learning in a population of wild brown-headed cowbirds. *Anim Behav 46,* 293–304.

Payne, R. B. (1982). Ecological consequences of song matching: breeding success and interspecific song mimicry in indigo buntings. *Ecol 63,* 401–411.

Payne, R. B. (1983). The social context of song mimicry: song matching dialects in indigo buntings (*Passerina cyanea*). *Anim Behav 31,* 788–805.

Payne, R. B. (1996). Song traditions in indigo buntings: origin, improvisation, dispersal and extinction in cultural evolution. In D. E. Kroodsma & E. L. Miller (Eds.), *Ecology and evolution of acoustic communication in birds* (pp. 198–220). Cornell, Ithaca, NY.

Payne, R. B. & Payne, L. L. (1993). Song copying and cultural transmission in indigo buntings. *Anim Behav, 46,* 1045–1065.

Payne, R. B., Payne, L. L. & Doehlert, S. M. (1988). Biological and cultural success of song memes in indigo buntings. *Ecol 69,* 104–117.

Searcy, W. A. (1984). Song repertoire size and female preferences in song sparrows. *Behav Ecol Sociobiol 14,* 281–286.

Searcy, W. A. (1992). Song repertoires and mate choice in birds. *Am Zool 32,* 71–80.

Searcy, W. A. & Andersson, M. (1986). Sexual selection and the evolution of song. *Annu Rev Ecol Syst 17,* 507–533.

Searcy, W. A. & Yasukawa, K. (1990). Use of the song repertoire in intersexual and intrasexual contexts by red-winged blackbirds. *Behav Ecol Sociobiol 27,* 123–128.

Smith, W. J. (1991). Singing is based on two markedly different types of signaling. *J Theor Biol 152,* 241–253.

Stoddard, P. K., Beecher, M. D., Campbell, S. E. & Horning, C. (1992). Song-type matching in the song sparrow. *Can J Zool 70,* 1440–1444.

Stoddard, P. K., Beecher, M. D., Horning, C. L. & Campbell, S. E. (1991). Recognition of individual neighbors by song in the song sparrow, a species with song repertoires. *Behav Ecol Sociobiol 29,* 211–215.

Stoddard, P. K., Beecher, M. D., Horning, C. H. & Willis, M. S. (1990). Strong neighbor-stranger discrimination in song sparrows. *Condor 97,* 1051–1056.

Stoddard, P. K., Beecher, M. D. & Willis, M. S. (1988). Response of territorial male song sparrows to song types and variations. *Behav Ecol Sociobiol 22,* 125–130.

Verner, J. (1976). Complex song repertoires of long-billed marsh wrens in eastern Washington. *Living Bird 14,* 263–300.

Woods, J. L. (1995). *Micro- and macrogeographic patterns of song sharing in bobolinks*. Unpublished doctoral dissertation, University of Michigan, Ann Arbor.

Chapter 4

REFERENTIAL SIGNALS

Christopher S. Evans

Department of Psychology
Macquarie University
Sydney, NSW 2109, Australia
chris.evans@mq.edu.au

ABSTRACT

Recent research on mechanisms of animal communication has been concerned largely with identifying the factors that have influenced the design of sexually selected signals. Work in this area has produced exciting advances that integrate proximate and ultimate levels of analysis. It has, however, neglected a whole class of signals that encode information in addition to attributes of the sender. Birds and primates are now known to have specific calls that allow companions to predict environmental events, such as the discovery of food or the appearance of a particular type of predator. These signals are functionally referential.

I propose a framework for recognizing referential signals, together with a strategy for the experimental analysis of such systems. Work on referential signalling is then reviewed, concentrating on the most obvious gaps in current models. I begin by considering the behavior of callers, assessing production specificity, the evidence for developmental plasticity in call usage and some problems introduced by nomenclature. The factors responsible for transitions between different signal classes have been identified in some systems, but little is known about fine-grained variation in signal morphology. Conclusions about the information encoded in animal signals will be sensitive to the level of analysis selected because relatively gross changes, such as variation in call type, can reflect external events while more subtle variation in structure remains affective in nature. The issue of signal design is then discussed. The factors responsible

Perspectives in Ethology, Volume 12
edited by Owings *et al.*, Plenum Press, New York, 1997

for determining the physical form of referential signals are much less well described than those that have shaped signals reflecting properties such as fighting ability or parasite load.

I suggest that there are aspects of signalling behavior that cannot be understood by focussing only upon issues of meaning, but rather will require consideration of additional factors, including the likelihood of tonic communication and of signalling to predators. The contribution of contextual cues has probably been underestimated. Social context plays an important role in determining whether signals are produced and is thus an essential component in any model that seeks to predict calling behavior. It is likely that contextual information also modulates the responses of receivers, both by providing additional information synchronous with the signal and by building up associations with particular signal types over the course of development.

I then assess critically some aspects of the relationship between communication and cognitive processes. Current evidence does not require us to conclude either that referential signals evoke representations of the eliciting event in the minds of receivers or that they are used deceptively. Finally, I consider the evolution of referential signals and suggest that ecological factors may have played an important role.

MECHANISMS OF ANIMAL COMMUNICATION

The recent resurgence of interest in proximate questions concerning animal communication has been in large part driven by the realization that perceptual processes have influenced the design of animal signals (reviews by Guilford and Dawkins, 1991, 1993; Dawkins, 1993; Pagel, 1993; Guilford, 1995; Dawkins and Guilford, 1996). Although we have known for some time that the structure of songs and calls can be accounted for, at least in part, by considering the physical characteristics of the signalling environment (e.g., Wiley and Richards, 1982; Forest, 1994) it is only quite recently that we have begun to understand the role played by the properties of signal receivers. Analyses of sexually dimorphic 'advertisement' signals, such as the conspicuous visual ornaments of birds, have been particularly revealing (e.g., M. Andersson, 1982, 1994; Møller, 1988, 1990; Barnard, 1990; S. Andersson, 1992). Such signals are likely to have arisen as a consequence of sexual selection operating through female mate choice, and this is thought to have affected not only the size of the structures, but also their degree of bilateral symmetry (e.g., Møller, 1992; Møller and Pomiankowski, 1993). In some cases, male signals appear to have evolved in response to preexisting biases in female receivers. Systematic comparative analyses provide compelling support for the hypothesis that the evolution of the receiver preferences predates those of the male traits (Basolo, 1990, 1995a,b;

Ryan et al., 1990; Ryan and Rand, 1993). These examples illustrate the considerable benefits that may accrue from a research strategy that integrates analyses of proximate factors, traditionally the province of experimental psychology and neuroscience, with an exploration of the functional problems that have always assumed central importance in behavioral ecology. Such an approach is more likely to provide a comprehensive account of behavior than a narrow focus on the issue of adaptive significance (Dawkins, 1989; Stamps, 1991).

Theoretical models of animal communication increasingly stress the concept of 'honest signalling' (Zahavi, 1975). This view suggests that many animal calls and displays are costly to produce, and that they consequently provide reliable information about attributes such as resource holding potential (e.g., Clutton-Brock and Albon, 1979) or resistance to parasites (Hamilton and Zuk, 1982; Zuk et al., 1990a,b). Formal mathematical and neural net models have been developed to describe the evolution of handicaps and ornaments (Arak and Enquist, 1993; Johnstone, 1994).

REFERENTIAL SIGNALS

The theoretical and empirical advances described above have been almost exclusively concerned with displays that reflect the physical characteristics of the sender. This is not the only thing that animals signal about. Recent observational and experimental studies have demonstrated that some vocalizations encode not only individual attributes such as species, size, and motivational state, but also information about environmental events. Research on these systems, collectively termed 'referential signals,' has been motivated by a very different set of theoretical concerns from those that provide the underpinning for studies of sexually selected displays. The issues addressed have included the evolution of human language (e.g., Snowdon, 1993a; Evans and Marler, 1995) and the possibility that communication affords particular insights into the cognitive processes of nonhuman animals (e.g., Griffin, 1981, 1984, 1992; Cheney and Seyfarth, 1990, 1992). Analyses of referential signalling have proceeded along a parallel path, largely unaffected by recent developments in behavioral ecology. I suggest that this is unfortunate, both because studies of referential signals have not typically been concerned with issues of design, and because theoretical work on signal structure has largely ignored problems requiring an understanding of meaning.

I will not attempt a comprehensive summary of work on referential signalling, as this topic has been the subject of a number of recent reviews (Marler et al., 1992; Macedonia and Evans, 1993; Snowdon, 1993a,b; Evans and Marler, 1995; Hauser, 1996). Instead, I shall concentrate on what we do not know. My goal is not to denigrate the work that has so far been conducted, but rather to identify lacunae in our understanding of referential signals, in the hope that this will encourage additional research.

Traditional models of animal communication suggest that signals principally encode motivational information (e.g., Rowell and Hinde, 1962; Bastian, 1965; Lancaster, 1965; Premack, 1975; Luria, 1982). This theoretical position implies that variation in the sender's internal state will be reflected by continuous gradation in the physical properties of the signal produced. Signals will be evoked under a very wide range of environmental circumstances and will consequently only be interpretable with the aid of contextual information.

There are now several examples of call systems with quite different properties. Vervet monkeys (Struhsaker, 1967), ring-tailed lemurs (Macedonia, 1990), and chickens (Gyger et al., 1987), all have structurally distinct alarm calls that are predator-class specific. Playback presentations of these alarm calls evoke adaptive responses from conspecific receivers (Seyfarth et al., 1980a,b; Macedonia, 1990; Evans et al., 1993a). These findings are consistent with the idea that the calls encode relatively specific information about the eliciting event (i.e., that they have the property of external reference) and are plainly incompatible with models relying on variation in motivational state, at least if this is conceived of in terms of general arousal.

It might, however, be possible to modify a motivational model to accommodate the specific eliciting conditions identified for the alarm calls of some primates and birds. This would be a challenging endeavor, because motivational accounts will have to explain the finding that call type is relatively insensitive to variation in the immediacy of threat posed by a predator. Vervets produce 'eagle alarms' to raptors, both when they are almost invisible in the distance, and when they are in the last phases of attack (Cheney and Seyfarth, 1990). Similarly, lemurs respond to avian predators with aerial alarm calls, and to carnivores with ground alarm calls, despite considerable variation, both natural and experimentally simulated, in the danger posed (Pereira and Macedonia, 1991). Call usage in chickens also appears to be dependent upon the physical properties of predator stimuli, rather than upon apparent distance (Evans et al., 1993b). For a motivational model to accommodate these results, it is necessary to allow a proliferation of stimulus-type-specific affective states—so that, for example, vervets would be held to have three qualitatively distinct fear responses (corresponding to leopards, eagles, and snakes, respectively), that are evoked despite considerable variation in general arousal and subsequent behavior. The existence of human phobias (Myers et al., 1984) demonstrates that such highly specific fear responses are not inconceivable. It seems more parsimonious to suggest that vervet calls encode information about predator type than to rely on a complex model in which predators evoke discrete motivational states, which then evoke corresponding calls, but it is hard to imagine an empirical test that would distinguish between these alternatives.

A rather different theoretical position considers all of the information that is encoded in an animal signal. Smith (1981, 1991, p. 214) suggests a relatively broad definition of the term 'referent' including "(a) several kinds of behavior (plus their probabilities and other variables), (b) physical characteristics of the signaller (e.g., its species and other identities), and (c) for some signals, external stimuli to which the signaller is responding." The more restrictive definition favored by some authors, which focuses especially on the possibility of external referents (e.g., Seyfarth et al., 1980b; Marler, 1985; Marler et al., 1992), approximates Smith's third category. It is important to note that if a class of signals were to encode highly specific information about the sender's subsequent behavior (category a), and if playback presentations were sufficient to elicit corresponding behavior in respondents, then it would be difficult to distinguish such signals from those that had external referents (category c). I suggest, however, that it is not necessary to treat behavioral referents and external referents as mutually exclusive interpretations. Vocalizations such as alarm calls make available many types of information about the sender; a partial list might include caller size, individual identity and affective state. It seems probable that such signals allow conspecific receivers some success in predicting the subsequent behavior of the sender. This is perhaps especially likely to be true in species that have stable social groups with individually distinctive vocalizations. There are a number of other complications. For example, it is particularly difficult to determine whether signals should be thought of as denotative (i.e., as labels for stimulus categories) or imperative (i.e., as instructions describing appropriate responses) (Cheney and Seyfarth, 1990, 1992; Baron-Cohen, 1992; Marler et al., 1992). Attempts to explore the 'meaning' of animal signals also tempt us to address difficult philosophical issues, such as the level of intentionality required to explain the observed behavior (e.g., Dennett, 1983; Cheney and Seyfarth, 1990) and whether animals are aware of their own knowledge or that of their companions (Allen, 1992; Armstrong, 1992; Schull and Smith, 1992; Snowdon, 1992). Problems of this kind are not unique to the study of animal communication; they are also characteristic of work on the behavior of preverbal human infants (Marler et al., 1992).

I shall focus instead on questions that are clearly accessible to experimental investigation. Systematic studies of animal signal systems can only establish that our subjects behave *as if* their vocalizations encode information about events in the external environment. The term 'functional reference' has been coined to describe this property. It acknowledges the constraints inherent in analyses of animal signals, including the difficult distinctions described above (Marler et al., 1992), and is intended to be neutral about philosophical issues that are not addressed directly by empirical evidence.

RECOGNIZING FUNCTIONAL REFERENCE

The discovery of referential signals in the natural behavior of nonhuman primates and birds invites comparative and developmental studies. An essential prerequisite for such a program is the development of agreed criteria for recognizing the property of functional reference. Recent theoretical papers have suggested that this should involve consideration both of the caller's behavior and of the effects of the signal on companions (Marler et al., 1992; Macedonia and Evans, 1993). Studies of signal production and perception thus assume equal importance.

The key considerations with regard to production are that referential signals should be structurally discrete and that they should have a degree of stimulus-specificity. Eliciting stimuli should belong to a coherent category, although the absolute size of this category, and hence the degree of referential specificity, could vary considerably. Variation of this sort is also characteristic of human speech, which provides paradigmatic examples of referential signalling. We are able to denote individuals and also to discuss much larger groups that are delineated by characteristics such as age or occupation. Despite differences in the number of possible eliciting stimuli, the terms 'Mary' and 'university professor' are both unambiguously referential. The key point is thus not the absolute level of specificity, but rather the relationship between event class and signal type. We would not expect the same class of referential signal to be produced in response to stimuli that are clearly drawn from qualitatively distinct categories.

The importance of this distinction is illustrated by work on California ground squirrels. These sciurid rodents have a complex series of alarm calls which form a continuum from broad-band 'chatters' to tonal 'whistles' (Owings and Virginia, 1978). Whistles are usually produced in response to raptors, whereas chatter calls are evoked by terrestrial carnivores. However, there are exceptions to this pattern of usage which suggest that the ground squirrel call system does not denote predator type. Squirrels being closely pursued by carnivores sometimes produce whistles, and chatters are given to distant hawks (Owings and Virginia, 1978; Leger et al., 1980; Owings and Leger, 1980). Similar patterns of call usage have been described in Belding's ground squirrels (Robinson, 1980, 1981; Sherman, 1985) and marmots (Blumstein, 1995a,b; Blumstein and Arnold, 1995; Blumstein, in press). The data on alarm call production in sciurids are consistent with the idea that these signals do not describe predator classes directly, but rather encode differences in response urgency perceived by the caller. This will vary quite reliably with predator type, as fast-moving hawks usually afford less time for escape than the relatively slow approach of a carnivore. Ground squirrels and marmots thus have call systems that are exquisitely well-matched to the hunting tactics of their two principal classes of predator; the signals are designed to allow receivers to judge the time available for fleeing to a burrow refuge, which is arguably information of greater functional importance than a taxonomic description of the approaching threat.

Signals with the property of functional reference must also meet a perception criterion. They should be sufficient, in the absence of the eliciting stimulus and of other normally available cues, to permit receivers to select appropriate responses. This property has been termed 'context independence' (Evans et al., 1993a; Macedonia and Evans, 1993). The most common technique for assessing perception of putative referential signals is playback experiments, in which recorded sounds are presented to conspecific receivers (e.g., Seyfarth et al., 1980b; Macedonia, 1990; Evans et al., 1993a). This approach, by design, strips away the contextual cues that might normally be provided by the nonvocal behavior of the sender. Appropriate responses reveal that such information is not essential, but they do not demonstrate that it is unimportant (see below).

Theoretical papers discussing the issue of external reference in animal signals have traditionally employed a simple dichotomous classification scheme in which signals are considered to be either affective or referential (Marler, 1977, 1978, 1984; Gouzoules et al., 1985). More recent work has modelled the properties of animal signals as points falling along a continuum, with signals that principally reflect the motivational state of the sender, such as the distress calls of precocial birds (Collias and Joos, 1953; Abraham, 1974) and the cries of human infants (Lester, 1985), at one end, and affect-free referential signals, such as machine-generated speech, at the other. Categorizing a signal as functionally referential is equivalent to postulating a threshold value on such an underlying continuum, and then demonstrating empirically that the properties of the signal are such that it is exceeded. There is a degree of unavoidable arbitrariness inherent in partitioning any sort of variation in this fashion. It is also true that whether or not a signal meets the criteria described above will be dependent not only upon the characteristics of the system being studied, but also upon extraneous factors such as the number of animals available for study (and hence the level of statistical power), and the sensitivity of the response assays employed. The advantage of this approach is that it allows us to address issues, such as the development and phylogenetic distribution of referential signals, that would otherwise be intractable.

A STRATEGY FOR EXPERIMENTAL ANALYSIS

It is important to note that no single observation or experimental test is sufficient to identify the property of functional reference. Instead, we require a program of research that involves studies of both production and perception (Fig. 1). It is consequently not straightforward to obtain compelling evidence for referential signalling and this, combined with the very small number of systems that have so far been studied, suggests that we may have underestimated the incidence of this phenomenon.

The strategy that I have outlined for exploring the properties of a system of putative referential signals involves mapping systematically the relationship between eliciting events (Fig. 1a) and signal morphology (Fig. 1c), and then

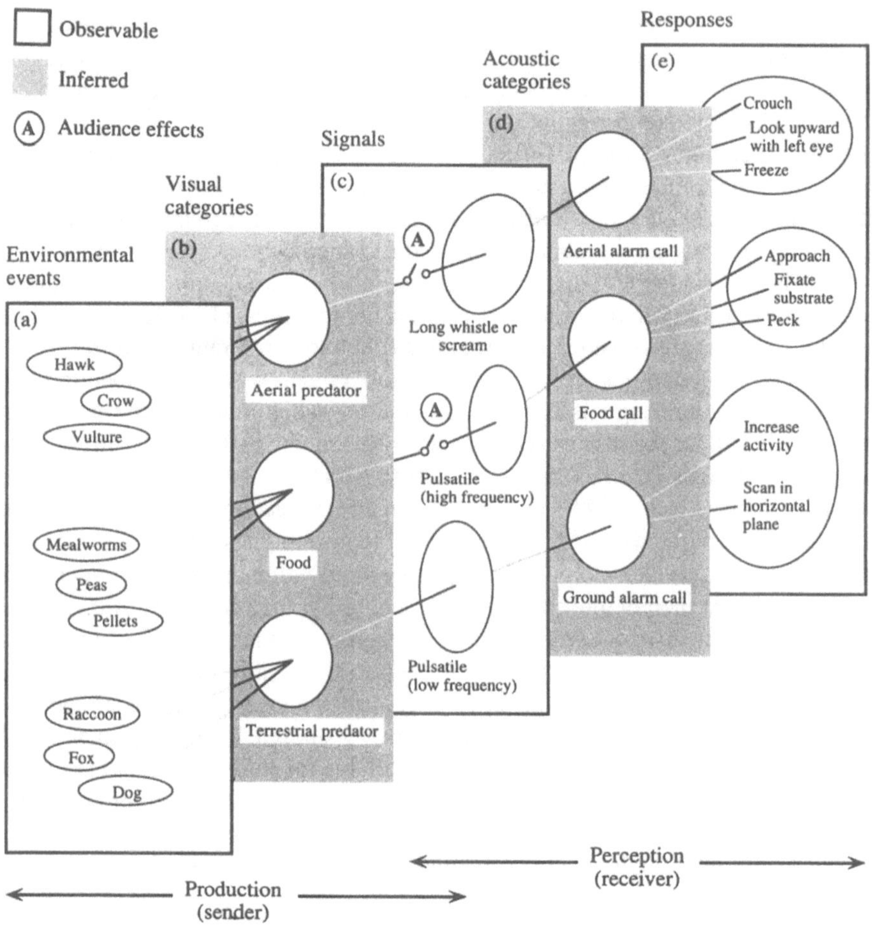

Fig. 1. Functional reference in vocal signals. Examples summarize findings obtained from experimental analyses of the production and perception of chicken calls. The system has both observable and inferred properties (see text for details).

assessing the effects of variation in signal type (Fig. 1c) on receiver response (Fig. 1e). In addition to these three observable levels (stimulus characteristics, signal structure, and receiver response), I have included two hypothetical levels that are necessarily hidden and have properties that can only be inferred. I am assuming for heuristic purposes that visual stimuli do not evoke call production directly, but rather that they are first recognized as members of a category (Fig. 1b) which is then linked to a particular call type. There is some evidence for this idea, as call production is not obligatory on presentation of stimuli (see below). I have also included an analogous level in which categorization of call type by receivers occurs (Fig. 1d). There is evidence for such processing in both birds (Nelson and Marler, 1989; Dooling et al., 1990) and mammals (May et al., 1989; but see Ehret, 1992; Hopp et al., 1992).

This scheme for exploring the properties of referential signals can be illustrated by recent work on the vocal behavior of chickens. Like all gallinaceous birds, chickens have a large vocal repertoire (Collias, 1987) which includes calls produced by both males and females upon discovering food and two structurally distinct alarm calls, one associated with the approach of aerial predators and the other with terrestrial predators (Konishi, 1963; Collias, 1987). Initial observational studies of the vocal behavior of chickens under naturalistic conditions confirmed earlier descriptions of alarm call usage (e.g., Schjelderup-Ebbe, 1922; Collias and Joos, 1953; Baeumer, 1962; Konishi, 1963). Aerial alarm calls were evoked by a relatively broad class of objects moving overhead, while ground alarm calls were evoked primarily by stimuli such as humans and dogs (Gyger et al., 1987).

The structure of the two alarm call types is qualitatively distinct and can perhaps be regarded as antithetical. Aerial alarm calls are made up of an initial pulse, followed by a second component of much greater duration which can be either broad-band (scream-like) or narrow-band (whistle-like). Ground alarm calls, in contrast, are short, broad-band pulses, typically delivered in long bouts (Fig. 4 in Evans et al., 1993a). Studies of chicken alarm calling thus suggested that these signals might satisfy the production criterion for functional reference. They were evoked by coherent and nonoverlapping sets of visual stimuli (Fig. 1a) and had discrete acoustic structure (Fig. 1c).

Subsequent laboratory experiments used videorecorded and computer-generated images of predators to confirm the relationship between predator class and the type of alarm call elicited (Evans et al., 1993a). The link between call type and receiver response was then assessed in a playback experiment. Upon hearing ground alarm calls, hens assumed an unusually erect, 'alert' posture and began scanning back and forth in the horizontal plane. In contrast, playback of aerial alarm calls evoked running toward cover, crouching with the feathers sleeked, and looking upward with one eye (Evans et al., 1993a). The hens thus behaved as though the call type predicted something about the circumstances under which the sound had originally been recorded. The responses evoked were precisely those that would facilitate detection of ground predators on the one hand, and of aerial predators on the other.

A peculiarity of the chicken visual system suggests that the type of information encoded, at least in aerial alarm calls, may be quite specific. There is a degree of hemispheric lateralization in the processing of visual stimuli such that the 'left eye system' and the 'right eye system' have different and complementary capabilities. The left eye system is superior for tasks involving spatial location, while the right eye system is better at some categorization tasks, such as the recognition of food (Andrew, 1988; Rashid and Andrew, 1989; Workman and Andrew, 1989). Following playback of aerial alarm calls, hens were much more likely to fixate upward with their left eye (Evans et al., 1993a). By doing so, they were bringing to bear the system best suited for locating a small rapidly moving target such as an aerial predator. The pattern of results from studies of both production and perception is hence consistent with the idea that chicken alarm calls are functionally referential.

Male chickens produce distinctive 'food calls,' both when they discover food during natural foraging behavior (Gyger and Marler, 1988) and when they are provided with edible objects under controlled laboratory conditions (Marler et al., 1986a,b; Evans and Marler, 1994). Recent experiments using instrumental conditioning techniques to control access to food demonstrate that calling can also be evoked by an artificial stimulus (a red light) that reliably predicts feeding opportunities (Evans and Marler, 1994), suggesting that associative learning may be important in determining the range of events responded to (see below). The duration of food calls, like that of ground alarm calls, is brief, but food calls are easy to identify because they have distinctive frequency modulation and a reliably higher dominant frequency than ground alarm calls (Fig. 2 in Evans and Marler, 1994). Food calls thus appear to meet the production criterion of functional reference, although there are reports of 'deceptive' calling in the absence of food (Marler et al., 1986b; Gyger and Marler, 1988) that will require further investigation. I shall discuss the issue of deception in more detail below.

Published accounts of food call production do not, however, compel an interpretation of functional reference because they do not include complementary analyses of perception. Hens respond to food calls by rapidly approaching the male, but we cannot exclude the possibility that this response is mediated by other aspects of the male's behavior, such as visual displays, or by motivational factors quite unrelated to food, such as sexual receptivity. Interpretation of the data currently available on food-associated calls in chimpanzees (Hauser and Wrangham, 1987; Hauser et al., 1993; but see Clark and Wrangham, 1993,1994), golden lion tamarins (Benz et al., 1992; Benz, 1993), toque macaques (Dittus, 1984) and rhesus macaques (Hauser and Marler, 1993a) is similarly constrained by the lack of playback experiments. The idea of functional reference has the merit of making a clear theoretical prediction about the behavior of hens, which is that presentation of food calls in the absence of other cues will be sufficient to elicit anticipatory feeding movements. Recently completed experiments have confirmed this prediction. Hens approach a loudspeaker playing food calls, moving their heads repeatedly downward to fixate frontally on the substrate and occasionally pecking at the floor. These responses are significantly more probable during playback of food calls than during playback of either ground alarm calls (which are structurally similar) or contact calls (which are produced under similar social circumstances) (Evans, in prep.). The data from studies of production and perception thus suggest that chickens have a total of at least three functionally referential calls, which encode information about the discovery of food and about the appearance of predators (Fig. 1).

SPECIFICITY

Analyses of referential signalling have always been concerned with the issue of specificity, that is, with the size of the stimulus set (Fig. 1a) corresponding to a particular call type (Fig. 1c). Analogies between the natural communication of animals and properties of human language (e.g., Cheney and Seyfarth, 1990; Evans and Marler, 1995) will be most compelling when animal signals are shown to be evoked by a small class of environmental events. Unfortunately, we have data on so few referential systems that it is not yet possible to evaluate sensibly patterns of specificity. It is, however, clear that in some species the external stimuli evoking calls are quite narrowly defined. For example, the 'eagle alarms' produced by adult vervet monkeys are elicited principally by martial eagles (Seyfarth and Cheney, 1980). Comparisons between the morphology of martial eagles and those of the other raptors to which vervets give eagle alarms suggest that the monkeys may be responding not only to the overall form of potential predators, but also to quite subtle details, such as melanic markings beneath the head and under the wings, and perhaps also to dynamic cues such as characteristic differences in flight pattern.

Studies of the natural vocal behavior of chickens provide a clear contrast to the vervet system. Chicken aerial alarm calls are associated with a relatively large set of airborne objects (Gyger et al., 1987). Laboratory experiments, in which birds were presented with computer-generated animations, have permitted the systematic manipulation of stimulus characteristics. The results demonstrate that aerial alarm calls are dependent upon attributes such as size and apparent speed (Evans et al., 1993b). Spatial location and shape are also important (Evans and Marler, in prep). These studies have provided a precise mapping of the stimulus parameters corresponding to call production and, together with measurements of non vocal behavior, they allow us to infer something about the way in which visual stimuli are recognized and categorized (Fig. 1b). Work of this kind is thus logically complementary to studies of visual categorization in animals using instrumental (e.g., Herrnstein et al., 1976; Herrnstein, 1984, 1991; Lea, 1984) or neurophysiological techniques (Kendrick and Baldwin, 1987; Perrett et al., 1987).

Developmental Plasticity

The specificity of calls produced by adult animals is likely to be a consequence not only of selection pressures operating over evolutionary time, but also of developmental factors. Vervet infants clearly have a predisposition to respond to aerial objects with eagle alarm calls, but they initially respond to a much more diverse array of events than adults, so that their calls are associated not only with the appearance of potential predators, but also with innocuous birds such as storks and even with inanimate objects such as falling leaves (Cheney and Seyfarth, 1980). There is a marked increase in specificity over the course of development, and it may be relevant that the calls produced by juveniles in response to genuine predators are much more likely to be followed by adult calls than are their responses to nonthreatening events. Observational data of this kind cannot, however, separate the effects of such experience from those of simple maturation.

Recent experimental studies of the development of alarm calling in chickens provide clear evidence for an effect of early experience (Palleroni and Marler, in prep). Chicks were reared under four experimental conditions. The first group was housed in an aviary together with the maternal hen and adult males. They received controlled exposure to a representative aerial predator (a trained falcon) and to a terrestrial predator (a ferret). The chicks were thus able to experience not only the morphology and distinctive hunting tactics of these two classes of predator, but also the calling behavior and other responses of adult conspecifics. Three other groups were reared under conditions that lacked one or both of these types of experience. One group of chicks was exposed to predators but with no adults present, another group was reared with adult birds but had no experience of predators, and a third group was deprived of both types of experience. When they were six months old, the chicks from each of the four groups were presented with video-recorded sequences depicting both aerial and terrestrial predators. The chicks that had both predator experience and an opportunity to view the antipredator behavior of adults had normal responses, but the calling behavior of each of the other three groups was deficient. These results suggest that development of fully competent adult antipredator behavior (i.e., responding to raptors with aerial alarm calls and to ferrets with ground alarm calls) is facilitated both by direct experience of the two predator classes and by exposure to the responses of adults.

Studies of development in both vervets and chickens hence provide evidence of developmental plasticity in the link between signal type and eliciting characteristics. There is, however, no suggestion of experiential effects on the physical characteristics of these signals. Vocal learning does not appear to occur either in the chicken (Konishi, 1963) or in nonhuman primates (Snowdon, 1990; Owren et al., 1992). These two model systems thus have only one of the two types of developmental plasticity characteristic of human language, in which both signal structure and the mapping between signal type and referent are dependent on the child's early environment (e.g., Locke, 1993). This is, nevertheless, a greater degree of malleability than might be expected in animal signalling systems, with the conspicuous exception of song development in oscine birds. It will be particularly interesting to determine whether the role of experience is to modify the relationship between eliciting stimuli and categorization (Fig. 1a-b), as is suggested by studies of predator recognition in other birds (Curio, 1993) and primates (Mineka, 1987; Mineka and Cook, 1987), or whether the link between category and signal type (Fig. 1b-c) may also be influenced.

Terminology

The problem of assessing referential specificity is complicated by issues of nomenclature. It is often necessary, for ease of exposition, to select labels for animal vocalizations that summarize the range of putative referents. The danger is that such terms bring with them a great deal of linguistic baggage and, perhaps unavoidably, a degree of ambiguity. For example, vervet 'leopard alarms' are also evoked by other animals, including eagles that are attacking (Cheney and Seyfarth, 1990) or feeding upon a dead vervet (Struhsaker, 1967), and pythons (M. Hauser, pers. comm.). They are also produced, although rarely, in the context of intergroup conflicts (Struhsaker, 1967; Cheney and Seyfarth, 1990). It is thus possible to argue that the term 'leopard alarm' is misleading, because it connotes a greater degree of referential specificity than can be supported by the empirical data. Such criticism would, however, be unfair, because it is clear that the term 'leopard' was intended as a shorthand description for a category of eliciting stimuli of which large spotted cats are, if not the prototypical exemplar, at least the most common (Seyfarth et al., 1980b). Nevertheless, there is the real danger that the terms selected to describe eliciting conditions may reveal more about human linguistic conventions and about the categorization behavior of the investigators than about the information content of the signals being studied. Such labels, when applied prematurely, may also affect the choice of future experiments (Snowdon, 1990). If a class of signals is labelled 'food calls' then we are perhaps less likely to conduct studies involving qualitatively different stimuli, even though they might reveal behavior that would fundamentally alter our interpretation, such as calling during affiliative social interactions in the absence of food. Labels are rarely neutral; they reflect our intuition about the properties of the system being studied and they have the potential to constrain the direction of our research.

There are several possible solutions to the labelling problem. The first is suggested by the study of animal signals that do not have the property of functional reference. There is consequently no temptation to apply linguistic terms. The songs of birds or the calls of anuran amphibians are identified with labels that are either completely arbitrary or perhaps reflect some aspect of acoustic structure. This has, of course, been the traditional practice in studies of primate vocal behavior, so that we read of 'rough grunts' in chimpanzees (Marler, 1976), 'threat-alarm-barks' in vervets (Struhsaker, 1967) and 'isolation peeps' in squirrel monkeys (Snowdon et al., 1985). It is difficult, however, to employ structural terms when faced with a vocal repertoire containing sounds for which we cannot readily coin onomatopoeic descriptions. The chicken's vocal repertoire includes a number of pulsatile sounds that do not lend themselves to economical prose description, although the eliciting conditions (food, social affiliation, and approach of predators) do. One way to avoid this problem would be to use completely arbitrary terms that were neutral with regard to issues of call meaning—so that we might speak of pulsatile call 1, pulsatile call 2, and pulsatile call 3. The obvious cost of such an approach is that our prose would become opaque and probably quite incomprehensible to the nonspecialist. The demands of terminological precision are thus to some degree incompatible with those of eloquence.

The most straightforward solution might be to inform the reader at the outset about the way in which terms are to be used. We will then be able to discriminate between: (i) 'leopard' as a shorthand description for a category of calls with particular acoustic characteristics, elicited by, among other things, leopards, and (ii) 'leopard,' a signal evoking mental representations of spotted cats in the minds of receivers, and sharing with some words the property of denoting a precisely defined category of external objects. Clearly these two usages make very different claims about the cognitive underpinnings of the signal being studied, and they will be evaluated according to correspondingly different logical and empirical criteria.

UNDERSTANDING VARIATION IN SIGNAL STRUCTURE: THE IMPORTANCE OF LEVELS

I have so far been discussing analyses that are concerned with variation at the level of call type. These involve identification of the eliciting event and of the factors mediating transitions from one signal class to another. Much less is known about the factors responsible for variation in signal structure within call type (Marler et al., 1992), although efforts have been made to ensure that such variation is represented in the exemplars chosen for use in playback experiments (e.g., Seyfarth et al., 1980b; Evans et al., 1993a). I suggest that the distinction between within-type and between-type variation is an important one, because systems may well prove to be referential at one level but not at the other.

Consider the properties of chicken food calls. These signals appear to be evoked by food items and by stimuli that reliably predict the availability of food. They are clearly not simply dependent upon social or sexual interaction (Evans and Marler, 1994). Analyses at the level of call type thus suggest that food calls predict events in the external environment (i.e., that they are functionally referential). There is also variation in the temporal characteristics of call bouts. The rate at which males produce food calls reflects their preference for a food stimulus, so that it is highest for live invertebrates and lower for less preferred items, such as their regular ration (Marler et al., 1986a). This relationship was initially considered to be evidence for highly specific referential signalling about object characteristics (Marler et al., 1986a), but subsequent analyses have demonstrated that the call rate also covaries with the speed at which males perform an operant task to obtain food. Males food called and key pecked vigorously at the beginning of test sessions, but both of these responses decayed with repeated food deliveries (Evans and Marler, 1994). This correlation between instrumental performance and food calling suggests a more parsimonious explanation for the earlier results, which is that temporal variation may simply reflect the male's motivational state with regard to food. Analyses of food-associated calls in cotton-top tamarins suggest that this system may have similar properties. While 97% of tamarin calls were produced in the presence of food (Snowdon, 1993b), the rate of calling was dependent on idiosyncratic individual preferences for particular food types, and therefore most likely reflected a motivational response (Elowson et al., 1991).

Determining whether food-associated calls are functionally referential proves to be more complicated than we might at first suppose. I suggest that the answer obtained will be, in part, dependent upon the level of analysis selected. It is possible for transitions between call types to encode stimulus class while within-category variation in signal structure remains principally dependent upon motivational state. There is thus the very real danger that results from studies of fine-grained variation will appear to contradict those obtained from analyses of differences between call types when, in fact, these two data sets are not comparable because they concern quite different phenomena.

The challenge for future studies of fine-grained variation in referential signals will be to determine whether there are structural parameters that vary with stimulus characteristics, yet are independent of those that serve as vehicles for affect (Marler et al., 1992). Addressing this issue will require the systematic exploration both of the consequences of manipulating stimulus attributes and of variation in the internal state of the sender. Such experiments would begin to redress the relative neglect of motivational or 'connotative' factors in studies of referential signalling (Owings, 1994).

DESIGN OF REFERENTIAL SIGNALS

Analyses of the information content of referential signals have revealed that some systems have a surprising degree of specificity. This finding does much to explain both the calling behavior of senders and the responses of receivers. Such work does not yet, however, allow us to make predictions about signal design. Clearly, we would expect calls that encode information about very different classes of events to have contrasting structure, and this is often true. But analyses of meaning provide little else in the way of insights into the physical form of referential signals. Why, for example, are vervet leopard alarms made up of a rapidly repeated series of pulses with a relatively low fundamental frequency, while snake alarms are delivered at a lower rate and have a much higher dominant frequency, sometimes exceeding 16 kHz (Seyfarth et al., 1980b)? No one has yet attempted a comprehensive model of the relationship between stimulus category and signal structure along the lines of Morton's (1977) motivational-structural rules, and given how little we still know about the details of referential signalling, such an effort would almost certainly be premature.

I suggest that a full account of any system of referential signals will need to consider issues in addition to that of call meaning, which has been the principal focus of most research conducted to date. Attempts to understand the information content of calls have necessarily been concerned with the correspondence between signal type and conditions of production, but there are a number of aspects of signalling behavior that are not captured by such analyses. This point can be illustrated by comparing the properties of two alarm call systems in the chicken (Table 1). Aerial alarm calls and ground alarm calls differ qualitatively in structure. Aerial alarm calls are low-amplitude vocalizations consisting of an initial pulse followed by a relatively long second component. Ground alarm calls are high-amplitude short pulses given in long bouts with an occasional 'scream' element. These characteristic differences in signal amplitude, which can exceed two orders of magnitude (Evans, unpub. data), suggest that there will be reliable differences in the active space of the signals, with aerial alarm calls audible over a relatively small area, and ground alarm calls audible over a much larger one. This difference is particularly pronounced when we examine the structure of aerial alarm calls given after the initial response to a predator. These sounds are almost invariably narrow-band (whistle-like), typically lack the introductory pulse that may facilitate localization of the caller by conspecifics, and are so soft

Table 1. A Comparison of Alarm Call Design

Property	Aerial alarm calls	Ground alarm calls
Structure	Pulse + second component (wide or narrow band)	Repeated broad-band pulses
Amplitude	Low (after initial pulse) Very low after first call	Consistently high
Active space (detectability)	Small (especially second and subsequent calls)	Large
Localizability	Poor (especially second and subsequent calls)	Excellent
Audience effect	Yes	No
Duration of calling	Brief	Prolonged
Potential receivers	Nearby social companions	Conspecifics over large area + potential predator

as to be almost inaudible. Ground alarm calls, in contrast, have acoustic characteristics, including rapid amplitude rise-time and broad frequency bandwidth, that should facilitate localization of the sender (Marler, 1955).

These differences in call structure are mirrored by differences in the duration of calling, and in nonvocal behavior. Aerial alarm calls are typically given on first sighting a predator and are accompanied by 'crouching' and 'freezing' with prolonged immobility (Table 1). Ground alarm calls have a very different timecourse: call rate builds up quite gradually after spotting the predator and calling may continue for some tens of seconds or even minutes after the predator has disappeared (Fig. 2 in Evans et al., 1993a). Long bouts of ground alarm calling are associated with increased motor activity, with the bird typically walking up and down vigorously in an erect posture (Table 1).

In summary, both acoustic structure and the time course of signalling suggest that aerial alarm calls are designed to allow the sender to remain cryptic, while ground alarm calls likely make the sender more conspicuous (Evans et al., 1993a). There are also reliable differences in the effects of an audience on these two call types. Aerial alarm calling is potentiated by the presence of conspecifics, regardless of age and sex (Karakashian et al., 1988), but there is no such audience effect on the production of ground alarm calls (Table 1). Isolated males confronted with a terrestrial predator call at a rate indistinguishable from that of males with social companions (Evans and Marler, in prep.).

It seems unlikely that considerations of call meaning will be sufficient to explain the signal characteristics and other behavior summarized above. Certainly the continued production of ground alarm calls, long after a predator stimulus had disappeared, grossly exceeds the level of signalling required to warn social companions. This may be an example of tonic communication (Schleidt, 1973; Owings et al., 1986), in which signals function to maintain a state of vigilance appropriate to a predator that, although no longer visible, may still be nearby—as is typically the case with slow-moving carnivores. It is also possible that the overall pattern of results, incorporating differences in signal structure, time course, and sensitivity to social context, reflects signalling to different potential receivers. Aerial alarm calls are ideally structured to alert flock members foraging nearby. The structure of ground alarm calls and the duration of calling are both consistent with the idea that although these signals are clearly salient to companions, they may also be designed to deter potential predators (e.g., Klump and Shalter, 1984; Caro, 1986a,b; Hasson, 1991; Caro et al., 1995). Assessing this idea will require quantitative estimates of the active space and localizability of alarm calls, not only for conspecifics but also for representative predators (e.g., Klump and Shalter, 1984). Observational studies of wild or feral populations will be necessary to determine whether animals engaging in ground alarm calling and associated conspicuous movements are indeed less likely to be attacked. Work of this kind has the potential to explain aspects of referential signal design that have so far been neglected.

THE ROLE OF CONTEXT

Signal Production: Audience Effects

Classical treatments of animal signalling behavior tend to assume that call production is essentially reflexive (e.g., Lyons, 1972). That is, that when a sufficient stimulus (e.g., a predator model or a food item) is presented, then the appropriate call, together with other responses (antipredator behavior or feeding) will necessarily be evoked. Recent work on a taxonomically diverse array of species demonstrates that such simple models are inadequate. Ground squirrels (Sherman, 1977; Owings et al., 1986), marmots (Blumstein et al., in press), vervet monkeys (Cheney and Seyfarth, 1985), downy woodpeckers (Sullivan, 1985), and chickens (Marler et al., 1986b; Karakashian et al., 1988; Evans and Marler, 1991, 1992, 1994) all modulate their vocal behavior according to social context.

In some systems at least, these 'audience effects' appear to be specific to vocal behavior. Chickens respond with antipredator behavior when presented with simple, hawk-shaped models of the kind used in the classic experiments of Lorenz and Tinbergen (Tinbergen, 1948), or computer-generated animations of aerial predators (Evans and Marler, 1992; Evans et al., 1993a,b), and this has allowed exploration of the effects of social context under controlled conditions. Alarm calls are produced by cocks when there are conspecifics present, but not when they are alone (Karakashian et al., 1988; Evans and Marler, 1991). Comparisons of the nonvocal responses to simulated predators reveal no apparent differences. Behavior such as crouching down, sleeking the feathers and fixating on the stimulus appears to be insensitive to social context, while call production is profoundly affected (Karakashian et al., 1988; Evans and Marler, 1991, 1992).

Studies of food calling provide opportunities for particularly sensitive comparisons between the effects of an audience on signalling and those on nonvocal behavior because animals will readily perform instrumental tasks to obtain food. If the food-delivery apparatus is computer-controlled, then it is possible to evaluate, moment by moment, the rate at which the subject works, perform synchronous analyses of signal structure, and then assess changes in both of these measures as a function of social context. This approach has recently been employed to explore the audience effect on food calling in chickens (Evans and Marler, 1994). Cocks were trained to peck a key for food reinforcement and then tested both in social isolation and with a female audience. Males produced significantly more food calls when a hen was confined in an adjacent cage than when they were alone, but the rate at which they keypecked was unaffected (Evans and Marler, 1994). At least under the conditions of this experiment, the presence of an audience potentiated calling in a specific fashion. It is therefore possible to reject the suggestion that the audience effect is simply a manifestation of social facilitation (Dewsbury, 1992), as this would be expected to affect both call production and feeding behavior (e.g., Zajonc, 1965).

The discovery of audience effects implies that any model seeking to predict the signalling behavior of animals will require consideration not only of the obvious environmental events, but also of social context. The presence of companions can be conceived of as modulating, either attenuating or enhancing, the link between stimulus categorization and vocal response (Fig. 1b-c).

Signal Perception: Immediate and Historical Context

The finding that some animal vocalizations appear to encode specific information about environmental events has stimulated quite extensive work focussing on the information content of calls. It is perhaps inevitable that this has led to the relative neglect of contextual cues (Leger, 1993). Contextual factors are, however, likely to play an important role in the perception of signals (Fig. 1c-e) (Smith, 1965, 1977, 1991; Leger, 1993). Playback experiments, which assess the responses evoked by a call in the absence of cues that are normally provided by the nonvocal behavior of the sender, strip away information that is potentially provided by 'immediate context' (Smith, 1991; Leger, 1993). If conspecifics are nevertheless able to select appropriate responses, then it follows that such cues are not essential, but this does not suggest that the animals being studied are insensitive to nonsignalling behavior when it can be monitored. Indeed, accounts of experiments in which referential calls have been presented to members of a social group typically describe behavioral responses, such as looking toward the loudspeaker (Seyfarth et al., 1980b), that are consistent with an attempt to evaluate the sender's nonvocal behavior, and perhaps to learn more about the environmental situation in which the call was produced. The relatively loud and easily locatable pulse element at the beginning of chicken alarm calls (Gyger et al., 1987) reliably evokes orienting responses from the other members of a group, and birds may well then be able to acquire much more specific information about eliciting conditions, because the caller is typically crouching and fixating upon the approaching predator (Evans, unpub. data).

Experimental investigations of the role played by immediate context have also been constrained by the technical difficulty of manipulating nonvocal behavior in a controlled way. The recent finding that a number of species respond socially to video-recorded images of conspecifics (Clark and Uetz, 1990, 1991, 1993; Evans and Marler, 1991; McQuoid and Galef, 1993; Macedonia et al., 1994; Macedonia and Stamps, 1994; Rowland et al., 1995a,b; Rosenthal et al., 1996) suggests a strategy for overcoming this obstacle. It should be possible to conduct experiments in which both signal and nonsignal behavior are systematically varied, with the goal of identifying the way in which immediate context interacts with call morphology to determine receiver response.

Animals are also sensitive to 'historical' context (Smith, 1991), that is, to the prior relationships between signals and environmental events. This is suggested by responsiveness to factors such as variation in sender 'reliability' (i.e., the correlation between individually distinctive calls and occurrence of a particular class of eliciting stimuli; Cheney and Seyfarth, 1988). The recent finding of developmental plasticity in the alarm calling behavior of chickens exposed to different regimes of predator experience (Palleroni and Marler, in prep.) similarly demonstrates an effect of historical event-signal relationships.

It might prove revealing to examine variation in the responsiveness of animals to functionally referential signals as a consequence of environmental cues. For example, do birds and primates become more sensitive to playback of alarm calls when they are in a part of their home range in which they have previously experienced attacks by the same class of predator that evoked the call? It is easy to imagine habitat features that predict a higher probability of attack by particular types of predator (e.g., areas of dense cover are more likely to conceal carnivores than open grassland). If associations of this type prove to be an important determinant of the level of responsiveness to alarm signals, then there will be implications for the interpretation of playback experiments. In field studies, it would be necessary to add controls for subject location at the moment of call presentation, or at least to systematically sample the responses evoked in different parts of a home range. In laboratory experiments, consideration will have to be given to the possibility that animals placed in a completely novel environment might have attenuated responses to conspecific signals, because the test situation provides none of the historical context cues that would normally facilitate selection of appropriate behavior. It follows that such experiments may be particularly conservative tests for functional reference.

REPRESENTATIONS: ARE THEY IN OUR MINDS OR THE ANIMALS'?

Highly specific animal signals appear to share with some words the property of identifying environmental events. This analogy might, however, be less exact than it at first appears. Words correspond to mental representations instead of denoting objects directly. The term 'functionally referential' explicitly acknowledges that, although animals behave as though their signals provide information about external stimuli, and although this is consistent with the responses being mediated by mental representations, it does not compel such an interpretation. One way to consider representations is as an intervening variable (e.g., Hinde, 1974) so that instead of signals eliciting behavior directly (signal

→ behavior), there is an intervening layer (signal → representation of eliciting conditions → behavior). We would be inclined to reject the latter account for reasons of parsimony unless there was some direct evidence for an intermediate level of processing.

Cheney and Seyfarth have conducted a series of field playback experiments designed expressly to determine whether the responses of vervet monkeys to conspecific alarm calls are mediated by mental representations (Cheney and Seyfarth, 1988). A subsequent experiment, which I will not discuss in detail, compared processing of vervet and superb starling alarm calls (Seyfarth and Cheney, 1990). Both studies employed a habituation/dishabituation paradigm that had previously been used in research on preverbal human infants: Vervets were first played a single exemplar of one call type (baseline). They then received a series of eight presentations of a different call (habituation) followed by a second presentation of the call heard initially (test). The critical comparison was of the duration of responses following the initial baseline presentation with those evoked by the final test presentation.

The experiments exploring the vervets' responses to calls produced by their companions (Cheney and Seyfarth, 1988) involved systematic manipulations of the relationship between the baseline/test stimulus and the sounds used in the intervening habituation series, varying acoustic structure, individual identity of the caller, and the putative referent. There was a significant decrement in the response to a 'chutter' call (characteristic of intergroup interactions) when the baseline and test presentations were separated by repeated presentations of the same individual's 'wrr' (another call type characteristic of interactions with neighboring groups). No such decrement was obtained when the habituation series was a 'wrr' recorded from another individual. Analogous experiments assessed response to leopard or eagle alarms, with a habituation series made up of the other type of alarm call recorded either from the same individual as the baseline/test stimulus, or from a different individual. There were no statistically significant differences between the responses to baseline and test presentations in these trials.

Cheney and Seyfarth suggest that the results from their habituation/dishabituation studies demonstrate that the principal determinant of the vervets' responses was the meaning of the calls presented, rather than their physical structure, and argue that this provides the strongest evidence available for *representational* signalling (Cheney and Seyfarth, 1988, 1990, 1992; Seyfarth and Cheney, 1992, 1993). Specifically, it is claimed that vervet monkeys "appear to process information at a semantic level, and not just according to acoustic similarity" (Cheney and Seyfarth, 1988), and that "monkeys have some representation of the objects and events denoted by different call types and that they compare and respond to vocalizations on the basis of these representations" (Seyfarth and Cheney, 1993). If true, these claims represent an important advance in our understanding of the cognitive capabilities of nonhuman primates, extending substantially the parallels between referential signalling and human speech.

The case for representations has additional theoretical importance because it is part of the foundation upon which subsequent arguments are constructed. These include consideration of more complex cognitive processes, such as the possibility that vervets attribute mental states (e.g., knowledge or belief) to their companions (Cheney and Seyfarth, 1990, 1992), and can thus be said to have a 'theory of mind.'

The conclusion that the responses of vervet monkeys to alarm calls are mediated by mental representations is based upon two assertions: (i) that the results obtained cannot be accounted for simply in terms of physical differences between the stimuli, as in traditional habituation/dishabituation studies (e.g., Nelson and Marler, 1989), and (ii) that differences in the duration of response between baseline and test presentations are attributable to the intervening habituation experience. I shall discuss each of these issues in turn.

Cheney and Seyfarth provide descriptive statistics for a selection of call characteristics such as total duration, dominant frequency, fundamental frequency, and voicing. They conclude that the structural properties of the intergroup calls (wrr and chutter) were not more similar than those of the leopard and eagle alarms, and suggest that the acoustic characteristics of the playback stimuli cannot therefore account for the results obtained (Cheney and Seyfarth, 1988). Quantitative analyses of this kind are an essential first step in exploring the recognition and categorization of animal signals, although the data provided are not sufficient to enable a comprehensive mapping of the acoustic space occupied by the calls in the vervet repertoire, as has been possible for analogous descriptions of other signals (e.g., Marler and Pickert, 1984; Nelson and Marler, 1990). A more serious concern, however, is that there is no way of assessing the perceptual significance of the acoustic features selected. Systematic playback experiments involving either titration along acoustic continuua (e.g., Gerhardt, 1978; Zoloth et al., 1979; Evans, 1993), or the use of a very large number of natural sounds together with multidimensional scaling analysis (e.g., Dooling et al. 1990) would be necessary to establish the way in which vervets partition the acoustic space occupied by their vocalizations. Ideally, perception of the calls would be analyzed in sufficient detail that the 'weighting' assigned to each acoustic feature could be estimated (e.g., Becker, 1982; Gaioni and Evans, 1986; Nelson, 1988; Weary, 1990, 1991; Gerhardt, 1992). There would then be an empirical basis for comparing the differences (not physical, but perceived) between the pairs of sounds that produced significant decrements in habituation experiments and those that did not. In the absence of such analyses, there is no compelling reason to reject a traditional interpretation based purely upon the physical properties of the playback stimuli.

A separate concern about the habituation/dishabituation technique employed by Cheney and Seyfarth (1988) is that the design lacks essential controls. Recall that the experiments involved presenting one type of sound (a1), followed by repeated presentation of another sound (b1–b8), and, finally, by a test presentation of the first sound heard (a2). Statistical comparisons were between the responses evoked by the initial and final presentations (i.e., a1 vs a2). A significant decrement in responding is described as "transfer of habituation" and is taken as evidence that the representations for sounds from classes 'a' and 'b' are similar. But this is the wrong logical comparison. What is required is an assessment of the response to a second presentation of sound 'a' when this has followed repeated playback of sound 'b' with the response to a second presentation of sound 'a' in the absence of such an intervening experience. This would require a parallel series of experiments in which some animals heard the sounds, as described by Cheney and Seyfarth, while a control group heard an initial playback of sound type 'a' (A1) followed, after the same period of elapsed time as in the habituation series, by a second presentation (A2):

Habituation a1	b1 - b8	a2
Control	A1	A2

Comparisons between the responses evoked by a2 and A2 would, if significant, suggest that the intervening experience with 'b' had affected responsiveness. The comparisons reported are uninterpretable, because the possibility cannot be excluded that a second presentation of 'a' would have evoked a smaller response *regardless of the intervening experience*. Assuming that more straightforward explanations based solely upon the physical structure of the sounds could be ruled out, this modified design would have the potential to provide evidence for representational signalling.

Note that nothing in the preceding discussion demonstrates that vervet alarms do not evoke mental representations in the minds of their companions. It is entirely possible that they do. I suggest, however, that the results reported so far do not require us to make such an inference.

Some Alternative Approaches

Work on instrumental conditioning suggests a different strategy for assessing whether the effects of referential signals are dependent upon representations of the eliciting event. Experiments in which the 'value' of a reinforcer (food or sucrose liquid) was diminished, either by pairing it with a toxin or by satiation, demonstrate that these manipulations reduced the rate at which rats performed an instrumental response (bar-pressing or chain-pulling). It follows that the rats' behavior was influenced by a stored representation of reinforcer properties (Colwill and Rescorla, 1985).

Similar logic could be applied in experimental analyses of responses to referential signals. If calls evoke particular behavioral responses in an essentially reflexive fashion, then changing the animal's experience with a class of visual stimuli (e.g., food or predators) should not alter the effects of presenting recorded sounds. If, on the other hand, calls affect behavior by representing a particular class of events, then increasing or decreasing the salience of these events should also perturb the response to the corresponding signals. For example, if chicken food calls cause hens to approach because they evoke a representation of palatable objects, then an experience that reduces the value of such objects should also attenuate responsiveness to the calls. This argument generates a strong prediction, which is that birds should approach a loudspeaker playing back food calls more rapidly when they are hungry than when they are satiated. The effects of manipulating hunger should also be specific to food calls, and should not affect responsiveness to other signals in the vocal repertoire.

Conceptually similar experiments could be conducted with alarm calls. It should be possible to increase or decrease the salience of simulated predators by using conditioned fear (e.g., Mineka, 1987; Mineka and Cook, 1987) and habituation paradigms. In systems where there is more than one type of alarm call, there is the opportunity to design a highly sensitive assay, because animals could be habituated to one set of putative referents (e.g., terrestrial predators) while leaving their experience of the other set (e.g., aerial predators) unaffected. The prediction in this case would be a selective reduction in responsiveness to ground alarm calls, with no change in responsiveness to aerial alarm calls. Control groups could be exposed to the opposite pattern of experiences (i.e., habituated to aerial predators). In both cases, presentations of alarm calls provide a particularly specific test, because it would be possible to discriminate a reduction in responsiveness to one class of predator from a more generalized diminution in fearfulness.

A recent report describing social interactions between junglefowl cocks and hens confined in a small arena suggests that manipulating hunger might not produce the predicted change in responsiveness to food calls (Van Kampen, 1994). Hens that had been food-deprived for 24 hours were no more likely to approach a male than when they had experienced *ad lib* access to food. This result is said to demonstrate that "hunger state is not a factor in determining the reaction of females to male food-calling." Unfortunately, no quantitative analyses of male vocal behavior are provided; it is consequently not feasible to verify that the sounds produced by cocks in Van Kampen's (1994) experiments are comparable with those that have been the subject of other studies (e.g., Evans and Marler, 1994). It is also impossible to assess the effects of control sounds (none were presented), to compare female responses to calling males with those to silent males, or even to partition out the contribution of vocal signals relative to other male behavior such as courtship displays. A similar manipulation of sexual motivation, which compared the behavior of hens that had been isolated for six days with that of hens maintained in a social group with a male, also failed to reveal any differences in the probability of approach (Van Kampen, 1994). It is quite possible that the negative results obtained in both of these experiments are simply attributable to the small sample size employed. I suggest that the relationship between the responses evoked by food calls and the motivational state of receivers might repay further investigation.

Experiments like those outlined above have the potential to provide strong evidence that referential signals affect receiver behavior by evoking representations. Note, however, that the results obtained would only establish the existence of 'nominal representations' (i.e., that the calls stand for something in the environment; Gallistel, 1990). This would be an important finding, but it should be noted that nominal representations are the most basic type; they are a relatively impoverished cognitive phenomenon compared with the 'computational representations' thought to be involved in processing information about time and spatial location (Gallistel, 1990). The use of the term 'representational' to describe the properties of animal signals certainly need not connote the sophistication implied by the linguistic use of the same term, in which representations have predicates (e.g., the brown dog that has fleas).

There are paradigms in which there is a consensus that representations are necessary constructs for explaining behavior. One clear example is filial imprinting (e.g., Hollis et al., 1991; Johnson, 1992; Bateson and Horn, 1994). Here there is compelling evidence that chicks store information about the characteristics of the maternal hen, and the neurophysiological basis of this process is now understood in some detail (Horn, 1985, 1990; Honey et al., 1995). The considerable timelag that may elapse between exposure to an imprinting object and successful performance on a discrimination task makes the inference that behavior is dependent on a stored representation unavoidable. This work suggests another possible approach for evaluating whether or not functionally referential signals are also representational. Experiments could be designed to assess whether exposure to referential signals significantly affects subsequent behavior measured hours or days later. For example, if birds were exposed to aerial alarm calls in one context, perhaps arranged to be visually distinctive, would they then be persistently more likely to engage in behavior that functions to detect aerial predators (e.g., scanning upward) than in other environments of a similar type? This strategy for detecting representations relies upon tests to determine whether changes in behavior caused by signals are based on memories stored for longer periods than the seconds or minutes that normally elapse in playback experiments. Intriguingly, there is anecdotal evidence to suggest that playback of snake alarms affected the behavior of vervet monkeys passing through the same area some hours later (Cheney and Seyfarth, 1990) but systematic experimental tests would be required to verify this.

THE PROBLEM OF DECEPTIVE SIGNALLING

There is probably no single issue that has generated more philosophical interest and debate than the possibility that animals are capable of behaving deceptively (Griffin, 1981, 1984, 1992; Ristau, 1983; Mitchell and Thompson,

1986; Whiten and Byrne, 1988; Byrne and Whiten, 1990, 1991). This is perhaps because successful deception requires a degree of cognitive complexity and flexibility in behavior that is qualitatively distinct from that envisaged both in traditional ethological accounts and in behaviorist analyses. As such, strong evidence for deception in the vernacular sense narrows considerably the gap between the cognitive properties of nonhuman animals and those that we impute to other humans. Like the issue of language, the problem of deception bears directly on the degree of continuity between humans and nonhuman animals and on the question of human uniqueness.

I will focus especially on deceptive signalling in the specific sense of transmitting false information about external events. I shall argue that analyses of referential signals make a critical contribution to understanding whether this phenomenon occurs in the natural behavior of animals.

It is important to be rigorous in separating the cognitive use of 'deception' from a purely functional one (Mitchell, 1986). The Mullerian mimicry of the viceroy butterfly provides a classic example of functional deception, but this hardly encourages us to speculate about its mental state. It is not, however, always straightforward to deduce which of these senses is intended in published accounts of deceptive behavior (see below).

There are several types of evidence for deception in animals. The first relies on the systematic collection and analysis of unique social interactions. These anecdotes are then assembled and examined to determine whether there are consistent trends (e.g., Whiten and Byrne, 1988). The second involves studies of interspecific communication, focussing upon signals that are designed to affect the behavior of potential predators (e.g., Ristau, 1983, 1991). And the third, which I will concentrate on here, involves intraspecific communication and the selective production of signals that are normally evoked by the approach of predators or the discovery of food. It is logical to separate these latter two data sets because there are likely to be important differences between inter- and intraspecific signals, both in the contingencies controlling production from moment to moment and in the selection pressures that have defined signal properties over evolutionary time. Evidence for deceptive intraspecific signals can be further sub-divided into instances of animals 'withholding' signals under conditions in which they are usually produced, and examples of signalling in the absence of the putative referent. I shall refer to the former as 'passive deception' and to the latter as 'active deception.'

Passive Deception

It is often suggested that remaining silent under conditions where a signal would normally be given is likely to be the most common form of deception in animal communication because the odds of retribution from companions are low (e.g., Cheney and Seyfarth, 1990; Hauser, 1992; Hauser and Marler, 1993a,b). There are, however, serious difficulties in interpreting the absence of behavior (Snowdon, 1992). It is always difficult to exclude the possibility that the normal eliciting conditions for the signal are not sufficiently well-understood and that some critical factor is simply absent on those occasions when signals are not evoked. In order to make a more compelling case for the inhibition of signalling, it would be necessary to demonstrate that the state of the sender when no signal is produced is otherwise identical with that when signals are given. This will require additional response assays, perhaps including physiological measures, and will be a methodologically challenging enterprise. Even if such data were available, the inference of deception would still be vulnerable on logical grounds, because it would involve asserting the null hypothesis. It will always be possible to suggest (and this argument will have some force) that a failure to demonstrate differences in physiological state, or in other nonsignalling responses to an eliciting event, is simply an artifact of assay insensitivity. I believe that it will consequently be difficult to make a truly compelling case for 'passive deception.'

An additional complication is introduced if we consider the information content of a signal in a broader sense. For example, food calls are likely to encode information not only about the availability of food but also about the subsequent behavior of the sender (Evans and Marler, 1994). Male chickens that food call at high rates typically refrain from ingesting the food item themselves and allow hens that approach to take it instead. The call thus predicts not only the presence of a palatable object but also a low probability of aggression by the sender. A striking contrast is apparent when this is compared with that of the behavior of the same cock in the presence of a rival male. Under these conditions, call production is almost completely abolished (Marler et al., 1986b). If our analyses were concerned solely with characteristics of the eliciting stimulus, then we might infer that this was a case of passive deception and that the silent males were concealing from competitors the presence of a valuable and limited resource. If, however, we consider other information in the signal, so that the meaning is not simply "food" but "food + a preparedness to share it," then there is no need to posit deception because the call is an honest commentary on the male's likely subsequent behavior (see Smith, 1991).

Active Deception

There is clearly the potential to construct a more logically compelling case for deception in situations where signals are produced in the absence of the events that normally elicit them. In such instances, we are dealing with overt behavior and are no longer handicapped by the requirement to demonstrate that a signal 'should' have been produced, as in the case of passive deception. However, it is often very difficult to determine whether a case is being made for deception in a cognitive sense, suggesting parallels with the most sophisticated aspects of human social behavior, or whether the term is simply intended in a less controversial functional sense. I shall illustrate this distinction with an example based loosely on work by Munn (1986a,b) and by Møller (1988).

Consider the problem of interpreting the following observation (Fig. 2). Bird A is approaching bird B, which is feeding on a rare and highly preferred food item. Bird A suddenly produces an alarm call, normally associated with the appearance of a hunting raptor. Bird B responds by dropping the food item and flying into cover nearby. Bird A then picks up and ingests the food. This is clearly a case of deception in the functional sense: bird A has altered the behavior of bird B in such a way that it benefits and bird B does not. There are parallels between such uses of vocal signals and traditional examples of deceptive signalling that involve morphological characters, as in the case of Batesian mimicry. This analogy raises a series of intriguing questions about the factors selecting for deceptive signals and for maintaining them in the population over evolution-

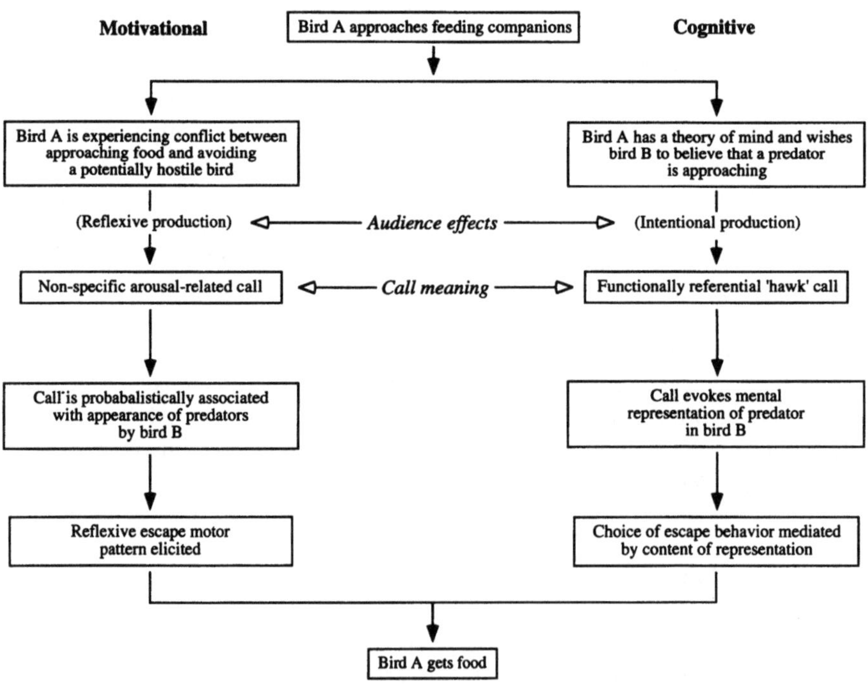

Fig. 2. Two alternative explanations for an observation of 'deceptive' alarm calling. The motivational model depends upon reflexive responses (perhaps attributable to associative learning) and changes in arousal. The cognitive model postulates sophisticated mental processing by both sender and receiver. Discriminating between these mechanisms requires analyses of call meaning and audience effects.

ary time (Guilford and Dawkins, 1991, 1993). It does not however, require us to infer that the alarm call is also deceptive in a cognitive sense. There are several possible proximate explanations for bird A's behavior. I have outlined two quite different accounts, deliberately selected so as to illustrate the contrast between them.

The first explanation, which draws on ideas from classical ethology, is principally concerned with motivational factors. Bird A approaches feeding companions and experiences an approach/avoidance conflict because it is attracted to the food but fearful of the potentially hostile individual currently in possession of it. The motivational conflict triggers a reflexive alarm call. This vocalization is a nonspecific signal reflecting bird A's internal state (i.e., high levels of arousal), which is usually associated with the appearance of predators. This contingency is responsible for the call evoking an escape response from bird B.

The alternative cognitive account could be outlined as follows: bird A has a theory of mind and wishes bird B to believe that a predator is approaching. To create this belief in bird B, bird A intentionally produces a highly specific referential alarm call. This evokes a mental representation of an approaching predator in bird B, which, in turn, mediates the choice of escape behavior. Bird B elects to flee to cover, which is the most appropriate response for avoiding an aerial predator.

Both of these models describe signalling that is deceptive in the functional sense, and if we were concerned solely with the fitness consequences of behavior, we would have no need to choose between them. Only the cognitive account would constitute deceptive behavior in the vernacular sense. Note that discriminating between these alternatives depends heavily upon understanding the type of information encoded in the signal. I do not believe that a convincing case can be made for deception in the cognitive sense if the signal does not meet the criteria for functional reference, because it would then be simply an accurate commentary on the internal state of the sender. Other components of the cognitive account may be less important and intermediate positions are surely possible. For example, we might choose to define cognitive deception in such a way that call production need not be intentional. A decision on this point will depend upon a more detailed understanding of the nature of audience effects and of whether or not they reflect volitional signalling. We may similarly be able to dispense with the requirement that birds have a 'theory of mind,' or perhaps be able to take a neutral position on this issue, because it is likely to be intractable to experimental investigation.

The issue of call meaning, however, remains central. In many recent accounts of deceptive signaliing, it is not clear whether the authors intend the functional or the cognitive sense, although reviews often assume the latter (e.g., Ristau, 1991; Griffin, 1992; Gould and Gould, 1994). There are typically not sufficient data to make a case for referential signalling, so the functional interpretation might be preferred on grounds of parsimony. For example, in Møller's analysis of alarm calling by great tits (Møller, 1988), and Munn's description of alarm calling in shrikes (Munn, 1986a,b), we do not have the necessary information about production and perception to establish call meaning. This is not a serious deficiency in either account, because the authors are principally concerned with the functional consequences of signalling.

Some Logical Difficulties in the Interpretation of Apparent Deception

Descriptions of food calling in chickens (Gyger and Marler, 1988) provide an intriguing potential case of active deception. Cocks were reported to food call while holding inedible objects, and to do so preferentially when hens were distant and hence less likely to detect the prevarication. Laboratory studies analyzing the conditions of food call production have demonstrated that, at least under controlled conditions, this vocalization is elicited specifically by food (Evans and Marler, 1994). Playback experiments have revealed that food calls are sufficient to elicit anticipatory feeding movements. Food calls thus have properties that satisfy both production and perception criteria for functional reference. Are we then required to conclude that selective calling by chickens provides an unambiguous example of deception in the cognitive sense?

There are both technical and logical deficiencies in the Gyger and Marler (1988) analysis that suggest such an interpretation would be premature. The data were obtained from opportunistic recordings of five cocks maintained in large pens. Statistical analyses were then performed with degrees of freedom determined by the total number of calling bouts (108) rather than the number of subjects (5). This is an example of the 'pooling fallacy' (Machlis et al., 1985; Martin and Bateson, 1993) in which the degrees of freedom are substantially inflated and the risk of Type I error is likely to be higher than the stated alpha level (but see Leger and Didrichsons, 1994). More conservative analyses would not have sufficient statistical power to demonstrate that the males, as a group, food called 'deceptively' more often when hens were distant than when they were close. An additional concern about this data set is that each of the subjects did not contribute equally to the analysis: two of the males yielded almost 76% of the data. What remains is thus essentially an anecdotal account describing the vocal behavior of two birds. Like other such descriptions of deceptive behavior (Whiten and Byrne, 1988; Byrne and Whiten, 1990, 1991), this is a thought-provoking example, but it does not establish that deceptive food calling is a property of chickens as a species, or even that it was a reliable phenomenon in the group under study.

Descriptions of natural vocal behavior also illustrate some of the logical problems that confront us in making a case for cognitive deception. Consider the observation of a cock food calling with an apparently inedible object such as a twig. Human observers would clearly be disposed to class this as a nonfood object and to regard the calling as a potential case of deception. There is, however, a real risk that this decision reflects more about human language, culture, and categorization by the observer than about the way in which objects are classified by the birds. It will be important to determine how cocks treat the apparently 'inedible' objects that they sometimes manipulate while food calling. If such objects are subsequently ingested, even though they have modest nutritive value, it will be reasonable to assert they are categorized as palatable by the birds and that, even though they may be less preferred than other objects, such as live invertebrates, they nevertheless qualify as 'food.' Much more thorough descriptions of the feeding behavior of galliformes will be required before we can discriminate 'honest' from 'deceptive' calling. It will be essential for this distinction to be based upon the animals' categorization scheme rather than those that human observers bring to the problem.

A second logical difficulty with interpreting putative instances of deceptive signalling is suggested by the literature on animal learning. Recent experiments employing instrumental conditioning techniques suggest strongly that contextual conditioning of food calling occurs. Birds were trained to work for food by pecking a hinged plastic key, which was illuminated from behind with a red light (Evans and Marler, 1994). In the first part of each test session, this light was switched off and the key was inactive. After two minutes, the light was switched on, signalling to the bird that pecks delivered to the key would produce food on a fixed ratio schedule (each bout of 16 pecks triggered a delivery of 3 small food pellets). Birds were initially unresponsive to the red light, which they had not previously experienced. Over the course of training, they rapidly learned to approach and to begin pecking as soon as the key was illuminated. After several such sessions, birds would engage in obvious foraging behavior (e.g., scratching and pecking at the ground) as they moved toward the response panel. They also food called as soon as the light was turned on. Later still, birds began food calling when first introduced into the test cage. Some individuals even began food calling when they were removed from their home cage and placed in a canvas bag to be weighed at the beginning of each test. None of these procedures had elicited food calling prior to training. These observations suggest that environmental cues reliably associated with the discovery of food can become sufficient to evoke food calling.

The possibility of contextual conditioning complicates the interpretation of food calling in the absence of food under natural conditions, because birds may well have been vocalizing in a location where food had previously been discovered. The implications of this are most easily illustrated with a hypothetical example of human behavior. Imagine two people (A and B) on a long-distance drive. They have not eaten for several hours. Person A spots a roadside billboard advertising a fast-food restaurant at the next motorway exit and exclaims, "Food!" Clearly A does not have a food item, but rather is responding to an environmental cue that reliably predicts feeding opportunities. It is possible that

B would entertain doubts about A's gastronomic judgment, but it is unlikely that they would consider their companion to be behaving *deceptively*. The implication of this logic is that a compelling case for deception will require an analysis not only of the immediate circumstances of signal production, but also of historical context (Smith, 1977, 1981, 1991), that is, of the animal's prior experience of pairings between the environmental events that normally elicit a call and the other stimuli reliably present at the time of production.

A final point that arises from the Gyger and Marler (1988) study concerns the frequency of 'deceptive' signalling. It was claimed that fully 45% of food calls were deceptive. This estimate is based on a flawed statistical analysis (see above), and the actual value may well be somewhat lower. Nevertheless, there is a striking contrast with classical theoretical analyses of animal signalling which suggest that 'deceptive' usage should be rare (Dawkins and Krebs, 1978). It is often assumed that this is a general rule in animal communication, so that claims for deceptive signalling at an appreciable rate are inherently not credible. Many of the original analyses of deceptive signalling were, however, based upon game theory models of aggressive interactions in which the costs of deception (severe physical injury) are potentially very high (Maynard Smith and Price, 1973; Maynard Smith, 1982). The payoff matrices describing signalling behavior such as food calling are almost certainly quite different. It is intriguing to note that in instrumental conditioning paradigms, animals will continue to work on very 'lean' schedules of reinforcement. Under these conditions, less than one response in a hundred might produce a delivery of food. This analogy suggests that, under more natural conditions, animals might be prepared to expend energy (e.g., by approaching a sender) even if the link between the sender's behavior and feeding opportunities has been quite tenuous. There is a real need for systematic studies of signal reliability with the goal of establishing the range of values that animals naturally experience for the correlation between signal production and presence of the putative referent.

ULTIMATE QUESTIONS

This will be a short section. I have so far been concerned almost exclusively with proximate questions, not because I regard them as more interesting or more important than evolutionary ones, but because this has been the principal focus of most studies conducted to date. Work on referential signalling has been comparative only in the weak sense (Wasserman, 1993) of evaluating the characteristics of animal signals to determine whether they have properties in common with language. This strategy has generated a number of important findings, some of which call into question traditional assumptions about the degree to which humans are unique. Dogmatic statements about the special attributes of language have perhaps been particularly tempting targets because they can successfully be attacked with a very modest data set; logically a single contradictory result will suffice. But there have been no studies of referential signalling that are comparative in the strong sense, providing a series of systematic analyses of closely related species that would permit us to reconstruct the evolution of a trait, as has been done successfully in studies of sexually selected signals (e.g., Ryan and Rand, 1993; Basolo, 1995a,b).

Evolution of Referential Signals

Some clues about the way in which comparative studies might proceed are provided by comparisons between the alarm call systems of ring-tailed and black-and-white ruffed lemurs. These two species have quite different behavior and ecology. Ring-tailed lemurs are highly terrestrial and occupy open habitat (e.g., Jolly, 1966; Tattersall, 1982), while ruffed lemurs are arboreal and live in dense rainforest (Pereira et al., 1988). Ring-tailed lemurs produce highly specific alarm calls, both during natural encounters with predators and during experimental simulations of such events. Playback experiments demonstrate that call type encodes sufficient information for companions to select appropriate responses (Macedonia and Polak, 1989; Macedonia, 1990; Macedonia and Yount, 1991; Pereira and Macedonia, 1991). Ruffed lemurs react to approaching predators with a graded series of calls which, like the alarm calls of ground squirrels and marmots, have only a probabilistic association with predator class. Consistent with these studies of call usage, playback experiments with ruffed lemurs do not reveal qualitatively different responses to different types of alarm call.

Ring-tailed lemurs share with vervet monkeys the traits of living in an open habitat and of spending a great deal of time on the ground. The transition from an arboreal to a largely terrestrial existence probably produced a substantial increase in predation risk. This factor is, however, clearly not sufficient to account for the characteristics of the ring-tailed lemur alarm call system, because there are many other species of small mammals (e.g., sciurid rodents) that are also exposed to frequent attacks from terrestrial predators but have rather less specific alarm calls. It has been suggested that one selection pressure for the evolution of referential alarm calls may have been the use of qualitatively distinct and incompatible strategies for avoiding different classes of predator (Macedonia and Evans, 1993). Squirrels avoid both coyotes and hawks by running toward a burrow, and their alarm call system seems to provide information that allows receivers to assess the time available for such a response. Vervets and ring-tailed lemurs have the shared characteristic that responses to avian predators involve movement into dense cover, while the safest refuge from ground predators is the outermost branches of trees (a location where vulnerability to attack from raptors is actually increased). Such ecological factors may have played a role in the evolution of alarm call systems that designate predator class (Macedonia and Evans, 1993).

Habitat Characteristics and Predator Recognition

Differences in ecology have also been implicated as one determinant of signal specificity. Studies of alarm call production under naturalistic conditions suggest that chickens respond to a fairly broad array of overhead objects (e.g., Gyger et al., 1987). Subsequent laboratory studies revealed that the size and speed of a simulated predator (Evans et al., 1993b) play an important role and that there is also a degree of sensitivity to shape and spatial location. Alarm call production in chickens is nevertheless substantially less highly specific than that of some other birds. For example, studies of alarm calling by lapwings (Walters, 1990) demonstrate that both South American and African species make subtle discriminations between raptors that are visually very similar. Lapwings inhabit open terrain and this probably affords them the opportunity to examine approaching raptors for some time before producing an alarm. In contrast, red junglefowl, which are the ancestral species for domesticated chickens (Fumihito et al., 1994), live in forest and dense brush where visibility is typically limited (Collias and Collias, 1967) and the time available for responding to potential avian predators will consequently be brief. This ecological constraint may have selected for a simple rule of thumb for predator recognition which, although it entails some loss in accuracy, facilitates rapid response. Other reports describing the anti-predator behavior of forest-dwelling birds also suggest a fairly high frequency of false alarms (Trail, 1987).

Although the evidence remains fragmentary, the results so far available are consistent with the idea that the specificity of alarm calls has been shaped by habitat characteristics. It is possible that the range of events evoking alarm calls is the product of a trade-off between the rate of Type I errors (i.e., calling when the approaching bird is not dangerous) and Type II errors (i.e., failing to respond to an approaching predator) over evolutionary time (Evans et al., 1993b). Birds living in open habitats may have been subject to selection for low Type I error rates, as the cost of the frequent false alarms that would otherwise be produced in an environment where aerial objects are visible much of the time would be prohibitive. This process would give rise to relatively specific alarm calls. Species inhabiting habitats where visibility is restricted and response times must be short may have been selected to reduce Type II error rates and their alarm calls may be less specific as a consequence. The logic of this argument is closely analogous to that of classical signal detection theory (e.g., Swets, 1961).

The theoretical ideas outlined above about the factors responsible for the evolution of referential signal systems (Macedonia and Evans, 1993) and about the relationship between habitat characteristics and signal specificity (Evans et al., 1993b) remain highly speculative. They do, however, both make clear predictions and are thus amenable to empirical test. This will require comparative studies with a focus upon groups of species selected either to have reliable differences in their strategies for avoiding predation, or to have similar anti-predator behavior, but different habitat characteristics.

CONCLUSIONS

We have learned a great deal about referential signals in the three decades since Struhsaker's pioneering studies of vervet monkeys (Struhsaker, 1967). In striking contrast to traditional assumptions about the affective nature of animal communication, it is now clear that both nonhuman primates and birds are capable of producing relatively specific alarm calls in response to approaching predators and that the information encoded in such signals is sufficient for companions to select appropriate responses. These systems have the property of functional reference. The discovery of referential signalling extends the parallels between the properties of human language and those of the natural behavior of nonhuman animals (Marler et al., 1992; Snowdon, 1993a; Evans and Marler, 1995). This is valuable because it provides an empirical basis for discriminating between those aspects of language that really are uniquely human and others that are more widely shared, and possibly quite ancient.

The discovery that nonhuman primates possess a communicative system with a surprising degree of flexibility and complexity has encouraged exploration along the lines of Griffin's suggestion (Griffin, 1981, 1984, 1992) that signalling behavior provides a potential window on the mental life of animals. This work has involved grappling with a series of difficult philosophical problems, such as the extent to which animal signals may usefully be viewed as deceptive, and whether monkeys attribute knowledge or ignorance to their companions (Cheney and Seyfarth, 1990). While these are undoubtedly important issues, I suggest that they are likely to prove recalcitrant. Specifically, I have argued that it will be difficult to build a logically impregnable case for cognitive deception, and that the current evidence does not require us to conclude that the effects of signals are mediated by mental representations of the eliciting event.

The focus on communication as a vehicle for analyzing complex cognitive processes has necessarily meant that a number of other issues have remained unaddressed. Analyses have been principally concerned with variation at the level of signal class. We have quite detailed descriptions of the factors determining signal type during production, and of the responses evoked, but we know much less about the significance of fine-grained variation in the structure of referential signals. It is likely that this reflects the interplay of stimulus attributes and motivational state (Marler et al., 1992), and it will be particularly important to determine whether there are aspects of signal morphology that reflect either of these factors independently.

The design of referential signals cannot be accounted for by considering just the physical properties of eliciting stimuli. This is most obviously true if we are concerned with explaining morphology in a broad sense, taking into account not only short-term variation in acoustic characteristics but also the time course of calling. Signal properties are likely to have been influenced by a long list of factors. These include the physical environment, the sensory attributes of receivers (which dictate the structural correlates of the requirement that signals encoding very different types of information should be readily discriminable), and design for effect over distances that may be quite short in some cases (e.g., aerial alarm calls) and considerable in others (e.g., ground alarm calls). In addition, we must consider the possibility that the various signals in a given repertoire may be designed to influence the behavior of different sets of receivers, so that some calls address only close companions while others also target heterospecifics such as an approaching predator.

Any complete account of communication will need to incorporate the effects of contextual factors. Analyses that examine only environmental events, such as the discovery of food or the approach of a potential predator, will fail to predict accurately the probability of calling in systems where signal production is sensitive to the nature and presence of an audience. The responses of receivers are almost certainly also influenced by contextual cues, which are available both from the nonsignalling behavior of the sender synchronous with call production and from relationships between signals and environmental events over much longer periods. The responses that we observe in natural social interactions are thus likely to reflect the integration of information from signals and from the contexts in which they are produced, but there have been very few experimental analyses of this process. If contextual cues prove to be important determinants of the responses evoked by referential signals, then this will imply that associative learning plays a larger role than is envisaged by current theory.

We do not currently have a comprehensive model of any system of referential signals.

I suggest that providing one is a worthwhile goal. This will not be a trivial undertaking because it will require a series of complementary analyses at different levels (Fig. 1). It will be necessary to consider not only the meaning of call types, but also fine-grained variation within signal class. We will also have to expand the time scale of our studies, so as to encompass behavior other than the initial response to an event, taking into account the possibility of long-term (tonic) effects. The development of referential signals will require further study, building on the evidence for plasticity in signal production and inquiring whether this is also true of the responses evoked. Finally, it will be essential to understand the effects of those social factors that modulate both the probability of calling and the behavior of receivers.

There are at least partial answers to some of these questions, but they have been obtained in relative isolation. We now have an opportunity to begin integrating the available findings, while at the same time identifying and addressing the remaining lacunae in our knowledge. By proceeding in this way we will be able to place referential signalling in the wider context of other social behavior, to account for signal design, and, in particular, to explore the fitness implications of some of the phenomena that we now understand at the level of mechanism.

ACKNOWLEDGMENTS

I am grateful to the following people for comments that substantially improved earlier drafts of this manuscript: D. T. Blumstein, K. C. Y. Cheng, M. Coltheart, L. Evans, J. M. Macedonia, D. H. Owings, N. S. Thompson. I also thank the editors for their patience. Preparation of this paper was supported by a grant from the Australian Research Council.

REFERENCES

Abraham, R. (1974). Vocalizations of the mallard (*Anas platyrhynchos*). *Condor, 76*, 401–420.

Allen, C. (1992). Monkeys mind. *Behav. Brain Sci., 15*, 147.

Andersson, M. (1982). Female choice selects for extreme tail length in a widowbird. *Nature, 299*, 818–820.

Andersson, M. (1994). *Sexual selection.* Princeton: Princeton University Press.

Andersson, S. (1992). Female preference főr long tails in lekking Jackson's widowbirds: Experimental evidence. *Anim. Behav., 43*, 379–388.

Andrew, R. J. (1988). The development of visual lateralization in the domestic chick. *Behav. Brain Res., 29*, 201–209.

Arak, A., and Enquist, M. (1993). Hidden preferences and the evolution of signals. *Phil. Trans. R. Soc. Lond. B, 340*, 207–213.

Armstrong, D. M. (1992). Monkeys and consciousness. *Behav. Brain Sci., 15*, 147–148.

Baeumer, E. (1962). Lebensart des Haushuhnes, dritter Teil - über seine Läute und allgemaine Ergänzungen. *Z. Tierpsychol., 19*, 394–416.

Barnard, P. (1990). Male tail length, sexual display intensity and female sexual response in a parasitic African finch. *Anim. Behav., 39*, 652–656.

Baron-Cohen, S. (1992). How monkeys do things with "words." *Behav. Brain Sci., 15*, 148–149.

Basolo, A. L. (1990). Female preference predates the evolution of the sword in swordtail fish. *Science, 250*, 808–810.

Basolo, A. L. (1995a). Phylogenetic evidence for the role of a pre-existing bias in sexual selection. *Proc. R. Soc. Lond. B, 259*, 307–311.

Basolo, A. L. (1995b). A further examination of a pre-existing bias favoring a sword in the genus Xiphophorus. *Anim. Behav., 50*, 365–375.

Bastian, J. (1965). Primate signaling systems and human languages. In DeVore, I. (ed.), *Primate behavior: Field studies of monkeys and apes* (pp. 585–606). New York: Holt, Rinehart and Winston.

Bateson, P., and Horn, G. (1994). Imprinting and recognition memory: A neural net model. *Anim. Behav., 48,* 695–715.

Becker, P. H. (1982). The coding of species-specific characteristics in bird sounds. In Kroodsma, D. E., and Miller, E. H. (eds.), *Acoustic communication in birds: Vol. 1. Production, perception and design features of sounds* (pp. 213–252). New York: Academic Press.

Benz, J. J. (1993). Food-elicited vocalizations in golden lion tamarins: Design features for representational communication. *Anim. Behav., 45,* 443–455.

Benz, J. J., Leger, D. W., and French, J. A. (1992). Relation between food preference and food-elicited vocalizations in golden lion tamarins (*Leontopithecus rosalia*). *J. Comp. Psychol., 106,* 142–149.

Blumstein, D. T. (1995a). Golden marmot calls: I. The production of situationally specific vocalizations. *Ethology, 100,* 113–125.

Blumstein, D. T. (1995b). Golden marmot calls: II. Asymmetrical production and perception of situationally specific vocalizations? *Ethology, 101,* 25–32.

Blumstein, D. T. (in press). Alarm calling in yellow-bellied marmots: I. The meaning of situationally variable alarm calls. *Anim. Behav.*

Blumstein, D. T., and Arnold, W. (1995). Situational specificity in alpine-marmot alarm communication. *Ethology, 100,* 1–13.

Blumstein, D. T., Steinmetz, J., Armitage, K. B., and Daniel, J. C. (in press). Alarm calling in yellow-bellied marmots: II. Kin selection or parental care? *Anim. Behav.*

Byrne, R. W., and Whiten, A. (1990). Tactical deception in primates: The 1990 database. *Primate Report, 27,* 1–101.

Byrne, R. W., and Whiten, A. (1991). Computation and mindreading in primate tactical deception. In Whiten, A. (ed.), *Natural theories of mind: Evolution, development and simulation of everyday mindreading* (pp. 127–141). Oxford: Basil Blackwell.

Caro, T. M. (1986a). The functions of stotting: A review of the hypotheses. *Anim. Behav., 34,* 649–662.

Caro, T. M. (1986b). The functions of stotting in Thompson's gazelles: Some tests of the predictions. *Anim. Behav., 34,* 663–684.

Caro, T. M., Lombardo, L., Goldizen, A. W., and Kelly, M. (1995). Tail-flagging and other antipredator signals in white-tailed deer: New data and synthesis. *Behav. Ecol., 6,* 442–450.

Cheney, D. L., and Seyfarth, R. M. (1980). Vocal recognition in free-ranging vervet monkeys. *Anim. Behav., 28,* 362–367.

Cheney, D. L., and Seyfarth, R. M. (1985). Vervet monkey alarm calls: Manipulation through shared information? *Behaviour, 94,* 150–166.

Cheney, D. L., and Seyfarth, R. M. (1988). Assessment of meaning and the detection of unreliable signals by vervet monkeys. *Anim. Behav., 36,* 477–486.

Cheney, D. L., and Seyfarth, R. M. (1990). *How monkeys see the world.* Chicago: University of Chicago Press.

Cheney, D. L., and Seyfarth, R. M. (1992). Précis of *How monkeys see the world. Behav. Brain Sci., 15,* 135–182.

Clark, A. P., and Wrangham, R. W. (1993). Acoustic analysis of wild chimpanzee pant hoots: Do Kibale Forest chimpanzees have an acoustically distinct food arrival pant hoot? *Am. J. Primatol., 31,* 99–109.

Clark, A. P., and Wrangham, R. W. (1994). Chimpanzee arrival pant-hoots: Do they signify food or status? *Int. J. Primatol., 15,* 185–205.

Clark, D. L., and Uetz, G. W. (1990). Video image recognition by the jumping spider *Maevia inclemens* (Araneae: Salticidae). *Anim. Behav., 40,* 884–890.

Clark, D. L., and Uetz, G. W. (1991). Morph-dependent mate selection in a dimorphic jumping spider: Demonstration of movement bias in female choice using video-controlled courtship behaviour. *Anim. Behav., 43,* 247–254.

Clark, D. L., and Uetz, G. W. (1993). Signal efficacy and the evolution of male dimorphism in the jumping spider, *Maevia inclemens. Proc. Natl. Acad. Sci. USA, 90,* 11954–11957.

Clutton-Brock, T. H., and Albon, S. D. (1979). The roaring of red deer and the evolution of honest advertisement. *Behaviour, 69,* 145–169.

Collias, N. E. (1987). The vocal repertoire of the red junglefowl: A spectrographic classification and the code of communication. *Condor, 89,* 510–524.

Collias, N. E., and Collias, E. C. (1967). A field study of the red jungle fowl in north-central India. *Condor, 69,* 360–386.

Collias, N. E., and Joos, M. (1953). The spectrographic analysis of sound signals of the domestic fowl. *Behaviour, 5,* 175–188.

Colwill, R. M., and Rescorla, R. A. (1985). Postconditioning devaluation of a reinforcer affects instrumental responding. *J. Exp. Psychol. Anim. Behav. Proc., 11,* 120–132.

Curio, E. (1993). Proximate and developmental aspects of antipredator behavior. *Adv. Study Behav., 22,* 135–238.

Dawkins, M. S. (1989). The future of ethology: How many legs are we standing on? In Bateson, P. P. G., and Klopfer, P. H. (eds.), *Perspectives in ethology: Vol. 8. Whither ethology?* (pp. 47–54). New York: Plenum Press.

Dawkins, M. S. (1993). Are there general principles of signal design? *Phil. Trans. R. Soc. Lond. B, 340,* 251–255.

Dawkins, M. S., and Guilford, T. (1996). Sensory bias and the adaptiveness of female choice. *Am. Nat., 148,* 937–942.

Dawkins, R., and Krebs, J. R. (1978). Animal signals: Information or manipulation? In Krebs, J. R., and Davies, N. B. (eds.), *Behavioural ecology: An evolutionary approach* (pp. 282–309). Oxford: Blackwell.

Dennett, D. C. (1983). Intentional systems in cognitive ethology: The 'Panglossian paradigm' defended. *Behav. Brain Sci., 6,* 343–390.

Dewsbury, D. A. (1992). Surplusage, audience effects and George John Romanes. *Behav. Brain Sci., 15,* 152.

Dittus, W. P. (1984). Toque macaque food calls: Semantic communication concerning food distribution in the environment. *Anim. Behav., 32,* 470–477.

Dooling, R. J., Brown, S. D., Park, T. J., and Okanoya, K. (1990). Natural perceptual categories for vocal signals in budgerigars (*Melopsittacus undulatus*). In Stebbins, W.C., and Berkley, M. A. (eds.), *Comparative perception: Vol. 2. Complex signals* (pp. 345–374). New York: John Wiley and Sons.

Ehret, G. (1992). Categorical perception of mouse-pup ultrasounds in the temporal domain. *Anim. Behav., 43,* 409–416.

Elowson, A. M., Tannenbaum, P. L., and Snowdon, C. T. (1991). Food-associated calls correlate with food preference in cotton-top tamarins. *Anim. Behav., 42,* 931–937.

Evans, C. S. (1993). Recognition of the spectral characteristics of conspecific signals in ducklings: Evidence for a simple perceptual process. *Anim. Behav., 45,* 1071–1082.

Evans, C. S., Evans, L., and Marler, P. (1993a). On the meaning of alarm calls: Functional reference in an avian vocal system. *Anim. Behav., 46,* 23–38.

Evans, C. S., Macedonia, J. M., and Marler, P. (1993b). Effects of apparent size and speed on the response of chickens, *Gallus gallus*, to computer-generated simulations of aerial predators. *Anim. Behav., 46,* 1–11.

Evans, C. S., and Marler, P. (1991). On the use of video images as social stimuli in birds: Audience effects on alarm calling. *Anim. Behav., 41,* 17–26.

Evans, C. S., and Marler, P. (1992). Female appearance as a factor in the responsiveness of male chickens during anti-predator behaviour and courtship. *Anim. Behav., 43,* 137–143.

Evans, C. S., and Marler, P. (1994). Food calling and audience effects in male chickens, *Gallus gallus*: Their relationships to food availability, courtship and social facilitation. *Anim. Behav., 47,* 1159–1170.

Evans, C. S., and Marler, P. (1995). Language and animal communication: Parallels and contrasts. In Roitblat, H. L., and Arcady-Meyer, J. (eds.), *Comparative approaches to cognitive science* (pp. 341–382). Cambridge, MA: MIT Press.

Forest, T. G. (1994). From sender to receiver: Propagation and environmental effects on acoustic signals. *Am. Zool., 34,* 644–654.

Fumihito, A., Miyake, T., Sumi, S. I., Takada, M., Ohno, S., and Kondo, N. (1994). One subspecies of the red junglefowl (*Gallus gallus gallus*) suffices as the matriarchic ancestor of all domestic breeds. *Proc. Natl. Acad. Sci. USA, 91,* 12505–12509.

Gaioni, S. J., and Evans, C. S. (1986). Perception of distress calls in mallard ducklings (*Anas platyrhynchos*). *Behaviour, 99,* 250–274.

Gallistel, C. R. (1990). *The organization of learning.* Cambridge, MA: MIT Press.

Gerhardt, H. C. (1978). Discrimination of intermediate sounds in a synthetic call continuum by female green treefrogs. *Science, 199,* 1089–1091.

Gerhardt, H. C. (1992). Multiple messages in acoustic signals. *The Neurosciences, 4,* 391–400.

Gould, J. L., and Gould, C. G. (1994). *The animal mind.* New York: Scientific American Library.

Gouzoules, H., Gouzoules, S., and Marler, P. (1985). External reference in mammalian vocal communication. In Ziven, G. (ed.), *The development of expressive behavior: Biology-environment interactions* (pp. 77–101). New York: Academic Press.

Griffin, D. R. (1981). *The question of animal awareness: Evolutionary continuity of mental experience.* New York: Rockefeller University Press.

Griffin, D. R. (1984). *Animal thinking.* Cambridge, MA: Harvard University Press.

Griffin, D. R. (1992). *Animal minds.* Chicago: University of Chicago Press.

Guilford, T. (1995). Animal signals: All honesty and light? *Trends Ecol. Evol., 10,* 100–101.

Guilford, T., and Dawkins, M. S. (1991). Receiver psychology and the evolution of animal signals. *Anim. Behav., 42,* 1–14.

Guilford, T., and Dawkins, M. S. (1993). Receiver psychology and the design of animal signals. *Trends Neurosci., 16,* 430–436.

Gyger, M., and Marler, P. (1988). Food calling in the domestic fowl, *Gallus gallus*: The role of external referents and deception. *Anim. Behav., 36,* 358–365.

Gyger, M., Marler, P., and Pickert, R. (1987). Semantics of an avian alarm call system: The male domestic fowl, *Gallus domesticus. Behaviour, 102,* 15–40.

Hamilton, W. D., and Zuk, M. (1982). Heritable true fitness and bright birds: A role for parasites? *Science, 218,* 384–387.

Hasson, O. (1991). Pursuit-deterrent signals: Communication between prey and predator. *Trends Ecol. Evol., 6,* 325–329.

Hauser, M. D. (1992). Costs of deception: Cheaters are punished in rhesus monkeys (*Macaca mulatta*). *Proc. Natl. Acad. Sci. USA, 89,* 12137–12139.

Hauser, M. D. (1996). *The evolution of communication.* Cambridge, MA: MIT Press.

Hauser, M. D., and Marler, P. (1993a). Food-associated calls in rhesus macaques (*Macaca mulatta*): I. Socioecological factors. *Behav. Ecol., 4,* 194–205.

Hauser, M. D., and Marler, P. (1993b). Food-associated calls in rhesus macaques (*Macaca mulatta*): II. Costs and benefits of call production and suppression. *Behav. Ecol., 4,* 206–212.

Hauser, M. D., Teixidor, P., Fields, L., and Flaherty, R. (1993). Food-elicited calls in chimpanzees: Effects of food quantity and divisibility. *Anim. Behav., 45,* 817–819.

Hauser, M. D., and Wrangham, R. W. (1987). Manipulation of food calls in captive chimpanzees: A preliminary report. *Folia Primatol., 48,* 207–210.

Herrnstein, R. J. (1984). Objects, categories and discriminative stimuli. In Roitblat, H. L., Bever, T. G., and Terrace, H. S. (eds.), *Animal cognition* (pp. 233–261). Hillsdale, NJ: Erlbaum.

Herrnstein, R. J. (1991). Levels of categorization. In Edelman, G. M., Gall, W. E., and Cowan, W. M. (eds.), *Signal and sense* (pp. 385–413). New York: Wiley-Liss.

Herrnstein, R. J., Loveland, D. H., and Cable, C. (1976). Natural concepts in pigeons. *J. Exptl. Psychol. Anim. Behav. Proc., 2,* 285–311.

Hinde, R. A. (1974). *Biological bases of human social behaviour.* New York: McGraw-Hill.

Hollis, K. L., ten Cate, C., and Bateson, P. P. G. (1991). Stimulus representation: A subprocess of imprinting and conditioning. *J. Comp. Psychol., 105,* 307–317.

Honey, R. C., Horn, G., Bateson, P., and Walpole, M. (1995). Functionally distinct memories for imprinting stimuli: Behavioral and neural dissociations. *Behav. Neurosci., 109,* 689–698.

Hopp, S. L., Sinnott, J. M., Owren, M. J., and Petersen, M. R. (1992). Differential sensitivity of Japanese macaques (*Macaca fuscata*) and humans (*Homo sapiens*) to peak position along a synthetic coo call continuum. *J. Comp. Psychol., 106,* 128–136.

Horn, G. (1985). *Memory, imprinting and the brain: An inquiry into mechanisms.* New York: Oxford University Press.

Horn, G. (1990). Neural bases of recognition memory investigated through an analysis of imprinting. *Phil. Trans. R. Soc. Lond. B, 329,* 133–142.

Johnson, M. H. (1992). Imprinting and the development of face recognition: From chick to man. *Curr. Direct. Psychol. Sci., 1,* 52–55.

Johnstone, R. A. (1994). Female preference for symmetrical males as a by-product of selection for mate recognition. *Nature, 372,* 172–175.

Jolly, A. (1966). *Lemur behavior.* Chicago, IL: University of Chicago Press.

Karakashian, S. J., Gyger, M., and Marler, P. (1988). Audience effects on alarm calling in chickens (*Gallus gallus*). *J. Comp. Psychol., 102,* 129–135.

Kendrick, K. M., and Baldwin, B. A. (1987). Cells in temporal cortex of conscious sheep can respond preferentially to the sight of faces. *Science, 236,* 448–450.

Klump, G. M., and Shalter, M. D. (1984). Acoustic behaviour of birds and mammals in the predator context. I. Factors affecting the structure of alarm signals. II. The functional significance and evolution of alarm signals. *Z. Tierpsychol., 66,* 189–226.

Konishi, M. (1963). The role of auditory feedback in the vocal behavior of the domestic fowl. *Z. Tierpsychol., 20,* 349–367.

Lancaster, J. (1965). *Primate behavior and the emergence of human culture.* New York: Holt, Rinehart and Winston.

Lea, S. E. G. (1984). In what sense do pigeons learn concepts? In Roitblat, H. L., Bever, T. G., and Terrace, H. S. (eds.), *Animal cognition* (pp. 263–276). Hillsdale, NJ: Erlbaum.

Leger, D. W. (1993). Contextual sources of information and responses to animal communication signals. *Psychol. Bull., 113,* 295–304.

Leger, D. W., and Didrichsons, I. A. (1994). An assessment of data pooling and some alternatives. *Anim. Behav., 48,* 823–832.

Leger, D. W., Owings, D. H., and Gelfand, D. L. (1980). Single-note vocalizations of California ground squirrels: Graded signals and situation-specificity of predator and socially evoked calls. *Z. Tierpsychol., 52,* 227–246.

Lester, B. M. (1985). Introduction. There's more to crying than meets the ear. In Lester, B. M., and Boukydis, C. F. Z. (eds.), *Infant crying* (pp. 1–27). New York: Plenum Press.

Locke, J. L. (1993). *The child's path to spoken language.* Cambridge, MA: Harvard University Press.

Luria, A. (1982). *Language and cognition.* Cambridge, MA: MIT Press.

Lyons, J. (1972). Human language. In Hinde, R. A. (ed.), *Nonverbal communication* (pp. 49–85). Cambridge: Cambridge University Press.

Macedonia, J. M. (1990). What is communicated in the antipredator calls of lemurs: Evidence from antipredator call playbacks to ringtailed and ruffed lemurs. *Ethology, 86,* 177–190.

Macedonia, J. M., and Evans, C. S. (1993). Variation among mammalian alarm call systems and the problem of meaning in animal signals. *Ethology, 93,* 177–197.

Macedonia, J. M., Evans, C. S., and Losos, J. B. (1994). Male *Anolis* lizards discriminate video-recorded conspecific and heterospecific displays. *Anim. Behav., 47,* 1220–1223.

Macedonia, J. M., and Polak, J. F. (1989). Visual assessment of avian threat in semi-captive ringtailed lemurs (*Lemur catta*). *Behaviour, 111,* 291–304.

Macedonia, J. M., and Stamps, J. A. (1994). Species recognition in *Anolis grahami* (Sauria, Iguanidae): Evidence from responses to video playbacks of conspecific and heterospecific displays. *Ethology, 98,* 246–264.

Macedonia, J. M., and Yount, P. L. (1991). Auditory assessment of avian predator threat in semi-captive ringtailed lemurs (*Lemur catta*). *Primates, 32,* 169–182.

Machlis, L., Dodd, P. W. D., and Fentress, J. C. (1985). The pooling fallacy: Problems arising when individuals contribute more than one observation to the data set. *Z. Tierpsychol., 68,* 201–214.

Marler, P. (1955). Some characteristics of animal calls. *Nature, 176,* 6–8.

Marler, P. (1976). Social organization, communication and graded signals: The chimpanzee and the gorilla. In Bateson, P. P. G., and Hinde, R. A. (eds.), *Growing points in ethology* (pp. 239–280). Cambridge: Cambridge University Press.

Marler, P. (1977). Primate vocalization: Affective or symbolic? In Bourne, G. (ed.), *Progress in ape research* (pp. 85–96). New York: Academic Press.

Marler, P. (1978). Affective and symbolic meaning: Some zoosemiotic speculations. In Sebeok, T. A. (ed.), *Sight, sound, and sense* (pp. 112–123). Bloomington, IN: Indiana University Press.

Marler, P. (1984). Animal communication: Affect or cognition? In Scherer, K. R. and Ekman, P. (eds.), *Approaches to emotion* (pp. 345–365). Hillsdale, NJ: Erlbaum.

Marler, P. (1985). Representational vocal signals of primates. *Fortschr. Zool., 31,* 211–221.

Marler, P., Dufty, A., and Pickert, R. (1986a). Vocal communication in the domestic chicken: I. Does a sender communicate information about the quality of a food referent to a receiver? *Anim. Behav., 34,* 188–193.

Marler, P., Dufty, A., and Pickert, R. (1986b). Vocal communication in the domestic chicken: II. Is a sender sensitive to the presence and nature of a receiver? *Anim. Behav., 34,* 194–198.

Marler, P., Evans, C. S., and Hauser, M. D. (1992). Animal signals: Motivational, referential, or both? In Papousek, H., Jürgens, U., and Papousek, M. (eds.), *Nonverbal vocal communication: Comparative and developmental approaches* (pp. 66–86). Cambridge: Cambridge University Press.

Marler, P., and Pickert, R. (1984). Species-universal microstructure in the learned song of the swamp sparrow (*Melospiza georgiana*). *Anim. Behav., 32,* 673–689.

Martin, P., and Bateson, P. (1993). *Measuring behaviour: An introductory guide*. Cambridge: Cambridge University Press.

May, B., Moody, D. B., and Stebbins, W. C. (1989). Categorical perception of conspecific communication sounds by Japanese macaques, *Macaca fuscata*. *J. Acoust. Soc. Am., 85,* 837–847.

Maynard Smith, J. (1982). *Evolution and the theory of games*. Cambridge: Cambridge University Press.

Maynard Smith, J., and Price, G. R. (1973). The logic of animal conflict. *Nature, 246,* 15–18.

McQuoid, L. M., and Galef, B. G. (1993). Social stimuli influencing feeding behaviour of Burmese fowl: A video analysis. *Anim. Behav., 46,* 13–22.

Mineka, S. (1987). A primate model of phobic fears. In Eysenck, H., and Martin, I. (eds.), *Theoretical foundations of behavior therapy* (pp. 81–111). New York: Plenum Press.

Mineka, S., and Cook, M. (1987). Social learning and the acquisition of snake fear in monkeys. In Zentall, T. and Galef, G. (eds.), *Social learning: Psychological and biological perspectives* (pp. 51–73). Hillsdale, NJ: Erlbaum.

Mitchell, R. W. (1986). A framework for discussing deception. In Mitchell, R. W., and Thompson, N. S. (eds.), *Deception: Perspectives on human and nonhuman deceit* (pp. 3–40). New York: State University of New York Press.

Mitchell, R. W., and Thompson, N. S. (eds.). (1986). *Deception: Perspectives on human and nonhuman deceit.* New York: State University of New York Press.

Møller, A. P. (1988). Female choice selects for male sexual tail ornaments in the monogamous swallow. *Nature, 332,* 640–642.

Møller, A. P. (1990). Male tail length and female mate choice in the monogamous swallow *Hirundo rustica. Anim. Behav., 39,* 458–465.

Møller, A. P. (1992). Female swallow preference for symmetrical sexual ornaments. *Nature, 357,* 238–240.

Møller, A. P., and Pomiankowski, A. (1993). Fluctuating asymmetry and sexual selection. *Genetica, 89,* 267–279.

Morton, E. S. (1977). On the occurrence and significance of motivation structural rules in some bird and mammal sounds. *Am. Nat., 3,* 855–869.

Munn, C. A. (1986a). Birds that 'cry wolf.' *Nature, 319,* 143–145.

Munn, C. A. (1986b). The deceptive use of alarm calls by sentinel species in mixed-species flocks of neotropical birds. In Mitchell, R. W. and Thompson, N. S. (eds.), *Deception: Perspectives on human and nonhuman deceit* (pp. 169–175). New York: State University of New York Press.

Myers, J. K., Weissman, M. M., Tischler, G. L., Holzer, C. E., Leaf, P. J., Orvaschel, H., Anthony, J. C., Boyd, J. H., Burke, J. D., Kramer, M., and Stoltzman, R. (1984). Six-month prevalence of psychiatric disorders in three communities. *Arch. Gen. Psychiatry, 41,* 959–967.

Nelson, D. A. (1988). Feature weighting in species song recognition by the field sparrow (*Spizella pusilla*). *Behaviour, 106,*158–182.

Nelson, D. A., and Marler, P. (1989). Categorical perception of a natural stimulus continuum: Birdsong. *Science, 244,* 976–978.

Nelson, D. A., and Marler, P. (1990). The perception of birdsong and an ecological concept of signal space. In Stebbins, W. C., and Berkley, M. A. (eds.), *Comparative perception: Vol. 2. Complex signals* (pp. 443–478). New York: John Wiley and Sons.

Owings, D. H. (1994). How monkeys feel about the world: A review of *How monkeys see the world. Lang. Commun. 1,* 15–30.

Owings, D. H., Hennessy, D. F., Leger, D. W., and Gladney, A. B. (1986). Different functions of 'alarm' calling for different time scales: A preliminary report on ground squirrels. *Behaviour, 99,* 101–116.

Owings, D. H., and Leger, D. W. (1980). Chatter vocalizations of California ground squirrels: Predator- and social-role specificity. *Z. Tierpsychol., 54,* 164–184.

Owings, D. H., and Virginia, R. A. (1978). Alarm calls of California ground squirrels (*Spermophilus beecheyi*). *Z. Tierpsychol., 46,* 58–70.

Owren, M. J., Dieter, J. A., Seyfarth, R. M., and Cheney, D. L. (1992). Food calls produced by adult female rhesus (*Macaca mulatta*) and Japanese (*M. fuscata*) macaques, their normally-raised offspring, and offspring cross-fostered between species. *Behaviour, 120,* 218–231.

Pagel, M. (1993). The design of animal signals. *Nature, 361,* 18–20.

Pereira, M. E., and Macedonia, J. M. (1991). Response urgency does not determine antipredator call selection by ringtailed lemurs. *Anim. Behav., 41,* 543–544.

Pereira, M. E., Seeligson, M. L., and Macedonia, J. M. (1988). The behavioral repertoire of the black-and-white ruffed lemur (*Varecia variegata variegata*). *Folia Primatol., 51,* 1–32.

Perrett, D. I., Mistlin, A. J., and Chitty, A. J. (1987). Visual neurones responsive to faces. *Trends Neurosci., 10,* 358–364.

Premack, D. (1975). On the origins of language. In Gazzaniga, M.S., and Blackmore, C.B. (eds.), *Handbook of psychobiology* (pp. 591–605). New York: Academic Press.

Rashid, N., and Andrew, R. J. (1989). Right hemisphere advantage for topographical orientation in the domestic chick. *Neuropsychologia, 27,* 937–948.

Ristau, C. A. (1983). Intentionalist plovers or just dumb birds? *Behav. Brain Sci., 6,* 373–375.

Ristau, C. A. (1991). Before mindreading: Attention, purposes and deception in birds? In Whiten, A. (ed.), *Natural theories of mind: Evolution, development and simulation of everyday mindreading* (pp. 209–222). Oxford: Basil Blackwell.

Robinson, S. R. (1980). Antipredator behaviour and predator recognition in Belding's ground squirrels. *Anim. Behav., 28,* 840–852.

Robinson, S. R. (1981). Alarm communication in Belding's ground squirrels. *Z. Tierpsychol., 56,* 150–168.

Rosenthal, G. G., Evans, C. S., and Miller, W. L. (1996). Female preference for dynamic traits in the green swordtail, *Xiphophorus helleri. Anim. Behav., 51,* 811–820.

Rowell, T. E., and Hinde, R. A. (1962). Vocal communication by the rhesus monkey (*Macaca mulatta*). *Proc. Zool. Soc. Lond., 138,* 279–294.

Rowland, W. J., Bolyard, K. J., and Halpern, A. D. (1995a). The dual effect of stickleback nuptial coloration on rivals: Manipulation of a graded signal using video playback. *Anim. Behav., 50,* 267–272.

Rowland, W. J., Bolyard, K. J., Jenkins, J. J., and Fowler, J. (1995b).Video playback experiments on stickleback mate choice: Female motivation and attentiveness to male colour cues. *Anim. Behav., 49,* 1559–1567.

Ryan, M. J., Fox, J. H., Wilczynski, W., and Rand, A. S. (1990). Sexual selection for sensory exploitation in the frog *Physalaemus pustulosus. Nature, 343,* 66–67.

Ryan, M. J., and Rand, A. S. (1993). Sexual selection and signal evolution: The ghost of biases past. *Phil. Trans. R. Soc. Lond. B, 340,* 187–195.

Schjelderup-Ebbe, T. (1922). Beiträge zur Sozialpsychologie des Haushuhnes. *Z. Tierpsychol., 88,* 225–252.

Schleidt, W. M. (1973). Tonic communication: Continual effects of discrete signs in animal communication systems. *J. Theoret. Biol., 42,* 359–386.

Schull, J., and Smith, J. D. (1992). Knowing thyself, knowing the other: They're not the same. *Behav. Brain Sci., 15,* 166–167.

Seyfarth, R. M., and Cheney, D. L. (1980). The ontogeny of vervet monkey alarm calling behavior: A preliminary report. *Z. Tierpsychol., 54,* 37–56.

Seyfarth, R. M., and Cheney, D. L. (1990). The assessment by vervet monkeys of their own and another species' alarm calls. *Anim. Behav., 40,* 754–764.

Seyfarth, R. M., and Cheney, D. L. (1992). Meaning and mind in monkeys. *Sci. Am., 267,* 122–128.

Seyfarth, R. M., and Cheney, D. L. (1993). Meaning, reference, and intentionality in the natural vocalizations of monkeys. In Roitblat, H. L., Herman, L. M., and Nachtigall, P. E. (eds.), *Language and communication: Comparative perspectives* (pp. 195–219). Hillsdale, NJ: Erlbaum.

Seyfarth, R. M., Cheney, D. L., and Marler, P. (1980a). Vervet monkey alarm calls: Semantic communication in a free-ranging primate. *Anim. Behav., 28,* 1070–1094.

Seyfarth, R. M., Cheney, D. L., and Marler, P. (1980b). Monkey responses to three different alarm calls: Evidence of predator classification and semantic communication. *Science, 210,* 801–803.

Sherman, P. W. (1977). Nepotism and the evolution of alarm calls. *Science, 197,* 1246–1253.

Sherman, P. W. (1985). Alarm calls of Belding's ground squirrels to aerial predators: Nepotism or self-preservation? *Behav. Ecol. Sociobiol., 17,* 313–323.

Smith, W. J. (1965). Message, meaning and context in ethology. *Am. Nat., 99,* 405–409.

Smith, W. J. (1977). *The behavior of communicating.* Cambridge, MA: Harvard University Press.

Smith, W. J. (1981). Referents of animal communication. *Anim. Behav., 29,* 1273–1274.

Smith, W. J. (1991). Animal communication and the study of cognition. In Ristau, C. A. (ed.), *Cognitive ethology: The minds of other animals* (pp. 209–230). Hillsdale, NJ: Erlbaum.

Snowdon, C. T. (1990). Language capacities of nonhuman animals. *Yearbook of Physical Anthropology, 33,* 215–243.

Snowdon, C. T. (1992). The sounds of silence. *Behav. Brain Sci., 15,* 167–168.

Snowdon, C. T. (1993a). Linguistic phenomena in the natural communication of animals. In Roitblat, H. L., Herman, L. M., and Nachtigall, P. E. (eds.), *Language and communication: Comparative perspectives* (pp. 175–194). Hillsdale, NJ: Erlbaum.

Snowdon, C. T. (1993b). A comparative approach to language parallels. In Gibson, K. R., and Ingold, T. (eds.), *Tools, language and cognition in human evolution* (pp. 109–128). Cambridge, London: Cambridge University Press.

Snowdon, C. T., Coe, C. L., and Hodun, A. (1985). Population recognition of infant isolation peeps in the squirrel monkey. *Anim. Behav., 33,* 1145–1151.

Stamps, J. A. (1991). Why evolutionary issues are reviving interest in proximate behavioral mechanisms. *Am. Zool., 31,* 338–348.

Struhsaker, T. T. (1967). Auditory communication among vervet monkeys (*Cercopithecus aethiops*). In Altmann, S. A. (ed.), *Social communication among primates* (pp. 281–324). Chicago, IL: University of Chicago Press.

Sullivan, K. A. (1985). Selective alarm calling by downy woodpeckers in mixed-species flocks. *Auk, 102,* 184–187.

Swets, J. A. (1961). Is there a sensory threshold? *Science, 134,* 168–177.

Tattersall, I. (1982). *The primates of Madagascar.* New York: Columbia University Press.

Tinbergen, N. (1948). Social releasers and the experimental method required for their study. *Wilson Bull., 60,* 6–51.

Trail, P. W. (1987). Predation and antipredator behavior of Guianan cock-of-the-Rock leks. *Auk, 104,* 496–507.

Van Kampen, H. S. (1994). Courtship food-calling in Burmese red junglefowl: I. The causation of female approach. *Behaviour, 131,* 261–275.

Walters, J. R. (1990). Anti-predatory behavior of lapwings: Field evidence of discriminative abilities. *Wilson Bull., 102,* 49–70.

Wasserman, E. A. (1993). Comparative cognition: Beginning the second century of the study of animal intelligence. *Psychol. Bull., 113,* 211–228.

Weary, D. M. (1990). Categorization of song notes in great tits: Which acoustic features are used and why? *Anim. Behav., 39,* 450–457.

Weary, D. M. (1991). How great tits use song-note and whole-song features to categorize their songs. *Auk, 108,* 187–190.

Whiten, A., and Byrne, R. W. (1988). Tactical deception in primates. *Behav. Brain Sci., 11,* 233–273.

Wiley, R. H., and Richards, D. G. (1982). Adaptations for acoustic communication in birds: Sound transmission and signal detection. In Kroodsma, D. E., and Miller, E. H. (eds.), *Acoustic communication in birds: Vol. 1. Production, perception and design features of sounds* (pp. 131–181). New York: Academic Press.

Workman, L., and Andrew, R. J. (1989). Simultaneous changes in behaviour and in lateralization during the development of male and female domestic chicks. *Anim. Behav., 38,* 596–605.

Zahavi, A. (1975). Mate selection: A selection for a handicap. *J. Theoret. Biol., 53,* 205–214.

Zajonc, R. B. (1965). Social facilitation. *Science, 149,* 269–274.

Zoloth, S., Petersen, M. R., Beecher, M. D., Green, S., Marler, P., Moody, D. B., and Stebbins, W. (1979). Species-specific perceptual processing of vocal sounds by monkeys. *Science, 204,* 870–873.

Zuk, M., Johnson, K., Thornhill, R., and Ligon, J. D. (1990a). Parasites and male ornaments in free-ranging and captive red jungle fowl. *Behaviour, 114,* 232–248.

Zuk, M., Thornhill, R., Ligon, J. D., Johnson, K., Austad, S., Ligon, S. H., Thornhill, N. W., and Costin, C. (1990b). The role of male ornaments and courtship behavior in female mate choice of red jungle fowl. *Am. Nat., 136,* 459–473.

Chapter 5

SEXUAL SELECTION AND THE EVOLUTION OF ADVERTISEMENT SIGNALS

Michael D. Greenfield

Department of Entomology
University of Kansas
Lawrence, Kansas 66045
email greenfie@kuhub.cc.ukans.edu

ABSTRACT

Evolution of the advertisement signals of male animals may be influenced by various sexual selection forces. Female choice mechanisms, male–male competition, and mate recognition may all initiate and maintain the signaling systems involved in sexual advertisement. The initiation of signaling systems may also be constrained phylogenetically; in particular, ancestral biases in perception may serve as preadaptations that select for specific signaling properties.

Whereas the various factors that purportedly influence signaling system evolution can be framed as operationally distinct hypotheses, testing and distinguishing these hypotheses empirically has proven difficult. Thus, it would not be prudent to propose at this time that any one factor has contributed more than another to signaling system evolution among animals. Signaling systems characterized by long-range acoustic or photic advertisements offer some advantages for empirical study. Advances in digital technology allow these signals to be characterized, simulated, modified, and played back. Moreover, by testing the responses of receivers toward simulated signals, the influences of male–male competition and female choice may be separated. Nonetheless, a survey of several acoustic and photic signaling systems in insects demonstrates that any attempt to unravel the many interdependent factors potentially influencing

Perspectives in Ethology, Volume 12
edited by Owings *et al.*, Plenum Press, New York, 1997

signaling system evolution must incorporate intensive genetic, environmental (field), and phylogenetic study in addition to basic behavioral assays.

INTRODUCTION

The evolution of sexual advertisement signals, first treated by Darwin (1871), remains one of the more intriguing and central problems in animal communication today. This is in spite of a recent flurry of research on sexual selection which has yielded a wealth of models and observational and experimental findings pertinent to signal evolution. In this chapter, I review these models and findings and assess whether we may expect any unifying principles on the evolution of advertisement signals among male animals to emerge.

Traditionally, the evolution of sexual advertisement signals has been attributed to mate (female) choice, intrasexual (male-male) competition, and mate (species) recognition. The source favored in the literature shifts according to prevailing paradigms in evolutionary thought, and current treatments suggest that certain female choice mechanisms have played the major role. For example, various compilations show that conspicuousness and exaggeration characterize male advertisement signals in many species (e.g., Ryan and Keddy-Hector 1992, Andersson 1994) and that females selectively favor males whose signals contain high levels of energy and/or power (see Ryan 1988). These generalizations might lead one to assume that advertisement signals serve as "reliable" indicators to females of a male's heritable viability or survivorship. However, another body of findings, obtained via comparative methods, suggests the rather different female choice mechanism in which signals that "exploited" preexisting sensory biases were selected (see Ryan and Keddy-Hector 1992). Moreover, these two mechanisms are not mutually exclusive or the only possibilities that may underlie the advertisement signals observed in many animals. This chapter shall emphasize that a multiplicity of explanations can yet be proffered for every signaling system and that some carefully focused work will be needed to reveal any unifying principles that may exist.

SIGNALING SYSTEM EVOLUTION VIA FEMALE CHOICE

By their very definition, advertisement signals are the means by which males exhibit their presence to females, and the evolution of these displays via female choice is therefore likely. According to the conventional scheme (Kirkpatrick and Ryan 1991), signal evolution via female choice may arise along either of two pathways: (1) choice of male signals based on associated material or somatic (nongenetic) benefits or (2) choice based on associated genetic factors. Whereas choice of male signals associated with a material or somatic benefit, such as the likelihood that paternal care will be provided, appears straightforward, explaining choice in which the female derives only genetic benefits—specific gametes—by virtue of attraction toward and mating with a particular male signaler has proven more elusive. Nonetheless, field observations on a wide variety of species suggest that the latter phenomenon is commonplace. It is perhaps most evident among vertebrates exhibiting special "lek" behaviors, in which females enter a traditional male congregation area where resources are notably absent, choose and copulate with a mate, and then quickly depart alone (Bradbury and Gibson 1983). However, evidence for female choice based on genetic factors may be found among many other species wherein pair formation is facilitated by long-range male advertisement signals, with some signals being more attractive or influential than others (see Kirkpatrick 1987). This more general scenario, exemplified by the mating activities of many arthropods, shall be our primary interest here.

Choice Based on Genetic Factors

Signal/Preference Covariance

Fisher (1915, 1958) first explained how female choice based on male genetic factors could operate via a process termed "runaway selection." Here, preference for "attractive" males occurs solely because the female's sons are likely to resemble their fathers and hence succeed in the "sexual selection arena" during the following generation. Unfortunately, the explanation was terse and solely verbal, and the proposed mechanism remained largely ignored.

At the beginning of the 1980s several theoreticians (e.g., O'Donald 1980, Lande 1981, Kirkpatrick 1982) revisited the problem of female choice based on genetic factors and applied formal modeling to the task. Their models essentially demonstrated that Fisher's proposed mechanism was possible. While details vary, the models begin with the assumption that both male features, e.g., advertisement signal characters, and female preferences for particular male features vary and are heritable. These conditions may lead to a genetic covariance in which males who produce the preferred signals will carry alleles for preference for such signals, which would be expressed later in their daughters. Likewise, females with strong preferences for certain male signals may carry alleles for those signals, which would be expressed in their sons. Via these mutual reinforcing processes, populations in which a majority of males exhibit extreme values of a signal character should come to include a preponderance of females who strongly prefer signals with those extreme values, and vice versa.

Arbitrary and Sensory Bias Models

Fisher's (1915, 1958) original conception of female choice based on genetic factors assumed that the process started due to an initial natural selection advantage accrued by a male bearing the preferred feature. Significantly, the models of Lande (1981) and Kirkpatrick (1982) showed that chance alone could initiate the process; e.g., random drift can lead to accumulation, within a given subpopulation, of alleles for strong preference for a particular feature. Henceforth, the process has sometimes been called the "arbitrary model." More recently, several authors (e.g., Basolo 1990, Ryan et al. 1990, Proctor 1991) have claimed that female preferences can also represent preexisting sensory biases that arose in contexts other than sexual selection, e.g., evading predators or locating food, and that male signals are simply those that best exploited such preferences. Signaling systems evolving via this "sensory bias model" would differ from those evolving via the arbitrary model in that signals and preferences would not coevolve at their inception. Rather, the signal would merely track the preference. In the context discussed above, mechanisms in both the arbitrary and sensory bias models refer only to how female choice based on genetic factors originated (see Christy 1995). Once initiated, however, the female choice process may be maintained by the same or different mechanisms. For example, signals that originated because of a preexisting sensory bias for such stimuli may then continue to undergo modification via the coevolutionary mechanism of the arbitrary model.

A critical stipulation in both models above is that the preferred male phenotypes remain independent of any natural selection advantage. Rather, males who produce preferred advertisement signals may actually suffer reduced viability in some cases because of the energy or conspicuousness associated with the advertisement. This stipulation implies that females who mate with preferred males would not be expected to produce more viable offspring, and the sole reward earned by these females would be aesthetically pleasing sons. These sons would be more attractive than the offspring of females mating with average males, though, and this sexual selection advantage accrued by them would offset any natural selection disadvantage suffered. If the sexual selection advantage provides exact compensation for the natural selection disadvantage, the population would be expected to assume evolutionary equilibrium; i.e., the signal and the form of the preference function for the signal remain unchanged.

Good Genes Model

Divergent from the arbitrary model, another body of theory has developed around the notion that female choice may be based on heritable viability of selected males, which would lead to enhanced viability in both male and female offspring of the discriminating female (see Kodric-Brown and Brown 1984).

This is often termed the "good genes model," and various mechanisms exist by which it could operate.

Several good genes mechanisms are variations of the "handicap principle" originally proposed by Zahavi (1975). According to this principle, attractive signals reduce—handicap—a male's viability but at the same time serve as a phenotypic indicator of his viability (Maynard Smith 1991). While initial modeling efforts suggested that the handicap principle was unlikely to work (e.g. Kirkpatrick 1986), recent amendments to the principle have shown that certain versions are indeed possible. A basic condition in these newer versions is that viability, the handicap feature, and female preference for the handicap feature are three separate, heritable traits (Pomiankowski 1988, Grafen 1990b). Additionally, the handicap trait must be "reliable" (see Zahavi 1993) in that (1) only males with high viability can fully exhibit the trait or (2) a male's viability can be accurately assessed only while he is exhibiting the trait.

Reliability of the first kind is termed a "conditional handicap," and it would be characterized by signals that are conditional upon a high level of viability (Zahavi 1977, Nur and Hasson 1984, Andersson 1986). For example, a male displaying a signal that is a conditional handicap may be handicapped by his expenditure of energy, but he would also be demonstrating that he had sufficient viability to acquire much energy, i.e., food, at an earlier juncture. Here, a male's expression of the signaling trait would be conditional upon bearing viability alleles as well as trait alleles; those males bearing only trait alleles would be prevented from fully exhibiting the trait because of inadequate development or a basic failure to acquire the necessary energy reserves.

Reliability of the second kind is termed a "revealing handicap," and it would be characterized by a signal that reveals a male's actual viability, which would otherwise remain concealed from females (Pomiankowski 1987). Here, a male bearing only trait alleles would develop the signal fully but would be unattractive to females, as would be any male bearing only viability alleles. In both kinds of handicaps reliability is a necessary component, as its absence would leave the mating system susceptible to "cheaters" — low-viability males who bear the handicap trait. The presence of cheaters would cause selection for female preference for the handicap to cease.

The genetic mechanisms on which the arbitrary and good genes models are based require a certain amount of additive genetic variance for viability, the signal, and the preference. Early during the development of modern sexual selection theory, recognition of this requirement led various authors to point out a serious difficulty: The additive genetic variance of traits subject to strong selection, as exerted by female choice, would probably be quite low (Williams 1975, Maynard Smith 1978). However, it has also been shown theoretically that factors such as recurrent mutation, migration, and cyclical selection pressures, none of which are implausible, would provide the prescribed variance for maintaining these genetic mechanisms (Charlesworth 1987, Heywood 1989). Recently, considerable attention has focused on the possibility that the last mentioned factor, cyclical selection pressures, can be generated by parasites (Hamilton and Zuk 1982). According to this proposal, signal quality reveals a male's level of infection by parasites, and resistance to infection is a heritable viability trait reliably indicated by the signal. If multiple genotypes exist both for resistance in the host and for virulence in the parasite, the host-parasite system may cycle regularly through the selection pressures imposed by the various genotypes.

Handicaps may also be nonheritable traits produced by random environmental factors (Dominey 1983). Assuming that when fully developed such handicaps reliably indicate heritable viability, female preference for them would be selected because male offspring inherit viability but not the handicap. Thus, the preference for certain signals can be selected even when the preferred signals themselves cannot.

Good genes mechanisms need not necessarily rely on signals that are handicaps, as any advertisement signal that reliably indicates the male's heritable viability may function. For example, a signal character that reliably increases with age may be more attractive in its exaggerated form because it could reflect an enhanced probability of survivorship (e.g., see Ryan 1980). That is, most males attaining an advanced age class may owe their achievement to high viability.

Other types of reliable signals may operate in good genes mechanisms by indicating heritable "developmental stability." Morphological symmetry may serve as such an indicator. Prevailing theory and some observations among various animals suggest that females may choose males whose bilateral symmetry is nearly perfect, because even slight asymmetrical deviations may reflect a heritable lack of developmental homeostasis in the presence of environmental stress (e.g., see Moller 1990a; but see Enquist and Arak 1993 and Johnstone 1993 who suggest that preference for symmetry may simply reflect a fundamental sensory bias). In the context of indirect female choice based on male signal characters, females might be attracted toward signals that reliably indicate a minimum amount of morphological asymmetry (Helversen and Helversen 1994; see Ryan et al. 1995 for a test yielding negative results).

Avoidance of Specific Genes

Owing to current fascination with the sensory bias and good genes models, several other genetic factors that potentially affect mate choice are often overlooked. These include avoidance of inbreeding (e.g., Ode et al. 1995) and of certain specialized forms of genetic incompatibility (e.g., see Gilburn et al. 1993, Gilburn and Day 1994). The latter may be immunologically based (see Zuk 1994). If selection against such pairings is sufficiently strong, females may evolve the ability to recognize the signals of those males that ought to be avoided; alternatively, females who mate promiscuously may employ postcopulatory choice of sperm to ensure that their eggs are properly fertilized (Zeh and Zeh 1994). Conceivably, selection may also lead to signal distinctiveness among males, as males who would fail to be recognized as closely related or genetically incompatible might inseminate inappropriate females. As outlined here, female preference would screen out a particular class of males (signals) rather than select in favor of a narrow category, which was implicit in the processes discussed previously.

Choice Based on Somatic Factors

Females may benefit materially in various ways through attraction toward and mating with a particular male based on his signaling, or at least by avoiding certain signalers. A male's signal characters might indicate presence of nutritional spermatophores, other nuptial gifts, or the likelihood that paternal care will be provided. And even in mating systems where such somatic factors are routinely absent, females may yet benefit materially in subtle ways through choice of specific signalers.

Ease of Signaler Localization

During the pair forming process, attracted females may be exposed to risks and energy demands that approach or even equal those to which signaling males are subject. These pressures, generally unappreciated in the past, may render the choice of and movement toward an easily located male critical (e.g., Hedrick and Dill 1993). Consequently, preferences for particular signals may reflect nothing more than energy constraints and the minimization of risk, and attractive signals may simply be those that are more easily localized (see Parker 1983, Andersson 1986). Further, males who generate high-amplitude signals that are broadcast over a wide area may encounter and mate with numerous females due to "passive attraction": These males would merely be detected by a greater sample of females. Production of a high-amplitude signal or one that is otherwise easily localizable may be a nonheritable trait that results solely from stochastic events, e.g., access to quality nutrition during early development. Under these circumstances, only choice based on somatic factors could occur.

Ejaculate Quality

Other potential somatic benefits that females may obtain via choice of specific males include minimizing the chance of acquiring a sexually transmitted disease or inadequate sperm (see Grafen 1990a). By not consorting with particular males, a female reduces not only her likelihood of choosing a mate whose heritable resistance to pathogenic infection is low but also of directly acquiring an infection herself (Borgia and Collis 1989, Moller 1990b). Because such infection could diminish current and future reproductive potential, the ability to recognize signal characteristics, e.g., low amplitude or irregular production, that reveal infection could be strongly selected in females. In turn, healthy males may be selected to exaggerate those signal characters that reliably indicate their lack of infection. Similarly, certain age classes, e.g., young or advanced categories, may be distinguished by a sperm supply insufficient to fertilize a female's eggs or by an excess amount of immotile or α nucleate sperm. Again, if such males can be recognized by their signals (Ritchie et al. 1995), females may be selected to screen them out.

In the examples given above, selection is expected to act simultaneously on somatic benefits and signals: The same factor that reduces a male's well-being will invariably affect his signaling as well. And while the state of well-being and the signal are not heritable traits themselves, the reliable association between the male's state and his signal is. This principle was invoked earlier in explaining how nonheritable handicaps may function.

Associated Resources

The general process linking signals with somatic benefits should also apply to situations in which territorial males signal in the vicinity of defended resources. Under such circumstances, which typify resource-defense polygyny systems, it is often presumed that females are choosing resources and not males per se. Nonetheless, regular access to high-quality resources may engender the production of intense, easily localizable male signals that offer passively attracted females reduced energy and risk during movement. A "shortcut," circumventing arduous trial-and-error localization, toward a valuable site may be offered as well, in which case females might actively monitor male signals for the telltale signs of acquired resources (see also Stamps 1987, Walker and Masaki 1989, and Shelly and Greenfield 1991 on using conspecific cues to assess habitat quality). Of course, the vigor and ability needed to acquire and successfully defend a prime location may be a heritable trait, in which case females who choose, either passively or actively, male signals that indicate an association with valuable resources may also receive genetic benefits. This possibility underscores the importance of anticipating that several mechanisms, e.g., choice based on both genetic and somatic factors, act jointly in the evolution of signals and preferences in a signaling system.

SIGNALING SYSTEM EVOLUTION VIA MALE-MALE COMPETITION

A male's success in competing with other males for female encounters may be influenced by the nature of his advertisement signals as well as by the way in which he adjusts his signals relative to those of his neighbors. This section shall examine the various ways in which male-male competition can effect the evolution of sexual advertisement signals and signaling systems as a whole. Close inspection reveals, however, that many such influences are interwoven with female choice.

Males as Signal Transmitters and Receivers

Because males may increase their fitness through monitoring the activities of neighboring males, they may be expected to perceive as well as transmit advertisement signals. Examination of signaling systems, particularly acoustic and visual ones, indicates that this prediction generally holds. In some species, the development of sensory organs and perceptual sensitivity in males even exceeds that in females. The general parity between female and male perception could reflect an underlying exploitative feature in the evolution of signaling systems: Basic sensory ability originates initially in the nonsexual contexts of avoiding natural enemies, finding food, assessing habitat, etc., and only later do male transmitters exploit the perception that is already common to both sexes. This evolutionary sequence is suspected in various groups of acoustic animals wherein hearing, found in both males and females, is believed to have preceded male advertisement songs (e.g., Otte 1977).

Adjustments in Signaler Location

Regular Spacing

Given sensory perception (and the coincidental ability to monitor the advertisement signals of neighbors), a male may benefit by moving away from a neighbor whose signal amplitudes are perceived to be excessive, as the high level would indicate a neighbor within a critical radius. Spatial adjustments of this sort would afford males exclusive signaling areas, and a regular spatial distribution among the population would result. Provided that males are equivalent in signaling prowess and females are initially distributed randomly within the preferred habitat, ideal free distribution theory predicts that regularly distributed males will achieve higher female encounter rates than males not spaced as such (e.g., Arak et al. 1990). This would occur because males who remain in exclusive signaling areas do not expend energy and time signaling close to calling neighbors, males with whom they must necessarily share the nearby females.

Aggregation

Under other circumstances, opposite spatial adjustments may be favored. If males do differ in signaling prowess, as probably occurs in most populations, inferior signalers may position themselves adjacent to superior ones and thereby benefit from the numbers of females that the latter attract (e.g., Arak 1988). In some cases, the inferior male may be thwarted from signaling by aggressive advances of the superior (dominant?) signaler. Such nonsignaling males are termed "satellites." It is reasoned that they achieve higher fitness through satellite behavior adjacent to an attractive signaler than they would by signaling weakly in a solitary location.

Signaler aggregation might also be favored when females evaluate available males more accurately by means of a simultaneous as opposed to a sequential comparison (Janetos 1980; but, see Real 1990 for another expectation). Here, females may only approach and mate with males found in aggregations, regardless of the signaling quality of solitary males, and all males, both inferior and superior signalers alike, would be forced to join aggregations. Such aggregations may be considered leks (Alexander 1975).

A similar outcome might occur when signaling males are associated with defended resources. Because a male's signal quality could be influenced by available food resources, males may be attracted toward a particularly powerful signaler. By using conspecific cues in this manner, the attracted male is afforded a shortcut to food. And if the site is sufficiently valuable, the benefit in resources may outweigh the drawback of frequent aggressive encounters with the other signaler(s). In turn, female preferences for signals that reveal the presence of several males may reinforce aggregation behavior. As discussed above, females attracted toward congregated signalers would be permitted a simultaneous comparison of available mates and, in this case, an easy path toward valuable resources as well. Thus, valuable sites may become "sinks" at which signaling males continue to accumulate until the per capita female encounter and mating rates level off and eventually decline (Greenfield 1997).

Adjustments in Signal Timing

Signal Competition and Precedence Effects

Female preferences may drive competing males to adjust the timing of their signals as well as their signaling locations (Greenfield 1994a). Among acoustic animals, the "precedence effect," a psychophysical phenomenon in which an animal is more sensitive to the first of several spatially separated stimuli occurring close in time (Wallach et al. 1949), may be an important influence in this regard. If females are strongly attracted to leading calls, timing mechanisms by which males can avoid following their neighbors' calls, and perhaps improve their chances of leading them too, would be selected. In species that call rhythmically, males may adopt an "inhibitory-resetting mechanism" for this purpose. Here, a male is inhibited from calling upon hearing a neighbor's signal and resets his call rhythm generator to the basal level. But he then releases his rhythm generator at the end of the neighbor's call and rebounds quickly from inhibition to call just prior to the neighbor's next anticipated call. Many cases of collectively synchronized and alternated calls in acoustic animals may result as incidental by-products of this mechanism (Greenfield and Roizen 1993, Greenfield 1994b).

Intergroup Signal Competition

Conceivably, intergroup competition among males may also drive collective signal interactions (Greenfield 1994a). In cases where females are influenced strongly by peak signal amplitude, a cluster of males—congregated for whatever reason—who cooperatively synchronize their signals may fare better than a cluster who do not. This should occur because the energy emanating from synchronized signals, integrated over a brief interval at the exact moment of synchrony, would exceed the energy of any isolated broadcast by a member of a nonsynchronizing cluster, thereby reaching females within a greater radius. And among rhythmic signalers, synchrony may also preserve a species-specific signaling rate that females must perceive before responding positively toward males in the locality; i.e., any male failing to synchronize would be committing an unexpected, spiteful act. In yet other cases, females may be attracted by a time-averaged value of signal amplitude, and a cluster of males who cooperatively alternate, rather than synchronize, their signals may fare better. This result could occur in communication involving light and sound, because the energy emanating from n synchronized acoustic and visual signals would be less than, or at most equal to, n times the energy of a single isolated signal (Bradbury 1981). Of course, once females arrive at a male group, competition between individual males would occur. The signal interactions expected to result from these individual level contests may or may not coincide with those selected by intergroup competition.

Interloping and Concealment

Opportunities for more elaborate competitive signal interactions exist in species where male/female dialogues are part of the pair-forming process (e.g., Otte and Smiley 1977, Buck 1988, Galliart and Shaw 1991). For example, a male may "interlope" by timing his signals so that they mask the replies of a female toward a rival male, thereby prolonging the availability of that female as a potential mate for himself. Or a male may detect and locate females by their replies toward his neighbor's signals.

Males who are subordinate in status may rely on other forms of "deceitful" signal interactions. A satellite male may time his signals such that they are short and occur entirely within the intervals during which the nearby dominant male's signals are generated. Thus, the satellite may signal, albeit weakly, in the vicinity of a strong signaler and yet remain undetected—and unmolested (Zelick 1986, Greenfield 1994a). His ability to conceal himself temporally behind the dominant male's signals would rely on perceptual impairment during signal generation (e.g., see Hedwig 1990, Narins 1992, Greenfield and Minckley 1993), a general feature in certain modes of communication such as sound and electricity.

Measurement of Aggressive Ability and Motivation

The influences of male-male competition on the evolution of advertisement signals are most obvious where measures of such signals, rather than specialized "graded aggressive signals," indicate aggressive ability and/or motivation to other males (e.g., see Schatral et al. 1984). In these situations, males generally evaluate their neighbor's advertisement signals and then adjust their own signaling accordingly. For example, a male may move away from signals that indicate superior aggressive ability, whereas indications of inferior ability may elicit an attack. As in the good genes model of female choice, reliability is expected to be an important component of the evaluation of neighbors' signals by males: Such evaluation is presumably carried out in lieu of actual fighting, which is probably fraught with time and energy costs and a risk of injury.

Where advertisement signals are evaluated by neighboring males as indications of strength and motivation, both intra- and intersexual selection are likely to influence signal evolution. These two influences probably act additively on signal characters, but not necessarily. The outcome should depend on whether female choice and male assessment are driven by comparable selection mechanisms leading to evaluation of similar features.

SIGNALING SYSTEM EVOLUTION VIA MATE RECOGNITION

When females respond and orient toward a male advertisement signal, they are expected at least to recognize the signal as a conspecific one. This has led to the view that mate recognition and sexual selection represent a unitary problem in pair formation (Ryan and Rand 1993). That is, attraction toward a heterospecific and toward an "inappropriate" conspecific could be functionally equivalent. Consequently, selection to reduce heterospecific attraction may have resulted in the evolution of certain interspecific differences among advertisement signals. Such evolved differences may be maintained by stabilizing selection exerted by female choice.

The above perspective notwithstanding, the actual influence of mate (species) recognition on signal evolution has been a controversial topic in animal communication. On the one hand, observations that the signals of sympatric, synchronic species usually differ in one or more characters were traditionally construed as evidence that selection for enhancing species recognition influenced signal evolution and diversification. However, various authors have stressed that interspecific signal divergence may have originated by chance or by natural or sexual selection in allopatric populations, without any reinforcement of interspecific differences once sympatry was regained (e.g., West-Eberhard 1984).

Several studies have attempted to distinguish the above possibilities by comparing signaling systems in different populations and screening for the presence of reproductive character displacement. Such displacement, in which a species' signal diverges from a particular heterospecific signal in areas where the two species are sympatric but not where the focal species occurs alone, could be construed as evidence for signal differences evolving under selection for enhanced mate recognition. Here, examining female preferences in addition to male signals should be critical: Owing to stronger selection acting on the female receiver than the male signaler, displacement in preferences may be more likely (e.g., Gerhardt 1994a). In some species this stronger selection pressure may be expressed as an aversion by mate-seeking females toward particular heterospecific signals, which supplements the females' specific preference for conspecific ones (Gwynne and Morris 1986).

Reproductive character displacement may also be detected by examining signal characters of the various species in a community that use the same signaling mode and then comparing the observed distribution of character values with that predicted by a random simulation (Otte 1989). Given the random simulation and specific ranges of character values in a community, a certain number of species pairs ought to have similar signals based on chance alone. Less frequent overlap between species than the random prediction may indicate that signal differences reflect, in part, selection for mate recognition.

Heterospecific signals not only represent potential sources of attraction but may also interfere with an animal's ability to recognize and localize a mate. Signal interference has been investigated most thoroughly in acoustic communication, and the usual responses to it appear to be short- and long-term shifts in signal timing (e.g., Greenfield 1988).

CASE STUDIES – ACOUSTIC AND PHOTIC SIGNALING SYSTEMS

In this section I describe four case studies illustrating different evolutionary factors that can influence advertisement signals and some of the approaches that can be used to discern which factors operate in a signaling system. Whereas these studies all provide some measure of explanation, each also reveals the rather formidable problems one may expect when attempting to identify these evolutionary factors. I have chosen to concentrate on studies of "long-range" acoustic and photic signaling systems in insects. As discussed below, many of these systems offer unusual opportunities to probe signal evolution.

Why Sound and Bioluminescence?

Pair-forming systems characterized by long-range male advertisements and female replies or orientation toward these advertisements are most appropriate for studies of advertisement signal evolution. This largely reflects the opportunities to examine female choice and male-male competition in isolation from each other and, in some cases, to do so via "playback" experiments (see Harvey and Bradbury 1991). In recent years the playback technique for investigating acoustic communication in particular has been greatly augmented by advances in digital technology that allow sounds to be physically described, edited and/or simulated, and played back with precision. Playback allows responses, either phonotaxis or acoustic replies to male advertisements, to be analyzed objectively over a wide range of values of each signal character. Moreover, the responses to each character of an advertisement signal can usually be studied independently of one another and of any other characteristics of the animal. These digital techniques have also been adapted for photic signals (Branham and Greenfield 1996), and bioluminescent species can now be studied in similar fashion.

Among taxa that rely heavily on acoustic communication, anurans and various acoustic insects are perhaps most suitable for playback experiments designed to bioassay female choice (see Gerhardt 1991, 1994b, Helversen and Helversen 1994). Because these groups are largely nocturnal, responses to advertisement sounds are modified little by visual cues, which may be difficult to control or simulate. Further, female phonotaxis toward males comprises the major part of mate choice and pair formation in many species.

Usually, phonotaxis is studied via choice tests in laboratory or field arenas, but in some acoustic insects devices that rely on locomotory compensation can be used (Bailey 1991). These devices retain a walking insect at a fixed location with respect to two stimuli (Kramer 1975). Thus, an "open loop" situation is maintained in which the insect's rate and direction of movement can be observed while the stimulus amplitudes that it perceives remain unchanged. The open loop situation provides more extensive and accurate measurements of phonotaxis and orientation, and it may be useful for assessing the strength of female preference.

Where female responses consist of reply signals to a signaling/searching male, less powerful single-stimulus (nonchoice) tests must be used (see Doherty 1985). Such techniques have been used successfully with both acoustic orthopterans (Helversen and Helversen 1994) and bioluminescent lampyrid beetles (Branham and Greenfield 1996).

Ultrasonic Signaling in Lesser Waxmoths: Reliable Viability Indicators?

Lesser waxmoths (*Achroia grisella*; Lepidoptera: Pyralidae) are highly unusual moths on several accounts. The larvae are obligate symbionts of honeybees, and pair formation in the adults is mediated by male ultrasonic signals that attract females up to several meters distant (Greenfield and Coffelt 1983, Spangler et al. 1984). These signals are generated by wing fanning and consist of pairs of approximately 60–130 μs pulses of 70–130 kHz sound. A brief (< 400 μs) silent interval normally separates the two pulses in a pair, and 80–120 pairs are produced per second (at 25°C). The males fan their wings at 40–60 complete cycles s^{-1} while remaining stationary on substrates near the entrances to honeybee colonies. During each upstroke and downstroke of the wings a pair of tymbals at the bases of the forewings are struck such that they resonate and emit a pair of ultrasonic pulses. The silent interval within a pair of pulses emitted at a given upstroke or downstroke occurs because resonations of the left and right tymbals, which represent the individual pulses, are slightly asynchronous.

Evolution of a Sex-Role Reversal and a Signaling Modality Switch

What factors could have led to evolution of the curious *A. grisella* signaling system? Ultrasonic advertisement most likely owes its occurrence to the prior existence of hearing, and specifically ultrasonic hearing, among moths (Spangler 1988). Ultrasonic hearing is a general feature in various moth families, including Pyralidae, and it probably evolved as a defense mechanism against insectivorous bats. When certain ecological circumstances, e.g., use of resources distributed in very limited patches, arose that selected for males to signal at predictable locations rather than to search—the male role in pair formation among the vast majority of moths—males may have adopted acoustic signaling rather than pheromones, the cues used by female signalers in typical moths (Greenfield 1981). Acoustic signals may be more easily localized by receivers, and while sound production probably demands more energy and is more attractive to natural enemies than pheromones, males are expected to expend more energy and take greater risks than females during sexual activities. These principles should apply to both attraction of females and male-male competitive interactions. The adoption of acoustic signaling would have been facilitated by the prior existence of (ultrasonic) hearing, and the specifically ultrasonic nature of the acoustic signals would be expected based on the small size of the animals. That is, sounds are emitted from a vibrating structure more efficiently by resonation than radiation, and the resonant frequency of a structure is inversely proportional to its size (*A. grisella* males are less than 1 cm in length, and their tymbals are approx. 0.3 mm in diameter) (see Ryan 1988). Whereas ultrasonic signals attenuate rapidly in the atmosphere, the aggregated distribution of *A. grisella* near honeybee colonies eliminates a need to communicate over great distances.

Female Preference Based on Signal Energy

How can we account for the specific attributes of ultrasonic advertisement in *A. grisella*? Signal character measurements (Jang et al. 1997), playback trials (Jang and Greenfield 1996), and energy measurements (Reinhold et al. 1997) provide some limited explanations. Coefficients of variation of pulse rate, pulse length, and silent interval among individual males are relatively high (> 10%; Fig. 1), suggesting that these are "dynamic characters" (sensu Gerhardt 1991) influenced by female choice other than mate recognition. This was confirmed by playback of digitized, edited male signals: *A. grisella* females preferentially oriented toward calls that were louder, delivered with higher pulse rates and pulse lengths, and included longer silent intervals within pairs of pulses than average calls in the male population (Fig. 1). These differences between the preference and the average signal indicate that female choice may exert strong directional selection on signal characters. Such directional selection is reported in many signaling systems relying on various communication modalities, and the selection consistently favors signals with higher, rather than lower, levels of energy or power than the average signal (Ryan and Keddy-Hector 1992, Andersson 1994). This tendency could imply that good genes mechanisms of female choice operate widely and lead to the evolution of extravagant advertisement signals in many species, including *A. grisella*.

However, finding that female preferences may exert directional selection is, by itself, insufficient evidence for rejecting several other evolutionary factors. For example, operating under the influence of chance alone, the arbitrary model of female choice could lead to preference for extravagant signals. Here, the degree of male signal elaboration might be held at an equilibrium value, well below that preferred by females, by phonotactic predators. Females could also derive a somatic benefit, reduced energy expenditure and risk while en route to a mate, by choosing males whose signals contain the higher energy and power features.

Measurements of the energy expended by signaling *A. grisella* males add some support for the good genes model. Respirometric apparatus that simultaneously recorded a male's VCO_2 and his song demonstrated that males who produced ultrasonic calls with attractive features (and who had earlier demonstrated their attractiveness to females in choice tests) always expended more energy than males producing nonattractive calls (Reinhold et al. 1997). But energy expenditure alone was insufficient to ensure attractiveness, as many nonattractive callers also expended high amounts of energy. This may reflect the contributions to overall attractiveness by the various signal characters, some of which (e.g., pulse rate) were correlated with energy expenditure and some of which (e.g., silent interval length) were not. However, signal characters that are uncorrelated with current energy expenditure may nonetheless indicate energy acquired and allocated during an earlier phase of development. Thus, studies of

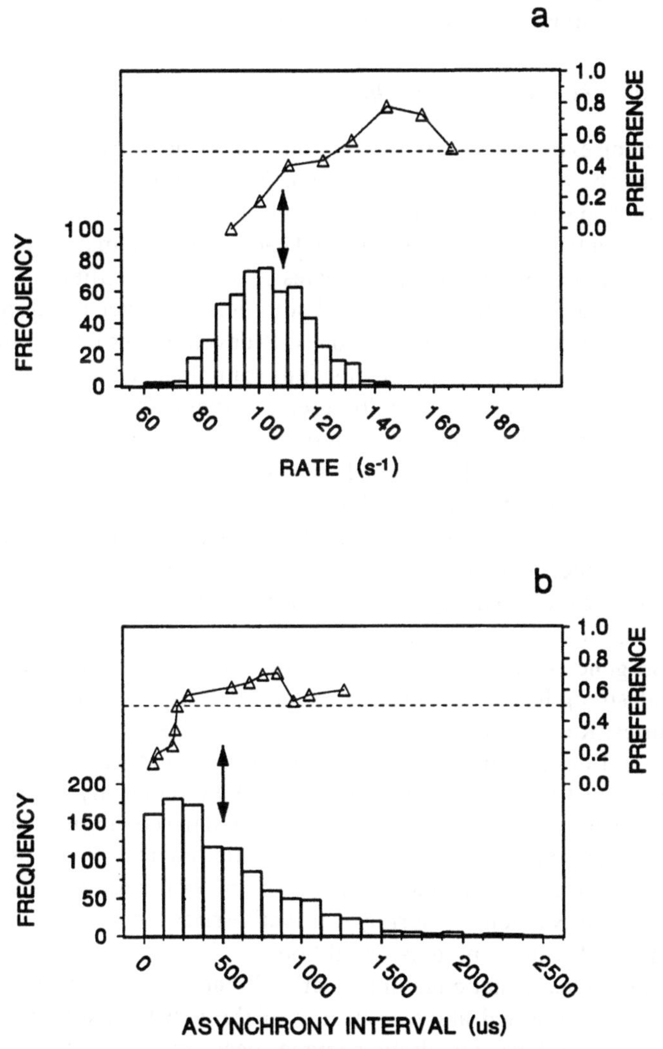

Fig. 1. (a) Frequency distribution of (double) pulse rates in a population of male *Achroia grisella* (Lepidoptera: Pyralidae) and female preference function for signals with various rates. Preference is the proportion of females that select, in a two-choice test, a signal broadcast with the given rate over one with the mean rate in the population (shown by arrow). (b) Frequency distribution of asynchrony intervals (time from onset of first pulse to onset of second pulse within a double pulse) in a population of male *A. grisella* and female preference function for signals with various asynchrony intervals. Preference is the proportion of females that select, in a two-choice test, a signal broadcast with the given asynchrony interval over one with the mean interval in the population (shown by arrow). Adapted from Jang and Greenfield (1996).

signaling energetics may clarify some points, but they cannot resolve the critical questions regarding signal evolution and female choice based on genetic factors.

Female Preference Yields Viable Offspring

Breeding experiments can advance our understanding in some systems and may determine whether a good genes mechanism underlies, at least in part, female choice and signal evolution. In *A. grisella*, attractive and nonattractive males were identified via choice tests, and each identified male was mated with three different randomly selected females (Jia and Greenfield 1997). Those offspring sired by attractive fathers had significantly faster development rates than offspring sired by nonattractive ones. If development rate confers greater fitness in some natural populations, this finding implies that females who choose and mate with attractive callers enjoy greater fitness. This does not negate the possibility that environmental factors too influence signal development and the benefits—somatic ones—that a female would gain by mating with an attractive caller, but it does show that female choice leads to genetic benefits. The presence of such benefits and their correlation with certain assessed features (pulse rate, gap length, loudness) of advertisement signals in *A. grisella* indicate that elaboration of these features may have evolved because they reliably indicate a male's heritable viability.

Good genes mechanisms are based on additive genetic variance for the male signal, strength of female preference, and viability, and covariance between these features. Recent breeding analyses of *A. grisella* showed that male signal characters and developmental indices are heritable and genetically correlated with each other (R.D. Collins, unpublished data). Further analyses are under way to measure the heritability of female preference and the covariance between male signal characters and female preference. Potential interactions between signaling, viability, and environmental quality are being examined as well. For example, higher viability among offspring of attractive sires seems to occur only under certain (favorable) environmental conditions.

Assuming that genetic factors underlie female choice and signal evolution in *A. grisella* as implied above, can any of the actual factors be identified? Polymorphisms for basic metabolic enzymes occur in Lepidoptera, and in some systems the different morphs experience varying levels of mating success (e.g., Watt et al. 1986). This observation and the suggestion that developmental rate is the key aspect of heritable viability in *A. grisella* recommend that loci controlling such metabolic enzymes be initial subjects for examination.

Are Generalizations Warranted?

Because directional selection imposed by female preference for male signals containing higher than average energy or power has been noted in many systems, one may be tempted to infer that signal reliability and good genes mechanisms are widespread. However, several considerations caution against dismissing alternative explanations, e.g., the arbitrary model, at this time. Whereas the chance process operating in the arbitrary model of female choice would generate an equal number of preferences for high and low signal energy among species, the observed bias toward high values could be spurious. That is, while evolving toward low values via the arbitrary model, signal characters might simply disappear without a trace (Ryan and Keddy-Hector 1992). Additionally, aesthetic sensibilities may lead biologists to study those systems containing the most extravagant signals.

While many studies do report female preferences for exaggerated values of a signal character, very few studies examine enough characters to confirm that overall signal energy and/or power are favored. Moreover, signal energy is not metabolic energy expended during signaling—presumably, but not necessarily, a currency with which viability could be assessed—and even fewer studies measure this latter energy quantity (Prestwich 1994).

Thus, generalizing on good genes mechanisms in signal evolution would be premature, and concentrating on these mechanisms may divert attention away from other evolutionary factors that should not be overlooked. For example, in *A. grisella* males fight over opportunities to occupy advantageous locations for encountering females, and they do so while generating their ultrasonic advertisement signals (Cremer and Greenfield 1997). They must also signal in the presence of heterospecific sounds, e.g., those of other ultrasonic moths and of honeybees, that may interfere with hearing by conspecifics. Presently, the influences of such male-male competition and potential heterospecific interference on signal evolution in *A. grisella* remain unknown.

Photic Signaling in Fireflies: Do Conservative Elements of Neural Design Influence Signaling Systems?

Pair formation in most species of fireflies (Coleoptera: Lampyridae) is effected by searching, signaling males that approach stationary females who reply to the male signals (Case 1984). As in (conventional) moths, ecological factors may select for this pair-forming protocol rather than the "textbook model" in which males signal and females search and/or are attracted; e.g., crickets, frogs, and peacocks. Nonetheless, signals in firefly systems could be influenced by the same evolutionary factors suggested to operate in the so-called textbook systems. For example, females may visibly reply only to males whose signals contain high photic energy because such signals are reliable viability indicators or are associated with somatic benefits. These expectations were investigated for the first time in recent studies of the Nearctic firefly *Photinus consimilis*.

Female Preference Determined by Photic Playback Experiment

Male *P. consimilis* advertise while flying slowly, 1–3 m above the ground (Branham and Greenfield 1996). Their signals are "trains" of 2–7 flashes delivered at 2.5–4.0 trains min^{-1}. Each flash in a train is approx. 70 ms long, and the flash rate varies from 2.5 to 3.7 s^{-1} among individuals (at 23°C)(Fig. 2). Receptive females reply with a train of dim flashes beginning 6–10 s following the end of the male's flash train.

Female replies were tested using a novel photic playback experiment in which simulated flashes were presented via a light-emitting diode (LED) driven by digitized signals stored on a digital tape recorder (DAT)(Branham and Greenfield 1996). These tests showed that females preferred male signals whose flash rates were significantly higher than the mean rate in the male population but whose flash lengths approximated the mean (Fig. 2). Thus, female choice imposes both directional and stabilizing selection on male signals, and females do not choose signals by simply integrating photic energy. This finding emphasizes the need to measure various signal characters before reaching conclusions that female preference may be based solely on signal energy.

Which evolutionary factors influence the *P. consimilis* signaling system? Although females do not prefer signals containing maximum photic energy, a good genes mechanism cannot be ruled out and is actually quite likely. That is, the rapid rhythm characterizing preferred signals may demand maximum energy that only highly viable males can expend. Alternatively, an arbitrary mechanism may operate. Here, an equilibrium level of male signal elaboration maintained by predation pressure from phototactic aerial predators, i.e., *Photuris* fireflies (see Lloyd and Wing 1983), is conceivable. Unfortunately, breeding experiments on fireflies cannot be realistically contemplated, and distinguishing these possibilities is not foreseeable.

Evolutionary Conservatism?

The combination of directional selection on flash rate and stabilizing selection on flash length indicates the presence of a perceptual filter yielded by phasic neural responses to flash onsets, followed shortly thereafter by adaptation and inhibition. A filter such as this may simply represent the vehicle through which genetic or somatic benefits are chosen and heterospecific signals are avoided. However, some remarkably similar cases of female choice for rapid rhythms are observed among diverse species pulsing acoustic and visual signals. These observations raise the possibility that evolutionarily conservative mechanisms in neural design (Dumont and Robertson 1986; see also Ryan 1994) influence the evolution of signaling systems; e.g., given equal signal (stimulus) power, neural excitation may be achieved more readily by repeated phasic responses to pulsed stimuli than by a prolonged tonic response to a continuous

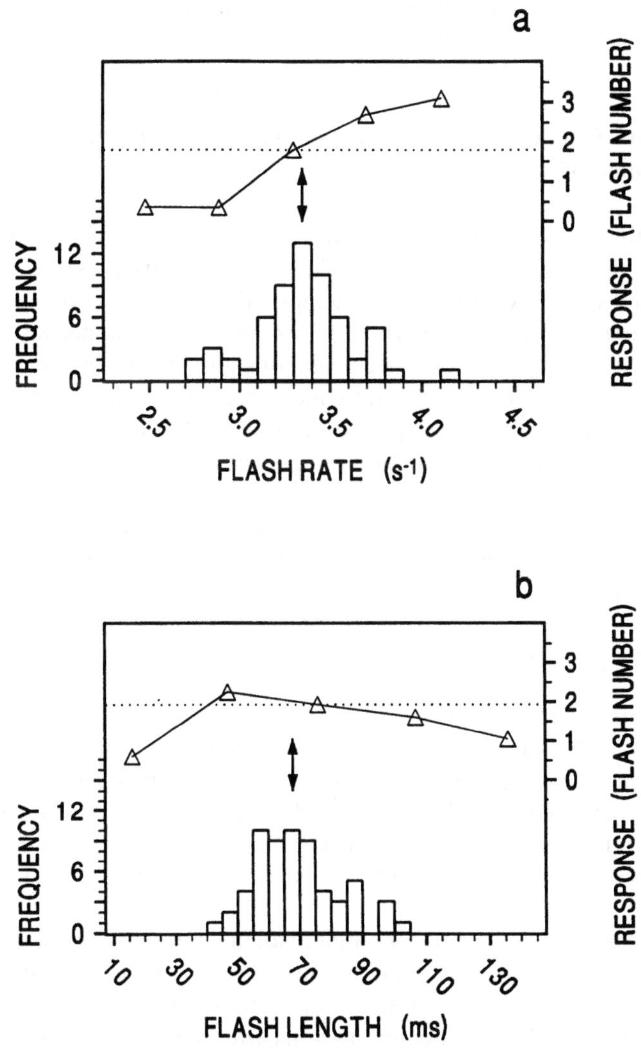

Fig. 2. (a) Frequency distribution of flash rates in a population of male *Photinus consimilis* (Coleoptera: Lampyridae) and female response function for simulated signals with various flash rates. Response is the proportion of tested females that flashed in reply to flash trains comprised of flashes delivered at the given rate. Arrow indicates mean rate in the male population. (b) Frequency distribution of flash lengths in a population of male *P. consimilis* and female response function for simulated signals with various flash lengths. Response level is the proportion of tested females that flashed in reply to flash trains comprised of flashes of the given length. Flash rate was held constant (3.3 s^{-1}) in all tests. Arrow indicates mean length in the male population. Adapted from Branham and Greenfield (1996).

stimulus. Neuroethological investigation and comparative phylogenetic analysis may help evaluate the potential contribution of this factor to signal evolution.

Synchronous Chorusing in Coneheaded Katydids: Adaptation or Epiphenomenon?

Signaling features include interactions between neighboring signalers in addition to the various measures of the signals themselves, i.e., amplitude, temporal, and spectral characters. The temporally structured choruses of anurans and orthopteran insects and the synchronous bioluminescent displays of *Pteroptyx* fireflies are among the more spectacular of signaling features that are defined by collective interaction rather than by measurement of an individual's signals. Are such collective interactions per se adaptive or do they merely represent an emergent property of the signaling strategies of individuals? Investigations of synchronous chorusing in the coneheaded katydid *Neoconocephalus spiza* (Orthoptera: Tettigoniidae) show that the latter is possible. These investigations also reveal additional ways in which conservative neural design might influence signal evolution.

Inhibitory Resetting Mechanism

The advertisement signals of male *N. spiza* are rhythmically produced "chirps" (Greenfield and Roizen 1993). Each chirp is approximately 50 ms long and may be heard by males and females up to 30 m distant. Chirp rates vary from 1.8 to 4.1 s^{-1}, but a solo individual usually maintains a given rate for several minutes. Neighboring males always synchronize their chirps, although all males occasionally drop out of the chorus for one or more chirp periods and then reenter in phase.

Playback experiments using single, isolated stimulus chirps showed that signal interactions between neighboring males were regulated by an inhibitory-resetting mechanism (Fig. 3) (Greenfield and Roizen 1993). When a male perceives a neighbor's chirp beginning *d* ms after the onset of his last chirp, he increases his concurrent chirp period by a delay interval slightly less than *d* ms. If both males have comparable chirp rates, synchrony arises by default. Otherwise, the faster individual does most of the signaling, because the call rhythm generator of the slower one is continually inhibited to the basal level before his chirps are triggered.

Additional playback experiments showed that a strong precedence effect influences female preference in *N. spiza*. As in many signaling systems, female *N. spiza* prefer male signals that are loud and longer than average. These energy-based signal characters are ignored, however, when the onset of one chirp precedes another by 10–75 ms: Females will choose the initial chirp even if

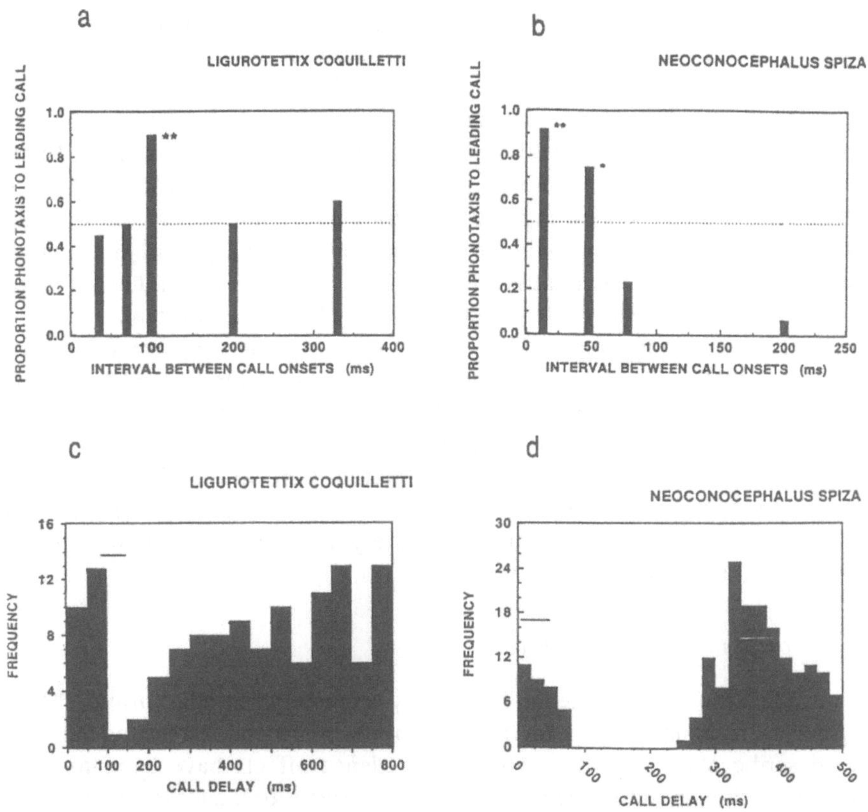

Fig. 3. (a and b) Precedence effects in *Ligurotettix coquilletti* (Orthoptera: Acrididae) and *Neoconocephalus spiza* (Orthoptera: Tettigoniidae). Bars show proportions of females orienting toward the earlier of two identical stimulus calls whose onsets were separated by the given delay interval. (c and d) Inhibitory-resetting mechanisms in the rhythmic calling of male *L. coquilletti* and *N. spiza*. Histograms show the frequency of calls produced at given delay intervals following the onset of a stimulus call. The delay intervals during which *L. coquilletti* males are inhibited from calling coincide with the precedence effect in female perception; thus, males do not produce calls that ineffectively follow a neighbor's calls. The delay intervals during which *N. spiza* males are inhibited from calling occur after the precedence effect in female perception; thus, males do not avoid production of calls that ineffectively follow a neighbor's concurrent call, but they do increase their chance of producing effective leading calls during the subsequent call period.

shorter and up to 4 dB lower in perceived amplitude (Fig. 3). Consequently, male *N. spiza* are under considerable selection pressure to avoid following a neighbor's call and to precede neighbors' calls by a brief interval; these are not identical objectives. A Monte Carlo simulation model showed that males would achieve the latter objective via an inhibitory-resetting mechanism provided that the mechanism includes a "relativity adjustment" for the velocity of signal transmission and "selective attention" toward only a subset of neighbors (Greenfield et al. 1997). The mechanism in *N. spiza* incorporates both predicted features.

Thus, synchronous chorusing in *N. spiza* may be an epiphenomenon emerging from individual males competing to jam each other's signals. Explanations relying on the adaptiveness of synchrony per se in the context of intergroup competition (intragroup cooperation) do not appear valid in this system. The males do not maintain a species-specific chirp rhythm at a given temperature, and they do not assemble in tight clusters that could benefit from maximizing signal energy integrated over a very brief time interval.

The epiphenomenon model relies solely on a precedence effect (and selective attention) in perception and a resettable oscillator controlling a rhythmic signal. Given these features, an inhibitory-resetting mechanism would be evolutionarily stable and may be selected. This model may explain chorusing phenomena in many acoustic insects and anurans (Greenfield 1994 a,b). Precedence effects and inhibitory-resetting are found in various species, whereas evidence that firmly supports explanations relying on intergroup competition is not reported in a single empirical study.

Evolution of the Precedence Effect

While the epiphenomenon model may accurately describe various chorusing events, it cannot account for the origin of the precedence effect and thus does not describe chorus evolution fully. Precedence effects have been found in mammals (though not in the context of sexual selection), anurans, and insects, and such diverse occurrence may reflect multiple origins and adaptive bases. Possibly, precedence effects prevent animals from responding to echoes of signals (see Wyttenbach and Hoy 1993). However, in some acoustic insects precedence effects last for 2–3 s following onset of an initial signal, and it is difficult to imagine any interfering echoes after a delay of that length. Alternatively, the precedence effect may represent another consequence of evolutionarily conservative design in the neural mechanisms, e.g., forward masking (see Sobel and Tank 1994), that recognize and respond to pulsed stimuli.

Do precedence effects confer genetic and/or somatic benefits to a female? As noted above, the initial signal in a sequence may be more easily recognized by neural mechanisms and thus precedence effects may be facilitating pair formation. However, precedence effects appear to differ fundamentally from all female preferences discussed previously: preferred signals are not necessarily distinguished by any physical characters, and it is not immediately clear how a chosen male would differ from an avoided one.

This perplexity dissipates somewhat when the competitive situations in which males often signal are considered. Females may be selected to choose pulsed signals based on their rate, but among a dense aggregation of signalers such choice may be perceptually impossible. If a precedence effect is already regulating female response, though, and males are thereby selected to exhibit inhibitory-resetting, females should be able to assess the relative signaling rates of local males readily. The faster males would produce most of the (effective) calls, while the slower ones would call very little owing to a cruel bind: Ignoring the rules of inhibitory-resetting would lead to production of ineffective following calls, while adhering to the rules would lead to repeated inhibition. Thus, the precedence effect would inadvertently magnify small differences in an energy-based signal character for which female preference is expected. This shows how a trait that may be selectively neutral at its inception in a signaling system can later acquire considerable adaptiveness through a feedback loop, ensuring its maintenance.

Acoustic Signaling in Territorial Grasshoppers: Advertisement of Resource and Male Quality?

Advertisement signals in resource-based mating systems can pose problems as intractable as any described in the previous cases. These problems arise because both resource and male quality can influence signaling, and potential feedback loops between resource and male quality can obscure distinction between these two influences. Investigations of the mating systems of *Ligurotettix* grasshoppers (Orthoptera: Acrididae), territorial species found on desert shrubs in southwestern North America, illustrate these difficulties well.

Signals and Settlement Patterns

Ligurotettix coquilletti (the "desert clicker") and *L. planum* primarily feed and reside on *Larrea tridentata* and *Flourensia cernua* shrubs, respectively (Greenfield 1990). In both systems specific antiherbivore properties of the foliage vary considerably such that most shrubs are avoided and only a few are preferred as food. Males are aggressive and defend preferred shrubs as mating territories. Because the relative abundance of preferred shrubs is low, several males may co-occupy a given shrub despite mutual aggression.

Females show less site fidelity than males and regularly move among the host shrubs. Playback experiments showed that their arrival at a shrub is largely influenced by acoustic signaling (clicking) of the resident male(s), with loudness representing the most influential signal character (Shelly et al. 1987, Shelly and Greenfield 1991). However, their subsequent long-term settlement is influenced by foliage quality, with the resident male(s) and its acoustic and visual signaling representing uncertain reinforcement.

Ligurotettix populations are markedly protandrous. Those males who eclose to adults earliest during the summer usually occupy the most valuable shrubs (Wang et al. 1990). Males maturing later either occupy shrubs of lower quality or accumulate on the valuable, previously occupied shrubs, where they may assume subordinate roles. Acoustic signaling both repels and attracts males to a shrub. That is, the signaling indicates that a site is already occupied as a defended territory, but it also indicates that a site is a valuable one. A late-eclosing male faced with the alternatives of occupying a low-quality site versus joining other males on a valuable one may respond to signaling's latter indication (Muller 1995).

Shortcuts and Feedback Loops

Numerous factors may influence evolution of the *Ligurotettix* signaling system (Greenfield 1997). Loud acoustic signals facilitate pair formation where the ratio of host shrubs to males is high. Thus, females who are attracted, either passively or actively, toward loud, regular male calls gain the somatic benefit of reduced risk and energy expenditure when orienting toward a mate. Because males seldom call from low-quality shrubs, females also gain a shortcut to a valuable resource patch. *Ligurotettix* cannot assess shrub quality from a distance (Greenfield et al. 1989), and females who orient toward loud, regular calls can reach a valuable shrub without an arduous trial-and-error search. Consequently, males would be under strong selection pressure to call loudly and regularly and to do so at high-quality shrubs.

Despite the expectation of vigorous calling, *Ligurotettix* males vary greatly in call amplitude and rate. Such variation could reflect underlying genetic and nutritional differences. Genetic factors associated with vigorous calling may also be associated with the vigor necessary to occupy a valuable shrub early and acquire superior nutrition, and so a feedback loop may connect genetic and nutritional influences on signaling. Thus, a female may gain genetic as well as somatic benefits via attraction toward loud, regular calling, which would be a reliable indication of heritable viability. Moreover, the feedback loop would accentuate any heritable differences in vigor and calling and thereby increase a female's opportunity to gain whatever genetic benefits may exist. This proposed scenario (Fig. 4)—entirely speculative regarding *Ligurotettix*—implies that good genes mechanisms may be anticipated in resource-defense mating systems wherein signal quality is influenced strongly by resource quality.

EPILOGUE

Have recent investigations opened a Pandora's box of unsolvable problems found by Darwin in 1871? This chapter may leave the impression that signaling system evolution is unlikely to be understood by present means and that each

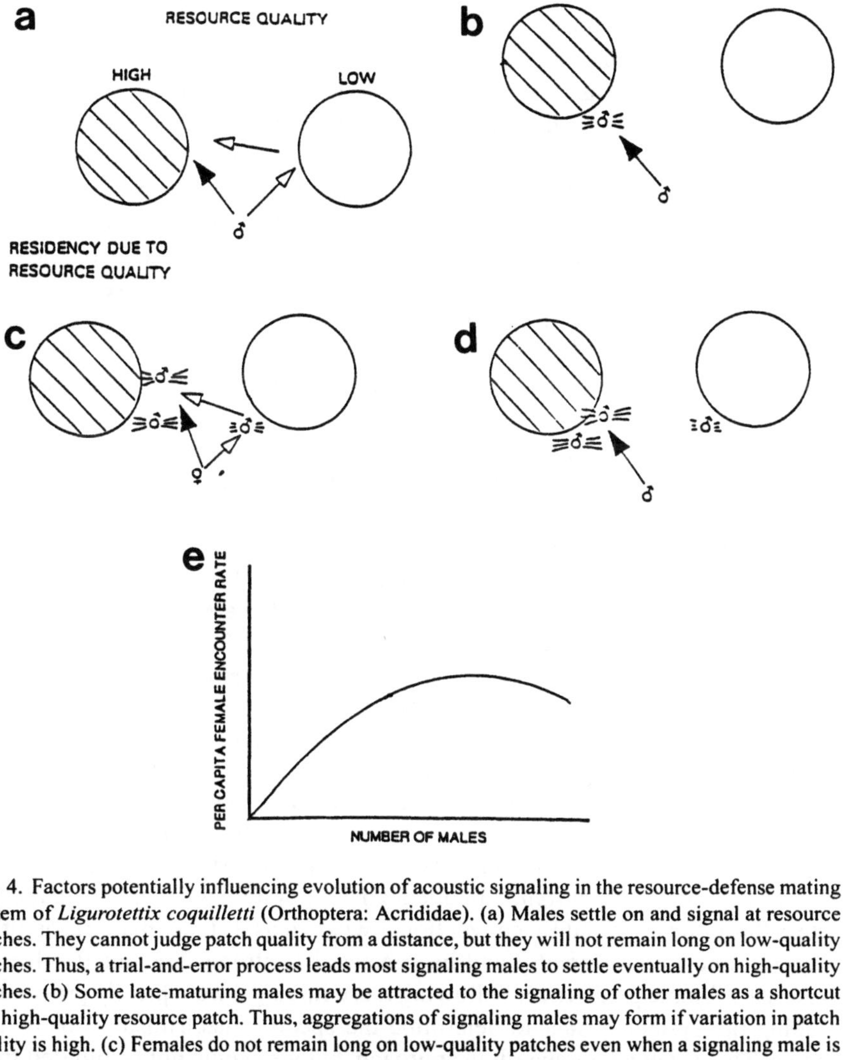

Fig. 4. Factors potentially influencing evolution of acoustic signaling in the resource-defense mating system of *Ligurotettix coquilletti* (Orthoptera: Acrididae). (a) Males settle on and signal at resource patches. They cannot judge patch quality from a distance, but they will not remain long on low-quality patches. Thus, a trial-and-error process leads most signaling males to settle eventually on high-quality patches. (b) Some late-maturing males may be attracted to the signaling of other males as a shortcut to a high-quality resource patch. Thus, aggregations of signaling males may form if variation in patch quality is high. (c) Females do not remain long on low-quality patches even when a signaling male is present. Females may be attracted to the signaling of males on high-quality patches because resource quality may influence signal characters. Additionally, females may be attracted to the signaling of male aggregations per se. Thus, females may be afforded a simultaneous comparison of available mates, a shortcut to a high-quality resource patch, and any genetic factors associated with a male's ability to locate and remain on a high-quality patch in the company of other males. (d) Additional feedback loops may lead males to orient toward aggregated signalers per se. Thus, males may be afforded a particularly effective shortcut to a high-quality resource patch. Females are expected to gain from orientation toward aggregations of signaling males, and this factor should enhance the benefits males acquire from joining a group of signalers. (e) Consequently, signaling males may continue to accumulate at high-quality resource patches until the per capita female encounter rate ceases to increase.

advance invariably reveals previously unrecognized problems. Full explanations for any given system and unifying principles accounting for the general evolution of advertisement signals appear equally unattainable.

This grim forecast may partly reflect the systems that have been chosen for study and the approaches used. Only by studying species in which breeding experiments and genetic analyses are possible and then implementing these techniques can we expect to identify factors in the evolution of a signaling system definitively. Unfortunately, many species with the most interesting behaviors and most extravagant signals—and which have attracted the most attention—cannot be approached with such techniques. But we should not overlook the possibility that some of these textbook systems, e.g., peacock plumage displays (see Petrie 1994), may be more amenable to breeding and genetic analyses than imagined and might yield significant information if correctly probed.

On the other hand, we should not expect (quantitative) genetic analyses alone to resolve all central problems in signal evolution. For example, finding support for a good genes mechanism in the *Achroia grisella* ultrasonic signaling system was extremely fortuitous, as lack of support could only have been interpreted as inconclusive: The possibility that viability differences exist between offspring of attractive and nonattractive sires but are only expressed under specific natural conditions could not have been rejected. Thus, major efforts must be made to study signaling systems in the context of natural environments, and in the full range of environmental variation.

Additionally, the ability of females to gain somatic benefits by orienting toward a more localizable male signal must not be forgotten. This is a most difficult problem, which can only be evaluated under natural conditions.

Recent work shows that female choice is undoubtedly a significant factor in the evolution of many signaling systems. This is particularly clear in cases such as the *Photinus consimilis* photic signaling system wherein a signal character (flash rate), viewed by conventional wisdom as a static, mate recognition cue, is found to vary greatly and be subjected to strong directional selection exerted by female preference. Following these findings, we typically proceed by probing the system further to determine the specific mechanism(s) of female choice. This is not an unrewarding approach, but we should not be blinded from the possibility that male-male competition, mate recognition, and ecological and phylogenetic factors may also influence signal evolution.

Because phylogenetic factors such as evolutionarily conservative neural mechanisms may be important influences on signal evolution, a greater diversity of species must be investigated than has been traditional. Taxa selected for investigation should be carefully distributed to allow testing of specific hypotheses; e.g., do precedence effects, and hence chorusing, reflect phylogenetic affinity? This approach has been used effectively in several tests of the sensory bias mechanism of female choice, and it should be applied more widely.

Presently, we are unable to account fully for the evolution of any one signaling system or to identify unifying principles pertaining to signal evolution among animals. Moreover, our investigations reveal that the various evolutionary factors potentially influencing signaling systems are more numerous and complex than previously considered. Nonetheless, these challenging problems in signal evolution need not remain unraveled provided that the above points and caveats are kept in mind.

ACKNOWLEDGMENTS

Final preparation of this chapter benefited greatly from critical reviews by Don Owings, Klaus Reinhold, Andy Snedden, and Nick Thompson.

REFERENCES

Alexander, R. D. (1975). Natural selection and specialized chorusing behavior in acoustical insects. In D. Pimentel (ed.), *Insects, Science, and Society* (pp. 35–77). New York: Academic Press.

Andersson, M. (1986). Evolution of condition-dependent sexual ornaments and mating preferences: Sexual selection based on viability differences. *Evolution, 40,* 804–816.

Andersson, M. (1994). *Sexual Selection*. Princeton, NJ: Princeton University Press.

Arak, A. (1988). Callers and satellites in the natterjack toad: Evolutionarily stable decision rules. *Animal Behaviour, 36,* 416–432.

Arak, A., T. Eiríksson, and T. Radesäter. (1990). The adaptive significance of acoustic spacing in male bushcrickets *Tettigonia viridissima*: A perturbation experiment. *Behavioral Ecology & Sociobiology, 26,* 1–7.

Bailey, W. J. (1991). *Acoustic Behaviour of Arthropods*. London, England: Chapman & Hall.

Basolo, A. (1990). Female preference predates the evolution of the sword in swordtails. *Science, 250,* 808–810.

Borgia, G. and K. Collis. (1989). Parasites and bright male plumage in the satin bowerbird (*Ptilonorhynchus violaceus*). *American Zoologist, 30,* 279–285.

Bradbury, J. W. (1981). The evolution of leks. In R. D. Alexander and D. W. Tinkle (eds.), *Natural Selection and Social Behavior: Recent Research and New Theory* (pp. 138–169). New York: Chiron Press.

Bradbury, J. W. and R. M. Gibson. (1983). Leks and mate choice. In P. Bateson (ed.), *Mate Choice* (pp.109–138). Cambridge: Cambridge University Press.

Branham, M. A. and M. D. Greenfield. (1996). Flashing males win mate success. *Nature, 381,* 745–746.

Buck, J. (1988). Synchronous rhythmic flashing in fireflies. II. *Quarterly Review of Biology, 63,* 265–289.

Case, J. F. (1984). Vision in mating behaviour of fireflies. In T. Lewis (ed.), *Insect Communication* (pp. 195–222). New York: Academic Press.

Charlesworth, B. (1987). The heritability of fitness. In J. W. Bradbury and M. Andersson (eds.), *Sexual Selection: Testing the Alternatives* (pp. 21–40). New York: Wiley.

Christy, J. H. (1995). Mimicry, mate choice, and the sensory trap hypothesis. *American Naturalist, 146,* 171–181.

Cremer, S. and M. D. Greenfield. (1997). Partitioning the components of sexual selection: Attractiveness and competitive behavior in male wax moths. *Ethology 103*, in press.

Darwin, C. (1871). *The Descent of Man and Selection in Relation to Sex*. London, England: John Murray.

Doherty, J. A. (1985). Phonotaxis in the cricket, *Gryllus bimaculatus* DeGeer: Comparisons of choice and no-choice paradigms. *Journal of Comparative Physiology A, 157,* 279–289.

Dominey, W. (1983). Sexual selection, additive genetic variance and the "phenotypic handicap." *Journal of Theoretical Biology., 101,* 495–502.

Dumont, J. P. C. and R. M. Robertson. (1986). Neuronal circuits: An evolutionary perspective. *Science, 233,* 849–853.

Enquist, M. and A. Arak. (1993). Symmetry, beauty and evolution. *Nature, 372,* 169–172.

Fisher, R. A. (1915). The evolution of sexual preference. *Eugenics Review, 7,* 184–192.

Fisher, R. A. (1958). *The Genetical Theory of Natural Selection* (2nd ed.) New York: Dover.

Galliart, P. L. and K. C. Shaw. (1991). Effect of intermale distance and female presence on the nature of chorusing by paired *Amblycorypha parvipennis* (Orthoptera: Tettigoniidae) males. *Florida Entomologist, 74,* 559–568.

Gerhardt, H. C. (1991). Female mate·choice in treefrogs: Static and dynamic criteria. *Animal Behaviour, 42,* 615–635.

Gerhardt, H. C. (1994a). Reproductive character displacement on female mate choice in the grey treefrog *Hyla chrysoscelis. Animal Behaviour, 47,* 959–969.

Gerhardt, H. C. (1994b). The evolution of vocalization in frogs and toads. *Annual Review of Ecology & Systematics, 25,* 293–324.

Gilburn, A. S. and T. H. Day. (1994). Evolution of female choice in seaweed flies: Fisherian and good genes mechanisms operate in different populations. *Proceedings of the Royal Society of London B, 255,* 159–165.

Gilburn, A. S., S. P. Foster, and T. H. Day. (1993). Genetic correlation between a female mating preference and the male preferred character in the seaweed fly, *Coelopa frigida. Evolution, 47,* 1788–1795.

Grafen, A. (1990a). Sexual selection unhandicapped by the Fisher process. *Journal of Theoretical Biology, 144,* 473–516.

Grafen, A. (1990b). Biological signals as handicaps. *Journal of Theoretical Biology, 144,* 517–546.

Greenfield, M. D. (1981). Moth sex pheromones: An evolutionary perspective. *Florida Entomologist, 64,* 4–17.

Greenfield, M. D. (1988). Inter-specific acoustic interactions among katydids *Neoconocephalus*: Inhibition-induced shifts in diel periodicity. *Animal Behaviour, 36,* 684–695.

Greenfield, M. D. (1990). Territory-based mating systems in desert grasshoppers: Effects of host plant distribution and variation. In R. F. Chapman and A. Joern (eds.), *Biology of Grasshoppers* (pp. 315–335). New York: Wiley–Interscience.

Greenfield, M. D. (1994a). Cooperation and conflict in the evolution of signal interactions. *Annual Review of Ecology & Systematics, 25,* 97–126.

Greenfield, M. D. (1994b). Synchronous and alternating choruses in insects and anurans: Common mechanisms and diverse functions. *American Zoologist, 34,* 605–615.

Greenfield, M. D. (1997). Sexual selection in resource defense polygyny: Lessons from territorial grasshoppers. In J. C. Choe and B. Crespi (eds.), *The Evolution of Mating Systems in Insects and Arachnids* (pp. 75–88). Cambridge: Cambridge University Press.

Greenfield, M. D., Alkaslassy, E., Wang, G.-y. and Shelly, T. E. (1989). Long-term memory in territorial grasshoppers. *Experientia, 45,* 775–777.

Greenfield, M. D. and Coffelt, J. A. (1983). Reproductive behaviour of the lesser waxmoth, *Achroia grisella* (Pyralidae: Galleriinae): Signalling, pair formation, male interactions, and mate guarding. *Behaviour, 84,* 287–315.

Greenfield, M. D. and Minckley, R. L. (1993). Acoustic dueling in tarbush grasshoppers: Settlement of territorial contests via alternation of reliable signals. *Ethology, 95,* 309–326.

Greenfield, M. D. and Roizen, I. (1993). Katydid synchronous chorusing is an evolutionarily stable outcome of female choice. *Nature, 364,* 618–620.

Greenfield, M. D., Tourtellot, M. K. and Snedden, W. A. (1997). Precedence effects and the evolution of chorusing. *Proceedings of the Royal Society of London.* 264, in press.

Gwynne, D. T. and Morris, G. K. (1986). Heterospecific recognition and behavioral isolation in acoustic Orthoptera (Insecta). *Evolutionary Theory, 8,* 33–38.

Hamilton, W. D. and Zuk, M. (1982). Heritable true fitness and bright birds: A role for parasites. *Science, 218,* 384–387.

Harvey, P. H. and Bradbury, J. W. (1991). Sexual selection: In J. R. Krebs and N. B. Davies (eds.), *Behavioural Ecology: An Evolutionary Approach* (3rd ed., pp. 203–233). Oxford: Blackwell Press.

Hedrick, A. V. and Dill, L. M. (1993). Mate choice by female crickets is influenced by predation risk. *Animal Behaviour, 46,* 193–196.

Hedwig, B. (1990). Modulation of auditory responsiveness in stridulating grasshoppers. *Journal of Comparative Physiology A, 167,* 847–856.

Helversen, O. von and Helversen, D. von. (1994). Forces driving coevolution of song and song recognition in grasshoppers. In K. Schildberger and N. Elsner (eds.), *Neural Basis of Behavioural Adaptations* (pp. 253–284). Stuttgart: Gustav Fischer Verlag.

Heywood, J. S. (1989). Sexual selection by the handicap mechanism. *Evolution, 43,* 1387–1397.

Janetos, A. C. (1980). Strategies of female choice: a theoretical analysis. *Behavioral Ecology & Sociobiology, 7,* 107–112.

Jang, Y., Collins, R. D., and Greenfield, M. D. (1997). Variation and repeatability of ultrasonic sexual advertisement signals in *Achroia grisella* (Lepidoptera: Pyralidae). *Journal of Insect Behavior, 10,* 87–98.

Jang, Y. and Greenfield, M. D. (1996). Ultrasonic communication and sexual selection in waxmoths: Female choice based on energy and asynchrony of male signals. *Animal Behaviour, 51,* 1095–1106.

Jia, F.-y. and Greenfield, M. D. (1997). When are good genes good? Variable outcomes of female choice in wax moths. *Proceedings of the Royal Society of London B, 264,* 1057–1063.

Johnstone, R. (1993). Female preference for symmetrical males as a by-product of selection for mate recognition. *Nature, 372,* 172–175.

Kirkpatrick, M. (1982). Sexual selection and the evolution of female choice. *Evolution, 36,* 1–12.

Kirkpatrick, M. (1986). The handicap mechanism of sexual selection does not work. *American Naturalist, 127,* 222–240.

Kirkpatrick, M. (1987). Sexual selection by female choice in polygynous animals. *Annual Review of Ecology & Systematics, 18,* 43–70.

Kirkpatrick, M. and Ryan, M. J. (1991). The paradox of the lek and the evolution of mating preferences. *Nature, 350,* 33–38.

Kodric-Brown, A. and J. H. Brown. (1984). Truth in advertising: The kinds of traits favored by sexual selection. *American Naturalist, 124,* 309–323.

Kramer, E. (1975). Orientation of the male silkmoth to the sex attractant bombykol. In D. A. Denton and J. P. Coghlan (eds.), *Olfaction and Taste Vol. 5* (pp. 329–335). New York: Academic Press.

Lande, R. (1981). Models of speciation by sexual selection on polygenic traits. *Proceedings of the National Academy of Sciences, U.S.A., 78,* 3721–3725.

Lloyd, J. E. and Wing, S. R. (1983). Nocturnal aerial predation of fireflies by light-seeking fireflies. *Science, 222,* 634–635.

Maynard Smith, J. (1978). *The Evolution of Sex*. Cambridge: Cambridge University Press.

Maynard Smith, J. (1991). Theories of sexual selection. *Trends in Ecology & Evolution, 6,* 146–151.

Moller, A. P. (1990a). Fluctuating asymmetry in male ornaments may reliably reveal male quality. *Animal Behaviour, 40,* 1185–1187.

Moller, A. P. (1990b). Effects of parasitism by a haematophagous mite on reproduction in the barn swallow. *Ecology, 71,* 2345–2357.

Muller, K. L. (1995). *Habitat settlement in territorial species: The effects of habitat quality and conspecifics*. Ph.D. dissertation. University of California, Davis.

Narins, P. M. (1992). Reduction of tympanic membrane displacement during vocalization of the arboreal frog, *Eleutherodactylus coqui. Journal of the Acoustical Society of America, 91,* 3551–3557.

Nur, N. and O. Hasson. (1984). Phenotypic plasticity and the handicap principle. *Journal of Theoretical Biology, 110,* 275–297.

Ode, P. J., Antolin, M. F. and Strand, M. R. (1995). Brood-mate avoidance in the parasitic wasp *Bracon hebetor* Say. *Animal Behaviour, 49,* 1239–1248.

O'Donald, P. (1980). *Genetic Model of Sexual Selection*. Cambridge: Cambridge University Press.

Otte, D. (1977). Communication in Orthoptera: In T. A. Sebeok (ed.), *How Animals Communicate* (pp. 334–361). Bloomington, IN: Indiana University Press.

Otte, D. (1989). Speciation in Hawaiian crickets: In D. Otte and J. Endler (eds.), *Speciation and its Consequences* (pp. 489–526). Sunderland, MA: Sinauer Associates.

Otte, D. and J. Smiley. (1977). Synchrony in Texas fireflies with a consideration of male interaction models. *Biology of Behavior, 2,* 143–158.

Parker, G. A. (1983). Mate quality and mating decisions: In P. Bateson (ed.), *Mate Choice* (pp. 141–164). Cambridge: Cambridge University Press.

Petrie, M. (1994). Improved growth and survival of offspring of peacocks with more elaborate trains. *Nature, 371,* 598–599.

Pomiankowski, A. N. (1987). Sexual selection: The handicap principle does work—sometimes. *Proceedings of the Royal Society of London B, 231,* 123–145.

Pomiankowski, A. N. (1988). The evolution of female mate preferences for male genetic quality. *Oxford Surveys in Evolutionary Biology, 5,* 136–184.

Prestwich, K. N. (1994). The energetics of acoustic signaling in anurans and insects. *American Zoologist, 34,* 645–643.

Proctor, H. C. (1991). Courtship in the water mite *Neumania papillator*: Males capitalize on female adaptations for predation. *Animal Behaviour, 42,* 589–598.

Real, L. (1990). Search theory and mate choice: I. Models of single-sex discrimination. *American Naturalist, 136,* 376–405.

Reinhold, K., Greenfield, M. D., Jang, Y. and Broce, A. (1997). Energetic costs of sexual attractiveness: Ultrasonic advertisement in wax moths. *Animal Behaviour, 54* in press.

Ritchie, M. G., Couzin, I. D. and Snedden, W. A. (1995). What's in a song? Female bushcrickets discriminate against the song of older males. *Proceedings of the Royal Society of London B, 262,* 21–27.

Ryan, M. J. (1980). Female mate choice in a neotropical frog. *Science, 209,* 523–525.

Ryan, M. J. (1988). Energy, calling and selection. *American Zoologist, 28,* 885–898.

Ryan, M. J. (1994). Mechanisms underlying sexual selection. In L. A. Real (ed.), *Behavioral Mechanisms in Evolutionary Ecology* (pp. 190–215). Chicago, IL: University of Chicago Press.

Ryan, M. J., Fox, J., Wilczynski, W., and Rand, A. S. (1990). Sexual selection for sensory exploitation in the frog *Physalaemus pustulosus. Nature, 343,* 66–67.

Ryan, M. J. and A. Keddy-Hector. (1992). Directional patterns of female choice and the role of sensory biases. *American Naturalist, 139,* S4-S35.

Ryan, M. J. and A. S. Rand. (1993). Species recognition and sexual selection as a unitary problem in animal communication. *Evolution, 47,* 647–657.

Ryan, M. J., Warkentin, K. M., McClelland, B. E., and Wilczynski, W. (1995). Fluctuating asymmetries and advertisement call variation in the cricket frog, *Acris crepitans. Behavioral Ecology, 6,* 124–131.

Schatral, A., Latimer, W., and Broughton, B. (1984). Spatial dispersion and agonistic contacts of male bush crickets in the biotope. *Zeitschrift für Tierpsychologie, 65,* 201–214.

Shelly, T. E. and Greenfield, M. D. (1991). Dominions and desert clickers (Orthoptera: Acrididae): Influences of resources and male signaling on female settlement patterns. *Behavioral Ecology & Sociobiology, 28,* 133–140.

Shelly, T. E., Greenfield, M. D., and Downum, K. R. (1987). Variation in host plant quality: Influences on the mating system of a desert grasshopper. *Animal Behaviour, 35,* 1200–1209.

Sobel, E. C. and Tank, D. W. (1994). In vivo Ca^{2+} dynamics in a cricket auditory neuron: An example of chemical computation. *Science, 263,* 823–826.

Spangler, H. G. (1988). Moth hearing, defense, and communication. *Annual Review of Entomology, 33,* 59–81.

Spangler, H. G., Greenfield, M. D., and Takessian, A. (1984). Ultrasonic mate calling in the lesser wax moth. *Physiological Entomology, 9,* 87–95.

Stamps, J. A. (1987). Conspecifics as cues to territory quality: A preference of juvenile lizards (*Anolis aeneus*) for previously used territories. *American Naturalist, 129,* 629–642.

Walker, T. J. and Masaki, S. (1989). Natural history. In F. Huber, T. E. Moore, and W. Loher (eds.), *Cricket Behavior and Neurobiology* (pp. 1–42). Ithaca, NY: Cornell University Press.

Wallach, H., Newman, E. B. and Rosenzweig, M. R. (1949). The precedence effect in sound localization. *American Journal of Psychology, 62,* 315–336.

Wang, G.-y., Greenfield, M. D. and Shelly, T. E. (1990). Inter-male competition for high-quality host plants: The evolution of protandry in a territorial grasshopper. *Behavioral Ecology & Sociobiology, 27,* 191–198.

Watt, W. B., Carter,P. A. and Donohue, K. (1986). Females' choice of "good genotypes" as mates is promoted by an insect mating system. *Science, 233,* 1187–1190.

West-Eberhard, M. J. (1984). Sexual selection, competitive communication and species-specific signals in insects. In T. Lewis (ed.), *Insect Communication* (pp. 283–324). New York: Academic Press.

Williams, G. C. (1975). *Sex and Evolution.* Princeton, NJ: Princeton University Press.

Wyttenbach, R. A. and Hoy, R. R. (1993). Demonstration of the precedence effect in an insect. *Journal of the Acoustical Society of America, 94,* 777–784.

Zahavi, A. (1975). Mate selection—a selection of a handicap. *Journal of Theoretical Biology, 53,* 205–214.

Zahavi, A. (1977). Reliability in communication systems and the evolution of altruism. In B. Stonehouse and C. M. Perrins (eds.), *Evolutionary Ecology* (pp. 253–259). London, England: Macmillan.

Zahavi, A. (1993). The fallacy of conventional signalling. *Philosophical Transactions of the Royal Society of London B, 340,* 227–230.

Zeh, J. A. and Zeh, D. W. (1994). Last-male sperm precedence breaks down when females mate with three males. *Proceedings of the Royal Society of London B, 257,* 287–292.

Zelick, R. (1986). Jamming avoidance in electric fish and frogs: Strategies of signal oscillator timing. *Brain, Behavior & Evolution, 28,* 60–69.

Zuk, M. (1994). Immunology and the evolution of behavior. In L.A. Real (ed.), *Behavioral Mechanisms in Evolutionary Ecology* (pp. 354–368). Chicago, IL: University of Chicago Press.

Chapter 6

DETERMINANTS OF CONFLICT OUTCOMES

William J. Hamilton III

Division of Environmental Studies
University of California, Davis
Davis, California 95616

John W. McNutt

Wild Dog Research Project
Private Bag 13
Maun, Botswana

ABSTRACT

Analysis of individual differences in competitive abilities is an essential part of the interpretation of unequal access to resources. Here we develop and apply a model applicable when dyadic contest outcomes determine access to resources, including mates. Individual differences in fighting ability may be decisive in determining contest outcomes. The components of fighting ability (FA) include inheritance (FAi), condition (FAc), experience (FAe) and development (FAd) to and throughout senescence (FAs). Present fighting ability (FAp) is a dynamic entity that varies as the value of these components change, waxing and waning as costs are incurred in contests. Fighting ability contributes to the probability that opponents will enter into and persist with patterns of conflict behavior and prevail as contest winners.

Simulations identify the probabilities of gaining access to status and resources when FA determines contest winners following dyadic contests. A sequence of interactions between opponents produce frequency distributions of fighting ability within populations which contain and may provide information to contestants about their fighting ability compared with that of other population

Perspectives in Ethology, Volume 12
edited by Owings *et al.*, Plenum Press, New York, 1997

members. Individual win:loss records may be incorporated into individual experience, resulting in effective behavioral responses to the probabilities of winning subsequent encounters. The probabilities of improving rank by transferring from one to another linear dominance hierarchy are determined given the assumption that FA determines rank. We identify the relationship of fighting ability to contests for resources in a competitive arena where some individuals are excluded from resources. These simulations emphasize the value of diachronic observations of individual contest histories and outcomes.

When fighting ability is not decisive in determining contest outcomes and the distribution of resources, the relative effect of other parameters in doing so is enhanced. The analysis of competitive behavior promotes identification of separable parameters which sometimes have opposing effects. Determinants of these additional processes occurring during animal conflict are identified and their utility in separating behavioral relationships is considered. These determinants include social support (S) from additional parties in contests and resource value (V), a collection of several relevant parameters influencing motivation. Motivation (M), the propensity of individuals to enter into and persist with contests, is a diversely used term. But the processes and phenomena it incorporates cannot be discarded, because without its role in the analysis of individual differences, fitness consequences of behavior are necessarily analyzed exclusively as a function of V.

Studies of the relationship of the determinants of conflict outcome identified here have been and can be incorporated into general models of contest behavior. To determine their respective effect upon conflict behavior and contest outcomes, some components of fighting ability can be set equal by experimentation or ordering of field observations. We extend our model to groups of individuals arranged in linear dominance hierarchies and to the ideal despotic distribution where access to resources is contested by many individuals. We make predictions about the exact probabilities of winning under alternative circumstances. We suggest how animals might adjust their behavior if they were to use the estimates of personal fighting ability identified by simulations to estimate their probability of winning.

INTRODUCTION

Conflict behavior allows individual animals to maximize their probability of gaining access to resources including mates and to minimize their lifetime costs of doing so. There are several essential pieces of information a contender requires to accurately estimate its probability of winning a contest. These include assessment of personal ability compared with that of the opponent and assessment of the motivation of an opponent to contest access to the resource. While perfect information is seldom available, even imperfect information can provide an estimate of relative ability and motivation. Contests usually begin with communication in the form of visual, vocal or other displays. In some instances these signals may convey sufficient information to resolve conflict. Conventional conflict settlement in this sense may be derived from contestants' earlier experiences, and decisions in response to relevant communicated information may be based on a history of contests. However, in the absence of perfect information, estimates are and can only be probabilistic and opponents may decide to engage in physical contests to determine which will get access to resources. Several context and intrinsic variables may be involved in determining conflict behavior and contest outcomes, and these will vary among species, populations and even by context for an individual. Nevertheless there are fundamental common properties of contest behavior which, when identified, facilitate interpretation of the determinants of conflict behavior. We assume that fighting is energetically costly, and that animals would not opt to fight if resources were unlimited. They place a premium on communication and accurate assessment. We examine the set of general determinants of conflict behavior and identify possible mechanisms of assessment, decision making and fighting. We review past analyses of animal conflict behavior and expand on them to identify a complete set of variables required to specifically address the questions: When do animals fight for access to limited resources? What determines the outcome of a contest? The aim of this paper is: (1) to facilitate an empirical multivariate approach to understanding the mechanisms of conflict behavior by providing a theoretical construct, an idealized model, of the relationships between conflict variables and behavior and (2) to identify an informational context based on estimates of probability of winning contests within which individual animals are expected to operate.

The application of game theory to interpretations of animal conflict behavior provides a framework for understanding alternative contest strategies from an evolutionary perspective, and especially why, for example, some contests are settled without fighting. The game theory approach addresses contests at the level of the genetic population using simplified frequency-dependent alternative strategy sets (e.g., hawk vs. dove). Following the first applications of game theory to animal conflict behavior (Maynard Smith and Price 1973; Parker 1974), the game theory approach, with the concept of an evolutionarily stable strategy (ESS; Maynard Smith 1982a) as the linchpin of theoretical validation, became the dominating paradigm for empirical analyses of animal conflict. Although game theory has provided important insights into ultimate explanations of alternative strategies, this paradigm imposes constraints that inhibit further detailed analyses of the mechanisms involved in conflict at levels at which detailed empirical research is currently conducted. Current emphasis upon this single theoretical paradigm in animal conflict behavior merits critical review.

It was not the intention of early ESS models to identify mechanisms of animal conflict or to reveal specific determinants of contest outcomes. Nevertheless, behavioral observations of dyadic contests have been analyzed with game theory as the conceptual framework. Among the recent few quantitative empirical tests of ESS model predictions are cases which have explicitly failed to demonstrate a quantitative fit of observations to predictions (e.g., Brockmann et al. 1979; Hammerstein and Riechert 1988). When results do not match predictions from the models, the terms of the models may be adjusted (Brockmann et al. 1979; Parker and Rubenstein 1981) or additional terminology added to exclude a range of otherwise inexplicable results (e.g., paradoxical ESSs which have no biological explanation, but are said to account for a part of the observations, Hammerstein and Riechert 1988). When no explanation supports theory, results may be attributed to a lag in the adaptive process, with the conclusion that an ESS has not yet been reached (Riechert 1979) or cannot be attained by the subject population due to gene flow constraints (Hammerstein and Riechert 1988). Thus, inferences from observations failing to match predictions from models remain ambiguous, a problem identified by others for optimization approaches in general (e.g., Gould and Lewontin 1979). We suspect abandoning the constraint of evolutionary stability as a criterion for validity in empirical investigations of conflict behavior would facilitate analysis of the mechanisms of contest behavior within a more temporally proximate, complex and dynamic multivariate context.

Application of game theory to animal conflict behavior, because it is limited to population level analyses using a limited set of generalized variables, (e.g., resource holding potential (RHP) and resource value (V)) and a small number of alternatives within a strategy set, is insufficient to enable more detailed understanding of animal conflict behavior at the level of the individual, where empirical studies often focus. The condensed terms of game theory models are often insufficient to describe the several processes modifying contest behavior and determining contest outcomes, and it has become common practice to model conflict with the terms RHP, V and costs and then to discuss the findings in terms of additional parameters that might reconcile observations with a model. The problem with this approach is that there is no set of observations that cannot be fully interpreted. There is thus no possibility that a null hypothesis will be rejected. Furthermore, while determining whether or not a particular set of strategies is an ESS is an interesting question, it is beyond the scope and focus of most empirical studies (e.g., Olsson 1994). To investigate and understand the mechanisms underlying acquisition and control of contested resources from the perspective of the participants in a complex multivariate system requires a more complex set of determinants than those offered in game theory models. Here we review the variables of conflict behavior, including the generalized terms used in game theory models of conflict, to identify sources of confusion. Then we develop what we consider to be a minimum set of determinants required to analyze empirically the multivariate nature of animal conflict behavior.

We first focus on the intrinsic variable fighting ability (FA) and identify the set of determinants represented by this characteristic. Then we identify a simple conflict context based on probabilities within which animals may acquire, from experience, increasingly accurate information about their relative fighting abilities. This hypothetical information context relies on a few broad assumptions regarding distribution of ability within a group, and enables predictions for detailed analysis of conflict behavior based on probability of winning contests. In the following section we define and discuss the set of independent variables in addition to fighting ability that can affect the behavior and the outcome of contests. We combine these under the heading of motivation. A conceptual construct of the relationships between these variables and fighting ability is developed, and we discuss incorporating these relationships into analyses of conflict behavior. We conclude by applying predictions from the fighting ability model to various empirical examples of some commonly described patterns of animal conflict settlement such as linear dominance hierarchies, the residence effect and the outcome effect.

DETERMINANTS OF CONFLICT OUTCOMES – FIGHTING ABILITY

Fighting Ability

Resource holding potential was originally defined by Parker (1974) as absolute fighting ability, but subsequent usage of RHP varies (e.g., "inherent fighting ability" — Parker 1982; "intrinsic power" — Dunbar 1988; "the constellation of factors that influence fighting ability" — Krebs and Dawkins 1984). Usage frequently confounds extrinsic and intrinsic variables. RHP may be a dependent variable, equivalent to the probability of winning (PW; Petrie 1984; Freeman 1987), or be an independent (subject) variable describing the relationship between individual ability and the outcome of a contest, as Parker (1974) originally intended. Contrary to Parker's (1974; Maynard Smith and Parker 1976) original intention, RHP may be used simply as a label for the contest winner. This usage has little utility in analyses of the determinants of contest outcomes. Nevertheless, this alternative meaning, implied by 'resource holding potential' has become common usage. "...RHP is a measure of the capacity of an individual to hold a resource" (Maynard Smith 1982b). With this usage, the possibility that an opponent could lose a contested resource to an individual with lower RHP is counterintuitive, but that possibility was clearly considered by Maynard Smith and Parker (1976). Similarly, Dunbar's (1988; Dunbar et al. 1990) recent use of 'power' as another surrogate term for FA suffers from alternative meanings and its absence from other literature. Since RHP as currently used is inexact we find no reason to call fighting ability by any other name, and return to this earlier (e.g., Ginsberg and Allee 1942) and still widely used term. Wherever possible we have used dictionary definitions of terms to enhance the prospect for wide consideration of these concepts.

Abilities are capacities, inherited and gained by experience, to perform actions. Fighting ability is a dynamic phenotypic individual characteristic. As used here, it is the potential to prevail in physical contests against others for status, resources, and mates. We distinguish between contests involving fighting and those which do not. Fights are encounters involving attempts to defeat an opponent by physical contact. Fights usually produce winners and losers, but draws are possible. Displays, probing, sparring and other interactions are examples of contest encounters that are not fights but may include physical contact and convey information about relative FA. Limited encounters, not involving contact between contenders may also resolve access to resources. Grafen (1987) clarifies the distinction between probing behavior and fighting and its relationship to game theory models. The distinction between probing and fighting is based on differences in costs and function. The probability of winning several high-cost fights, for example, is low and can be quantitatively defined. Probing and its associated displays, for some species, has relatively low or no costs (spiders, Hammerstein and Riechert 1988) and may occur repetitively. The difference between fighting and sparring is readily distinguished, for example, in most ungulates (e.g., Estes 1969; Geist 1971, 1974; Barrette 1977; Hirth 1977; Wilson and Franklin 1985; Barrette and Vandal 1990). Among the chacma baboons we observe males display with loud calls (wahoos) repeatedly over a period of days or weeks. When a rare physical contest between these monkeys occurs the distinction between vocal sparring and its associated positional maneuvering and overt fighting is obvious. In other animals the distinction may be less obvious, as for some insects (e.g., Crespi 1986), though some contests described as fights between adult insects do have clear winners and losers (e.g., Hamilton et al. 1976; Davies 1978; Waage 1988; Marden and Waage 1990).

Differences in fighting ability between individuals are influenced by stage of development to and throughout senescence, condition and experience (Collias 1943; Parker 1974). These variables, combined with a heritable component, collectively determine the present (instantaneous) FAp (Table 1). FA may explain some, all or none of the variance in contest outcomes and access to contested resources. Abilities other than FA as, for example, foraging skills, may determine competitive potential, especially when contest competition is of limited or no importance to resource acquisition. FA is an individual characteristic which generally increases during ontogeny to some maximum individual potential, then declines as a result of costs incurred in contests and from senescence. Temporal patterns of FA vary by species and individual, both between and within populations. These variables collectively comprise the minimum set of components of fighting ability. Fighting ability contributes to the probability that opponents will enter into and persist with patterns of conflict behavior and prevail as contest

Table 1. Components of Fighting Ability and Some Examples of Factors Contributing to Them. These Positive and Negative Effects Collectively Determine an Individual's Present Fighting Ability (FAp)

FAi (Inherited characteristics)
Gender
Adult body size
Other inherited morphological and functional differences between individuals
Behavioral traits
FAd (Developmental state)
FAc (Condition)
Current mobilizable energy
Slowly mobilizable energy reserves
Stress state
Conditioning (exercise)
Parasite effects
Injury effects
Disease effects
FAe (Experience)
Acquired tactics
Degree of coordination of known behavior
Knowledge of context of effective behavior
Play
FAs (Senescence)
Wear effects
Design-based declines
Declines in capacity to recover

winners. To determine their respective effect upon conflict behavior and contest outcomes, some components of FA can be set equal by experimentation or ordering of field observations, but empirical analyses which fail to identify a complete set of FA components may reach invalid conclusions. We find no examples in the literature on fighting behavior or from personal experience that cannot be allocated to these specific components. A review of recent literature shows that some authors consider a more or less complete set of these components (Olsson 1994; Thorpe et al. 1995). Others do not.

Inheritance (FAi)

FAi is the heritable component of differences in fighting ability. In nature FA often is measured by the outcome of contests between adults. The contribution of inherited differences between individuals to differences in performance, while conceptually straightforward, may be empirically problematic. In many animal species gender has an enduring effect upon the expression of behavior and other attributes determining conflict behavior. For some species gender may be more important than any other variable in explaining the contest outcomes (e.g., Richner 1989). All components determining FA have an inherited component including, but not limited to, some determinants of body size and behavior patterns. Individual differences between rhesus macaques (*Macaca mulatta*) in response to stressors have inherited components (Suomi 1983). Artificial selection for characteristics modifying the ability and disposition of animals to fight (Reed 1969) has been highly successful within domesticated strains of dogs, fish and birds (Darwin 1871; summarized in Huntingford and Turner 1987). The results of artificial selection in modification of contest behavior and patterns of conflict outcomes (e.g., Van de Poll et al. 1982; Thornhill and Sauer 1992) are based upon variation in heritable characteristics, some of which determine FA. Some authors refer to the inheritance of fighting ability (e.g., Rohwer 1982). More often the terms intrinsic (cf. Maynard Smith and Parker 1976; Popp and DeVore 1979; Whitfield 1987) and inherent (Parker 1982) are used, implying but usually avoiding specific statements about the contribution of inheritance. These circumlocutions and fashionable deference to current social mores dealing with the concept of inheritance of behavior in any organism have been an impediment to the development of behavioral biology as a precise science. Surrogate terms for motivation, considered below, have also contributed to vagueness and a discontinuous discourse.

Motivation to enter into and persist with encounters is determined in part by inherited characteristics (Ginsburg and Allee 1942; Barlow et al. 1986). A game theoretic approach may profitably analyze the environmental, including demographic circumstances in which alternative levels of aggressiveness succeed. Here in addition we distinguish between disposition to attack, experience notwithstanding, which we call inherent aggressiveness, and that part of aggression which has been modified by experience. Inherited dispositions may determine contest outcomes (Barlow et al. 1986). Interspecific and intraspecific differences in aggressive behavior (Ginsburg and Allee 1942; Scott and Fuller 1965; Bekoff 1974) as well as interindividual differences within a population (Thornhill and Sauer 1992) have a heritable component the characteristics of which are probably related to patterns of conflict behavior and contest outcomes.

Centuries of artificial selection for fighting performance by cock and dog breeders (Hoff 1981) has produced subjects unsuited to unqualified analyses of ESSs or other studies of the adaptive features of conflict because the aggressiveness of domestic animals is not in harmony with maximization of lifetime reproductive success except as determined by breeders. Nevertheless, strain differences and differences in aggressiveness, termed gameness by breeders of fighting stock (Finsterbusch 1980), demonstrate that inherent aggressiveness is subject to selection, independent of ability. For pit bulls (as considered here, American Staffordshire Terriers) expression of innate aggressiveness is also subject to modification by experience (Clifford et al. 1990). Elaborate apparatus and training regimens such as those which feature the preparations of fighting domestic cocks (Finsterbusch 1980) are essential to maximize the probability of winning staged contests. The expression of inherent aggressiveness is thus responsive to modification by experience in these cases gained during the training process. The results of breeding for fighting ability in domestic animals nonetheless reveal the presence of inherited controls upon thresholds for the expression of aggression modified by artificial selection. These thresholds are treated here as determinants of motivation (see below) rather than ability.

During growth, most animals avoid fights with older, larger and more experienced individuals. Inherited patterns of responses to larger opponents may discourage developing individuals (lower FAd, below) from entering encounters with individuals who could easily defeat them. A single escalated encounter against a substantially larger and/or older more experienced individual could be costly. Avoidance of contests with larger individuals, independent of experience, may prevent costly premature fights. Inherited behavior patterns of this sort may be one basis for the widespread use of displays exaggerating apparent body size. These displays are more likely to be effective against younger and inexperienced than older and more experienced opponents. It is also probable among longer-lived animals that individuals learn the punishing costs of interacting with larger individuals during development. Dogs, monkeys, rats and mice reared in isolation are more prone to attack than socialized controls (Jackson 1988), suggesting that for these relatively well-studied animals development of appropriate levels of expression of aggression in contests is in part gained by experience. Recent analyses of conflict often ignore or do not make explicit the role of inheritance in determining the range of potential responses to competitors.

Development (FAd) and Senescence (FAs)

FAd is the effect of growth and maturation to adulthood upon FA. Differences in FAd within same-age and same-sex cohorts may be relatively small compared with other parameters determining contest outcomes. However, especially among fish such as salmonids (Newman 1956; Yamagishi 1962) and cichlids (Francis 1988), FAd changes as a function of nutritional and/or social environmental conditions, largely independent of age (see also anthophorid bees, *Centris pallida*, Alcock et al. 1977; naked mole rats, *Heteromeles*, Jarvis 1978, 1981). Within a natural social grouping of competitors there are often cohorts of both sexes and of broadly different ages and degrees of development (FAD) reflecting both social organization and demography. In such groups differences in FAd are often decisive in determining access to resources.

Fighting ability typically rises to some maximum value during development, then declines. Age may be viewed as a determinant of male competitive success (e.g., Dunbar et al. 1990) but is more appropriately identified as one of its correlates (Noë 1994). While age usually closely tracks development, the relationship of age to the probability of winning also includes experience and condition (FAe and FAc, below). In empirical studies it will be useful to identify the respective contribution of these components to age-correlated differences between individuals in fighting ability. Among mammals (cf. mountain sheep, *Ovis canadensis*, Geist 1971; chimpanzees, *Pan troglodytes*, Bygott 1979; savanna baboons, *Papio* spp., Packer 1979; Dunbar 1988; Noë 1994) and birds (cf. Robinson 1986a, b; Arcese 1987, 1989) individual FA characteristically rises to some maximum value (Fig. 1) correlated, but not necessarily synchronous, with the age at which adult size is attained.

Fighting ability then declines due to losses in condition and the effects of senescence (FA), as with mechanical senescence (Finch 1990) from wing wear in damselflies, *Calopteryx maculata* (Forsyth and Montgomerie 1987). In the chacma baboons (*P. ursinus*) we study, adult males attain their maximum probability of winning contests for alpha rank at about 100 months when their upper canines reach maximum length and their probability of attaining maximum rank (Hamilton and Bulger 1990) and access to fertile females (Bulger 1993) peaks. Weight gain at high rank may add to FA and contribute to the temporary maintenance of high rank. Infrequent but serious fights with other contenders

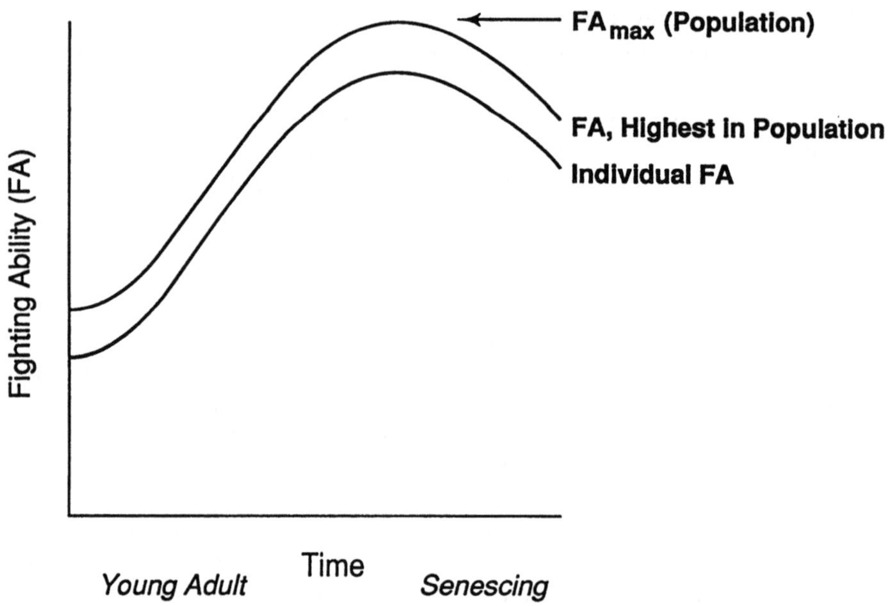

Fig. 1. The late ontogeny of fighting ability (FA), illustrating its increase to a maximum lifetime value followed by decline due to senescence (FAs) and irrecoverable losses in condition. The mean and maximum fighting ability (FAmax) for the population are identified to indicate variance from a mean and a maximum equivalent to FA = 100 in our model (pg. 195). Note that FA also varies on a shorter temporal scale from the ontogenetic trajectory owing to changes in condition (FAc) and/or experience (FAe).

may result in debilitating injuries (FAc if fully recoverable, FAs if not) including especially canine breakage. Our observations of aging adult males (>12 yrs) are few, but a general decline in capacity at this age is evident three years after males first attain alpha rank, and few males recover alpha or beta rank and access to fertile females after their first loss to a challenger. The analysis of senescence is a particularly challenging parameter to evaluate (Alexander 1987) since from the perspective of the approach advocated here it would be necessary to separate changes in condition from the cessation of increases in FAd. The relationship of FAi to FAs in Fig. 2 reflects Williams' (1957) pleiotropic theory of the evolution of senescence.

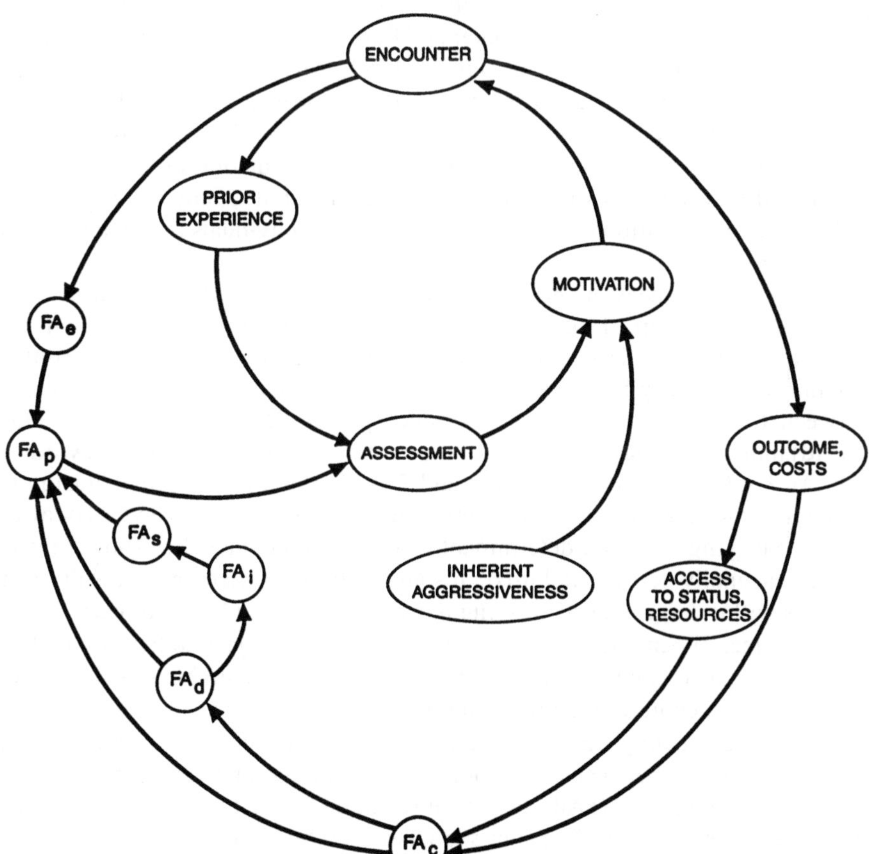

Fig. 2. Components contributing to fighting ability and their interrelationships. Additional parameters regulating aggression and providing information which might modify probabilities and patterns of contest behavior are shown in Figs. 5 and 6 below.

Body size, a commonly evaluated operational measure of FA, is not a component of FA here but we identify its contribution to several components of fighting ability. Body size has some measurable heritability (FAi, e.g., Cade 1984; Dixon and Cade 1986) but it is also an expression of development, regulated by the endocrine system, and of condition. Within some animal populations greater relative body size, measured by linear skeletal or other morphological measures (red-winged blackbirds, *Agelaius phoeniceus*, Searcy 1979; shrews, *Sorex araneus*, Barnard and Brown 1984; fish, *Salmo gairdneri*, Abbott et al. 1985) or mass (spiders, *Agelenopsis aperta,* Riechert 1978; *Euophrys parvula*, Wells 1988; frogs, *Uperoleia rugosa*, Robertson 1986), predicts contest outcomes (review in Enquist and Leimar 1983). However, within natural populations of other species, body size, however measured, often does not predict contest winners (e.g., red-winged blackbirds, Eckert and Weatherhead 1987; olive baboons, *P. anubis*, Packer and Pusey 1985; Bercovitch 1989; gelada baboons, *Theropithecus gelada*, Dunbar 1988; African elephants, *Loxodonta africana*, Poole 1989; carrion crows, *Corvus corone*, Richner 1989; velvet swimming crabs, *Necora puber*, Smith et al. 1994). The effects of body size range from slight differences accurately predicting contest outcomes (<2% by weight for cichlid fishes, *Cichlasoma citrinellum*, Barlow et al. 1986; Jakobsson et al. 1979) to the observation that substantially smaller individuals (elephants, Poole 1989a) may prevail.

Since differences in performance are not necessarily determined by body size, the relationship of body size to outcomes needs consideration within a multivariate context. Differences in body size may predict or determine contest outcomes, especially between young, less developed and older individuals, or between the sexes in sexually dimorphic species. But younger and smaller (<FAd) individuals may not be part of the contending cohort in nature (swordtail fish, *Xiphophorus helleri*, Ribowski and Franck 1993a) and their inclusion or exclusion from analyses will bias interpretations of observations toward or away from identifying size as a determinant of contest outcomes. For some species (e.g , seasonally territorial birds), resource apportionment may result from contests within large cohorts of same-age and same-sex adults, all nearly equal in body size. When such cohorts breed seasonally large numbers of mature individuals may compete for limited resources and a measurable size difference correlated with contest outcomes is not necessarily predicted.

When evaluating size as a determinant of contest outcomes some authors do not distinguish between differences due to skeletal or other morphological changes during growth (FAd) and those resulting in fluctuations in body mass about the skeletal frame (FAc, but see Olsson 1994). Mass of individuals as a measure of size and, therefore, as a predictor of FA, may explain much of the variance in contest outcomes for animals that are not resource limited, as for provisioned and captive animals. But mass or other measures of size are inadequate measures of fighting ability because they do not necessarily reflect FAc, which may be equally or more important than FAd in determining FA. Despite numerous observations that skeletal size and mass are often not closely correlated with contest outcomes, especially in seasonally breeding birds but also among adult primates, they continue to be used as operational or surrogate measures of FA (e.g., Andren 1990).

Condition (FAc)

Variation in FA due to differences in condition (FAc) may be induced by changes in nutritional state, molt status (birds, Collias 1943; crustaceans, Caldwell 1986; Adams and Caldwell 1990), levels of energy reserves (fowl, Collias 1943; spiders, Riechert 1988; African cape buffalo, *Syncerus caffer*, Prins 1989; Prins and Iason 1989), conditioning (exercise) and the effects of injuries (Packer 1979), parasitism (Sutherland and Parker 1985) and disease. FAc may change gradually or abruptly and at any age. Abrupt changes in condition may result from injury or exhaustion. Packer (1979) suggests that losses in FAc may result from the cumulative effects of injuries to olive baboons, but does not provide convincing evidence. Some losses in FAc, including some of those due to injuries and many of those due to loss of energy reserves, are fully recoverable. Other losses of FAc are permanent, especially for some injury and disease effects.

Confounding of FAd and FAc often results because of their interdependence. Among adult holometabolous insects individual differences in FA are due to the components FAd. The relationship of size to FA for emergent insects which continue to gain mass (FAc), cuticular material (FAd) and, as a result, competitive potential following eclosion (dragonflies, *Erythemis simplicicollis*, McVey 1988) reflects the phylogenetic pattern of development imposed by eclosion.

Differences between individuals in FAc which influence FA may accrue during intervals when no, or relatively little, contest competition occurs. Such changes may determine condition before animals arrive at breeding areas following migration (elephant seals, *Mirounga angustirostris*, Le Boeuf 1974; Le Boeuf and Reiter 1988; snow geese, *Chen hyperborea*, Rockwell et al. 1985), or before entering contests at the end of winter (fallow deer, *Dama dama*, Festa-Bianchet et al. 1990; moorhens, *Gallinula chloropus*, Petrie 1983; sage grouse, *Centrocercus urophasianus*, Gibson and Bradbury 1985).

Experience (FAe)

Experience may modify fighting ability by incorporating new or refining existing patterns of conflict behavior (Barash 1982). Experience may be gained during play (Groos 1898; Aldis 1975), sparring and probing (Grafen 1987) and from observation of contests between others (Freeman 1987). For example, male chacma baboons nearing maturation (>80 mos) often closely approach escalated contests by adults (>90 mos) coming so close that the encounters may appear to be triadic. However, these younger males do not participate in these contests except as close observers and probably their actions are used to gain experience through observation. FAE increases when individuals learn new fighting tactics and may continue to do so throughout an individual's lifetime. Nevertheless, at some point, individual FA will reach its maximum lifetime fighting ability (FAMAX), then decline as gains in experience do not keep pace with irremediable losses in FA from condition (FAC) and senescence (FAS), expressed in part as a slowing of the repair process following injury (Fig. 1), and in numerous other ways (Finch 1990).

Prior experience in conflict may also contribute to and modify motivation to engage in aggressive interactions through perception of one's probability of winning (Fig. 2). This relationship is the basis for the probability model for estimating relative FA from prior contest experience described below.

Summary of Fighting Ability

Individual fighting ability, as determined by the combined effects of the determinants identified above and their interactions, changes throughout an individual's lifetime. The rate of change of FA may differ with life history stage. Contests may have both negative and positive effects upon FA and its several components which also affect contest behavior (see below). The dynamic nature of FA through time has important consequences for conflict behavior. Due to continuous individual changes in FA we expect animals to reevaluate their estimate of their own FA and to change their contest behavior accordingly.

Differences in FA within a cohort of same-age and same-sex individuals are likely to be less than those among individuals who differ in age and/or sex, but it is incorrect to assume that FA is operationally equal for same-age and -sex individuals (Landau 1965). In an analysis of conflict in nature it is essential to distinguish individuals who are serious contenders for resources from those who are not (Grafen 1987). The relevance of studies of conflict to social animals in nature depends upon identification of the social and demographic composition of operationally relevant cohorts of competitors. If opponents are not same age and same sex or are otherwise not equivalents, large differences in FAd, gender or experience probably will determine contest outcomes.

Studies of conflict behavior of captive animals often pit same-age and same-sex adults against one another (Ginsburg and Allee 1942; McBride 1958; King 1965; primates, Bernstein and Gordon 1980; fish, Mosler 1985; Barlow et al. 1986; Keeley and Grant 1993). In these experiments the size of opponents is determined by the experimenter, not by demography and natural social groupings. Similarly, same-age and same-sex cohorts of domestic chickens frequently have been the subjects in investigations of dominance hierarchies (McBride 1958; King 1965; Chase 1982, 1985). These subjects are raised under relatively uniform nutritional and social conditions and the negative feedback relationship between contest outcome at resources and condition may be largely or entirely obscured by the conditions of captive management. Thus, differences between all FA components in this context with the possible exception of FAi would be expected to differ from natural social groupings of ancestral junglefowl (*Gallus gallus*). Even variance of FAi may have been modified by selection, as within poultry strains. In interactions within these breeds, individual differences in FA may be minimized or enhanced, depending upon breeding programs. Multivariate analyses of the contribution of the components of FA among such experimental cohorts may identify and emphasize determinants of conflict resolution not characteristic of or relatively unimportant to natural populations. When one or more FA components are controlled by experimental design, the effect of asymmetries in other FA components or other conflict determinants will increase. For example, we expect provisioned populations of animals, both in nature and captivity, to lose part or all of the differences between them in the expression of FAc, enhancing the probability that FAi, FAe, and FAd will determine contest outcomes.

Because the physical and experiential determinants of FA are dynamic, the contribution to contest outcomes will vary between individuals throughout their lifetimes. Feedback relationships between conflict outcomes and FA will result in varying effects of alternative parameters upon different individuals within populations. The importance of particular conflict parameters to animal populations can be elucidated by considering past and present individual circumstances, including demography of relevant social groupings. Separation of the components of fighting ability in models may present difficulties, but for empirical behavioral studies of contest outcome determinants, their separate analysis is essential before concluding, by elimination, that other parameters such as resource value and costs determine outcomes and access to resources.

PROBABILITIES OF GAINING ACCESS TO STATUS AND RESOURCES ON THE BASIS OF FIGHTING ABILITY

Among animals that contest one another for resources we know of no example for which untested behavior patterns are the sole basis for contest resolution. There appears to be no evidence demonstrating that territorial animals defeat challengers without having their fighting ability tested in the current or some previous encounter. Nor is there compelling evidence that exclusive access to contested resources is conceded to nonkin outside the context of personal ability and experience.

Throughout their lives individuals gain probabilistic information about their personal ability which can be related to the abilities of others. This

Table 2. Loser's Probability of Winning (PW) against Opponents of Stated Experience Based upon Previous Contest Outcomes. The Table Shows that a Few Consecutive Losses Result in a Predictable Decline in PW in Future Contests. Note that for the Loser Opposing an Unknown, the Best Estimate for PW Is Equivalent to that Derived from an Opponent with a 0:0 Record (i.e., No Information). If the Opponent Is a Territory Holder and the Loser a Challenger, the Ownership Asymmetry Allows Some Refinement of the Estimated PW (i.e., the Opponent Has Won at Least Once)

Winner's win:loss record	0:0	1:0	2:0	3:0	4:0
Loser's win:loss record					
0:0	1/2	1/3	1/4	1/5	1/6
0:1	1/3	1/6	1/10	1/15	1/20
0:2	1/4	1/10	1/20	3/100	2/125
0:3	1/5	1/15	3/100	2/125	7/1000
0:4	1/6	1/20	2/125	7/1000	1/250

Records are of wins and losses against the field. Outcomes are determined by relative FA. PW for winner is (1−PW) for loser.

information may become available to individuals competing for resources based upon direct experience in prior contests or derived from observation of interactions between others. Acquisition of limited information during contests has been incorporated into the analysis of fighting behavior in other models (Parker and Rubenstein 1981; Enquist and Leimar 1983, 1987, 1990; Leimar and Enquist 1984).

Using simulations we demonstrate probabilities of gaining access to status and resources when fighting ability alone determines dyadic contest outcomes (Table 2). Individual win:loss records are the substance of individual experience. We hypothesize that this experience results in modified responses to conflict situations based on increasingly accurate estimates of the probabilities of winning or losing additional encounters in particular contexts with particular individual opponents or categories of opponents. Simulations illustrate that a sequence of interactions between opponents produces frequency distributions of fighting ability within populations (Fig. 3) which contain information about contestants and their fighting ability and that of other population members. Our model can be extended to groups arranged in linear dominance hierarchies and can be applied to the ideal despotic distribution (Fretwell and Lucas 1969) where access to resources is contested by many individuals. An ideal despotic distribution of resources determined by FA, giving individuals with higher FA proportionately greater access to better resources, can be achieved in at least three ways. Opponents may determine by fighting which individual will gain access to a resource, they may assort themselves without fighting on the basis of prior knowledge of their FA and estimates of the probability of winning (PW) against prospective opponents, or they may act on inherited response patterns which limit the frequency and context of encounters they will contest. Our model

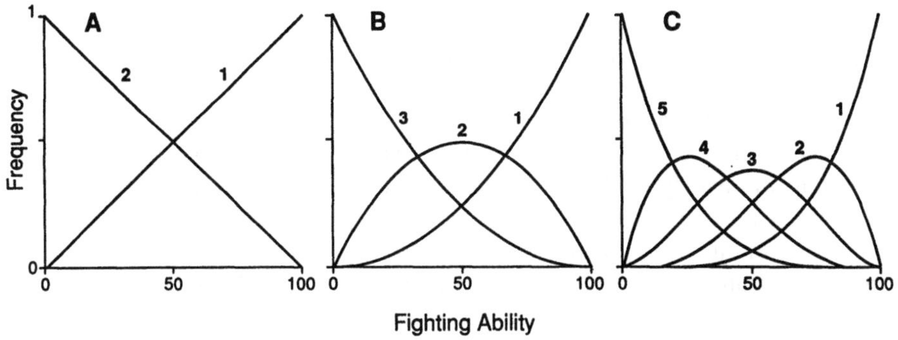

Fig. 3. The frequency distribution of fighting ability in a modeled group of individuals given assumptions stated in the text, A, after a single contest for (1) winners = 1/2 and (2) losers = 1/2. B, after two contests, the group divided to a record of (1) double winners = 1/4, (2) one win and one loss = 1/2 and (3) double losers = 1/4. C, after four contests, with a record of (1) all wins = 1/16, (2) three wins and one loss = 1/4, (3) two wins and two losses = 3/8, (4) one win and three losses = 1/4 and (5) four losses = 1/16. Distributions B and C are also those of individuals in a modeled linear dominance hierarchy of three and five individuals respectively.

suggests the quantitative criteria animals may use to incorporate features from both of these sources of information into responses to opponents.

Dyadic contests are determined by FA in our model: greater FA wins. The degree to which FA may explain patterns of contest behavior and outcomes is also modeled by Leimar and Enquist (1984). Their approach is logically similar to ours but we make predictions about the exact probabilities of winning under various generalized and alternative circumstances. Animals may adjust their behavior as if they were using estimates of their own FA, identified here by simulations, to estimate their PW on the basis of limited and imperfect information.

Fighting Ability and the Probability of Winning

Dyadic contests are simulated given the following broad assumptions:

a. all individuals have a fighting ability (FA), a dimensionless parameter whose relative value lies within a bounded range (0–100),
b. two individuals have identical FA,
c. relative FA determines outcomes in contests over limited resources.

To generate an initial distribution of relative FA in simulated cohorts we draw random numbers from 0 to 100, 100 representing the greatest FA for any population. Use of the limits 0 to 100 avoids the need to consider absolute FA or to identify a specific distribution of FA values within groups or populations. The

FA scale is relative. Because our simulated values are percentiles they are applicable to any FA distribution, regardless of the range or distribution of actual values. The probabilities of winning future contests vary predictably, depending upon prior individual contest history.

We simulate sequences of contests based upon multiple (N = 20,000) iterations to identify the PW given stated assumptions and prior contest history or equivalent experience. All contests are conducted independently, and all serial win:loss combinations for from zero to 10 contests are simulated independently. In our notation (Table 2) a 0:2 record means zero wins and two losses. Because greater FA determines contest winners, to select a 0:2 individual from a group of three randomly drawn individuals is equivalent to rank ordering the three according to their FA, then selecting the third-ranked individual. To simulate contests between a third-ranking individual in a group of three against the first-ranking individuals, those with 2:0 records, three more individuals are drawn at random, ranked, and the FA of 0:2 individuals is then compared with the FA of 2:0 individuals. The frequency with which 0:2 individuals win against 2:0 individuals is identified as the estimate of the PW, in this case 1/20. The result is a series of individuals in additional contests (Table 2).

Individuals may have no knowledge of their own relative FA except that gained from experience with former competitors. After each encounter they gain an increasingly accurate estimate of their own relative FA because the variance of the estimate decreases (Fig. 4). For any contestant, information and experience about personal FA and the PW can be gained from the win:loss record. This information may include the identity of individuals it has defeated or by whom it has been defeated. Individuals with no prior experience and no information about the experience of an opponent have a best estimate of 1/2 for PW (Landau 1951).

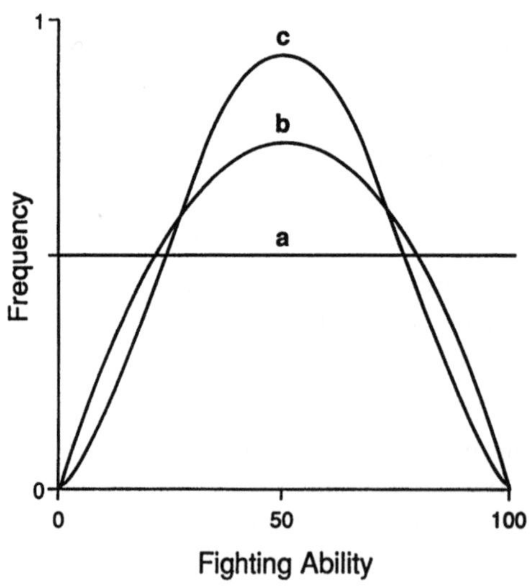

Fig. 4. Frequency distribution of individuals by fighting ability. a) with no record, b) individuals with a 1:1 win:loss record, and c) individuals with a 2:2 record. In all three cases the PW = 0.5 against an unknown opponent in the next contest.

After each contest the estimate of the probability of winning future contests is updated (Table 2). A one-time loser has only a 1/3 chance of winning the next contest against an unknown opponent. The PW is 2/3 for a winner of a single contest (Slater 1986, based upon the same logic). The PW for successive losers is also predictable. For individuals with no prior contest experience the PW against unknown opponents following successive losses, is 1/3, 1/4, etc. Similarly, the PW for a one-time loser winning against a one-time winner is 1/6. Values in Table 2 show that individuals need to participate in relatively few contests to identify their prospects against relevant opponents. A mixed record of wins and losses will yield some intermediate PW for a subsequent contest, and the more contests engaged in the better will be the potential to estimate relative FA.

In addition to personal contest history the precision of estimates of PW can be increased if the win:loss record of future opponents is observed or can be estimated. An inexperienced observer has information to predict that its PW is 1/3 against a one-time winner and 2/3 against a one-time loser (Table 2). Slater (1986) makes this argument as an alternative to Chase's (1985) interpretation of outcomes following initial encounters among triads of chickens. Extension of Table 2 predicts that after five consecutive losses an individual will have a 14% PW against an unknown opponent, and after five consecutive wins, an 86% PW. In experiments with field crickets (*Teleogryllus oceanicus*) Burk (1983) found values of 18% and 87% after five losses and five wins respectively, in close agreement with our predictions. If animals are responsive to, or behave as if they are sensitive to, the probability of winning contests with specific contest experiences, contest patterns (see "Costs" below) will reflect these probabilities.

In suggesting that animals estimate their probability of winning from previous contests we assume: (1) that animals make assessments based on available information regarding relevant conflict variables including relative FA, (2) that information is imperfect and probabilistic, (3) that information about PW is derived from direct experience, and (4) that decisions reflect this information, modifying subsequent contest behavior. By suggesting that the fundamental assessment of contest variables includes an individual's estimate of its PW based on experience, we mean only that their responses are expected to reflect the distributions of FA and PW based on the fact that variance (uncertainty) in PW decreases with increasing number of encounters (experience). This pattern of decreasing uncertainty with increasing experience provides probabilistic information about PW applicable to all fighting animals.

DETERMINANTS OF CONFLICT OUTCOMES – MOTIVATION

Fighting ability may or may not determine encounter outcomes. Here we consider social, environmental, and physiological determinants facilitating, regulating and inhibiting behavior when animals compete to determine their social status and to gain access to resources. We identify a minimum complete set of parameters and their components contributing to contest dynamics.

The extent to which individual animals may evaluate and utilize alternative sources of information, identified as determinants in this paper, varies by species and in some cases between (see Hammerstein and Riechert 1988; Worthman 1990) and within populations. Additional terms delineating the processes occurring during animal conflict are identified and their utility in identifying the determinants of contest outcomes considered.

Resource Value (V)

Resource value (V; Parker 1974; Maynard Smith and Parker 1976; Parker and Rubenstein 1981) is another fundamental parameter of game theory models which represents in various ways a contested resource and the payoff to the winner of securing that resource. As with RHP, usage of the parameter varies widely in the literature, primarily because it subsumes several context and state dependent variables. V can be a resource related value or the context and contestant specific value of winning. Hammerstein and Riechert (1988) identify V as the fitness value of winning less the fitness cost of not contending. Others (e.g., Wells 1988) use V to denote the relative value of a resource. In addition V is identified as assessed or estimated resource value (Parker and Rubenstein 1981). We adopt this latter usage here because it is best suited to interpretation of the behavior of contesting individuals acting on estimates of a resource's value. However, a single parameter may be inadequate to accommodate the wide range of phenomena it is intended to represent in behavioral analyses of the determinants of conflict outcomes.

One difficulty with V has resulted from a confounding of the concept of payoff asymmetries for the contestants. A resource, such as the caloric value of a food item or a potential mating, has some absolute value independent of the contestants. There is no asymmetry in the absolute value of any resource. Where absolute resource value varies in the environment, for example the quality of a territory, an asymmetry in information about, and therefore perception of V may exist as between a resident and an intruder (cf. Parker 1974; Riechert 1979; Ewald 1985; Beletsky and Orians 1987a, b; Stamps 1987).

Additional complexity in interpreting V is added by asymmetries in the relative value of a resource, as those due to differences in the residual reproductive value of contestants (Williams 1966; Fisher 1930; Trivers 1972; Pianka 1976), which may exist independent of asymmetries in contestants' perception of absolute resource value. Residual reproductive value, called "expected future reproductive success" (EFRS) by Grafen (1987), is expected to bear an inverse relationship to contest effort (Grafen 1987) or V. In seasonally semelparous animals (e.g., anthophorid bees, Alcock et al. 1977; speckled wood butterflies, *Pararge aegeria*, Davies 1978; digger wasps, *Sphex ichneumoneus*, Brockmann et al. 1981; spiders, *A. aperta*, Hammerstein and Riechert 1988) and anadromous salmonids, the effect of EFRS on V in conflict may be precluded, because deferring reproduction to another season is not an option. This 'desperado' behavior (Grafen 1987) generally is ignored in empirical analyses and interpretations of animal conflict and we know of no examples of desperado behavior in nature except among semelparous animals (e.g., marsupial mice, *Antechinus*, Cockburn and Lee 1988).

Costs

For most animals fighting has costs to both winners and losers. On the average costs are greater for losers than winners (captive salamanders, Jaeger 1981; captive fish, Barlow et al. 1986). Costs include injuries and/or losses of condition (e.g., energetic costs, Smith and Taylor 1993) and thus reduce present FA, thereby reducing the relative PW future contests for both contestants (Grafen 1987; Guhl and Allee 1944; Maynard Smith 1974; Jaeger 1981). Individuals may depreciate their perception of their FA following losses, further reducing the likelihood they will enter into or persist with additional contests. On multipurpose or feeding territories there may be a positive feedback relationship between successful and future outcomes because the condition of winners is enhanced by gaining access to limited resources (Barnard and Burk 1979). In other cases persistent defense of resources such as mates may reduce condition (FAc) and thus the probability of maintaining tenure (ungulates holding harems, Geist 1971; pinnipeds, Le Boeuf 1974; caciques, S. Robinson, personal communication, but see Thorpe et al. 1995). The feedback relationship between contest outcomes and growth (FAd) is a conspicuous feature of the development of fish, whose adult size may vary widely as a consequence of social dominance and access to resources, resulting in individual differences in growth rates (Magnuson 1962; Abbott et al. 1985).

In contests with high fitness cost, we expect losers to assess their relative ability and to avoid probable defeats. Even if an individual has a relatively low FA it usually will have some nonzero PW future contests and some positive EFRS (Grafen 1987). If there is little or no EFRS, fights are expected to be unconstrained and lethal (Enquist and Leimar 1990). Similarly, in contests with low or no costs, fighting for critical resources might continue indefinitely (Clutton-Brock and Harvey 1976), irrespective of PW, and constrained only by time (e.g., beetles, Marden 1987; Otronen 1988). In a population where contest competition is the basis for access to resources, there is some estimated personal FA threshold below which individuals are unlikely to win (Leimar and Enquist 1984) against the established cohort of resource holders. These individuals are unlikely to engage in further contests until their ability or the situation (fewer or weakened resource-holding opponents; reduced V because most breeding opportunities are gone, etc.) improves. The FA corresponding to this threshold will depend upon the distribution of available resources and the proportion of competitors excluded from those resources. A limited series of costly fights is likely to result in prompt settlement of respective tenure at resources (Leimar and Enquist 1984).

The nearly ubiquitous observation of restrained conflict over access to limited or potentially limiting resources, often called conventional settlement, is frequently the result of decisions made in part based upon prior experience and use of information so gained to respond in particular situations to the probability of winning contests against opponents having specific characteristics. Some assume or imply that animals do not compete for resources with other potential contenders to the limits of their ability, instead substituting conventional, ritualized or otherwise limited aggressive behavior (e.g., Maynard Smith and Price 1973; Davies 1978). One of the counterintuitive findings of ESS game theory models is that there may be conventionally settled encounters and fitness gains and losses which are unrelated to ability (e.g., "uncorrelated asymmetries," Maynard Smith and Parker 1976). Although interpretation of the most widely cited example of conventional settlement of contests (speckled wood butterfly, Davies 1978) has been challenged (Austad et al. 1979; Wickman and Wiklund 1983; Grafen 1987), it continues to be cited, without caveat, as an example of conventional conflict settlement relevant to interpretations of conflict in other species (e.g., Beletsky and Orians 1987b; Yokel 1989). An alternative explanation to the possibility that conflict is settled without physical contact between opponents is that costs limit the number and intensity of contests (Geist 1974; Parker 1974; Barnard and Burk 1979; Clutton-Brock et al. 1979; Abbott et al. 1985). Individuals may thus appear to be choosing conventional settlements because their behavior accurately reflects their assessment of their probability of winning.

Risk

Risk as used here is the product of potential costs and their probability of occurrence. Several authors note that risk is a parameter relevant to models of conflict behavior (Maynard Smith and Parker 1976; Hammerstein 1981; Parker and Rubenstein 1981; Rubenstein 1982; Enquist and Leimar 1983), but others fail to distinguish risk from cost (e.g., Andersson 1982; Montgomerie and Weatherhead 1988). Risk associated with fighting behavior varies among species, between populations and between and for individuals through time.

Assessment of risk may be a determinant of decisions to fight for a resource (Fig. 5). Rohwer (1982) describes a theoretical inverse curvilinear relationship between risk and the probability of winning. He concludes that at maximum risk both opponents have an equal probability of injuring the other and the probability of either prevailing is 0.50. Maynard Smith and Parker (1976) also concluded

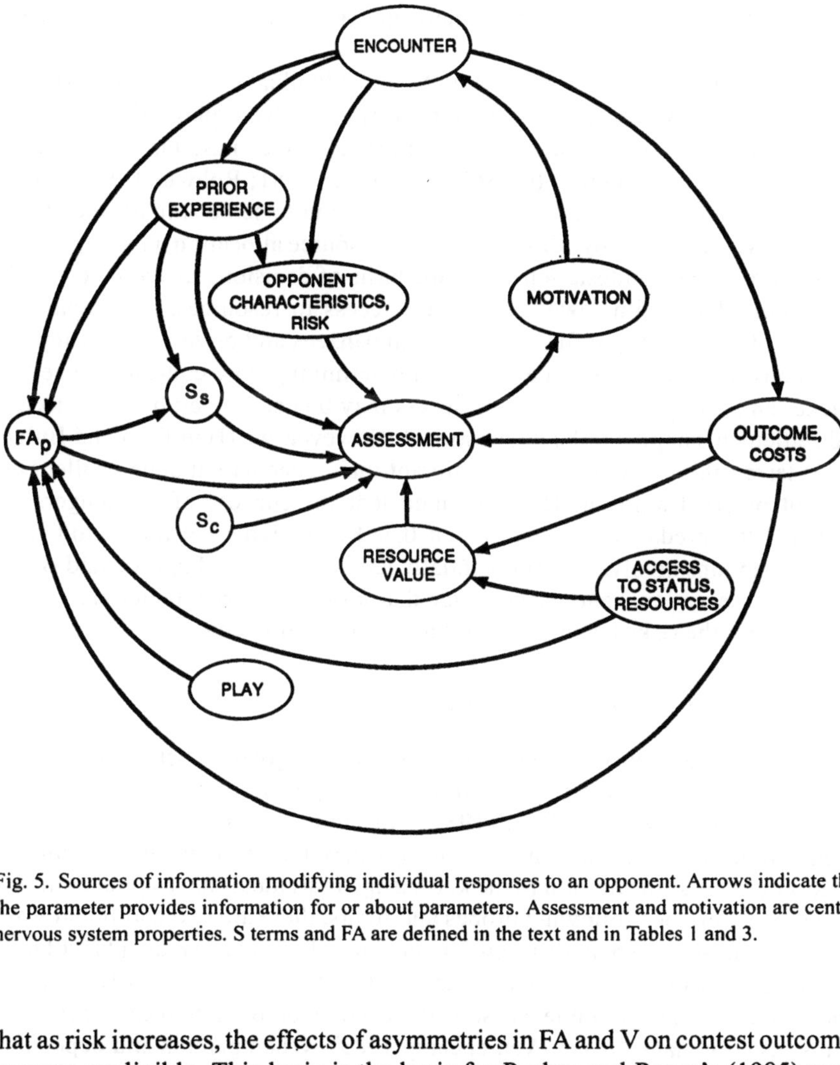

Fig. 5. Sources of information modifying individual responses to an opponent. Arrows indicate that the parameter provides information for or about parameters. Assessment and motivation are central nervous system properties. S terms and FA are defined in the text and in Tables 1 and 3.

that as risk increases, the effects of asymmetries in FA and V on contest outcomes become negligible. This logic is the basis for Packer and Pusey's (1985) comparative analysis of risk, noted below.

Costs of contests and their probability of occurrence may vary independently. The probability of costly injury, for example, from conflict may be negligible in some taxa, while the energetic costs of conflict may be great (e.g., beetles, *Physodesmia globosa*, Marden 1987; *Coprophanaeus ensifer*, Otronen 1988; hummingbirds, *Archilochus alexandri*, Ewald 1985; newts, *Notophtalmus viridescens*, Verrell 1986). Risk may be relatively high in other species (e.g., spotted hyenas, *Crocuta crocuta*, Kruuk 1972; lions, *Panthera leo*, Schaller 1972; American buffalo, *Bison bison*, Lott 1979; red deer, *Cervus elaphus*,

Clutton-Brock et al. 1979; North American cervids, Geist 1986; velvet swimming crabs, Thorpe et al. 1995).

As risk increases, the probability that an individual will enter contests may decline. When risks are high, as for animals armed with morphological weapons, and V is relatively low, assessments of prospects by contestants may result in settlements without fighting (Geist 1974; Berger 1981; Rohwer 1982; Rubenstein 1982). Thus, because of high risks of fighting to some animals the effects of large asymmetries in fighting ability upon resource allocation may be partially or fully obscured. Contests settled on the basis of decisions to avoid costs may allow individuals with lower FA to gain access to resources. A comparative analysis of the risk of conflict to lions, chimpanzees and olive baboons appears to demonstrate the importance of risk in determining how contests are settled (Packer and Pusey 1985). Resource access may be determined by prior access, and settlements appear to be conventional. However, we (Hamilton and Bulger 1990) tested this possibility for immigrant male chacma baboons challenging resident paternal alpha males, but found that the success of challengers far exceeded the predicted value of about 0.5. Risky fights between adult male chacma baboons involve use of potentially lethal weapons (Hamilton and Brain, personal observations) and we conclude that resident paternal males are unwilling to accept the risks taken by immigrants with no progeny.

Social Determinants (S) of Contest Outcomes

Early game theory models avoided analysis of contests between individuals known to one another, thus avoiding the complications introduced by prior acquaintanceship (Maynard Smith 1982a). In nature contestants often are known to each other and these social relationships may be important in determining contest outcomes. In addition to the knowledge individuals have about competitors from experience following probes and fights, there are two additional ways social relationships can modify contest outcomes. Individuals may: (1) withhold the full expression of aggressive potential (concessions, Sc) and (2) support others in contests (Ss) (Table 3). Sc is the concession of resources or access to others, usually mates or kin (Vehrencamp 1983). When parents and dependent offspring interact, contest outcomes may not reflect the full potential of partici-

Table 3. Terminology Used in This Chapter to Identify Social Effects in Conflict Situations

Term	Usage	Number of interactants
FA	Personal fighting ability	dyads
Sc	Concessions to relatives, mates and friends	dyads
Ss	Support by third and additional parties	triads

pants because it may not be advantageous to either party to impose high costs upon the other (Parker et al. 1989). We view these self-imposed limits as aspects of motivation (Fig. 5). In some cases Sc may be the outcome of assessment, in others it may be an innate feature of species typical behavior based upon the evolutionary history of contest contexts and consequences. Ss is the intrusion of third and additional parties in support of others, producing outcomes which might be determined otherwise in dyadic contests were no third parties to intrude. We use the term Ss to refer to social support for allies, whether or not they are relatives. Social settings may also occur which suppress the expression of aggressive behavior, as that of dominant ravens (*C. corax*) at carcasses (Heinrich and Marzluff 1991) and resident territorial spotted hyenas allowing commuters through their territory uncontested (Hofer and East 1993).

When social support (Ss) is a determinant of contest outcomes (Scott and Fuller 1965; Gouzoules 1975; Vehrencamp 1979, 1983; Hand 1986; Smuts 1987; Pereira 1989; Noë 1990), it needs to be separately identified and analyzed. In some cases social partners are decisive in determining contest outcomes, obscuring the effects of individual differences in FA. This is the case for some primate coalitions (Packer 1977; Dunbar 1988; Pope 1990; reviewed by Silk 1987; modeled by Noë 1994). Among Japanese macaques (*M. fuscata*, Kawai 1965; Walters and Seyfarth 1987) and yellow baboons (*P. cynocephalus*) adult females attain their mother's rank (Hausfater et al. 1982). Thus, resource access may be largely independent of individual FA, although Dunbar (1988) suggests that individual 'power' (FA) plays a role in determining rank within and between gelada baboon (*Theropithecus gelada*) matrilines. Group size also may be a determinant of outcomes in intergroup interactions among primates (Cheney 1987) and is widely characteristic of social carnivores (lions, Bygott 1979; McComb et al. 1994; Grinnell et al. 1995; spotted hyenas, Kruuk 1972; dwarf mongooses, *Helogale parvula*, Rood 1987; African wild dogs, *Lycaon pictus*, McNutt 1996). Larger family groups of geese tend to prevail in contests over food in nonbreeding situations (Boyd 1953; Hanson 1953; Raveling 1970). Among some territorial animals (fur seals, Getty 1981, 1987; song birds, Jaeger 1981) neighbors collaborate to evict intruders, perhaps stabilizing relationships between individuals in the neighborhood. In circumstances where conflict involves numerical superiority by groups over other groups or of collaborating dyads over single individuals, social support may mask the expression of individual FA such that relatively inferior individuals (<FA) prevail in contests. In this case patterns of settlement such as early settlement, for example, where social support is provided by a neighborhood effect, may be decisive in determining contest outcomes (see below). The fighting ability of savanna baboon coalition partners may, however, be a determinant of the outcome of contests (Noë 1990) and the assessed FA of potential partners may influence their choice of coalition partners (e.g., Pereira 1989; Noë 1994). We identify this relationship between FA and Ss in Fig. 5.

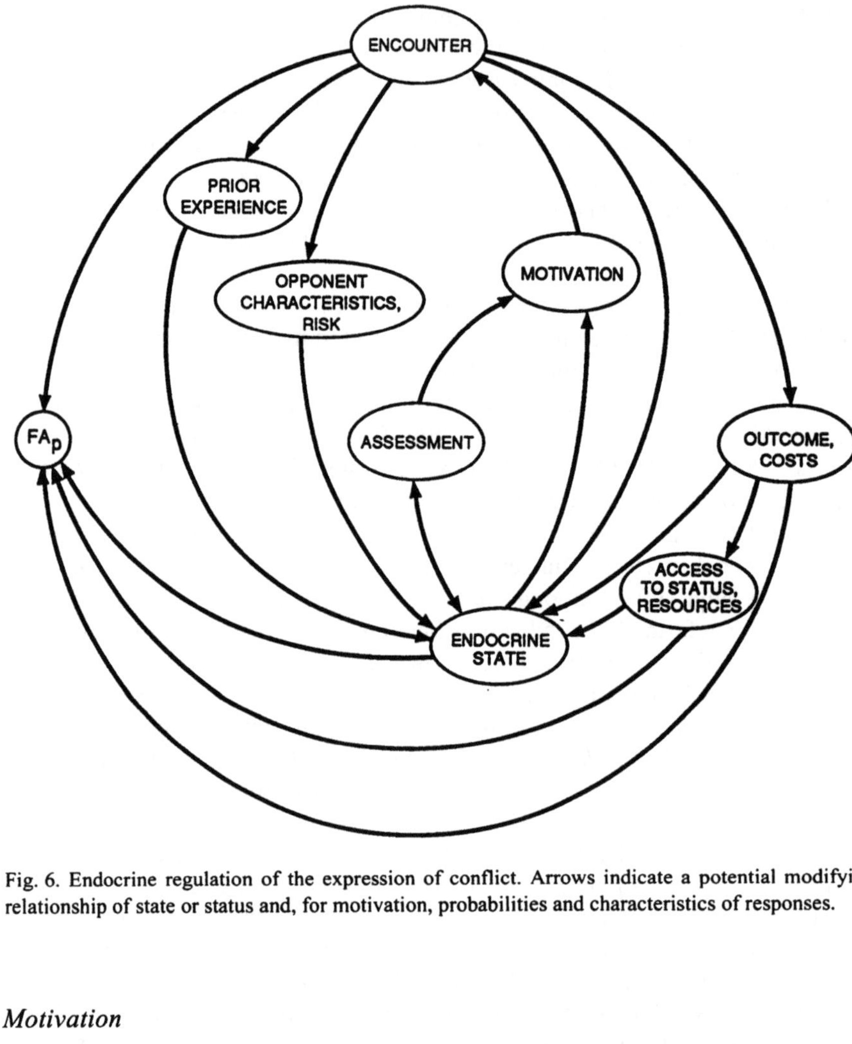

Fig. 6. Endocrine regulation of the expression of conflict. Arrows indicate a potential modifying relationship of state or status and, for motivation, probabilities and characteristics of responses.

Motivation

Motivation is defined here as willingness to engage in behavior. Many authors are at present unwilling to use the term motivation, but the array of surrogate terms such as tendencies, propensities and resource value are inadequate substitutes. Dunbar and colleagues (1990, p. 665) for example, substitute for motivation "...the extent to which a male is prepared to take on...." Excision of motivation from behavioral vocabulary and thinking has led to an ambiguity in some literature about the distinction between ability and motivation (e.g., Hand 1986). Motivation may change in response to changes in ability, but they are separable entities. In game theory the ability-motivation distinction is made between RHP (ability) and a payoff asymmetry in V. In some usage V has become a surrogate term for motivation. Motivation of participants determines when, how long and how vigorously individuals will pursue a dyadic contest they have neither won nor lost. In dyadic interactions, individuals may engage in a contest from similar or differing motivations, dispositions, or assessments of resource values. Multiple pathways leading to motivation (Fig. 5) suggest the need to evaluate outcome determinants from a multivariate perspective. Motivation may be modified by long-term associations with others, especially relatives, mates and coalition partners (Sc). The relative value of resources may add to or subtract from motivation, depending upon assessment of several parameters. Prior experience in contests, including play, sparring and previous fights provide a source of information for individuals about their probability of prevailing, which should affect motivation. Costs likely to accrue in contests also could modify motivation. Dynamic changes in motivation have been modeled by Parker and Rubenstein (1981) who note the difficulty of measuring changes in assessments of resource value (V). But motivation cannot be set equal to resource value because V, however defined, is only one of several determinants of motivation in contest behavior (Figs. 2, 5, 6).

Mobilization Responses

The timing and intensity of conflict are regulated by endocrine changes inducing changes in motivation and condition. Endocrine state changes correlated with and induced by hormones are also changes in condition (FAc). Since we wish to limit parameters insofar as possible, we include these changes within endocrine state and, if they modify ability, within FAc. Others may wish to add a term to the FA components (Table 1) to distinguish the effects of endocrine state upon contest outcomes. If so, it will be useful to distinguish between that part of the endocrine response which enhances ability and that which modifies motivation. This relationship is represented in Fig. 6 where the effect of endocrine state upon the expression of FA is noted.

Adaptations to the regulation of advantageous timing of maximum competitive ability include the following:

Endocrine Regulation. Endocrine regulation of motivation is achieved by neural and endocrine changes which may modify responsiveness in subsequent encounters and to particular challenges. These regulatory processes are tuned to seasonal opportunities for mating and modify the timing of competitive behavior (Worthman 1990; Wingfield et al. 1990).

Spatial Patterning of Movements. Building or rebuilding energy reserves may be achieved by managing one's social environment. For example, by leaving social groups, individuals can limit intraspecific competition (African cape buffalo, Prins 1989) and behavioral subordination and suppression of endocrine status (long-tailed macaques, *M. fascicularis*, van Noordwijk and van Schaik 1985; monkeys (*Saimiri sciurus*), DuMond and Hutchinson 1967).

Stress

Stress is a fundamental physiological process which can have profound consequences on contest behavior, and needs to be considered in an adaptive context. Often the suite of physiological parameters that comprise the stress response are viewed as a chronic state and a cost. The adaptive mobilization of resources through the reallocation of energy reserves to available energy has received less attention. Earlier group selection interpretations of the role of stress in the natural regulation of densities (Christian and Davis 1964) were rejected (e.g., Ricklefs 1979) but have not been replaced by adequate alternative hypotheses (but see Pianka 1988). The stress response, earlier described as the general adaptation syndrome (GAS; Selye 1971) initially increases availability of energy, resulting in prompt enhancement of performance (Sapolsky 1987). (Selye's (1971) physiological 'adaptation' term differs from adaptation in evolutionary biology.) In conflict situations, where opponents are highly competitive and have nearly equal FA, mobilization of available reserves and their conversion to available energy (Sapolsky 1987), adaptive in an evolutionary sense, changes condition (FAc) while imposing costs (Sapolsky 1987). The stress response is part of a suite of endocrine changes tuning motivation and available energy to adaptive lifetime outcomes.

Studies of the behavior and endocrinology of African elephants illustrate the limits of application of game theory to analysis of empirical conflict behavior. Male elephants undergo long-term changes in their motivation to fight (Poole 1987, 1989a), paralleling the behavioral (Poole 1987) and endocrine (Hall-Martin 1987) state of musth, which determines contest outcomes between musth and nonmusth contestants (Poole 1989a). Musth is initiated by individuals in good condition (Poole 1989b), and is correlated with high circulating testosterone levels (Hall-Martin 1987). However, prolonged periods of musth with costs of reproductive efforts, including exceptionally long treks (Hall-Martin 1987) and fights with other males (Hall-Martin 1987), result in deterioration of condition (Poole 1989a). Using game theory parameters RHP and V, Poole (1989a) interpreted her observations as a demonstration of the role of resource value (V) rather than RHP in determining contests because poorer-condition musth males prevail over nonmusth contestants (Poole 1989a). Shoehorning such detailed patterns of conflict into the limited relationships prescribed by game theory oversimplifies the already empirically achieved level of understanding. Changes in the social milieu, endocrine state and current condition all play a part in determining the motivation of elephants in contests (Poole 1987, 1989a, b). At the very least, elephant male-male conflict is too narrowly defined if the set of determinants used to analyze the behavior does not include other relatively proximate and temporally dynamic variables such as the endocrine state of the contestants.

APPLYING THE MODEL

Linear Dominance Hierarchies

There is uncertainty about the relationship of fighting ability (FA) to rank (Landau 1951) and the maintenance of linearity in hierarchies (Landau 1965; Silk 1987). Barnard and Burk (1979) treat hierarchies and their maintenance from the perspective of probable dyadic contest outcomes. Their analysis emphasizes the avoidance of contests by incompetent individuals. The relationship of numbers of individuals in hierarchies (Landau 1965; Chase 1984) to their maintenance and character has been unclear. Comparison of the FA of individuals in different linear dominance hierarchies (LDHs) has seemed impractical because ordinal rank of FA is not directly comparable to FA ranks in other hierarchies, although cardinal indices resolve this problem when outcomes are not fully linear (Boyd and Silk 1983). Use of randomly assigned FA values avoids the problem of estimating probabilities of success on the basis of rank-ordered abilities alone. To simulate the possibility of increasing rank as a function of changing social group, we make the following assumptions about LDHs:

1. There is an exact ordering of individuals by FA among group members. That is, there is a linear dominance hierarchy based upon FA. Individuals with higher FA hold higher rank.
2. The distribution of FA within a population or social group is, on the average, the same as that of neighboring populations.

The number of FA values randomly drawn from 0 to 100, as for dyadic contests above, equals the number of individuals in the group. Rank ordering of these values identifies individual ranks in LDHs. It also allows comparison of ranked individuals from one group with individuals from any other. From the results of these simulations we can address the question: What is the probability of acquiring a higher rank by joining a different group?

Given the stated assumptions, every individual has a finite probability of entering any other hierarchy at each rank, depending upon the number of individuals in that hierarchy and the FA of the contender (Fig. 7). Others have suggested that an individual might improve status in a hierarchically organized society by choosing to enter another group where weaker individuals comprise the membership (Clutton-Brock and Harvey 1976; Cheney and Seyfarth 1983; Pusey and Packer 1987). We evaluate this possibility using a group size of 10 individuals organized in an LDH on the basis of FA. By repeatedly comparing individuals from one random group with others, we simulate contests and establish the frequency that individuals can increase rank by moving to another group (a neighboring LDH). We use these frequencies to estimate PW an alpha

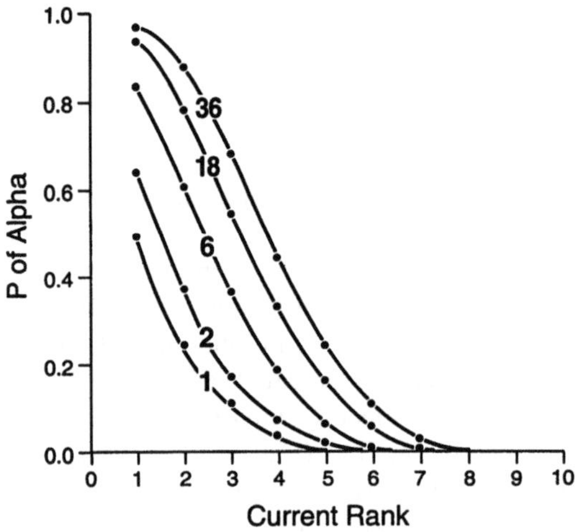

Fig. 7. The probability of an individual rising to alpha status by moving to another same-size hierarchy derived from simulations. Probability is a function of former rank (shown here for group size = 10) and number of opportunities (groups) for immigration. Ranks are based upon fighting ability (FA) and hierarchically organized groups are randomly drawn. (1) shows probability for one alternative hierarchy for initial ranks 1 - 10; (2) for two alternatives; (6) for six alternatives; (18) for 18 alternatives; and (36) for 36 alternatives.

rank elsewhere (i.e., the immigrant beats the highest rank in neighboring LDHs). The probability of an immigrant rising in rank by moving to another group is a function of its preemigration rank and the number of groups it is able to challenge. Figure 7 illustrates the increases in PW an alpha position with the number of LDHs challenged. For example, any fifth-ranking individual from a group of 10, challenging only one other LDH has a PW = 0.016 of rising to alpha status. This PW increases to 0.075 if the contestant could choose among six randomly drawn same-size groups. By comparison, a rank rise from second to first rank becomes highly probable (>0.66) if more than one group is contested. This same method will similarly identify the probability for successful entry from any rank to any other rank in variable-size hierarchies.

In practice, the increase in probability of promptly rising in rank by challenging a large number of cohorts is less than that specified by Fig. 7 because the immigrating individual may incur costs and lose FA during successive challenges. Values from the simulations show that choice of another same-size hierarchy in the same population seldom will offer more than a gain of one rank if the FA of all cohort members is independent and varies independently, i.e., if there is no social support (Ss) from other individuals during the process of rank establishment. Only immigration to a smaller group or some change in parameters other than FA determining contest outcomes will provide any substantial opportunity for rank rises by changing group affiliation. It is thus improbable that inferred substantial rank rises to alpha status by chacma baboons transferring between groups of similar size (Hamilton and Bulger 1990) can be explained by immigration into other less competitive groups. We suggest that the observations are most probably explained by differences in motivation. A choice of weaker groups is unlikely to provide a substantial gain in rank in hierarchically organized societies.

The Outcome Effect

An animal's contest behavior may reflect information about relative FA in two distinct ways: (1) how to fight (FAe) and (2) when to fight (motivation). Here we are concerned exclusively with the latter. Experience of contests won or lost may have positive or negative effects, modifying motivation and thus the probability of prevailing in subsequent contests (Ginsburg and Allee 1942; Alexander 1961; Burk 1983; Franck and Ribowski 1987, 1989; Jackson 1991; Ribowski and Franck 1993b). Such changes in motivation are not necessarily in FAE because they do not necessarily change ability. We view changes in motivation as adjustments of the criteria for deciding when to enter contests based upon perceived relative FA and an estimate of the probability of winning.

Changes in motivation in response to outcomes often involve changes in endocrine states (Baptista et al. 1987; Wingfield et al. 1987). The probabilities identified here suggest a relationship between behavior changes and correlated hormonal responses (Harding and Follett 1979; Harding 1981; Hannes et al. 1984; Wingfield et al. 1987) following changes in individual perceptions of social status (Worthman 1990), including those associated with resource ownership. When ownership of resources reflects prior contest history, i.e., that the owner has a higher FA than previous owners, contestants may behave in subsequent contests as if they had gained information about their relative FA, changes that might be correlated with, if not modulated by changes in endocrine status mediated by the central nervous system.

For many animals prior experience modifies contest behavior (Bronstein 1985; Wallen and Wojeiechowski-Metzlar 1985; Popp 1988). The consistent result of experiments in which trained or selected winners are pitted against equivalent losers in laboratory settings with same-age same-sex subjects is that consistent winners and losers behave as if they expect to win or lose subsequent contests (Ginsberg and Allee 1942; Franck and Ribowski 1987). If animals have evolved to respond to opponents on the basis of probable costs and outcomes, their current motivation will reflect these probabilities. A classical explanation for this experimentally identified prior outcome effect is that an attitude or state predicting and in part determining outcomes develops (Kahn 1951). The outcome effect, experimentally demonstrated and analyzed by psychologists, has not been incorporated into game theoretic studies. Incorporation of the fundamental probabilities of Table 2 into analyses and interpretations of conflict reconciles these earlier observations with cost/benefit analyses. Experimental demonstrations of outcome effects may result from an adaptive response to probable outcomes.

The Resident Effect

The resident effect is a corollary to the outcome effect. It is the greater than chance probability that an owner or resident will retain its residence when challenged by an intruder (Davies 1978; Krebs 1982; Rohwer 1982; Barnard and Brown 1984; Beletsky and Orians 1987b). The resident effect appears to be characteristic of most animal contests for space and will in part be based upon prior outcomes. If the outcome of dyadic contests is determined by relative FA, ownership of resources in successive encounters will increasingly correspond to the distribution of FA among competitors (Leimar and Enquist 1984).

To investigate the estimated role of FA upon resident effects we simulate a series of contests between all population members with half of all contestants excluded from residences and with contest outcomes determined by FA. A single first round of dyadic contests between randomly drawn contestants, produces half winners (residents) and half losers (nonresidents), for which, on the average, 2/3 of all eventual (after repeated contests between residents and nonresidents) residents are already situated at resources they will hold at the end of contests. If losers are randomly paired with a winner (resident) in a second round, their average PW = 1/6, and decreases following additional losses. Probabilities for successive rounds given the stated conditions are illustrated graphically in Fig. 8. The difference between the curves for successive rounds illustrates the decreasing probability of changing either resident or nonresident status in successive rounds, i.e., it produces a resident effect.

If FA is the basis for resolution of contest outcomes and if there is an ideal despotic distribution of resources (Fretwell and Lucas 1969), the minimum relative FA necessary to confer resident status will approximate the proportion of nonresidents in the population. If, for example, 50% of the physically mature population are nonbreeders as observed by Smith (1978) for the rufous-necked sparrow (*Zonotrichia capensis*) and 52% by Jenkins et al. (1967) for red grouse

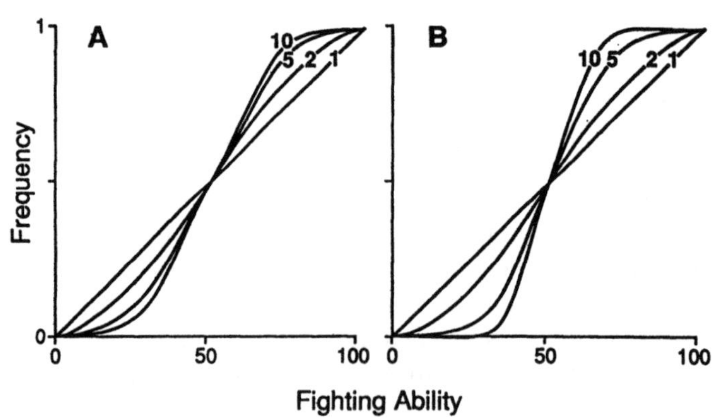

Fig. 8. Frequency distribution of fighting ability (FA) for winners after 1 through 10 rounds of encounters if resources are adequate to accommodate 50% of contestants. A, for 10 contestants and B, for 100 contestants. The frequency distribution curves are for winners after 1, 2, 5 and 10 contests. Frequency distribution curves for losers are mirror images of those shown here for winners.

(*Lagopus scoticus*), the initial probability prior to a contest, of establishing or maintaining a territory is 0.5; the FA necessary to acquire and retain a territory is on average ≥50. With varying ratios of nonresidents to residents, probabilities for initial settlement of individuals by FA on residences and their subsequent rearrangement will follow. There is a predictable error in the degree of correspondence between individual ability and realized access to contested resources (Leimar and Enquist 1984), which is a function of both deterministic and stochastic contest costs (which when combined with probability equals risk), which limit the number of contests and thus the availability of perfect information to competitors. While relative FA may not explain all resident effect observations, the model provides a quantitative basis for evaluating the extent to which it does.

One of the conditions of the model is that no two individuals have the same FA. However, given the lifetime schedule of FA (Fig. 1) it follows that two individuals must occasionally have identical FA when the curves for individuals with opposing trajectories intersect (Dunbar 1988) or as FA approaches its developmental asymptote (FAmax). The rate at which FA changes throughout the lifetime and according to individual lifetime experiences varies by species. Stability of resource apportionment is expected to reflect the relative stability of FA distributions. The degree of correspondence between FA and access to status and resources can be used to analyze empirical examples. Territory holders may be, on the average, individuals with higher FA than floaters. Among red grouse, contests in the fall determine which individuals will hold territories over winter and the following spring (Watson and Moss 1971). Nonbreeding rufous-necked sparrows do not hold territories but establish hierarchical rank relationships in the vicinity of particular territories, and the higher-ranking individuals fill breeding site (territory) vacancies (Smith 1978). Reversals between resident and floater status may take place during the breeding season in both red-winged blackbirds and song sparrows (*Melospiza melodia*, Eckert and Weatherhead 1987; Smith and Arcese 1989). In our model these reversals may be accounted for by a change in the FA of residents or opponents. Changes in FA may include: (1) a loss in condition (FAc) of replaced residents and/or (2) a gain in condition (FAc) or experience (FAe) by new winners during the breeding season. Observations of red-winged blackbirds and song sparrows suggest that outcomes are largely determined by condition and/or recent experience (Searcy 1979; Smith and Arcese 1989; Shutler and Weatherhead 1991, 1992).

Lack of a correlation between morphological characteristics (FAd) and outcomes does not exclude FA as a determinant of contest outcomes. Eckert and Weatherhead (1987) found that in one red-winged blackbird population floaters were morphologically indistinguishable from residents. Nor do song sparrows who gain residences appear to differ morphologically from those who do not (Smith 1988). Crucial dyadic contests for resources often take place within cohorts of same-sex competitors near their maximum fighting ability (FAmax).

In a population such individuals are expected to have nearly equal FA (e.g., Jones 1987). Thus we expect exclusion of individuals with lower FA from aggressive competitive arenas to reduce or eliminate measurable morphological differences between these competent competitors.

Despite extensive experimentation and observation the relationship of FA, resource value (V) and experience to territory ownership in red-winged blackbirds remains unresolved. Several authors agree that FA is not an obvious determinant of successful territory ownership (Beletsky and Orians 1987b; Shutler and Weatherhead 1991). First ownership status of experimentally removed territorial males conveys an advantage in contests for the original territory if the original male is released within two days (Beletsky and Orians 1987b; Shutler and Weatherhead 1992). When floaters occupy experimentally vacated territories, the basis for their success needs to be compared with the relative FA of individuals in the floating population (cf. Smith 1978). In the case of red-winged blackbirds the FA of residents and the males who replace them does not appear to differ sufficiently to overcome the experimentally reversed residence asymmetry (Shutler and Weatherhead 1991). An alternative explanation for manipulated resident-floater reversals, reflects the dynamic nature of estimates of FA. If the resident effect, in terms of the estimate of relative FA increasing because of ownership, takes effect over a period of time as a consequence of the residence position, then motivation to fight to retain a territory may increase the longer the territory is held.

In territorial systems for two neighboring residents the PW is 0.5 for each, should they contest. That estimate may be refined by observing the contest behavior of a neighbor against other neighbors or intruders (Freeman 1987) or from direct encounters with that neighbor. The extent to which there may be an ideal despotic distribution by FA will depend upon: (1) the relative FA of opponents and their win:loss records and (2) the ability of individuals to identify superior territories. We expect the frequency of contests with neighbors to be cost limited, and the relationship of relative FA to territory quality to be information limited and thus imperfect. A frequent explanation for the resident effect is that resources such as territories are more valuable to owners than they are to nonresidents because of an information asymmetry (e.g., Parker 1974; Riechert 1979; Rohwer 1982; Leimar and Enquist 1984; Ewald 1985; review in Enquist and Leimar 1987). Because successful reproductive opportunities often depend upon acquisition of limited and localized resources, it is not obvious why critical resources should be more valuable to owners than they are to same-age same-sex non-residents. In nature, respect for ownership and the resulting resident effect are inferred following episodic observations. To identify whether individual fighting ability determines access to limited resources (van den Berghe 1988) long-term diachronic observations identifying successive conflict contexts and contest outcomes are essential.

SUMMARY

Fights in nature are, for most species, relatively infrequent events. This has given rise to a widely held view, especially in popular literature and in narrations of filmed depictions of animal conflict, that contests are decided on a ceremonial or conventional basis. If escalated encounters are infrequent relative to an animal's lifetime, as with savanna baboons and African wild dogs, routine observation, even over fairly long intervals, may not yield direct observations of fighting. This may be particularly frustrating to enlightened analysis if fighting, or the possibility of fighting, is considered to be an important experience shaping the social behavior of a particular animal species. However, it is usually the case, at least among vertebrates, that responses to opponents are based in part upon prior experiences. If individual contest costs are high, individuals can afford relatively few fights. From a cost-benefit perspective, costly and high-risk fights are predicted to be infrequent. Our model of probability estimates for winning contests illustrates that relatively few fights are required for an individual to assess accurately its probability of winning, particularly if there is information regarding previous successes by resource-holding individuals.

In addition to identifying how prior contest experience can provide information about current relative fighting ability, we have identified other conflict variables which may modify the probability of entering contests by modifying motivation. These include variables such as the endocrine state and social support systems which fluctuate on varying temporal scales throughout an individual's life.

Game theory as a theoretical paradigm for analysis of conflict behavior has provided us with predictions from a set of testable hypotheses about animal conflict at the population level (Maynard Smith and Parker 1976; Parker and Rubenstein 1981), and the framework continues to guide the study of animal conflict. However, vague reference to a game theoretic model has become standard practice in many current studies of conflict behavior in spite of uncertainty in interpretation of discrepancies between theoretical predictions and empirical observations. These uncertainties are due in part to constraints imposed by limiting models to broadly generalized and thus imprecisely defined parameters. Simplified formulations of game theory may be inadequate to guide evaluation of the determinants of contest behavior and outcomes because the determinants of fighting ability and motivation are subsumed by terms which include separate and separable entities and may have opposing effects.

As in game theory, we assume that patterns of conflict behavior have an evolutionary history. But we do not impose the condition of evolutionary stability on observed conflict behavior because we assume that population level strategy sets derived from ESS modeling identify a minimum metric of variability that any given individual might utilize in contests depending on specific contexts, and we focus on investigating mechanisms of that behavior at the level of individual contestants. Proximate mechanisms of behavior mediate broadly flexible behavior of individuals on a temporal scale that ultimate explanations do not necessarily accommodate. Identification of parameters influencing competitive behavior has been a fundamental feature of the experimental analysis of animal behavior and the development of our current understanding of behavior, both from the perspective of evolutionary ecology and the study of mechanisms. Analyses of the determinants of behavior are proceeding effectively in disciplines which investigate complex parameters independently from game theoretic models of contests and their consequences. We find these studies to be effective in advancing our understanding of behavior and advocate a continuation of this approach to the analysis of conflict behavior. We have identified a complete set of parameters which may be used to evaluate contests even in complex contexts.

To model the relationship of fighting ability to patterns of status and resource acquisition among competing individuals, we describe frequency distributions of estimated personal FA and the resulting future probability of winning following a limited number of cost-free contests. These fundamental probabilities may pattern the behavioral responses of animals to their opponents. Costs of fighting constrain opportunities to gain experience and introduce error in matching ability and resource acquisition. We identify distribution patterns for the relationship of FA to status and tenure at resources. These probabilities provide a basis for evaluating individual conflict behavior.

Sequential interactions between opponents produce predictable frequency distributions of FA within populations which contain information about an individual's ability relative to that of other population members. Sequences of encounters result in individual win:loss records that are incorporated into an individual's experience and from which estimates of the PW in subsequent encounters can be made. Deviations from these quantitative predictions identify the degree to which other parameters (e.g., costs, risks, value asymmetries) determine outcomes. Contest strategies may be viewed as the consequences of developing and labile abilities and motivations whose components are conceptually and empirically separable. Studies of classical learning and hormonal responses (Worthman 1990) identify potential mechanisms mediating appropriate behavioral responses to probabilities of prevailing in future contests based upon information providing estimates of individual relative FA.

Analyses of conflict need to identify initial status of contestants and costs of encounters and the potential for probabilistic responses to them. It is likely that different patterns for the expression of FA components and other parameters will be found in different species and populations. Since social primates rely more upon cognition and less upon internal states in their responses to social companions, it follows that they may be able to make more externally informed assessments (Bercovitch and Goy 1990; Worthman 1990). We expect primates to be more precise and to consider more information than most other animals and so to be less constrained by prior experience and long-term states. We expect that social primates will respond in conflict situations based upon active decision making. They will assess more kinds of information than some other animals, especially those which respond reflexively, guided by comparatively few stimuli. Callithricid monkeys are more dependent upon endocrine control and less susceptible to cortical mediation than the larger primates in more complex social environments (Worthman 1990).

A probabilistic framework based upon the fundamental parameters of animal conflict combined with complete multivariate analysis of contests provides a more promising empirical approach to analysis of behavioral access to resources than paradigms which assume that FA is static or is operationally equal among competitors. Analyzing conflict behavior within the underlying context that individual decisions are based upon assessments of probable contest outcomes has broad applicability and unifies several classical behavior paradigms.

While our analysis here has been confined to fighting ability and to nonhuman animals, the relationships identified here are applicable to the analysis of all abilities, including those of humans.

ACKNOWLEDGMENTS

We thank Roderick Thompson for programming and running simulations. Anne Forcella edited and provided insightful comments on several drafts. We thank R. Bowen, J. Brown, C. Clark, G. Grether, M. Hauser, J. Henshel, C. Hom, Y. Hso, P. Marler, N. Pelkey, M. Periera, G. W. Salt, M. Seely, W. Shields, S. Robinson, K. Winkler, C. Worthman, and J. Yoshimura for their comments on drafts of this paper. We are indebted to the Harry Guggenheim Foundation and the Frankfurt Zoological Society for financial support during field studies. Without the long-term efforts of John Bulger in maintaining field observations of a focal baboon population in Botswana this study would not have been possible. We are grateful to the Office of the President of Botswana and to the Director of Wildlife and National Parks for permission to conduct field studies of baboons and wild dogs in Botswana.

REFERENCES

Abbott, J. C., Dunbrack, R. L., and Orr, C. D. (1985). The interaction of size and experience in dominance relationships of juvenile steelhead trout (*Salmo gairdneri*). *Behaviour, 92,* 241–253.

Adams, E., and Caldwell, R. L. (1990). Deceptive communication in asymmetric fights of the stomatopod crustacean *Gonodactylus bredini*. *Anim. Behav., 39,* 706–716.

Alcock, J., Jones, C. E., and Buchmann, S. L. (1977). Male mating strategies in the bee *Centris pallida* Fox (Anthophoridae: Hymenoptera). *Am. Nat., 111,* 145–155.

Aldis, O. (1975). *Play fighting*. New York: Academic Press.

Alexander, R. D. (1961). Aggressiveness, territoriality, and sexual behavior in field crickets (Orthoptera: Gryllidae). *Behaviour, 17,* 130–223.

Alexander, R. D. (1987). *Biology of moral systems*. New York: Aldine De Gruyter.

Andersson, M. (1982). Sexual selection, natural selection and quality advertisement. *Biol. J. Linn. Soc., 17,* 375–393.

Andren, H. (1990). Despotic distribution, unequal reproductive success, and population regulation in the jay *Garrulus glandarius L. Ecology, 71,* 1796–1803.

Arcese, P. (1987). Age, intrusion pressure and defence against floaters by territorial male song sparrows. *Anim. Behav., 35,* 773–784.

Arcese, P. (1989). Territory acquisition and loss in male song sparrows. *Anim. Behav., 37* 45–55.

Austad, S. N., Jones, W. T., and Waser, P. M. (1979). Territorial defence in speckled wood butterflies: why does the resident always win? *Anim. Behav., 27,* 960–961.

Baptista, L. F., DeWolfe, B. B., and Avery-Beausoleil, L. (1987). Testosterone, aggression and dominance in Gambel's white-crowned sparrows. *Wilson Bull., 99,* 86–91.

Barash, D. P. (1982). *Sociobiology and behavior,* (2nd ed.). New York: Elsevier.

Barlow, G. W., Rogers, W., and Fraley, N. (1986). Do Midas cichlids win through prowess or daring? It depends. *Behav. Ecol. Sociobiol., 19,* 1–8.

Barnard, C. J., and Brown, C. A. J. (1984). A payoff asymmetry in resident-resident disputes between shrews. *Anim. Behav., 32,* 302–304.

Barnard, C. J., and Burk, T. (1979). Dominance hierarchies and the evolution of "individual recognition." *J. Theor. Biol., 81,* 65–73.

Barrette, C. (1977). Fighting behaviour of muntjac and the evolution of antlers. *Evolution, 31,* 169–176.

Barrette, C., and Vandal, D. (1990). Sparring, relative antler size, and assessment in male caribou. *Behav. Ecol. Sociobiol., 26,* 383–387.

Bekoff, M. (1974). Social play in coyotes, wolves, and dogs. *Bio. Sci., 24,* 225–230.

Beletsky, L. D., and Orians, G. H. (1987a). Territoriality among male red-winged blackbirds. I. Site fidelity and movement patterns. *Behav. Ecol. Sociobiol., 20,* 21–34.

Beletsky, L. D., and Orians, G. H. (1987b). Territoriality among male red-winged blackbirds. II. Removal experiments and site dominance. *Behav. Ecol. Sociobiol., 20,* 339–349.

Bercovitch, F. B. (1989). Body size, sperm competition, and determinants of reproductive success in male savanna baboons. *Evolution, 43,* 1507–1521.

Bercovitch, F. B., and Goy, R. W. (1990). The socioendocrinology of reproductive development and reproductive success in macaques. In Ziegler, T. E., and Bercovitch, F. B. (eds.), *Socioendocrinology of primate reproduction* (pp. 59–93). New York: Wiley-Liss.

Berger, J. (1981). The role of risks in mammalian combat: zebra and onager fights. *Z. Tierpsychol., 56,* 297–304.

Bernstein, I. S., and Gordon, T. P. (1980). The social component of dominance relationships in rhesus monkeys (*Macaca mulatta*). *Anim. Behav., 28,* 1033–1039.

Boyd, H. (1953). On encounters between wild white-fronted geese in winter flocks. *Behaviour, 5,* 85–129.

Boyd, R., and Silk, J. B. (1983). A method for assigning cardinal dominance ranks. *Anim. Behav., 31,* 45–58.

Brockmann, H. J., Grafen, A., and Dawkins, R. (1979). Evolutionarily stable nesting strategy in a digger wasp. *J. Theor. Biol., 77,* 473–496.

Bronstein, P. M. (1985). Agonistic and reproductive interactions in *Betta splendens*. *J. Comp. Psychol., 98*, 421–431.

Bulger, J. B. (1993). Dominance rank and access to estrous females in male savanna baboons. *Behaviour, 127*, 67–103.

Burk, T. (1983). Male aggression and female choice in a field cricket (*Teleogryllus oceanicus*): The importance of courtship song. In Gwynne, D. T., and Morris, G. K. (eds.), *Orthopteran mating systems* (pp. 97–119). Boulder, CO: Westview Press.

Bygott, J. D. (1979). Agonistic behavior, dominance and social structure in wild chimpanzees of the Gombe National Park. In Hamburg, D. A., and McCown, E. R. (eds.), *The great apes* (pp. 405–427). Menlo Park: Benjamin/Cummings

Cade, W. H. (1984). Genetic variation underlying sexual behavior and reproduction. *Am. Zool., 24*, 355–366.

Caldwell, R. L. (1986). The deceptive use of reputation by stomatopods. In Mitchell, R. W., and Thompson, N. S. (eds.), *Deception: Perspectives on human and nonhuman deceit* (pp. 129–145). Albany, NY: State University of New York Press.

Chase, I. D. (1982). Dynamics of hierarchy formation: the sequential development of dominance relationships. *Behaviour, 80*, 218–240.

Chase, I. D. (1984). Social process and hierarchy formation in small groups: a comparative perspective. In Barchas, P. R. (ed.), *Social hierarchies* (pp. 45–80). Westport, CT: Greenwood.

Chase, I. D. (1985). The sequential analysis of aggressive acts during hierarchy formation: an application of the 'jigsaw puzzle' approach. *Anim. Behav., 33*, 86–100.

Cheney, D. L. (1987). Interactions and relationships between groups. In Smuts, B. B., et al. (eds.), *Primate societies* (pp. 267–281). Chicago, IL: University of Chicago Press.

Cheney, D. L., and Seyfarth, R. M. (1983). Nonrandom dispersal in free-ranging vervet monkeys: social and genetic consequences. *Am. Nat., 122*, 392–412.

Christian, J. J., and Davis, D. E. (1964). Endocrines, behavior, and population. *Science, 146*, 1550–1560.

Clifford, D. H., Green, K. A., and Watterson, R. M. (1990). *The pit bull dilemma*. Philadelphia, PA: Charles.

Clutton-Brock, T. H., Albon, S. D., Gibson, R. M., and Guinness, F. E. (1979). The logical stag: adaptive aspects of fighting in red deer (*Cervus elaphus L.*). *Anim. Behav., 27*, 211–225.

Clutton-Brock, T. H., and Harvey, P. H. (1976). Evolutionary rules and primate societies. In Bateson, P. P. G., and Hinde, R. A. (eds.), *Growing points in ethology* (pp. 195–237). Cambridge: Cambridge University Press.

Cockburn, A., and Lee, A. K. (1988). Marsupial femmes fatales. *Nat. Hist., 97*, 40–47.

Collias, N. E. (1943). Statistical analysis of factors which make for success in initial encounters between hens. *Am. Nat., 77*, 519–538.

Crespi, B. J. (1986). Size assessment and alternative fighting tactics in *Elaphrothrips tuberculatus* (Insecta: Thysanoptera). *Anim. Behav., 34*, 1324–1335.

Darwin, C. R. (1871). *The descent of man and selection in relation to sex*. London: John Murray.

Davies, N. B. (1978). Territorial defence in the speckled wood butterfly (*Pararge aegeria*): the resident always wins. *Anim. Behav., 26*, 138–147.

Dixon, K. A., and Cade, W. H. (1986). Some factors influencing male-male aggression in the field cricket *Gryllus integer* (time of day, age, weight and sexual maturity). *Anim. Behav., 34*, 340–346.

Dugatkin, L. A., and Biederman, L. (1991). Balancing asymmetries in resource holding power and resource value in the pumpkinseed sunfish. *Anim. Behav., 42*, 691–692.

DuMond, F. V., and Hutchinson, T. C. (1967). Squirrel monkey reproduction: the "fatted" male phenomenon and seasonal spermatogenesis. *Science, 158*, 1067–1070.

Dunbar, R. I. M. (1988). *Primate social systems*. Ithaca, NY: Comstock Publishing Associates.

Dunbar, R. I. M., Buckland, D., and Miller, D. (1990). Mating strategies of male feral goats: a problem in optimal foraging. *Anim. Behav., 40,* 653–677.

Eckert, C. G., and Weatherhead, P. J. (1987). Owners, floaters and competitive asymmetries among territorial red-winged blackbirds. *Anim. Behav., 35,* 1317–1323.

Enquist, M., and Leimar, O. (1983). Evolution of fighting behaviour: decision rules and assessment of relative strength. *J. Theor. Biol., 102,* 387–410.

Enquist, M., and Leimar, O. (1987). Evolution of fighting behaviour: the effect of variation in resource value. *J. Theor. Biol., 127,* 187–205.

Enquist, M., and Leimar, O. (1990). The evolution of fatal fighting. *Anim. Behav., 39,* 1–9.

Estes, R. D. (1969). Territorial behavior of the wildebeest (*Connochaetes taurinus* Burchell, 1823). *Z. Tierpsychol., 26,* 284–370.

Ewald, P. W. (1985). Influence of asymmetries in resource quality and age on aggression and dominance in black-chinned hummingbirds. *Anim. Behav., 33,* 705–719.

Festa-Bianchet, M., Apollonio, M., Mari, F., and Rasola, G. (1990). Aggression among lekking male fallow deer (*Dama dama*): territory effects and relationship with copulatory success. *Ethology, 85,* 236–246.

Finch, C. E. (1990). *Longevity, senescence, and the genome.* Chicago, IL: University of Chicago Press.

Finsterbusch, C. A. (1980). *Cock fighting.* Surrey: Hindehead, Saiga Publishing.

Fisher, R. A. (1930). *The genetical theory of natural selection.* Oxford: Clarendon Press.

Forsyth, A., and Montgomerie, R. D. (1987). Alternative reproductive tactics in the territorial damselfly *Calopteryx maculata*: sneaking by older males. *Behav. Ecol. Sociobiol., 21,* 73–81.

Francis, R. C. (1988). Socially mediated variation in growth rate of the Midas cichlid: the primacy of early size differences. *Anim. Behav., 36,* 1844–1845.

Franck, D., and Ribowski, A. (1987). Influences of prior agonistic experiences on aggression measures in the male swordtail (*Xiphophorus helleri*). *Behaviour, 103,* 217–240.

Franck, D., and Ribowski, A. (1989). Escalating fights for rank-order position between male swordtails (*Xiphophorus helleri*): effects of prior rank-order experience and information transfer. *Behav. Ecol. Sociobiol., 24,* 133–143.

Freeman, S. (1987). Male red-winged blackbirds (*Agelaius phoeniceus*) assess RHP of neighbors by watching contests. *Behav. Ecol. Sociobiol., 21,* 307–311.

Fretwell, S. D., and Lucas, H. L. (1969). On territorial behavior and other factors influencing habitat distribution in birds. *Acta Biotheor., 19,* 16–36.

Geist, V. (1971). *Mountain sheep: A study in behavior and evolution.* Chicago, IL: University of Chicago Press.

Geist, V. (1974). On fighting strategies in animal combat. *Nature, 250,* 354.

Geist, V. (1986). New evidence of high frequency of antler wounding in cervids. *Can. J. Zool., 64,* 380–384.

Getty, T. (1981). Competitive collusion: the presumption of competition during the sequential establishment of territories. *Am. Nat., 118,* 426–431.

Getty, T. (1987). Dear enemies and the prisoner's dilemma: why territorial neighbors form defensive coalitions. *Am. Zool., 27,* 327–336.

Gibson, R. M., and Bradbury, J. W. (1985). Sexual selection in lekking sage grouse: phenotypic correlates of male mating success. *Behav. Ecol. Sociobiol., 18,* 117–123.

Ginsburg, B., and Allee, W. C. (1942). Some effects of conditioning on social dominance and subordination in inbred strains of mice. *Physiol. Zool., 15,* 485–506.

Gould, S. J., and Lewontin, R. C. (1979). The spandrels of San Marco and the Panglossian paradigm: a critique of the adaptationist programme. *Proc. R. Soc. London B, 205,* 581–598.

Gouzoules, H. (1975). Maternal rank and early social interactions of infant stumptail macaques, *Macaca arctoides. Primates, 16,* 405–418.

Grafen, A. (1987). The logic of divisively asymmetric contests: respect for ownership and the desperado effect. *Anim. Behav., 35,* 462–467.

Grinnell, J., Packer, C., and Pusey, A. E. (1995). Cooperation in male lions: kinship, reciprocity or mutualism. *Anim. Behav., 49,* 95–105.

Groos, K. (1898). *The play of animals.* New York: Appleton.

Guhl, A. M., and Allee, W. C. (1944). Some measurable effects of social organization in flocks of hens. *Physiol. Zool., 17,* 320–347.

Hall-Martin, A. J. (1987). Role of musth in the reproductive strategy of the African elephant. *S. Afr. J. Sci., 83,* 616–620.

Hamilton, W. J., III, and Bulger, J. B. (1990). Natal male baboon rank rises and successful challenges to resident alpha males. *Behav. Ecol. Sociobiol., 26,* 357–362.

Hamilton, W. J., III, Buskirk, R. E., and Buskirk, W. H. (1976). Social organization of the Namib Desert tenebrionid beetle *Onymacris rugatipennis. Can. Ent., 108,* 305–316.

Hammerstein, P. (1981). The role of asymmetries in animal contests. *Anim. Behav., 29,* 193–205.

Hammerstein, P., and Riechert, S. E. (1988). Payoffs and strategies in territorial contests: ESS analyses of two ecotypes of the spider *Agelenopsis aperta. Evol. Ecol., 2,* 115–138.

Hand, J. L. (1986). Resolution of social conflicts: dominance, egalitarianism, spheres of dominance, and game theory. *Quart. Rev. Biol., 61,* 201–220.

Hannes, R. P., Franck, D., and Liemann, F. (1984). Effects of rank-order fights on whole-body and blood concentrations of androgens and corticosteroids in the male swordtail (*Xiphophorus helleri*). *Z. Tierpsychol., 65,* 53–65.

Hanson, H. C. (1953). Inter-family dominance in Canada geese. *Auk, 70,* 11–16.

Harding, C. F. (1981). Social modulation of circulating hormone levels in the male. *Am. Zool., 21,* 223–231.

Harding, C. F., and Follett, B. K. (1979). Hormone changes triggered by aggression in a natural population of blackbirds. *Science, 203,* 919–920.

Hausfater, G., Altmann, J., and Altmann, S. (1982). Long-term consistency of dominance relations among female baboons (*Papio cynocephalus*). *Science, 217,* 752–755.

Heinrich, B., and Marzluff, J. M. (1991). Do common ravens yell because they want to attract others? *Behav. Ecol. Sociobiol., 28,* 13–21.

Hirth, D. H. (1977). Social behavior of white-tailed deer in relation to habitat. *Wildl. Monog., 53,* 1–55.

Hofer, H., and East, M. L. (1993). The commuting system of Serengeti spotted hyaenas: how a predator copes with migratory prey. II. Intrusion pressure and commuters' space use. *Anim. Behav., 46,* 559–574.

Hoff, C. P. (1981). *Dogfighting in America: A national overview.* New York: ASPIC.

Huntingford, F., and Turner, A. (1987). *Animal conflict.* London: Chapman and Hall.

Jackson, W. M. (1988). Can individual differences in history of dominance explain the development of linear dominance hierarchies? *Ethology, 79,* 71–77.

Jackson, W. M. (1991). Why do winners keep winning? *Behav. Ecol. Sociobiol., 28,* 271–276.

Jaeger, R. G. (1981). Dear enemy recognition and the costs of aggression between salamanders. *Am. Nat., 118,* 972–974.

Jakobsson, S., Radesateer, T., and Jarvi, T. (1979). On the fighting behaviour of *Nannacara anomala* (Pisces, Cichlidae). *Z. Tierpsychol., 49,* 210–220.

Jarvis, J. U. M. (1978). Energetics of survival in *Heterocephalus glaber* (Ruppell), the naked mole-rat (Rodentia: Bathyergidae). *Bull. Carnegie Mus. Nat. Hist., 6,* 81–87.

Jarvis, J. U. M. (1981). Eusociality in a mammal: cooperative breeding in naked mole-rat colonies. *Science, 212,* 571–573.

Jenkins, D., Watson, A., and Miller, G. R. (1967). Population fluctuations in the red grouse (*Lagopus lagopus scoticus*). *J. Anim. Ecol., 36,* 97–122.

Jones, J. S. (1987). The heritability of fitness: bad news for 'good genes'? *TREE, 2,* 35–38.

Kahn, M. W. (1951). The effect of severe defeat at various age levels on the aggressive behavior of mice. *J. Genet. Psychol., 79,* 117–130.

Kawai, M. (1965). On the system of social ranks in a natural troop of Japanese monkeys, 1: Basic rank and dependent rank. In Imanishi, K., and Altmann, S. A. (eds.), *Japanese monkeys*. Atlanta, GA: Emory University Press.

Keeley, E. R., and Grant, J. W. A. (1993). Visual information, resource value, and sequential assessment in convict cichlid (*Cichlasoma nigrofasciatum*) contests. *Behav. Ecol., 4*, 345–349.

King, M. G. (1965). The effect of social context on dominance capacity of domestic hens. *Anim. Behav., 13*, 132–133.

Krebs, J. R. (1982). Territorial defence in the great tit (*Parus major*): do residents always win? *Behav. Ecol. Sociobiol., 11*, 185–194.

Krebs, J. R., and Dawkins, R. (1984). Animal signals: mind-reading and manipulation. In Krebs, R., and Davies, N. B. (eds.), *Behavioural ecology: An evolutionary approach*, (2nd ed., pp. 380–402). Oxford: Blackwell Scientific Publications.

Kruuk, H. (1972). *The spotted hyena: A study of predation and social behavior*. Chicago, IL: University of Chicago Press.

Landau, H. G. (1951). On dominance relations and the structure of animal societies: I. Effect of inherent characteristics. *Bull. Math. Biophys., 13*, 1–19.

Landau, H. G. (1965). Development of structure in a society with a dominance relation when new members are added successively. *Bull. Math. Biophys., 27*, 151–160.

Le Boeuf, B. J. (1974). Male-male competition and reproductive success in elephant seals. *Am. Zool., 14*, 163–176.

Le Boeuf, B. J., and Reiter, J. (1988). Lifetime reproductive success in northern elephant seals. In Clutton-Brock, T. H. (ed.), *Reproductive success* (pp. 344–362). Chicago, IL: University of Chicago Press.

Leimar, O., and Enquist, M. (1984). Effects of asymmetries in owner-intruder interactions. *J. Theor. Biol., 111*, 475–491.

Lott, D. F. (1979). Dominance relations and breeding rates in mature male American bison. *Z. Tierpsychol., 49*, 319–337.

Magnuson, J. J. (1962). An analysis of aggressive behavior, growth, and competition for food and space in medaka (*Oryzias latipes*, Pisces, Cyprinodontidae). *Can. J. Zool., 40*, 313–363.

Marden, J. H. (1987). In pursuit of females: following and contest behavior by males of a Namib Desert tenebrionid beetle, *Physadesmia globosa*. *Ethology, 75*, 15–24.

Marden, J. H., and Waage, J. K. (1990). Escalated damselfly territorial contests are energetic wars of attrition. *Anim. Behav., 39*, 954–959.

Maynard Smith, J. (1974). The theory of games and the evolution of animal conflicts. *J. Theor. Biol., 47*, 209–221.

Maynard Smith, J. (1982a). *Evolution and the theory of games*. Cambridge: Cambridge University Press.

Maynard Smith, J. (1982b). Do animals convey information about their intentions? *J. Theor. Biol., 97*, 1–5.

Maynard Smith, J., and Parker, G. A. (1976). The logic of asymmetric contests. *Anim. Behav., 24*, 159–175.

Maynard Smith, J., and Price, G. R. (1973). The logic of animal conflict. *Nature, 246*, 15–18.

McBride, G. (1958). The measurement of aggressiveness in the domestic hen. *Anim. Behav., 6*, 87–91.

McComb, K., Packer, C., and Pusey, A. (1994). Roaring and numerical assessment in contests between groups of female lions, *Panthera leo*. *Anim. Behav. 47*, 379–387.

McNutt, J. W. (1996). Sex-biased dispersal in African wild dogs (*Lycaon pictus*). *Anim. Behav., 52*, 1067–1077.

McVey, M. E. (1988). The opportunity for sexual selection in a territorial dragonfly, *Erythemis simplicicollis*. In Clutton-Brock, T. H. (ed.), *Reproductive success* (pp. 44–58). Chicago, IL: University of Chicago Press.

Montgomerie, R. D., and Weatherhead, P. J. (1988). Risks and rewards of nest defense by parent birds. *Quart. Rev. Biol., 63,* 167–187.

Mosler, H. (1985). Making the decision to continue the fight or to flee: an analysis of contests between male *Haplochromis burtoni* (Pisces). *Behaviour, 92,* 129–145.

Newman, M. A. (1956). Social behaviour and intra-specific competition on two trout species. *Physiol. Zool., 29,* 223–243.

Noë, R. (1990). Alliance formation among male baboons: shopping for profitable partners. In Harcourt, A. H., and de Waal, F. B. M. (eds.), *Coalitions and alliances in humans and other animals* (pp. 285–321). Oxford: Oxford University Press.

Noë, R. (1994). A model of coalition formation among male baboons with fighting ability as the crucial parameter. *Anim. Behav., 47,* 211–213.

Olsson, M. (1994). Nuptial coloration in the sand lizard, *Lacerta agilis*: an intra-sexually selected cue to fighting ability. *Anim. Behav., 48,* 607–613.

Otronen, M. (1988). Intra- and intersexual interactions at breeding burrows in the horned beetle, *Coprophanaeus ensifer. Anim. Behav., 36,* 741–748.

Packer, C. (1977). Reciprocal altruism in olive baboons. *Nature, 265,* 441–443.

Packer, C. (1979). Male dominance and reproductive activity in *Papio anubis. Anim. Behav., 27,* 37–45.

Packer, C., and Pusey, A. (1985). Asymmetric contests in social mammals: respect, manipulation and age-specific aspects. In Greenwood, P. J., Harvey, P. H., and Slatkin, M. (eds.), *Evolution: Essays in honour of John Maynard Smith* (pp. 173–186). Cambridge: Cambridge University Press.

Parker, G. A. (1974). Assessment strategy and the evolution of fighting behaviour. *J. Theor. Biol., 47,* 223–243.

Parker, G. A. (1982). Phenotype-limited evolutionarily stable strategies. In King's College Sociobiology Group (eds.), *Current problems in sociobiology* (pp. 173–201). Cambridge: Cambridge University Press.

Parker, G. A., Mock, D. W., and Lamey, T. C. (1989). How selfish should stronger sibs be? *Am. Nat., 133,* 846–868.

Parker, G. A., and Rubenstein, D. I. (1981). Role assessment, reserve strategy, and acquisition of information in asymmetric animal conflicts. *Anim. Behav., 29,* 221–240.

Pereira, M. E. (1989). Agonistic interactions of juvenile savanna baboons. II. Agonistic support and rank acquisition. *Ethology, 80,* 152–171.

Petrie, M. (1983). Female moorhens compete for small fat males. *Science, 220,* 413–415.

Petrie, M. (1984). Territory size in the moorhen (*Gallinula chloropus*): an outcome of RHP asymmetry between neighbours. *Anim. Behav., 32,* 861–870.

Pianka, E. R. (1976). Natural selection of optimal reproductive tactics. *Am. Zool., 16,* 775–784.

Pianka, E. R. (1988). *Evolutionary ecology,* (4th ed). New York: Harper and Row.

Poole, J. H. (1987). Rutting behavior in African elephants: The phenomenon of musth. *Behaviour, 102,* 283–316.

Poole, J. H. (1989a). Announcing intent: the aggressive state of musth in African elephants. *Anim. Behav., 37,* 140–152.

Poole, J. H. (1989b). Mate guarding, reproductive success and female choice in African elephants. *Anim. Behav., 37,* 842–849.

Pope, T. R. (1990). The reproductive consequences of male cooperation in the red howler monkey: paternity exclusion in multi-male and single-male troops using genetic markers. *Behav. Ecol. Sociobiol., 27,* 439–446.

Popp, J. W. (1988). Effects of experience on agonistic behavior among American goldfinches. *Behav. Proc., 16,* 11–19.

Popp, J. L., and DeVore, I. (1979). Aggressive competition and social dominance theory: synopsis. In Hamburg, D. A., and McCown, E. R. (eds.), *The great apes* (pp. 316–338). Menlo Park: Benjamin/Cummings.

Prins, H. H. T. (1989). Condition changes and choice of social environment in African buffalo bulls. *Behaviour, 108,* 297–324.

Prins, H. H. T., and Iason, G. R. (1989). Dangerous lions and nonchalant buffalo. *Behaviour, 108,* 262–296.

Pusey, A. E., and Packer, C. (1987). Dispersal and philopatry. In Smuts, B. B., et al. (eds.), *Primate societies* (pp. 250–266). Chicago, IL: University of Chicago Press.

Raveling, D. G. (1970). Dominance relationships and agonistic behavior of Canada geese in winter. *Behaviour, 37,* 291–319.

Reed, C. A. (1969). The pattern of animal domestication in the prehistoric Near East. In Ucko, P. J., and Dimbleby, G. W. (eds.), *The domestication and exploitation of domestic animals* (pp. 361–380). London: Duckworth.

Ribowski, A., and Franck, D. (1993a). Demonstration of strength and concealment of weakness in escalating fights of male swordtails (*Xiphophorus helleri*). *Ethology, 93,* 265–274.

Ribowski, A., and Franck, D. (1993b). Subordinate swordtail males escalate faster than dominants: A failure of the social conditioning principle. *Aggressive Behavior, 19,* 223–229.

Richner, H. (1989). Phenotypic correlates of dominance in carrion crows and their effects on access to food. *Anim. Behav., 38,* 606–612.

Ricklefs, R. E. (1979). *Ecology.* New York: Chiron.

Riechert, S. E. (1978). Games spiders play: Behavioral variability in territorial disputes. *Behav. Ecol. Sociobiol., 3,* 135–162.

Riechert, S. E. (1979). Games spiders play. II. Resource assessment strategies. *Behav. Ecol. Sociobiol., 6,* 121–128.

Riechert, S. E. (1988). The energetic costs of fighting. *Am. Zool., 28,* 877–884.

Robertson, J. G. M. (1986). Male territoriality, fighting and assessment of fighting ability in the Australian frog *Uperoleia rugosa. Anim. Behav., 34,* 763–772.

Robinson, S. K. (1986a). Competitive and mutualistic interactions among females in a neotropical oriole. *Anim. Behav., 34,* 113–122.

Robinson, S. K. (1986b). Benefits, costs, and determinants of dominance in a polygynous oriole. *Anim. Behav., 34,* 241–255.

Rockwell, R. F., Findlay, C. S., Cooke, F., and Smith, J. A. (1985). Life history studies of the lesser snow goose (*Anser caerulescens caerulescens*). IV. The selective value of plumage polymorphism: net viability, the timing of maturation, and breeding propensity. *Evolution, 39,* 178–189.

Rohwer, S. (1982). The evolution of reliable and unreliable badges of fighting ability. *Am. Zool., 22,* 531–546.

Rood, J. P. (1987). Dispersal and intergroup transfer in the dwarf mongoose. In Chepko-Sade, B. D., and Halpin, Z. (eds.), *Mammalian dispersal patterns: The effects of social structure on population genetics* (pp. 85–102). Chicago, IL: University of Chicago Press.

Rubenstein, D. I. (1982). Risk, uncertainty and evolutionary strategies. In King's College Sociobiology Group (eds.), *Current problems in sociobiology* (pp. 91–111). Cambridge: Cambridge University Press.

Sapolsky, R. M. (1987). Stress, social status, and reproductive physiology in free-living baboons. In Crews, D. (ed.), *Psychology of reproductive behavior: An evolutionary perspective* (pp. 291–322). Englewood Cliffs, NJ: Prentice-Hall.

Scott, J. P., and Fuller, J. L. (1965). *Genetics and the social behavior of the dog.* Chicago, IL: University of Chicago Press.

Schaller, G. B. (1972). *The Serengeti lion: A study of predator-prey relations.* Chicago, IL: University of Chicago Press.

Searcy, W. A. (1979). Morphological correlates of dominance in captive male red-winged blackbirds. *Condor, 81,* 417–420.

Selye, H. (1971). *Hormones and resistance.* New York: Springer-Verlag.

Shutler, D., and Weatherhead, P. J. (1991). Owner and floater red-winged blackbirds: Determinants of status. *Behav. Ecol. Sociobiol., 28,* 235–241.

Shutler, D., and Weatherhead, P. J. (1992). Surplus territory contenders in male red-winged blackbirds: Where are the desperados? *Behav. Ecol. Sociobiol., 31,* 97–106.

Silk, J. B. (1987). Social behavior in evolutionary perspective. In Smuts, B. B., et al. (eds.), *Primate societies* (pp. 318–329). Chicago, IL: University of Chicago Press.

Slater, P. J. B. (1986). Individual differences and dominance hierarchies. *Anim. Behav., 34,* 1264–1265.

Smith, I. P., Huntingford, F., Atkinson, R. J. A., and Taylor, A. C. (1994). Mate competition in the velvet swimming crab *Necora puber* — Effects of perceived resource value on male agonistic behaviour. *Mar. Biol., 120,* 579–584.

Smith, I. P., and Taylor, A. C. (1993). The energetic cost of agonistic behaviour in the velvet swimming crab, *Necora (= Liocarcinus) puber* (L.). *Anim. Behav., 45,* 375–391.

Smith, J. N. M. (1988). Determinants of lifetime reproductive success in the song sparrow. In Clutton-Brock, T. H. (ed.), *Reproductive success* (pp. 154–172). Chicago, IL: University of Chicago Press.

Smith, J. N. M., and Arcese, P. (1989). How fit are floaters? Consequences of alternative territorial behaviors in a nonmigratory sparrow. *Am. Nat., 133,* 830–845.

Smith, S. M. (1978). The "underworld" in a territorial sparrcw: adaptive strategy for floaters. *Am. Nat., 112,* 571–582.

Smuts, B. B. (1987). Gender, aggression, and influence. In Smuts, B. B., et al. (eds.), *Primate societies* (pp. 400–412). Chicago, IL: University of Chicago Press.

Stamps, J. A. (1987). The effect of familiarity with a neighborhood on territory acquisition. *Behav. Ecol. Sociobiol., 21,* 273–277.

Suomi, S. (1983). Models of depression in primates. *Psychol. Med., 13,* 465–468.

Sutherland, W. J., and Parker, G. A. (1985). Distribution of unequal competitors. In Sibly, R. M., and Smith, R. H. (eds.), *Behavioural ecology* (pp. 255–273). Oxford: Blackwell Scientific Publications.

Thornhill, R., and Sauer, P. (1992). Genetic sire effects on the fighting ability of sons and daughters and mating success of sons in a scorpionfly. *Anim. Behav., 43,* 255–264.

Thorpe, K. E., Taylor, A. C., and Huntingford, F. A. (1995). How costly is fighting? Physiological effects of sustained exercise and fighting in swimming crabs, *Necora puber* (L.) (Brachyura, Portunidae). *Anim. Behav., 50,* 1657–1666.

Trivers, R. L. (1972). Parental investment and sexual selection. In Campbell, B. (ed.), *Sexual selection and the decent of man, 1871–1971* (pp. 136–179). Chicago, IL: Aldine.

van den Berghe, E. P. (1988). Piracy as an alternative reproductive tactic for males. *Nature, 334,* 697–698.

Van de Poll, N. E., De Jonge, F., Van Oyen, H. G., and Van Pelt, J. (1982). Aggressive behaviour in rats: effects of winning or losing on subsequent aggressive interactions. *Behav. Proc., 7,* 143–155.

Van Noordwijk, M. A., and Van Schaik, C. P. (1985). Male migration and rank acquisition in wild long-tailed macaques (*Macaca fascicularis*). *Anim. Behav., 33,* 849–861.

Vehrencamp, S. L. (1979). The roles of individual, kin, and group selection in the evolution of sociality. In Marler, P., and Vandenbergh, J. G. (eds.), *Handbook of behavior and communication, Vol. 3* (pp. 351–394). New York: Plenum Press.

Vehrencamp, S. L. (1983). A model for the evolution of despotic versus egalitarian societies. *Anim. Behav., 31,* 667–683.

Verrell, P. A. (1986). Wrestling in the red-spotted newt (*Notophthalmus viridescens*): resource value and contestant asymmetry determine contest duration and outcome. *Anim. Behav., 34,* 398–402.

Waage, J. K. (1988). Confusion over residency and the escalation of damselfly territorial disputes. *Anim. Behav., 36,* 586–595.

Wallen, K., and Wojeiechowski-Metzlar, C. (1985). Social conditioning and dominance in male *Betta splendens*. *Behav. Proc.*, *11*, 181–188.

Walters, J. R., and Seyfarth, R. M. (1987). Conflict and cooperation. In Smuts, B. B., et al. (eds.), *Primate societies* (pp. 306–317). Chicago, IL: University of Chicago Press.

Watson, A., and Moss, R. (1971). Spacing as affected by territorial behavior, habitat and nutrition in red grouse (*Lagopus l. scoticus*). In Esser, A. H. (ed.), *The use of space by animals and men* (pp. 92–111). New York: Plenum.

Wells, M. S. (1988). Effects of body size and resource value on fighting behaviour in a jumping spider. *Anim. Behav.*, *36*, 321–326.

Whitfield, D. P. (1987). Plumage variability, status signaling and individual recognition in avian flocks. *TREE*, *2*, 13–18.

Wickman, P., and Wiklund, C. (1983). Territorial defense and its seasonal decline in the speckled wood butterfly (*Pararge aegeria*). *Anim. Behav.*, *31*, 1206–1216.

Williams, G. C. (1957). Pleiotropy, natural selection, and the evolution of senescence. *Evolution*, *11*, 398–411.

Williams, G. C. (1966). Natural selection, the costs of reproduction, and a refinement of Lack's principle. *Am. Nat.*, *100*, 687–690.

Wilson, P., and Franklin, W. L. (1985). Male group dynamics and inter-male aggression of guanacos in southern Chile. *Z. Tierpsychol.*, *69*, 305–328.

Wingfield, J. C., Ball, G. F., Dufty, A. M., Jr., Hegner, R. E., and Ramenofsky, M. (1987). Testosterone and aggression in birds. *Am. Sci.*, *75*, 602–608.

Wingfield, J. C., Hegner, R. E., Dufty, A. M., Jr., and Ball, G. F. (1990). The "challenge hypothesis": theoretical implications for patterns of testosterone secretion, mating systems and breeding strategies. *Am. Nat.*, *136*, 829–846.

Worthman, C. M. (1990). Socioendocrinology: key to fundamental synergy. In Ziegler, T. E., and Bercovitch, F. B. (eds.), *Socioendocrinology of primate reproduction* (pp. 187–212). New York: Wiley.

Yamagishi, H. (1962). Growth relation in some small experimental populations of rainbow trout fry, *Salmo gairdneri* Richardson with special reference to social relations among individuals. *Jpn. J. Ecol.*, *12*, 43–53.

Yokel, D. A. (1989). Intrasexual aggression and the mating behavior of brown-headed cowbirds: their relation to population densities and sex ratios. *Condor*, *91*, 43–51.

Chapter 7

INCIDENTAL EMISSIONS, FORTUITOUS EFFECTS, AND THE ORIGINS OF COMMUNICATION

Mark S. Blumberg

Department of Psychology
University of Iowa
Iowa City, Iowa 52242
e-mail: mark-blumberg@uiowa.edu

Jeffrey R. Alberts

Department of Psychology
Indiana University
Bloomington, Indiana 47405

ABSTRACT

In recent years, a small group of animal behaviorists has been calling for a renewed focus on proximate mechanisms in the study of behavioral evolution (Kennedy, 1992; Stamps, 1991). These calls have been made to counter the current view, implicit in most contemporary analyses of animal behavior, that we can understand ultimate causation without worrying about the mechanistic details. As noted by Stamps (1991), however, today's "students of ultimate causation in behavior have begun to 'rediscover' the importance of proximate mechanisms" (p. 342). Specifically, there is a growing realization that proximate mechanisms are not mere details—rather, they vitally shape our understanding of function and of the evolutionary origins of behavior and structure.

Communication is an area of animal behavior, we argue, that can especially benefit from a renewed focus on proximate mechanism. This argument is

Perspectives in Ethology, Volume 12
edited by Owings *et al.*, Plenum Press, New York, 1997

developed as follows: (1) Students of animal communication have tended to make preliminary assumptions of function when confronted with novel signals; (2) these assumptions of function have steered us away from studying the proximate mechanisms that underlie signal production; and (3) by assuming function and ignoring mechanism, we arrive at a distorted view of the communicatory relations between senders and receivers and, in addition, deny ourselves a path to understanding the evolutionary origins of communication. This argument is supported by examples of how appropriate emphasis on mechanism broadens and deepens our understanding of communication.

INTRODUCTION

> ...theories of behavior which ignore problems of origin can scarcely be called complete.
> —M. T. Ghiselin (1969, p. 208)

Judith Stamps (1991) has recently noted that a resurgence of interest in proximate causation is benefiting the study of ultimate causation in animal behavior. Such a resurgence in the study of mechanism, she argues, is necessary to balance the explosion of research in behavioral ecology and sociobiology of the last 20 years, research that has focussed predominantly on ultimate causation. Many examples are provided in her article that illustrate the ways in which the study of mechanism can further the study of evolution, but none is directly related to the topic of animal communication. The aim of this paper is to examine the applicability of Stamps' perspective to the field of animal communication.

At any given time, a particular organism possesses a variety of anatomical features, some of which are adaptive and some nonadaptive. The same is true for behavioral features, including vocalizations and other acoustic behaviors. Animal vocalizations are produced by a diversity of physiological and biomechanical mechanisms, and they are also diverse with respect to the degree to which they have been modified by the evolutionary process. At one extreme are vocalizations that affect the behavior of other animals and have been modified for communication. At the other extreme are many vocalizations that are emitted as incidental by-products of other mechanisms (e.g., sneezes, coughs), that are ignored by other animals, and have not been modified for communication.

Thus, one goal of the study of acoustic communication should be to understand the evolutionary processes by which incidental vocalizations are modified and incorporated into a communicatory system. To achieve that goal, we should follow Stamps' suggestion and elevate the importance of proximate mechanism in the study of communication. By doing so, we will correct for the tendency of current perspectives to distort the relative contributions of senders and receivers to communicatory systems. Specifically, senders (i.e., animals that are, at any given moment, emitting information) are often conceptualized today as the primary communicatory actors while receivers (i.e., animals that are, at any given moment, processing information) are conceptualized as passive participants. On the contrary, we argue, receivers are equal participants in communicatory systems who often act as selecting agents on the varied signals to which they are exposed.

Although Darwin founded the scientific study of the evolution of communication, he did not himself fall into the habit of assuming that every novel vocalization had a communicative function. In fact, he cautioned against such assumptions. For example, in *The Descent of Man, and Selection in Relation to Sex* (1871/1981) he writes: "Animals of all kinds which habitually use their voices, utter various noises under any strong emotion, as when enraged and preparing to fight; but this may merely be the result of their nervous excitement, which leads to the spasmodic contraction of almost all the muscles of the body, as when a man grinds his teeth and clenches his hands in rage" (p. 275).

Even today, Darwin's clarity and grasp of parsimony are remarkable. In effect, his appeal is for initial consideration of the possibility that a vocal emission is an acoustic by-product of physiological or biomechanical forces, rather than assuming that each sound has a communicatory function. Of course, considerations of function and mechanism are not mutually exclusive. Nonetheless, we can no longer sanction the current imbalance with regard to the study of function and mechanism (Kennedy, 1992). When we ignore mechanism, we blind ourselves to evolutionary history.

After Darwin, other theorists have cautioned against the "unwarranted uses of the concept of adaptation. This biological principle should be used only as a last resort. It should not be invoked when less onerous principles, such as those of physics and chemistry or that of unspecific cause and effect, are sufficient for a complete explanation" (Williams, 1966, p. 11).

Williams contrasts adaptations with what he variously calls "fortuitous relationships," "incidental consequences," and "by-products." Of course, many of these by-products confer a survival advantage, but that is not sufficient for them to qualify as adaptations: "In an individual organism an effect should be presumed to be the result of physical laws only, or perhaps the fortuitous effect of some unrelated adaptation, unless there is clear evidence that it is produced by mechanisms designed to produce it" (Williams, p. 261).

But what do we call those by-products that have come to be shaped by natural selection for a new function? For example, the feather insulation of birds has been referred to as a "preadaptation" for flight, a term whose teleology detracts from its explanatory usefulness. Gould and Vrba (1982) correct for this temporal distortion of language by suggesting that, in the above example, we should speak of feathers as "exaptations" for flight, by which they mean that although feathers are fit for their role in flight they were not originally selected for that role.

Forty years ago, Morris (1956) drew attention to the ways in which the feathers of birds are also used in a variety of ways as communicatory signals. In some cases, he argued, feather postures have become specialized as signals in that the "original motor patterns concerned have been modified in any way in connection with their new secondary function" (p. 82). In other words, feathers have not only been exapted for flight but, in some species, for communication as well.

The above discussion illustrates how a number of theorists have emphasized the importance of distinguishing between functions and effects. As we discuss in the next section, failure to make this distinction distorts the relative contributions of senders and receivers and thus makes the evolution of communicatory systems more difficult to understand.

THE DYNAMIC RELATIONS OF SENDERS AND RECEIVERS

Deciding on a definition of communication on which a majority of researchers can agree has proved to be a difficult task. Most broadly, we might, as Smith (1977) preliminarily suggests, "define communication as any sharing of information from any source" (p. 13). To this broad definition, any of a number of restrictions may be added (see Burghardt, 1970, for a review and analysis of such restrictions). For example, one might insist that a communicatory signal alter "the probability pattern of behavior in another organism... from what it would have been in the absence of the signal" (Wilson, 1975, p. 176). Burghardt (1970) argues that a communicatory act must be intended to communicate, where "intent can be looked at scientifically merely by considering that it is to the real or perceived advantage of the signaler or the signaler's group for it to get its message across to whatever organism is involved" (p. 12). In addition, one might insist that the signal be specialized for communication, or that the communicatory system be adaptive for the sender and/or the receiver. Finally, some may even wish to restrict communication to exchanges between members of the same species.

A consensus on a single definition of communication has not been reached for several reasons, one of which is that no unambiguous boundary on the continuum from incidental by-product to communicatory signal has yet been located. This ambiguity arises from the fact that the relations among senders and receivers are complex, comprising a dynamic system of multiple components interacting in diverse ways (Kelso, 1995; Thelen and Smith, 1994). As a result, the "cause" of a communicatory system cannot be localized in any single component of that system. Just as organismal development arises from the dynamic interactions between the organism and the environment (Oyama, 1985), communication arises from the dynamic interactions between sender and receiver.

Recent definitions of communication have tended to focus on the sender. But any attempt to define communication exclusively in terms of the sender is doomed because no acoustic signal can acquire an adaptive function unless a receiver can detect it and process the information in an appropriate way. For example, stingless bees emit an audible (to humans) hum while flapping their wings during warmup and, moreover, the frequency of the hum is directly correlated with thoracic temperature (cited in Heinrich, 1979, p. 56). In other words, the hum is informative. But is the information contained in this sound emission of biological interest to any animal other than the entomologist interested in the thoracic temperature of bees?

As another example, the rattle of rattlesnakes could potentially alert conspecifics to a dangerous situation. The rattle's sound energy, however, is concentrated above 2000 Hz while rattlesnakes are most sensitive to sounds below 700 Hz; these and other facts suggest that rattling is not important for conspecific communication (Fenton and Licht, 1990). In this case, the potential for communication exists, but conspecific receivers have proven ill-equipped to detect or process the information. On the other hand, rattling does incidentally provide valuable information about rattlesnake size and dangerousness to ground squirrels aggressively defending their young against a snake (Rowe and Owings, 1990, in press; Swaisgood, 1994).

Thus, while there are an infinite number and variety of signals being emitted at any given time, these signals can only begin to develop a communicatory function, and their communicatory significance can only be assessed, when they elicit responses from receivers. As an example of a well-accepted communicatory behavior, consider the "mating call" of male field crickets of the genus Gryllus. In response to this call, female crickets approach the male and mating ensues. This call alters the female's behavior, it confers a reproductive advantage to the male, and it is intraspecific. Thus, we might tentatively describe this interaction as the male cricket communicating to the female or, in the language of Krebs and Dawkins (1984), manipulating her.

One's comfort with such descriptions is challenged somewhat when one considers that parasitoid flies of the genus Ormia also move toward calling male crickets but, in their case, they deposit their offspring on or near the male. The offspring then burrow inside the cricket, grow, and eventually kill their host (Robert et al., 1992). As Krebs and Dawkins (1984) point out, "most people would not wish to say that the crickets were signalling to the flies... because most authors agree in wanting to exclude such incidental consequences" (p. 380). After all, it is argued, the male cricket did not intend (in Burghardt's sense) to attract the fly or, in other words, the male cricket's song was not designed by natural selection for attracting the fly.

From the female cricket's perspective, responding to the male's song provides her with an opportunity to mate and thus increase her reproductive fitness; she is taking advantage of information in her environment and manipulating the male cricket for her own ends. Similarly, the parasitoid fly is taking advantage of information in her environment and ensuring her own reproduction at the expense of those crickets that are unlucky enough to attract the flies by singing. But the fly's response is not mere happenstance. On the contrary, natural selection has provided her with a hearing organ that is tuned to the male cricket's song and is unique in its convergence upon a design that "more resembles a cricket's ear than a typical fly's ear" (Robert et al., 1992, p. 1135). In other words, the female fly's ear is specialized for detecting the male cricket's song.

If we attempt to define communication from the sender's perspective alone, we will be forced, like Krebs and Dawkins (1984), to argue that the male cricket is communicating with the female cricket but not with the female fly. But, on what objective basis do we conclude that the specialization of a sender's signal is indicative of communication while specialization of the receiver's detection apparatus is not? Are they not two equally effective means by which senders and receivers establish a communicatory relationship? We believe they are.

If it is acknowledged that senders and receivers both participate fully and actively in the communicatory process, it is then only a small step to acknowledge that specialization of either participant equally qualifies as an important feature of communication. Thus, we are arguing that communication is not the product of any single privileged participant but emerges from the relations and interactions among senders and receivers. Senders and receivers are both necessary for the formation and maintenance of communicatory systems, where maintenance includes both positive and negative forces acting on each contributor's behavior. Acceptance of this view will not, however, make us feel any less uncomfortable about stating that the male cricket is communicating to the fly; this lack of comfort is in part a result of our natural tendency to anthropomorphize and insist that no right-thinking cricket would actively call out to its executioner (see Kennedy, 1992, for a detailed discussion of these anthropomorphic tendencies). But, if we step back and view the system in its entirety, we see that female crickets and flies are acting similarly and, each in their own way, contributing to the dynamics of this communicatory system.

Specialization, of a sender's signal or a receiver's detection apparatus, can only occur after senders and receivers have entered into a relationship. Thus, although sounds may be produced incidentally at first (as in the stingless bee), in time they may come to be specialized for sound production. On the other side of the coin, but much less well examined or understood, are the incidental responses of receivers to various signals. For example, Tungara frogs (Physalaemus pustulosus) emit vocalizations composed of two components, a 'whine' and a 'chuck' (Ryan et al., 1990). Female Tungara frogs prefer males that emit chucks of lower frequencies; lower frequency chucks signal a male of larger body size. But females of a related species, P. coloradorum, also prefer calls that contain chucks even though the calls of conspecific males do not contain chucks.

Similarly, and in the visual modality, the preference of the females of one species of freshwater guppy (Xiphophorus helleri) for males with large swordtails may have originated in a bias of the sensory system of ancestral fish before the swordtail evolved (Basolo, 1990). This view is supported by the finding that females of a related but swordless species (X. maculatus) also prefer males with swordtails. Although there is disagreement regarding the phylogenetic relations between the swordless and sworded species (Pomiankowski, 1994), the role of sensory bias in signal evolution deserves far more attention than it has received thus far (Guilford and Dawkins, 1993).

ANSWERS FIRST, QUESTIONS SECOND

When Krebs and Dawkins (1984) state that "most of the sounds given off by [animals] are best interpreted as being adapted...to influence the behaviour of other animals" (p. 380), they are expressing a commonly held view among communication researchers. We have argued above that this assumption of communicatory function blinds us to mechanism and, in turn, to evolutionary history. As we argue below, this assumption has yet another, but related, drawback.

Practitioners of comparative psychology, ethology, behavioral ecology, and sociobiology share the view that naturalistic contexts give meaning and significance to behavioral phenomena. Thus, it is common practice today that when a vocalization is detected, a researcher next determines the contexts in which the vocalization is emitted and the behaviors that accompany it. Based on this information, a functional hypothesis is formulated that relates the emission of the vocalization to the context in which it is emitted. For example, a vocalization may be discovered that is emitted by males during the mating season, a discovery that would suggest that the vocalization functions as a sexual attractant. Similarly, a vocalization may be emitted when two males meet in an aggressive encounter, thus suggesting a role for this vocalization in the modulation of aggression. What is important here is that it requires very little imagination to hypothesize a communicatory function for a newly discovered vocalization.

As one example of the progression from the discovery of a vocalization to the suggestion of a communication hypothesis, consider the "cackle" of male Japanese macaques (Macaca fuscata): "Males cackle only during copulations. These rhythmic calls, occurring before, during and after thrusting at various times, and also while the female is cackling and copulating, may serve to strengthen the pair bond and possibly hasten the culmination of a mating sequence" (Green, 1981).

The existence of a stereotyped vocalization that can be recognized and categorized, coupled with its regular appearance in a particular setting (physical and social), is generally sufficient to inspire a communication hypothesis. In the above case, the discovery of a vocalization within a copulatory context implied a communicatory function for the vocalization and thus drove the suggestion of two possible functions.

This example provides insight into the question-and-answer process as it often exists in the study of animal communication. The suggestion that the cackle of the Japanese macaque "may serve to strengthen the pair bond and possibly hasten the culmination of a mating sequence" is based upon observation of the context in which the vocalization is produced—no observation is reported that these macaques form stronger pair bonds or mate faster than related species that do not cackle. Viewed as an analytic strategy, the researcher invented problems for the animal (i.e., need to strengthen the pair bond, need to hasten the culmination of the mating sequence) for which the vocalization was to be a solution.

The sneeze-like Snough vocalization of golden lion tamarins (Leontopithecus rosalia) provides another example. McLanahan and Green (1977) write that the Snough "sounds very similar to a sneeze. Its frequent occurrence at the end of feeding and locomotor bouts leads us to postulate that it may communicate completion of an activity" (p. 262). While golden lion tamarins may emit the Snough after completing particular activities, the potential benefit of communicating that fact is not addressed. Nor is any evidence provided that the behavior of conspecifics changed at the end of feeding or locomotor bouts. On the other hand, if such observational data had been offered, and if the behavioral changes seemed well-timed, then a search for such a timing mechanism would be reasonable.

In contrast, there are examples of vocalizations that, when discovered, have provided clear explanations for unexplained phenomena. For example, researchers noticed that large groups of Asian elephants often display coordinated behaviors despite the absence of a detectable signal. The riddle of this coordinated behavior was solved when it was discovered that these elephants emit infrasonic vocalizations (i.e., vocalizations below the range of human hearing) and that these vocalizations are emitted immediately prior to the coordinated movements of multiple elephants (Payne et al., 1986). Unfortunately, the significance of "the remarkable coordination" (Poole et al., 1988, p. 386) of elephants is still unclear. Nonetheless, this history provides a nice example of a clearly defined question generating the search for a solution.

These examples demonstrate the practical difference between searching for an answer to a question versus searching for a question to an answer. When questions are stated clearly (e.g., "What signal triggers the coordinated movements of elephants?"), we can imagine what the answer to the question will look like, and can map out a series of experiments with a clear terminus. In contrast, when we search for a question to an answer ("The Snough vocalization of golden lion tamarins is important for X"), our search may never terminate. Thus, when questions come second, there are serious consequences for the form of the scientific approach that emerges.

ANALYSES OF ACOUSTIC COMMUNICATION IN RODENTS

Thus far, we have detailed the potential pitfalls of assuming communicatory function and ignoring mechanism. Now the time has come to explore how considerations of mechanism broaden our understanding of the origins of communication. We will do this in the next section by providing examples of signals that are produced as incidental by-products of a diverse array of production mechanisms. In this section, we summarize our own work on the vocalizations of newborn and adult rats.

In our investigations of the vocalizations of Norway rats (Rattus norvegicus), we have set aside issues of communicatory function and have concentrated instead on the mechanisms responsible for the production of these vocalizations. To illustrate this approach and its benefits, we discuss three broad categories of rodent (mostly ultrasonic) vocalizations, all of which have received a great deal of experimental attention: the "distress calls" of newborn rats, the "short calls" of the adults of many rodent species during strenuous activities, and the post ejaculatory "long calls" of adults. In each case we show that attention to the mechanisms that produce and constrain these vocalizations enhances our understanding of their evolutionary history and communicatory significance.

"Distress Calls" of Rodent Young

The newborns of many rodent species emit ultrasonic vocalizations when they become cold following isolation from the nest (Zeppelius and Schleidt, 1956; Noirot, 1972). The fact that rodent young with limited thermoregulatory abilities emit these vocalizations when they are cold, and the fact that rodent mothers often respond to these vocalizations by retrieving the pup to the warm nest (Allin and Banks, 1972) or relocating the site of a disturbed nest (Brewster and Leon, 1980), have led many to conclude that the function of these "distress calls" is to elicit maternal care and protection. Given the remarkable fit between the behavior of the pups, its physiological needs, and the response of the mother, this seemed like a reasonable conclusion.

We noted, however, that although the effects of cold on this vocalization had been studied (Okon, 1970, 1971), there had been no attempt to relate ultrasound emission to the pup's other well-known physiological responses to cold. These responses include nonshivering thermogenesis via the activation of brown adipose tissue (BAT), an organ specialized for heat production. We hypothesized that if the vocalization is a component of the pup's overall response to cold exposure, then we might find that ultrasound emission and heat production by BAT are initiated at the same time.

Contemporaneous activation of BAT and ultrasound emission during cold exposure did occur, and was accompanied by an increase in oxygen consumption and the expression of a unique respiratory pattern characterized by prolonged expirations (Blumberg and Alberts, 1990; see also Blumberg and Stolba, 1996, for a recent update on this earlier work). The ultrasound occurred during these prolonged expirations, as had been demonstrated earlier (Roberts, 1972). This concurrence of ultrasound emission and prolonged expiratory duration suggested to us that the cold-exposed rat pup, like other mammalian newborns, employs a respiratory mechanism called laryngeal braking that helps maintain lung inflation. Laryngeal braking involves the constriction of the larynx following inspiration, resulting in prolonged expiratory duration and enhanced gas exchange in the lungs. In this context, any sound produced by the combination of laryngeal constriction and increased intrathoracic pressure is a by-product of expiration against a constricted larynx, as is the audible "grunt" emitted by human infants and lambs during laryngeal braking (Harrison et al., 1968; Johnson et al., 1977).

As we pursued this line of reasoning, new insights were gained. First, we used manipulations that presumably cause distress (i.e., starvation, hypoxia) but that inhibit physiological responding to cold (Blumberg and Alberts, 1991b). Cold-exposed pups that were starved or hypoxic did not activate nonshivering thermogenesis and they also did not ultrasound, suggesting that the vocalization is not simply a distress response. Second, with our physiological perspective, we revisited the issue of the vocalization's proximate stimulus and were able to demonstrate that the vocalization is specifically modulated by thermal and respiratory factors as opposed to factors more broadly defined as anxiety-related (Blumberg et al., 1992a, b). Remarkably, pups even emit the vocalization during recovery from deep hypothermia at body temperatures so cold that motor behavior is prevented (Hofer and Shair, 1992). These results suggest that extreme caution should be exercised by those suggesting the use of the ultrasound-producing rat pup as an animal model for human infant separation anxiety.

There is currently no evidence that the rat pup's vocalization or laryngeal apparatus is specialized for communication (Roberts, 1975); such evidence would provide support for the hypothesis that the vocalization is an adaptation or exaptation for communication. But we do know that while pups are deaf to their own vocalization at ages when they are emitting it most often, the mother's hearing curve is tuned specifically to the pup's vocalization frequency; this is also true for a number of other rodent species including the house mouse (Mus musculus) and the red-backed mouse (Clethrionomys glareolus; Brown, 1973). Thus, selective pressure on the mother's hearing sensitivity and propensity to retrieve pups may have been more significant than the selective pressure on pups to modify their ultrasonic emissions during isolation from the nest. This perspective does not rule out the possibility that ultrasound production elicited by cold exposure can be modulated by nonthermal factors such as olfactory and tactile cues from the mother. Such modulation may very well occur, especially in older pups and for brief durations (Hofer et al., 1993, 1994). More work is needed, however, in which the physiological consequences of such stimuli are more carefully measured and controlled.

We conclude that pup ultrasound is emitted as an incidental by-product, it is a reliable and informative indicator of a cold and metabolically active pup, adult rats can detect the vocalization, and the pup's mother often responds to this vocalization by retrieving the pup to the warm nest. Moreover, we suggest that this is a richer, more concrete, and less mentalistic description of the interaction between pup and mother than that which states that isolation from the nest causes anxiety and/or distress in the pup who is then motivated to cry out for maternal retrieval to the nest.

Vocalizations Associated with Arousal and Locomotion in Adult Rodents

Many rodent species emit ultrasonic or audible pulses during mating and other contexts involving high levels of arousal. For example, during copulatory behaviors, male and female collared lemmings (Dicrostonyx groenlandicus) emit an ultrasonic 'twitter' (Brooks and Banks, 1973) and male and female rats (Rattus norvegicus) emit a 40–70 kHz vocalization (Thomas and Barfield, 1985). For these species, and for many others, a number of communicatory functions have been hypothesized for these vocalizations; these hypothesized functions include facilitation of female receptive and/or proceptive behaviors and the inhibition of female aggressive behaviors during copulation.

A cursory examination of reports on these 'mating calls' reveals a curious association between these emissions and the animals' locomotor behaviors. This association has only been studied in detail for the ultrasonic vocalization of the Mongolian gerbil (Meriones unguiculatus; Thiessen and Kittrell, 1979; Thiessen et al., 1980). This vocalization is emitted during different modes of locomotion, all of which involve the landing of the forepaws on the ground. Thiessen and his colleagues attributed the emission of sound to the forcible expulsion of air through the larynx as a result of physical compression of the lungs during landing.

That vigorous locomotor behaviors, in which the forepaws land forcibly on the ground, are associated with the emission of sound is not surprising when one considers the biomechanics of locomotion in rodents and other mammals (Bramble, 1989; Blumberg, 1992). In addition, mammals typically time footfall patterns and respiration in such a way that expiration occurs as forelimbs strike the ground (Bramble and Carrier, 1983). These and other considerations suggest that, at least in some cases, the ultrasonic vocalizations associated with locomotion in sexually active rodents, as well as those vocalizations occurring in other contexts (e.g., during pelvic thrusting in mice and hind foot thumping in gerbils; Sales, 1972), are acoustic by-products of biomechanical stress and thoracic compression.

By neglecting the importance of the biomechanical constraints on ultrasound production, experiments designed to study the hormonal or neural bases of ultrasound production fail to control for confounding variables such as locomotion. For example, Dizinno and Whitney (1977) exposed castrated male mice (Mus musculus) to adult females and measured the latency to first detection of ultrasound. They found that these male mice had longer latencies to production of first ultrasound than either controls or castrates injected with testosterone. They conclude that "male androgen levels influenced the production of short latency ultrasounds by male mice" and that the "results are consistent with the courtship function hypothesized for these ultrasonic calls." Of course, castrates are also much slower to investigate a female and attempt intromission. These potential confounds are not considered, however, because ultrasound production is commonly viewed as an unconstrained behavior (Blumberg, 1992).

The possibility that a vocalization is an incidental by-product of locomotion, or any other behavior, does not preclude the possibility that the vocalization conveys information that can be used by conspecifics. For example, female receptive behaviors are facilitated by the ultrasonic vocalizations of male hamsters (Mesocricetus auratus; Floody and Pfaff, 1977). But finding such an effect of a vocalization is not a sufficient demonstration that it has a communicatory function (Blumberg and Alberts, 1992).

The 22-kHz Vocalization of Male Rats

Male rats (Rattus norvegicus) emit this "long call" following ejaculation, during aggressive encounters, as well as spontaneously during the day or night (Barfield and Geyer, 1972; Francis, 1977; Adler and Anisko, 1979). On the basis of contextual correlations, Francis (1977) writes that "it is unlikely that these calls are non-functional because they are so common, because they do not occur randomly, and because they are continually produced even though they could endanger the caller in natural conditions" (p. 238). Similarly, others write that "it occurs with temporal regularity and is very loud... . It would be difficult to accept that a behavior pattern with such an insistent quality would be without communicatory significance" (Barfield et al., 1979, p. 471).

But the search for a communicatory function of the 22-kHz vocalization has not been successful. For example, following the discovery that the vocalization is emitted by a defeated male after an aggressive encounter, it was suggested that the vocalization serves to inhibit further physical aggression by the dominant rat (Sales, 1972). Despite the finding that the vocalization correlates with aggressive behaviors, it has not yet been demonstrated that the vocalization has any effect on the aggressive behaviors of conspecifics. Similarly, attempts to demonstrate an affect of this vocalization on conspecifics during sexual encounters have not been successful (see Blumberg and Alberts, 1991a).

Failure to identify a role for the 22-kHz vocalization in modulating behavior has spawned additional, new hypotheses. For example, after one unsuccessful attempt to identify a communicatory effect of the 22-kHz vocalization in aggressive behavior, it was concluded that "even if domesticated rats are potentially similar to wild rats in the social regulation of aggressive behavior, they might acquire responsiveness to 22-kHz vocalizations only after they are reared in a more natural environment" (Takeuchi and Kawashima, 1986, p. 550). Similarly, the authors of another paper write that "the failure to demonstrate a role for ultrasonic vocalizations in the present study and [in another] study may reflect the constraints of the experimental procedure and testing conditions employed" (Takahashi et al., 1983, p. 211). They then suggest another communicatory role for the vocalization.

Negative results often generate new hypotheses that can be empirically tested. We see a problem, however, when the accumulation of negative evidence has little effect on the underlying assumption that the vocalization serves a communicatory function. Instead, as illustrated above, investigators explain negative evidence as the result of unspecified variables, termed "unnatural" conditions or inappropriate methods. If such suspicions arise—and they frequently do in studies of natural behavior under controlled and constrained conditions—the investigators may reasonably be expected to specify, if not alter and test, the critical variables. In the absence of specific explanations for unsuccessful experiments, however, it is appropriate to consider explicitly the possibility that the vocal emission does not serve a communicatory function; indeed, that possibility should be the null hypothesis.

Despite widespread treatment in the literature of the communicatory aspects of the rats' ultrasonic vocalizations, we concentrated instead on the physiological bases of this vocalization (Blumberg and Alberts, 1991a). Citing experimental evidence that the vocalization accompanies the increases in oxygen consumption and brain temperature that occur during the "chill phase" of fever (Blumberg and Moltz, 1987), and noting that laryngeal braking accompanies this phase of fever in lambs (Johnson and Andrews, 1990), we hypothesized that the 22-kHz vocalization may, like the vocalizations of infant rats, be the acoustic by-product of laryngeal braking. This hypothesis, although in need of direct testing, is consistent with the fact that the contexts in which this vocalization is emitted involve profound physiological activation (see Blumberg and Alberts, 1991a). Moreover, if supported, this hypothesis directs our attention toward those features that could make it informative to conspecifics. In other words, improved understanding of the physiological correlates of the vocalization will suggest ways in which conspecifics could potentially benefit from the information being provided. On the other hand, as we learn more about the vocalization, we may discover that the information provided is not valuable or not reliable, thus helping to explain the continued lack of success in identifying an effect of the vocalization on receivers.

MORE EXAMPLES OF INCIDENTAL SIGNAL PRODUCTION

Although the vast majority of animal signals have been investigated from the perspective of communicatory function, there have been a number of instances where researchers have attended to mechanism and, by doing so, have enriched our perspective.

Acoustic Signals

The sneezing behaviors of a number of species of New World monkeys have received more than a little attention, and a number of different explanations have been put forth for these behaviors, including roles as display and displacement behaviors, as well as responses to infection. For example, as described above, McLanahan and Green (1977) suggested that the sneeze-like Snough vocalization of golden lion tamarins (Leontopithecus rosalia) "may communicate completion of an activity."

In contrast, and in a beautifully reasoned analysis of this sneezing behavior, Schwartz and Rosenblum (1985) showed that previous explanations of sneezing were insufficient and unsupported, and they provided their own explanation for this behavior following a series of simple but novel experiments. As these investigators raised and lowered the air temperature inside a cage housing a squirrel monkey (Saimiri sciureus), they monitored the occurrence of sneezing. They found that the rate of sneezing increased as air temperature was raised and decreased as air temperature was lowered. Based on these observations, as well as considerations of hemodynamic changes in the head during a sneeze, they hypothesized that sneezing helps to regulate brain temperature during heat stress. Although their hypothesis has not been tested directly, their perspective dramatically alters the conventional approaches of previous investigators, and suggests that these sneezes may not be the controlled communicatory signals that some researchers have assumed.

Gans and Maderson (1973) discuss the many mechanisms underlying the wide variety of reptilian sound production. These authors are particularly sensitive to the incidental nature of many of these sounds, mentioning the faint whistling during expiration in turtles and the sounds accompanying the defensive fecal expulsions of many lizards and snakes. (They write (p. 1197): "Fecal discharge is of course a common defensive mechanism in squamates, but the associated sound is usually much less specific.") Gans and Maderson conclude their review with a passage that reflects on mechanism and historical origins:

> Such [reptilian] sounds might be initially produced by the convulsive expiration of air (the less the pulmonary filling, the greater the flexibility of the trunk). They might occur when portions of the body's keratinous cover are rubbed against each other in an agitated animal. They might,

> furthermore, arise when appendages involved in excitement vibrations contact leaves, twigs, or portions of the substratum. All of the patterns actually observed represent (relatively minor) amplifications upon such themes. (p. 1202)

Demski, Gerald, and Popper (1973) review sound production in teleost fish. They note, for example, that fish of various species make sounds while swimming, while grinding their teeth during eating, and when caught and taken out of the water. Fish sounds are also produced by the release of gas bubbles and by the muscular contraction of the swim bladder. Although, for example, the feeding sounds of predators feeding on prey are attractive to other predators and repellent to other prey, these authors state that "many of the sounds produced by fish may have no biological significance but may be incidental to other aspects of the fish's behavior" (p. 1142).

Olfactory Signals

As we have seen, vocal signals can be shaped by natural selection or merely incidental to physiological or biomechanical processes. Similarly, researchers of olfactory communication have struggled with the distinction between those olfactory signals that qualify as pheromones in the strict sense (i.e., involving species-specific chemical release and a well-defined response) and those that are nonspecific chemical signals (Beauchamp et al., 1976).

One attempt to deal with this conceptual distinction between pheromonal and incidental olfactory emission involved the olfactory signals of bullhead catfish (Ictalurus nebulosus; Bryant and Atema, 1987). These researchers were investigating whether catfish use body odors to detect the presence of newcomers to a territory and found that catfish exhibited increased aggression toward another fish fed a "strange" diet. Thus, catfish appear to be sensitive to relatively small changes in the olfactory stimuli in the waste products of other animals, and these stimuli are sufficient to regulate dominance and territorial relationships between these fish. Faced with this striking example of a signal that is an incidental by-product of digestion, Bryant and Atema suggest that "it is possible, perhaps likely, that many so-called pheromones in vertebrates will turn out to be rather nonspecific metabolites, exerting their influence by virtue of chemical habituation and familiarity" (p. 1658). The clear inference from this statement is that students of olfactory communication, like those in acoustic communication, may be invoking relatively complex explanations when more simple ones will suffice.

Galef (1986) similarly argues against the tendency among communication researchers to focus on ritualized or formalized displays and ignore the more subtle ways in which animals learn about their world. In his studies of the means by which weanling Norway rats (Rattus norvegicus) learn what foods to eat, Galef has shown that these young animals use olfactory cues on the breath of postprandial adults to orient them to nonpoisonous foods. Galef stresses that these breath cues are passively emitted by the adults as incidental by-products of ingestion, thus allowing these animals to adapt to different environments in which the identity of safe foods may vary.

Visual Signals

This form of communication, especially as regards facial expressions, has been the focus of attention before and since Darwin's classic contribution to the subject (Darwin, 1872/1965). Like other areas of communication, there has been a tendency to overinterpret the meaning of facial expressions. Ghiselin (1974) states that investigators of facial expression in humans and primates "tend to presuppose that emotional expressions are there for the sake of communication, ignoring Darwin's view that some have a communicative function and others do not" (p. 255). Although human facial expressions have been reinterpreted within a physiological, homeostatic framework (Zajonc, 1985), the assumption that facial expressions serve primarily as communicatory signals remains because of a lack of direct experimental evidence to the contrary.

Damselflies exhibit a behavior called wingclapping in which the wings are spread apart and then snapped together. Investigators have suggested that this behavior functions as a territorial declaration and/or as a signal between an ovipositing female and her mate that they are both present (Bick and Bick, 1978). Wingclapping, however, occurs in other contexts as well including feeding, grooming, and even when a female is alone (Erickson and Reid, 1989). Erickson and Reid hypothesized that wingclapping by damselflies (Calopteryx maculata) is a thermoregulatory behavior (perhaps by cooling the body surface convectively or through some other mechanism) and, in support of their hypothesis, showed that wingclapping increases during radiant heating.

The dynamics of pigeon (Columba livia) flock behavior provide a striking example of how receivers interpret incidental visual information in their environment. Davis (1975) investigated a phenomenon known as the "contagion of flight," in which an individual bird can induce flight in an entire flock. Others had tried, without success, to identify the "alarm signal" that induces this contagion. Studying this phenomenon in pigeons, Davis observed that under normal conditions these birds typically engaged in preflight behaviors before taking off. When, however, a pigeon took off suddenly and thus did not engage in these preflight behaviors, the other pigeons in the flock took off as well. Thus, it may be the absence of a behavior that induces contagion of flight in pigeons, not the presence of an alarm signal. This example demonstrates once again how receivers can use incidental information provided by conspecifics.

WHAT ABOUT BIRDSONG?

"In the adaptation of birds to an aerial environment, the evolution of feathers and a remarkably efficient respiratory system have incidentally enabled birds to develop complex systems of communication" (Hooker, 1968, p. 311).

As a group, birds emit more complex, rich, and varied vocalizations than do mammals. Related to this difference between avian and mammalian vocalizations, neuroethological approaches to birdsong have paid dividends far exceeding similar approaches in mammals; the notion of a young bird learning the local "dialect" from an adult bird is commonplace in the birdsong literature and is virtually unheard of in the mammalian literature. Functionally, male birds appear to accrue reproductive benefits from singing, although such benefits are admittedly "not well established" (Kroodsma and Byers, 1991). But even as various hypotheses regarding the functional importance of birdsong gain and lose support, we will still be left with the basic, vexing question, "Why birdsong and not, for example, ungulatesong?" Does the answer to this question lie in the ecology of birds, their physiology, or both?

Natural selection works on the components of characters that already exist. One of the initial barriers to acceptance of Darwin's theory was that of explaining the presence of complex and seemingly perfect structures for which intermediate structures were not evident. Understanding the perfection of the eye presented such a problem to Darwin, as did the geometric harmony of the bee's honeycomb. These are no longer considered difficult problems for Darwinism; a simple eye is better than no eye at all, and some exquisitely complex structures emerge from the application of remarkably few behavioral rules (e.g., wasp nests; Kugler and Turvey, 1987). In either case, basic biological materials and behavioral components must exist if natural selection is to have something on which to work.

Understanding acoustic communication in animals requires a similar approach. At the most obvious level, acoustic communicatory systems cannot evolve unless animals can make sounds. Some animals, such as insects, lack an active respiratory system in which air can be expelled through a vibrating structure; instead, crickets, for example, have developed stridulation, in which one wing is rubbed against another (Hoy et al., 1977). Mammals and birds, on the other hand, make noise when they breathe. As Darwin (1872/1965) noted, "Involuntary and purposeless contraction of the muscles of the chest and glottis ...may have first given rise to the emission of vocal sounds" (p. 84). Similarly, Spurway and Haldane (1953) wrote of animal vocalization as "a ritualisation of breathing."

Using the conceptual perspective that we are pursuing here, one might wonder about the physiological bases of birdsong and whether birds have respiratory mechanisms that have been exapted for communication. This perspective is captured by Hooker's statement quoted above, as well as Morris's

(1956) contention that, in birds, "Respiratory changes [have led to] alterations in the breathing rate, amplitude and regularity, which have evolved into vocalisation on the one hand, and inflation displays, on the other." Standing alone, however, these statements do not direct us toward a clear understanding of the evolution of birdsong. Such direction can only be provided by addressing the song production mechanisms of birds directly.

As is well known, birdsong is produced by a novel avian structure, the syrinx, that sits between the bronchi and the trachea. The syrinx (once called the "lower larynx" before Huxley (1877) renamed it, ostensibly to avoid confusion), has become synonymous with the vocal organ of birds. In fact, it was based primarily on syringeal anatomy that Muller, in 1878, first classified the passerine species (Ames, 1971). No doubt, the uniquely structured syrinx of the songbirds, coupled with the uniquely complex songs of these species, have had a profound impact on the nature of experimental investigations of this organ.

The syrinx is very similar in structure to the larynx, although the latter structure is found in both birds and mammals. The larynx consists of membranes controlled by muscles that constrict and adduct the membranes and thus close or open the upper respiratory airway, respectively. Of course, the larynx is recognized as a vocal organ but, as argued in depth by Negus over sixty years ago, several of its functions precede, both in terms of evolutionary time and physiological necessity, the vocal uses of that organ. As Negus (1929) stated it, "in the larynx an organ has been evolved, particularly by arboreal animals, to subserve functions of locomotion, prehension, olfaction and deglutition, and that by the various modifications brought in, an instrument has been provided capable of use for sound production in a highly efficient form" (p. 267). Similarly, the avian larynx serves many functions, including roles in protection against foreign bodies, respiration, swallowing, and modulation of sound (McLelland, 1989).

The syrinx has been treated very differently by comparative anatomists. Its primary role as a vocal organ has apparently never been seriously questioned and, because studies of avian respiration have proceeded without any notice being given to the syrinx, there has never been any reason to doubt that it plays an exclusive role in sound production. Nonetheless, the syrinx, like the larynx, might have evolved as an organ necessary for physiological and/or respiratory regulation and only later was exapted for sound production.

Before addressing this question directly, consider that the avian respiratory system is qualitatively different from that of mammals and must function adequately in environments (i.e., high altitudes) that have very low oxygen concentrations. First, the avian system is unique in that air flows unidirectionally through the bird's respiratory system, in contrast to the mammalian system of bidirectional flow into and out of the lungs (Bretz and Schmidt-Nielsen, 1971). The avian system is also different in that inspiratory and expiratory flow is generated by a series of air sacs. Consider also that passerine species (which have the most highly derived syringeal musculature) maintain metabolic rate and body temperature significantly higher than do mammals and nonpasserine species (Schmidt-Nielsen, 1984; Caputa, 1984). Moreover, a number of respiratory system mechanisms help protect the avian brain and body from the constant thermal threat posed by elevated body temperatures. These respiratory mechanisms include gular flutter and panting, during which, in some birds, respiratory frequencies can increase 20–30 times over resting values (Dawson, 1982).

We recognize that such discussions of the uniqueness of avian physiology do not bear directly on any possible nonvocal functions of the syrinx. Moreover, we have not found many scientific reports that address this possibility. However, some interesting evidence does exist. Specifically, in a paper devoted primarily to the neural control of sound production in chaffinches (Fringilla coelebs), Nottebohm (1971) reports the consequences of bilateral denervation of the branch of the hypoglossus nerve that innervates the syrinx. He writes:

> An unexpected result of bilateral section of the hypoglossus was the respiratory disorder which overtook the operated birds when placed under respiratory stress. The usual response to respiratory stress, as induced in intact birds by alarm and excess heat, is to hyperventilate... .Seven of the eleven birds died during the first month following the operation. When frightened, they would start producing a "wheezing" sound, drop from their perches, and lie on the floor of their cages. If the disturbance persisted, respiration slowed down and became more laborious until it stopped and the bird apparently died from asphyxia... .[T]he fact that the operated birds do not differ in their behavior from the intact animals as long as they are not disturbed, suggests that their "relaxed" respiratory rhythms are similar. Operated animals, furthermore, keep their weight and general condition well (pp. 235–236).

Nottebohm then provides an explanation for the devastating impact of bilateral denervation of the syrinx:

> During inspiration the soft medial walls, including the internal tympaniform membranes of each bronchus tend to collapse due to the Bernoulli effect... . Thus, when hyperventilation is called for, inspiration becomes more laborious and hypoventilation results. The more frantic the inspiratory effort, the less air reaches the lungs... . If this interpretation is correct, the syringeal musculature plays an active role during the inspiratory effort, so that bronchial walls are kept taut and the passage of air meets the least resistance (p. 236).

Youngren et al. (1974) make a similar observation in chickens (Gallus gallus), a nonpasserine species. These investigators examined the effects of bilateral hypoglossectomy on their birds during exercise. They write: "The hypoglossectomized birds quickly became exhausted, made low wheezing sounds during inspiration and expiration, and recovered very slowly during their exertions. The normal bird was difficult to exhaust, did not wheeze, and recovered almost immediately" (p. 412). Moreover, these same investigators found that activity of one syringeal muscle, the tracheolateralis, is tightly coupled to respiratory activity, again suggesting that at least part of the syringeal anatomy has a nonvocal function. Conversely, they also found that "nearly normal calls can be evoked when both the sternotrachealis and the tracheolateralis muscles are inactivated" (Youngren et al., 1974, p. 412). Clearly, there is room to doubt the widely held view of the syrinx as an organ with an exclusively vocal function (see also Brackenbury, 1989; Phillips and Peek, 1975).

The evidence just cited is too slim to allow us to draw any conclusions regarding the possible nonvocal functions of the syrinx. On the other hand, the mammalian larynx is an organ that is known to play important roles in respiratory and other physiological functioning and yet, in adult rats, bilateral denervation of either the inferior or superior laryngeal nerves is not a lethal surgical procedure (e.g., Mortola and Piazza, 1987; Thomas et al., 1981). Considering the uniquely complicated respiratory system of birds, perhaps we should not be surprised that denervation of the syrinx is lethal during arousal and respiratory activation. Thus, perhaps the syrinx, like the larynx, is more than simply a vocal organ. If this is indeed the case, the comparative study of avian sound production could be integrated with our substantial knowledge of avian physiology, and, in time, bring us a step closer to birdsong's evolutionary origins.

CONCLUSIONS

> **Desdemona:** Hark! Who is't that knocks?
> **Emilia:** It's the wind.
>
> —Shakespeare's *Othello* Act IV Scene III

In this chapter, we have argued for greater caution when confronted with a vocalization, or any other signal, about whose production mechanisms we know little. Ignorance of and disinterest in these mechanisms, coupled with the widely held assumption that, a priori, animal vocalizations serve communicatory functions, blurs critical distinctions between the diverse mechanisms that produce these sounds. By blurring these distinctions, we blind ourselves to a signal's information value, its potential for communication, and its evolutionary history. Even more fundamental, however, is that, even in the face of accumulating negative evidence against a signal's design for communication, it is rarely acknowledged that some animal signals have no communicatory function. In sum, increasing attention to the mechanisms that produce and constrain vocal and other communicatory behaviors has the beneficial effect of exposing the historical forces that may have shaped their evolution.

Communication is a dynamic and two-way process, but the current trend of attributing intentionality and other mentalistic categories to a signaling animal (e.g., Hauser and Nelson, 1991) degrades the role of the receiver as an active participant in communicatory systems: *Intention*, *deception*, and *manipulation* are terms that are nearly always used to describe the sender doing something to the receiver. Furthermore, these terms imply a host of concepts, such as control, planning, and conscious design, that are contrary to the empirical findings of experimental psychology and inconsistent with our understanding of the serendipity and contrivance of evolution (Blumberg and Wasserman, 1995). The debate over the meaning and applicability of mentalistic concepts as tools for explaining the vocal behaviors of some animals (especially primates) will continue. But such concepts should not become the dominant explanatory tools of animal communication when more parsimonious, less mentalistic, and more empirically fruitful approaches are available.

Given the infinite number of visual, acoustic, and olfactory signals emitted by animals throughout the day and throughout evolutionary history, it is only reasonable to expect that there will be a similar multitude of ways in which animals can interact on the basis of those signals. At one extreme, most signals go unheeded. At the other, some signals are specialized for their communicatory effects and will be received by conspecifics that have specialized detection systems. Between these two extremes, however, lies the diversity that we expect the complexities of animal behavior and the serendipity of the evolutionary process to produce. There are many possible paths from incidental emission to fortuitous effect to communicatory function. Focusing exclusively on communicatory function will only conceal these paths and thus prevent us from unveiling the evolutionary origins of communication.

ACKNOWLEDGMENTS

We are sincerely grateful to Nick Thompson and Don Owings for their remarkable dedication to a constructive and intellectually stimulating editorial process. Preparation of this chapter was supported in part by National Institute of Mental Health Grants MH50701 to M.S.B. and MH28355 to J.R.A.

REFERENCES

Adler, N. & Anisko, J. (1979). The behavior of communicating: An analysis of the 22 kHz call of rats (Rattus norvegicus). *Am. Zool., 19,* 493–508.

Allin, J. T. & Banks, E. M. (1972). Functional aspects of ultrasound production by infant albino rats (Rattus norvegicus). *Anim. Behav., 2,* 175–185.

Ames, P. L. (1971). The morphology of the syrinx in passerine birds. *Bulletin of the Peabody Museum of Natural History (Yale University), 37,* 1–194.

Barfield, R. J., Auerbach, P., Geyer, L. A., & McIntosh, T. K. (1979). Ultrasonic vocalizations in rat sexual behavior. *Am. Zool., 19,* 469–480.

Barfield, R. J. & Geyer, L.A. (1972). Sexual behavior: Ultrasonic postejaculatory song of the male rat. *Science, 176,* 1349–1350.

Basolo, A. (1990). Female preference predates the evolution of the sword in swordtail fish. *Science, 250,* 808–810.

Beauchamp, G. K., Doty, R. L., Moulton, D. G., & Mugford, R. A. (1976). The pheromone concept in mammalian chemical communication: a critique. In R. L. Doty, (Ed.), *Mammalian Olfaction, Reproductive Processes, and Behavior* (pp. 144–160). New York: Academic Press.

Bick, G. H. & Bick, J. C. (1978). The significance of wingclapping in Zygoptera. *Odonatologica, 7,* 5–9.

Blumberg, M. S. (1992). Rodent ultrasonic short calls: Locomotion, biomechanics, and communication. *J. Comp. Psychol., 106,* 360–365.

Blumberg, M. S. & Alberts, J. R. (1990). Ultrasonic vocalizations by rat pups in the cold: an acoustic by-product of laryngeal braking? *Behav. Neurosci., 104,* 808–817.

Blumberg, M. S. & Alberts, J. R. (1991a). On the significance of similarities between ultrasonic vocalizations of infant and adult rats. *Neurosci. Biobehav. Rev., 15,* 383–390.

Blumberg, M. S. & Alberts, J. R. (1991b). Both milk-deprivation and hypoxia diminish metabolic heat production and ultrasound emission by rat pups during cold exposure. *Behav. Neurosci., 105,* 1030–1037.

Blumberg, M. S. & Alberts, J. R. (1992). Functions and effects in animal communication: Reactions to Guilford and Dawkins. *Anim. Behav., 44,* 382–383.

Blumberg, M. S., Efimova, I. V., & Alberts, J. R. (1992a). Ultrasonic vocalizations by rat pups in the cold: the primary importance of ambient temperature and the thermal significance of contact comfort. *Dev. Psychobiol., 25,* 229–250.

Blumberg, M.S ., Efimova, I. V., & Alberts, J. R. (1992b). Thermogenesis during ultrasonic vocalization by rat pups isolated in a warm environment: a thermographic analysis. *Dev. Psychobiol., 25,* 497–510.

Blumberg, M. S. & Moltz, H. (1987). Hypothalamic temperature and the 22-kHz vocalization of the male rat. *Physiol. Behav., 40,* 637–640.

Blumberg, M. S., & Stolba, M. A. (1996). Thermogenesis, myoclonic twitching, and ultrasonic vocalization during moderate and extreme cold exposure. *Behavioral Neuroscience, 110,* 305–314.

Blumberg, M. S. & Wasserman, E. A. (1995). Animal mind and the argument from design. *American Psychologist, 50,* 133–144.

Brackenbury, J. H. (1989). Functions of the syrinx and the control of sound production. In A. S. King & J. McClelland (Eds.), *Form and Function in Birds, Volume 4* (pp. 193–220). San Diego, CA: Academic Press.

Bramble, D. M. (1989). Axial-appendicular dynamics and the integration of breathing and gait in mammals. *Am. Zool., 29,* 171–186.

Bramble, D. M. & Carrier, D. R. (1983). Running and breathing in mammals. *Science, 219,* 251–256.

Bretz, W.L. & Schmidt-Nielsen, K. (1971). Bird respiration: flow patterns in the duck lung. *J. Exp. Biol., 54,* 103–118.

Brewster, J. & Leon, M. (1980). Relocation of the site of mother-young contact: Maternal transport behavior in Norway rats. *J. Comp. Physiol. Psychol., 94,* 69–79.

Brooks, R. J. & Banks, E. M. (1973). Behavioral biology of the collared lemming [Dicrostonyx groenlandicus (Traill)]: an analysis of acoustic communication. *Anim. Behav. Monogr., 6,* 1–83.

Brown, A. M. (1973). High levels of responsiveness from the inferior colliculus of rodents at ultrasonic frequencies. *J. Comp. Physiol., 83,* 393–406.

Bryant, B. P. & Atema, J. (1987). Diet manipulation affects social behavior of catfish: importance of body odor. *J. Chem. Ecol., 13,* 1645–1661.

Burghardt, G. M. (1970). Defining "communication." In J. W. Johnston, D. G. Moulton, & A. Turk (Eds.), *Communication by chemical signals, Volume 1* (pp. 5–18). New York: Appleton-Century-Crofts.

Caputa, M. (1984). Some differences in mammalian versus avian temperature regulation: putative thermal adjustments to flight in birds. In J. R. S. Hales (Ed.), *Thermal physiology* (pp. 413–417). New York: Raven Press.

Darwin, C. (1871/1981). *The Descent of Man, and Selection in Relation to Sex*. Princeton: Princeton University Press.

Darwin, C. (1872/1965). *The Expression of the Emotions in Man and Animals*. Chicago, IL: University of Chicago Press.

Davis, J. M. (1975). Socially induced flight reactions in pigeons. *Anim. Behav., 23,* 597–601.

Dawson, W. R. (1982). Evaporative losses of water by birds. *Comp. Biochem. Physiol., 71A,* 495–509.

Demski, L. S., Gerald, J. W., & Popper, A. N. (1973). Central and peripheral mechanisms of teleost sound production. *Am. Zool., 13,* 1141–1167.

Dizinno, G. & Whitney, G. (1977). Androgen influence on male mouse ultrasounds during courtship. *Horm. Behav., 8,* 188–192.

Erickson, C. J. & Reid, M. E. (1989). Wingclapping behavior in Calopteryx maculata (P. de Beauvois) (Zygoptera: Calopterygidae). *Odonatologica, 18,* 379–383.

Fenton, M. B. & Licht, L. E. (1990). Why rattle snake? *J. Herpetol., 24,* 274–279.

Floody, O. R. & Pfaff, D. W. (1977). Communication among hamsters by high-frequency acoustic signals. III. Responses evoked by natural and synthetic ultrasounds. *J. Comp. Physiol. Psychol., 91,* 820–829.

Francis, R. L. (1977). 22-kHz calls by isolated rats. *Nature, 265,* 236–238.

Galef, B. G. (1986). Olfactory communication among rats: information concerning distant diets. In D. Duvall, D. Muller-Schwarze, & R. M. Silverstein (Eds.), *Chemical Signals in Vertebrates 4: Ecology, Evolution and Comparative Biology* (pp. 487–505). New York: Plenum Press.

Gans, C. & Maderson, P. F. A. (1973). Sound producing mechanisms in recent reptiles: Review and comment. *Am. Zool., 13,* 1195–1203.

Ghiselin, M. T. (1969). *The triumph of the Darwinian method*. Chicago, IL: University of Chicago Press.

Ghiselin, M. T. (1974). *The economy of nature and the evolution of sex*. Berkeley: University of California Press.

Gould, S. J., & Vrba, E. S. (1982). Exaptation — a missing term in the science of form. *Paleobiol., 8,* 4–15.

Green, S. M. (1981). Sex differences and age gradations in vocalizations of Japanese and lion-tailed monkeys (Macaca fuscata and Macaca silenus). *Am. Zool., 21,* 165–183.

Guilford, T. & Dawkins, M. S. (1993). Receiver psychology and the design of animal signals. *Trends Neurosci., 16,* 430–436.

Harrison, V. C., Heese, H., & Klein, M. (1968). The significance of grunting in hyaline membrane disease. *Pediatrics, 41,* 549–559.

Hauser, M.D. & Nelson, D. A. (1991). 'Intentional' signaling in animal communication. *Trends Ecol. Evol., 6,* 186–189.

Heinrich, B. (1979). *Bumblebee Economics*. Cambridge: Harvard University Press.

Hofer, M. A., Brunelli, S. A., & Shair, H. N. (1993). Ultrasonic vocalization responses of rat pups to acute separation and contact comfort do not depend on maternal thermal cues. *Developmental Psychobiology, 26,* 81–95.

Hofer, M. A., Brunelli, S. A., & Shair, H. N. (1994). Potentiation of isolation-induced vocalization by brief exposure of rat pups to maternal cues. *Developmental Psychobiology, 27,* 503–517.

Hofer, M. A. & Shair, H. N. (1992). Ultrasonic vocalization by rat pups during recovery from deep hypothermia. *Dev. Psychobiol., 25,* 511–528.

Hooker, B. I. (1968). Birds. In T. A. Sebeok, (Ed.), *Animal Communication* (pp. 311–337). Bloomington: Indiana University Press.

Hoy, R. R., Hahn, J., & Paul, R. R. (1977). Hybrid cricket auditory behavior: evidence for genetic coupling in animal communication. *Science, 195,* 82–84.

Huxley, T. H. (1877). *A manual of the anatomy of vertebrated animals*. London.

Johnson, P. & Andrews, D. (October, 1990). *Regulation during fetal and early postnatal period: effects of temperature and state*. Paper presented at the Animal Models of SIDS Workshop, Washington, D.C.

Johnson, P., Harding, R., McClelland, M., & Whyte, P. (1977). Laryngeal influences on lung expansion and breathing in lambs. *Ped. Res., 11,* 1025.

Kelso, J. A. S. (1995). *Dynamic patterns: The self-organization of brain and behavior*. Cambridge, MA: MIT Press.

Kennedy, J. S. (1992). *The new anthropomorphism*. Cambridge: Cambridge University Press .

Krebs, J. R. & Dawkins, R. (1984). Animal signals: Mind-reading and manipulation. In J. R. Krebs & N. B. Davies (Eds.), *Behavioral Ecology: An Evolutionary Approach* (pp. 380–402). Sunderland, MA: Sinauer Associates.

Kroodsma, D. E. & Byers, B. E. (1991). The function(s) of bird song. *Am. Zool., 31,* 318–328.

Kugler, P. N. & Turvey, M. T. (1987). *Information, natural law, and the self-assembly of rhythmic movement*. Hillsdale, NJ: Lawrence Erlbaum.

McClelland, J. (1989). Larynx and trachea. In A. S. King & J. McClelland (Eds.), *Form and function in birds, Volume 4* (pp. 69–103). San Diego, CA: Academic Press.

McLanahan, E. B. & Green, K. M. (1977). In D. G. Kleiman (Ed.), *The biology and conservation of the Callitrichidae* (pp. 251–269). Washington, D.C.: Smithsonian Institution Press.

Morris, D. (1956). The feather postures of birds and the problem of the origin of social signals. *Behavior, 9,* 73–113.

Mortola, J. P. & Piazza, T. (1987). Breathing pattern in rats with chronic section of the superior laryngeal nerves. *Respir. Physiol., 70,* 51–62.

Negus, V. E. (1929). *The mechanism of the larynx*. London: Wm. Heinemann.

Noirot, E. (1972). Ultrasounds and maternal behavior in small rodents. *Dev. Psychobiol., 5,* 371–387.

Nottebohm, F. (1971). Neural lateralization of vocal control in a passerine bird. I. Song. *J. Exp. Zool., 177,* 229–262.

Okon, E. E. (1970). The effect of environmental temperature on the production of ultrasounds by isolated non-handled albino mouse pups. *J. Zool. Lond., 162,* 71–83.

Okon, E. E. (1971). The temperature relations of vocalization in infant golden hamsters and Wistar rats. *J. Zool. Lond., 164,* 227–237.

Oyama, S. (1985). *The ontogeny of information*. Cambridge: Cambridge University Press.

Payne, K. B., Langbauer, W. R., Jr., & Thomas, E. M. (1986). Infrasonic calls of the Asian elephant (Elephas maximus). *Behav. Ecol. Sociobiol., 18,* 297–301.

Phillips, R. E. & Peek, F. W. (1975). Brain organization and neuromuscular control of vocalization in birds. In P. Wright, P. G. Caryl, & D. M. Vowles (Eds.), *Neural and Endocrine Aspects of Behaviour in Birds* (pp. 243–274). Amsterdam: Elsevier.

Pomiankowski, A. (1994). Swordplay and sensory bias. *Nature, 368,* 494–495.

Poole, J. H., Payne, K., Langbauer, W. R., Jr., & Moss, C. J. (1988). The social contexts of some very low frequency calls of African elephants. *Behav. Ecol. Sociobiol., 22,* 385–392.

Robert, D., Amoroso, J., & Hoy, R. R. (1992). The evolutionary convergence of hearing in a parasitoid fly and its cricket host. *Science, 258,* 1135–1137.

Roberts, L. H. (1972). Correlation of respiration and ultrasound production in rodents and bats. *J. Zool. Lond., 168,* 430–449.

Roberts, L. H. (1975). The functional anatomy of the rodent larynx in relation to audible and ultrasonic cry production. *Zool. J. Linn. Soc., 56,* 255–264.

Rowe, M. P. & Owings, D. H. (1990). Probing, assessment, and management during interactions between ground squirrels and rattlesnakes. Part 1: Risks related to rattlesnake size and body temperature. *Ethology, 86,* 237–249.

Rowe, M. P. & Owings, D. H. (in press). Probing, assessment, and management during interactions between ground squirrels and rattlesnakes. Part 2: Cues afforded by rattlesnake rattling. *Ethology*.

Ryan, M. J., Fox, J. H., Wilczynski, W., & Rand, A. S. (1990). Sexual selection for sensory exploitation in the frog Physalaemus pustulosus. *Nature, 343*, 66–67.

Sales, G. D. (1972). Ultrasound and mating behaviour in rodents with some observations on other behavioural situations. *J. Zool. Lond., 168*, 149–164.

Schmidt-Nielsen, K. (1984). *Scaling: Why is animal size so important?* Cambridge: Cambridge University Press.

Schwartz, G. G. & Rosenblum, L. A. (1985). Sneezing behavior in the squirrel monkey and its biological significance. In L. A. Rosenblum & C. L. Coe (Eds.), *Handbook of squirrel monkey research* (pp. 253–269). New York: Plenum Press.

Smith, W. J. (1977). *The behavior of communication: An ethological approach*. Cambridge: Harvard University Press.

Spurway, H. & Haldane, J. B. S. (1953). The comparative ethology of vertebrate breathing. I. Breathing in newts, with a general survey. *Behavior, 6*, 33–76.

Stamps, J. A. (1991). Why evolutionary issues are reviving interest in proximate behavioral mechanisms. *Am. Zool., 31*, 338–348.

Swaisgood, R. R. (1994). *Assessment of rattlesnake dangerousness by California ground squirrels*. Unpublished doctoral dissertation, University of California, Davis.

Takahashi, L. K., Thomas, D. A., & Barfield, R. J. (1983). Analysis of ultrasonic vocalizations emitted by residents during aggressive encounters among rats (Rattus norvegicus). *J. Comp. Physiol. Psychol., 97*, 207–212.

Takeuchi, H. & Kawashima, S. (1986). Ultrasonic vocalizations and aggressive behavior in male rats. *Physiol. Behav., 38*, 545–550.

Thelen, E. & Smith, L. B. (1994). *A dynamic systems approach to the development of cognition and action*. Cambridge, MA: MIT Press.

Thiessen, D. D. & Kittrell, E. M. W. (1979). Mechanical features of ultrasound emission in the Mongolian gerbil, Meriones unguiculatus. *Am. Zool., 19*, 509–512.

Thiessen, D. D., Kittrell, E. M. W., & Graham, J. M. (1980). Biomechanics of ultrasound emissions in the Mongolian gerbil, Meriones unguiculatus. *Behav. Neural Biol., 29*, 415–429.

Thomas, D. A. & Barfield, R. J. (1985). Ultrasonic vocalization of the female rat (Rattus norvegicus) during mating. *Anim. Behav., 33*, 720–725.

Thomas, D. A., Talalas, L., & Barfield, R. J. (1981). Effect of devocalization of the male on mating behavior in rats. *J. Comp. Physiol. Psychol., 95*, 630–637.

Williams, G. C. (1966). *Adaptation and natural selection*. Princeton: Princeton University Press.

Wilson, E. O. (1975). *Sociobiology: The new synthesis*. Cambridge: Harvard University Press.

Youngren, O. M., Peek, F. W., & Phillips, R. E. (1974). Repetitive vocalizations evoked by local electrical stimulation of avian brains. III. Evoked activity in the tracheal muscles of the chicken (Gallus gallus). *Brain Behav. Evol., 9*, 393–421.

Zajonc, R. B. (1985). Emotion and facial efference: a theory reclaimed. *Science, 228*, 15–21.

Zippelius, H. M. & Schleidt, W. M. (1956). Ultraschall-Laute bei jungen mausen. *Naturwissenschaften, 43*, 502.

Chapter 8

STUDYING HOW CETACEANS USE SOUND TO EXPLORE THEIR ENVIRONMENT

Peter L. Tyack

Biology Department
Woods Hole Oceanographic Institution
Woods Hole, Massachusetts 02543-1049

ABSTRACT

Many biologists implicitly assume that mechanisms for echolocation and communication are separate and compartmentalized. For example, the high-frequency vocal and auditory specializations of dolphins are typically only discussed in terms of echolocation and the low-frequency sounds of baleen whales are usually presented as signals for long-range communication. However, signals that evolved for one purpose may develop other functions. Some porpoises appear to use rhythmic patterns of "echolocation" clicks as communicative signals. When a whale makes a low-frequency sound for communication, the sound may echo from the seafloor and possibly provide the whale with important information about its environment. Research on the evolution of echolocation in marine mammals suffers from a dearth of studies of ecological function and from a lack of broad comparative reviews. If studies of marine mammal sonar included more analysis of the problems for which sonar may have evolved, we might discover fascinating new kinds of biosonar. For example, low-frequency sound is better suited than high frequency for long-range sonar in the sea, and many targets of great importance to marine mammals, such as large bathymetric features and fish with resonant swim bladders are also well suited to low-frequency sonar. Some marine mammals have the skills required to engage in bistatic sonar, in which one animal may

Perspectives in Ethology, Volume 12
edited by Owings *et al.*, Plenum Press, New York, 1997

listen to how the sounds of another individual are modified by the environment. Targets such as concentrations of fish with resonant swim bladders may absorb more energy than they scatter, leading to significant advantages for bistatic sonar in a forward propagation mode. These examples blend features typically associated with the domains of sonar and communication. I suggest that auditory and vocal skills evolved to function in one of these domains may preadapt animals for developing abilities in the other domain. Vocal learning, in particular, is required for many forms of sonar, and it also enables the evolution of very different communication systems than are possible when vocal output is unaffected by auditory input.

INTRODUCTION

In spite of the considerable successes of research on cetacean biosonar, there are several unexploited opportunities. Extensive research has defined the biosonar signals produced by bottlenose dolphins, the performance of their echolocation, and specializations of their auditory system for echolocation. Yet we are ignorant of how cetaceans use echolocation in the natural environment. We know little about the targets for which this remarkable sensory ability evolved. In this paper, I will explore several new directions for research on how cetaceans may use sound to explore their environment. Our understanding of cetacean echolocation also remains isolated from research on acoustic communication, and I will discuss several potential interactions between auditory and vocal abilities evolved for echolocation and for communication.

The basic facts of dolphin echolocation are well known to most ethologists and are well summarized in Au (1993). Bottlenose dolphins produce clicks of very short duration (tens of microseconds, or less than 0.0001 sec), with energy well above the highest frequencies we can hear. Humans cannot hear well above 20 kHz, but dolphin echolocation clicks have energy above 100 kHz. Bottlenose dolphins can hear well at high frequencies, up to about 150 kHz. The auditory system of bottlenose dolphins is also specialized for rapid processing of very short signals. Bottlenose dolphins can detect small targets several centimeters in diameter out to ranges of about 100 m (Murchison 1980). Bottlenose dolphins can measure the distance to a target by timing how long it takes for an echo to return from one of their clicks, and they tend to wait to produce a click until they have processed the echo from the last click. Dolphins can discriminate targets of different composition, shape, or size.

Cetacean biosonar has been studied almost exclusively in captive bottlenose dolphins echolocating on artificial targets. Some of the first research on dolphin echolocation used fish as targets (e.g., Kellogg 1961). While these may be important targets for dolphins, their acoustic properties are complicated, and were not quantified in these early studies. The use of artificial targets with simple and quantified acoustic properties represented an important advance in studies of dolphin echolocation. The early focus on bottlenose dolphins as subjects for echolocation research made sense as these were the most common animals in captivity and they are readily trained for experimental work. It was also a reasonable research strategy to concentrate on understanding one species in depth before including potential variation in echolocation across species.

In four decades of research on marine mammal echolocation, it is remarkable how few studies have used biologically relevant targets or species other than bottlenose dolphins. While we know a great deal about dolphin echolocation in a narrow artificial context, we know very little about how cetaceans actually use echolocation in the wild to solve problems of orientation, obstacle avoidance, and prey detection.

I believe that dolphin echolocation is at a point described by Griffin (1980) retrospectively for bats:

> Something I should like to emphasize, because it may have implications for the future, is that after these basic facts [of bat echolocation] had been generally accepted there was what now seems in retrospect an incredible lack of interest in further studies of echolocation .... I cannot overemphasize the intellectual inertia and difficulty in justifying the necessary effort to continue studies of echolocation and inquire what further ramifications might develop after a wider variety of species had been studied under a wider range of natural conditions. We should be alert in the future for similar mental blocks that may restrict imagination and thus retard progress. (p. 4)

EVOLUTIONARY COMPARISONS AND STUDIES OF ECOLOGICAL FUNCTION HAVE BEEN CENTRAL TO OUR UNDERSTANDING OF ECHOLOCATION IN BATS

Echolocation in bats has become a classic topic in ethology, sensory biology, and neurobiology, while echolocation by marine mammals is usually mentioned in passing, if at all, by most texts. This stems not merely from the difficulty of studying marine mammals, for both aerial bats and aquatic cetaceans live in environments that are foreign to humans. Both groups are extremely mobile in the wild and are difficult to maintain in captivity. The neurophysiological research which has uncovered many of the neural mechanisms underlying echolocation in bats has been facilitated by the greater availability of these animals than marine mammals for invasive research. However, new noninvasive methods are becoming available for neurobiological research. I believe that the more significant obstacle comes from less integration of psychophysical and physiological studies of cetacean echolocation with behavioral and ecological studies.

Synergy between Studies of Behavioral Ecology and Sensory Physiology of Bat Biosonar

"From the beginning, the study of hearing in bats has been characterized by a synergistic partnership between behavioral and physiological approaches. The results have been nothing short of spectacular..." (Grinnell 1995, p. 1).

Studies of echolocation and communication benefit from synergy between naturalistic and experimental research. There are strong traditions of field studies of communication and experimental studies of echolocation in cetaceans, but there is little interchange between these approaches. Most studies of acoustic communication in cetaceans are observational studies, often in the wild, focusing on the acoustic structure and social functions of vocal signals as opposed to sensorimotor or cognitive mechanisms. Most studies of cetacean echolocation, as we have seen, involve experimental tests using captive dolphins and artificial targets. By contrast, many experiments in bat echolocation take as their starting point a detailed understanding of what signal the bat uses and what information it needs for a particular natural task.

For example, the earliest observations that stimulated interest in bat biosonar concerned the ability of bats to avoid large obstacles such as walls. Griffin (1974) initially studied obstacle avoidance in controlled experiments using evenly spaced wires. However, since many bats feed on insects flying at night, it seemed that bats might also use echolocation to detect their prey. Griffin and his colleagues therefore studied the sounds made by wild bats as they fed on flying insects. This research demonstrated that as bats pursued their prey, their biosonar signals became shorter with a more rapid repetition rate (Griffin 1974). Later work combined visual and audio recordings to define stages of predation as the big brown bat, *Eptesicus fuscus*, approached, tracked, and caught its prey, with each stage having characteristic frequency-modulated (FM) downsweep biosonar signals (e.g., Simmons & Kick 1983; Kick & Simmons 1984). The variation in biosonar signals with biosonar task helped to define how bats process these signals. The size, shape, and motion of prey targets coupled with the flight patterns of the bats also impose design constraints on bat biosonar that help narrow the range of appropriate biosonar mechanisms. The discovery of how echolocation is used to find prey fostered the development of experimental laboratory models of prey detection involving appropriate artificial or even synthetic models. These models have been critical for later research on biosonar signal processing by bats (e.g., Simmons et al. 1995).

Most flying insects that are preyed upon by bats flutter their wings at rates of roughly 10–100 times/second. This may seem to be a minor detail for bat echolocation, but this subtle feature of the bat's environment is critical for the evolution of several important features of bat echolocation. Several species of bats from different families have independently evolved a biosonar signal that differs from the short FM downsweep of *Eptesicus*. These other species have a long sound of a constant frequency (long CF) before ending the signal with an FM downsweep. These long CF/FM bats tend to feed on insects flying near wooded areas, where the surrounding objects are likely to return louder echoes than the prey of particular interest to the bat (Neuweiler 1990). The prey can be discriminated from this clutter, however, because their flying motion and fluttering wings change the frequency of the returning echo, just as a moving siren seems to change in frequency. This change in frequency due to the relative motion of source and receiver is called a Doppler shift. The long CF/FM bats have evolved a highly specialized biosonar system in order to resolve these small Doppler frequency shifts. Their auditory systems are specialized for detecting a narrow range of frequencies near the CF portion of their biosonar signal. This has been called an acoustic fovea, in analogy to the fovea of the eye, which is the area of most acute vision (Schuller & Pollak 1979). However, a feeding bat is moving as well as its prey, and its motion also causes a Doppler shift that could move the echo frequency outside of the foveal frequency band. These bats have evolved a precise compensation mechanism to shift the sound of their outgoing CF biosonar signals in order that the incoming echoes from prey will center on the optimal frequency for the auditory system. This compensation mechanism may lead to complex signal changes in the wild depending upon the relative velocities of bat and target. It was first demonstrated in an elegant lab experiment, using a stationary bat echolocating on a moving pendulum (Schnitzler 1968). The auditory systems of long CF/FM bats are incredibly sensitive to tiny shifts in frequency. For example the horseshoe bat, *Rhinolophus ferrumequinum*, can detect changes of < 50 Hz in an 83-kHz signal (Schuller et al. 1974) and the mustache bat, *Pteronotus parnellii*, can detect frequency modulations of as little as ± 10 Hz from a 60-kHz signal (Bodenhamer and Pollak 1983). The biosonar of these bats is sensitive enough to detect the wingbeat frequency of an insect. Von der Emde and Menne (1989) played synthetic wingbeat sounds to horseshoe bats, and found that these bats could detect changes as low as 2.8–4.6 Hz from a baseline wingbeat rate of 50 Hz. It has been suggested that they use this as a cue for discriminating insects with different wingbeat patterns (Schnitzler & Ostwald 1983).

We Need Closer Coupling between Field Observations and Experimental Studies of How Cetaceans Use Sound

This feedback between studies of how bats use biosonar in nature and studies of sensory and neural specializations for bat echolocation has been central for the remarkable progress in this field. Insight into the particular problems for which bat biosonar has evolved has enabled experiments that give us a much greater appreciation for the sophistication of bat echolocation. There has been much less of this kind of feedback in research on cetacean echolocation. For example, there has been so little field observation of echolocation that more than a decade after Norris and Møhl (1983) proposed that odontocetes can actually debilitate prey with their sonar signals, there still are no convincing data on the effects of clicks upon prey. While different biosonar signals produced by a bat species have been shown to be specialized for different biosonar problems, Au (1993) reports much less evidence that dolphin biosonar signals are similarly specialized for ecologically relevant problems. However, almost all of the experimental studies of dolphin echolocation involve stationary dolphins echolocating on stationary and artificial targets that bear little resemblance to the biologically relevant targets for which dolphin biosonar evolved. There is evidence in bottlenose dolphins (*Tursiops truncatus*), beluga whales (*Delphinapterus leucas*), and false killer whales (*Pseudorca crassidens*) of a correlation between the source level of an echolocation click and the peak frequency (Au et al. 1985, 1995, Brill et al. 1992). Dolphins have been shown to control the correlated source level or peak frequency of their clicks as a function of ambient noise (Au et al. 1985) or as a function of the echolocation task (Moore & Pawloski 1991), but these changes appear less specialized than those demonstrated for bat echolocation. Dolphin biosonar may in fact be much less specialized than bat biosonar. On the other hand, dolphin researchers may simply be unlikely to discover these specializations until they achieve a better integration of studies on the functions of biosonar in nature with psychophysical experiments of echolocation with captive animals and artificial tasks. For example, it will be particularly interesting to study whether there are any differences in how dolphins echolocate at different stages of searching, detecting, or pursuing prey, or when they or their targets are moving. Dolphins move and scan rapidly while echolocating. If they can process echoes equally rapidly, there may be tight coupling between a dolphin's search movements and its biosonar signal emission and processing.

We Need a Broader Comparative Perspective of Cetacean Echolocation

> ...echolocation sounds are consistent within a given species and differ in species-specific ways. Comparative studies show that these differences correlate with preferred habitats and hunting strategies. The differences also help establish the information-gathering value of different components of the echolocation sounds (Grinnell 1995, p. 2).

Studying the evolution of any specialized adaptation such as echolocation requires a broad comparative perspective (Harvey & Pagel 1991). Comparative studies also offer advantages to researchers interested in selecting an experimental subject for investigating biological mechanisms. Preliminary behavioral studies on diverse species may help direct experimenters to the best species for a particular problem. For example, when Griffin initially developed an interest in how bats echolocate to find prey, he went to a lake over which foraging bats would fly straight and then dive with many rapid turns apparently in pursuit of an insect. Of the two bats common in the area, the little brown bat, *Myotis lucifugus*, produced FM downsweeps that varied in repetition rate, but that otherwise did not change much. However, Griffin (1974) did notice changes in the biosonar signals of big brown bats, *Eptesicus fuscus*. As the big brown bats flew to the lake, they produced a slow series of FM downsweeps. As a bat pursued its prey, the downsweeps became shorter with a more rapid repetition rate (Griffin 1974). These early field observations suggested that big brown bats would be better subjects than little brown bats for studies of how bats change their biosonar signals at different stages of predation. If bat researchers had limited their work to little brown bats, bat and dolphin sonar might seem similarly unspecialized. Au (1993) in his comprehensive review of dolphin biosonar suggests that "there is no solid evidence of dolphins purposefully changing the spectral content and duration of their biosonar signals when approaching prey or a stationary target." However, this conclusion stems almost exclusively from data on the bottlenose dolphin, which is a generalist in terms of foraging and habitat use. One might not expect a highly specialized sonar signal in such a species, and Ketten (1994) describes this species as having relatively unspecialized high-frequency hearing. There has been so little work on species other than the bottlenose dolphin that we do not know whether this is typical of other dolphin species. Aside from occasional references to beluga whales, *Delphinapterus leucas*, and the false killer whale, *Pseudorca crassidens*, the comprehensive review of dolphin biosonar by Au (1993) devotes only four pages to "signals from other species." This small percentage accurately reflects the limited work on other species, particularly by American biologists.

The lack of a broader array of comparative data makes it difficult to judge whether differences between species merely involve trivial details, or whether different species have evolved different biosonars designed to solve different problems. For example, Turl and Penner (1989) report that the interclick interval of bottlenose dolphins is greater than the two-way travel time from the click to the sonar target, while beluga whales will emit packets of clicks with shorter intervals. This suggests that bottlenose dolphins do not emit a click until they have processed the preceding echo, and that beluga whales, which sometimes emit a click before hearing the preceding echo, may have a different echolocation strategy. Among bats, there are clearly much more profound differences in the biosonar systems, for example, of the bats that analyze Doppler shifts in long CF signals versus those that use short downsweeps as biosonar signals. Specializations for long-duration constant-frequency signals, finely tuned hearing curves, and Doppler compensation have evolved independently in several distantly related groups of bats. The convergence suggests that this complex of traits forms an integrated biosonar system, perhaps evolved for hunting in environments where the echoes of insects are masked by background clutter (Neuweiler et al. 1988).

There are some hints of differences in the biosonar systems among the cetaceans that may parallel the differences in biosonar among bats. Several cetacean taxa, porpoises of the family Phocoenidae and dolphins of the genus *Cephalorhynchus*, have more specialized hearing mechanisms than those of bottlenose dolphins and biosonar signals that are longer in duration and narrower in bandwidth. These taxa have been identified by Ketten (1994) as having inner ears that are particularly specialized for high-frequency audition above 100 kHz. As an illustration of the different echolocation signals, Figure 1 compares clicks from a bottlenose dolphin and from a harbor porpoise. The echolocation signals of bottlenose dolphins in open water are very short in duration (40–70 μsec), relatively broadband (80–154 kHz), and with high sound pressure levels (up to over 220 dB re 1 μPa at 1 m) (Au 1980). By contrast, the echolocation signals of phocoenid porpoises and *Cephalorhynchus* dolphins are 5–10 times longer (150–600 μsec), roughly half the bandwidth (120–160 kHz), and with sound pressure levels between 150 and 170 dB, several orders of magnitude weaker than the loudest bottlenose dolphin clicks (*Phocoena phocoena*: Amundin 1991, Kamminga & Wiersma 1981; *Phocoenoides dalli*: Hatekayama & Soeda 1990;

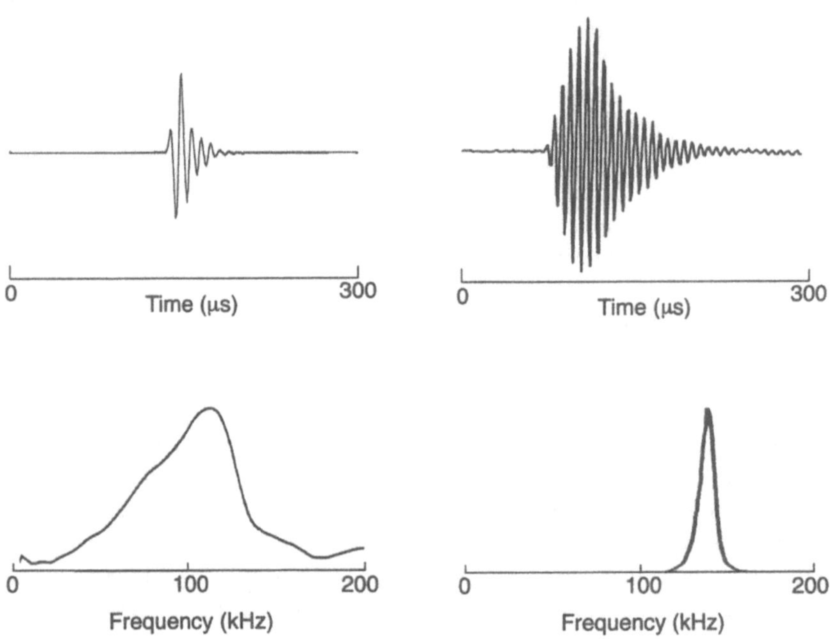

Fig. 1. Waveform (top) and spectra (bottom) of clicks from bottlenose dolphins, *Tursiops truncatus* (left) and harbor porpoise, *Phocoena phocoena* (right). Bottlenose dolphin figure is an average from an entire click train adapted from Au (1980). Harbor porpoise figure is from a single click from a young animal adapted from Kamminga (1988).

Cephalorhynchus commersonii: Kamminga & Wiersma 1982, Evans et al. 1988; *C. hectori*: Dawson & Thorpe 1990).

The analogy between porpoises and *Cephalorhynchus* with the long CF/FM bats can only be taken so far. The long CF/FM bats produce biosonar signals more than 80% of the time when they are echolocating (Fenton 1995). These bats can process information from an echo even while the bat is still producing the outgoing signal. This is presumably enabled by their Doppler shift compensation described above. On the other hand, dolphins, porpoises, and most other bats are actually producing outgoing clicks less than 20% of the time when they are echolocating. Most species tested adjust their outgoing pulses so as not to interfere with incoming echoes. The actual durations of porpoise and *Cephalorhynchus* echolocation signals are less than one tenth those of the long CF/FM bats, and most available information would suggest that they produce echolocation signals only for a low percentage of the time when they are echolocating. While porpoises and *Cephalorhynchus* appear to have inner ears tuned particularly for high frequencies and produce narrow band clicks, there have been no tests of whether they use Doppler information in their echolocation.

What I do want to suggest is that, as in the bats, the convergence of biosonar signals among distantly related animals suggests the evolution of a specialized biosonar system well worthy of further study. Watkins et al. (1977) and Dudok van Heel (1981) suggest that porpoises and *Cephalorhynchus* inhabit similar coastal habitats and have converged upon similar morphology and biosonar systems especially suited to this ecological niche. I am suggesting a convergence in echolocation systems of a sort that has been suggested by Endler (1992) for communication signals:

> Thus, sensory systems, signals, signaling behavior, and habitat choice are evolutionarily coupled. These suites of traits should coevolve in predictable directions. (Endler 1992, p. S125)

More research on a variety of cetacean species is needed to enable the kinds of comparisons that have proven so productive in studies of echolocation and communication in other taxa.

ACOUSTIC CHARACTERISTICS OF BIOLOGICALLY RELEVANT TARGETS FOR MARINE MAMMAL BIOSONAR

Our ignorance of how wild marine mammals use sound to explore their ocean environment contrasts markedly with the enormous effort of underwater acousticians in exploring how humans can use sound to find out about the ocean. Humans map the bathymetry of the sea using ever more sophisticated sonars. Where pelagic fishing once depended upon large doses of luck, fish-finding sonars have become so good as to take the luck out of this kind of fishing. Any predator-prey ecologist would immediately appreciate the sophistication of military sonars used for antisubmarine warfare, along with the counter measures employed against these sonars. The field of acoustic oceanography in the last few decades has developed the analytical tools and data to start to understand the opportunities and problems which ocean acoustics offered for marine mammals during their tens of million years of adaptation to life at sea. Acoustic oceanography allows biologists to develop models of how sound might be used to solve specific problems encountered by marine mammals. These models will usually specify design features for sounds used to solve the problem, as well as set bounds on what sounds, receivers, and signal processing are feasible or optimal. For example, a sonar designed for depth sounding is likely to have very different signals, receivers, and processors from a sonar designed to classify different species of plankton.

In order to broaden the scope of our thinking about how marine mammals use sound to explore their environment, I will therefore explore the acoustic properties of some biologically relevant targets. Underlying this examination is the assumption that the acoustic properties of targets are important sources of selection for animal biosonar systems. In analogy to bat research, are there marine equivalents to fluttering insects, targets whose acoustic properties may hold the key to how animals use sound to explore their environment? The implications of this discussion for animal biosonar will necessarily be speculative, but my goal is to direct research attention to potentially important problems that are not being studied. I will therefore particularly focus this discussion on animals and biosonar problems that have not been studied in captivity.

The high-frequency specializations of bats and dolphins form the basis for our understanding of echolocation. High-frequency sounds are particularly useful for detecting echoes from small targets or for increasing the resolving power of a sonar used with objects that are larger than the wavelength of the sonar signal. However, high frequencies of sound do not in general propagate as far as lower frequencies (Urick 1983). Fenton (1995) agrees with Griffin (1974) that bat echolocation typically operates at short ranges up to several tens of meters. The high-frequency echolocation system of coastal bottlenose dolphins has detection ranges of about 100 m (Murchison 1980). Cetaceans inhabit a broad range of habitats. Some may dive more than a kilometer below the surface, and pelagic species may migrate thousands of kilometers. Many of these species face problems where they would benefit from sensing features of their environment at ranges well beyond the demonstrated detection range for dolphin clicks. A variety of sensory modalities are available to animals to solve these problems. Deep-diving seals may rely upon vision that is sensitive to low light levels at depth (Schusterman 1972), while migrating animals may rely upon a remarkable variety of cues for orientation, including visual, acoustic, olfactory, and magnetic (Schmidt-Koenig 1975; Papi 1992). However, seafaring humans have quickly recognized the utility of acoustics as a distance sense in the sea where vision is usually limited to tens of meters. Many marine sonars developed by humans use frequencies well below those of bats or dolphins in order to operate at these greater ranges. Since we know that many cetacean species specialize in low-frequency acoustics for communication, it seems reasonable to explore whether they might also use lower-frequency sounds to explore their environment at greater ranges.

Many important marine targets will reflect low-frequency sounds of the sort usually thought to be used by cetaceans for social communication. Marine mammals with unspecialized hearing may be able to obtain important information about their environment by simply learning how to interpret echoes from these signals. Some reviews of the negative evidence concerning high-frequency echolocation in whales and seals clearly consider this possibility (e.g., baleen whales: Herman 1980, Moore 1980; pinnipeds: Schusterman 1981). In order to distinguish between highly specialized abilities for echolocation of the sort demonstrated for bats and dolphins, and the more generalized abilities to use audition to learn about the environment, I will follow the usage of Kinne (1975) and use the term biosonar to include these more general abilities. This matches the broad range of applications called sonar in engineering as well. In a later section of this paper, I will consider the possibility that animals may use several forms of sonar which are familiar to sonar engineers, but are very different from what biologists think of as echolocation. I would like to emphasize that when I discuss animal biosonar in this sense, the term does not presume a specialization of sensory ability. Rather I want to focus on the potential utility of sound as a distance sense for marine mammals to learn about their environment. If marine mammals can use general mammalian auditory abilities and nonspecialized vocal signals to obtain information about their environment, this may illuminate potential pathways for the evolution of more specialized adaptations for high-frequency echolocation.

Sensing the Physical Environment in Bottom and Surface Reverberation

Bottom Reverberation

Humans have devoted billions of dollars to developing techniques to use sound to explore the seafloor and sediments below it. This effort provides extensive theory and data on the design features of echosounding. Different frequencies of sound have different properties of reflecting off the seafloor or penetrating it and reflecting off inhomogeneities in the sediments below. The frequency range of sounds that penetrate sediments well is typically below 100 Hz. The amount of sound energy reflecting off a rigid object drops rapidly if the circumference of the object is less than the wavelength of the sound. The speed of sound in seawater, denoted by the variable c, is about 1500 m/sec. The relationship between wavelength, λ, speed, c, and frequency, f, is: $\lambda = c/f$. This means that the wavelength in seawater of a 15-Hz whale call would be 100 m; that of a 150-Hz call, 10 m. These long wavelengths would not be useful for detecting small objects, but do reflect off of large bathymetric features. Wavelengths smaller than the circumference of the target are still effective, so higher frequencies can also be used for detecting large targets. For example, the seafloor is a very large target, but depth sounding sonars often use frequencies of 3.5 or 12 kHz, with wavelengths of 43 cm and 12.5 cm. Echo sounders typically have a downward-directed pulse and measure the time until the first reflection. With a nominal sound speed of 1500 m/sec, if there were a 4-sec delay between the pulse output and the echo, then the seafloor would be 3 km below the ship.

Depth Sounding

The simplest of echo sounders simply detect a single echo in order to estimate range from the bottom. Some animal sounds have long been known to have the potential for this kind of depth sounding. Griffin (1955) analyzed what appeared to be calls and bottom echoes from an unknown marine animal dubbed the "echo fish." The "echo fish" recordings were made using a surface hydrophone in waters 5100 m deep. "Echo fish" calls were 500 Hz in frequency and 0.3–1.5 sec in duration. Each of these calls was followed by an apparent echo 1.47–1.77 sec later, with an echo intensity only 27%-56% the amplitude of the initial call. Since the speed of sound in the ocean is about 1500 m/sec, the difference in path length between the direct and apparent bottom reflected path was about 2400 m. The strength of both signal and echo led Griffin to speculate that the source was probably not at a great distance from the hydrophone. In the simplest case, if the "echo fish" were directly below the hydrophone, then it would have been 1200 m above the seafloor or 3900 m deep. This "echo fish" was likely also to have heard these echoes, and might also have been able to time the echo delay to estimate its distance from the seafloor.

An echo sounder would clearly also be useful to a marine mammal that dives rapidly and deeply. Sperm whales, for example, routinely dive from 300 to over 1000 m in dives that last around 40–50 minutes. Sperm whales tracked with a sonar descended at rates of 1–2 m/sec (Papastavrou et al. 1989). The rate of descent of sperm whales tagged with a transponder tag averaged about 1 m/sec with a maximum of 4 m/sec (Watkins et al. 1993). These tagged whales were within a few kilometers of the island of Dominica. If one of these whales had no way of detecting the bottom, it clearly would have run a risk of collision. However, when sperm whales start a dive, they are reported to start regular series of clicks as they reach depths of 150–300 m (Papastavrou et al. 1989). These clicks are similar enough to those from some depth sounders that Backus and Schevill (1966) report that sperm whale clicks actually can make false targets on a ship's depth sounder. Male sperm whales produce particularly loud clicks with a slow repetition rate (Weilgart & Whitehead 1988). In waters off Dominica, bottom reverberation has been recorded from these slow clicks. Figure 2 shows a spectrogram of one such click followed by reverberation as such a nondirectional click echoes off the surrounding bathymetry. These echoes were received 1.4 seconds after the direct click, indicating that the echo path was over 2100 m longer than the direct one. While the echoes from the slow click shown in the figure are more obvious at these ranges than those from the regular clicks that can be seen in the center of the figure, clearly a whale producing regular clicks would be likely to hear echoes from the seafloor as it approached within a few hundred meters of it. Whatever other reasons a whale might have for producing the clicks Weilgart and Whitehead, (1988) a diving whale would do well to heed such echoes as a warning of an approaching obstacle.

When you listen to the pattern of bottom reverberation shown in Figure 2, you can hear a pattern of several returns, presumably as the omnidirectional

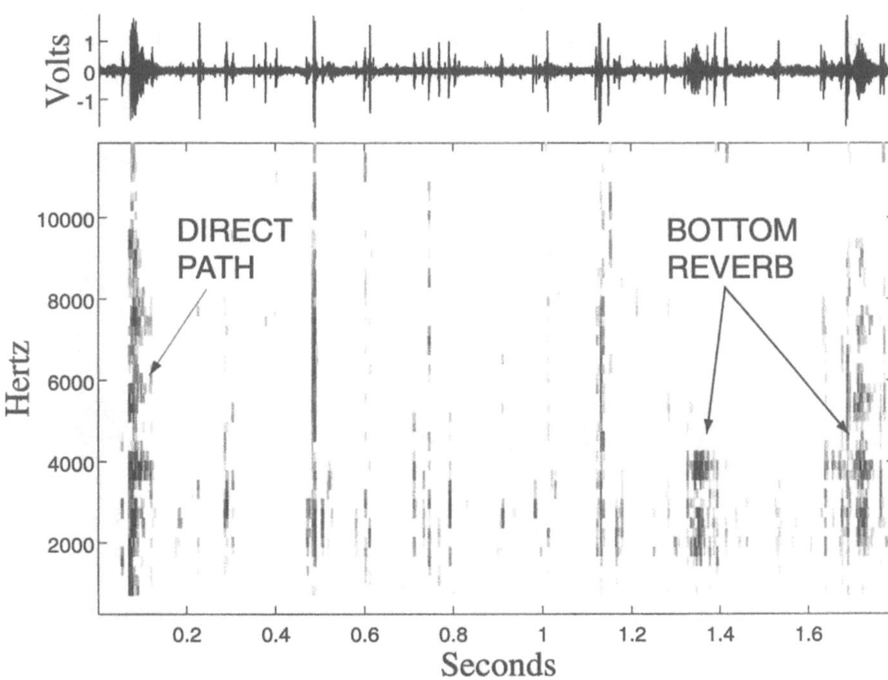

Fig. 2. Spectrogram of slow click from a sperm whale, *Physeter macrocephalus*, in waters near the island of Dominica in the Caribbean. The direct arrival of the click is visible on the left of the spectrogram, and echoes of the click reflecting off the seafloor are visible on the right of the spectrogram just before 1.4 and just after 1.6 seconds. The shorter clicks in the middle of the spectrogram are regular clicks from sperm whales.

sperm whale click reflects off a variety of bathymetric features in the vicinity. We do not know how well sperm whales can locate sounds such as echoes, but the ability of dolphins to locate sound sources has been tested. Bottlenose dolphins can differentiate signals separated by several degrees (Renaud & Popper 1975). If sperm whales have comparable abilities, then they might be able to separate the superposition of echo returns from different bearings and ranges shown in Figure 2 into some sort of bathymetric map or image. Pack and Herman (1995) have shown that bottlenose dolphins can use either echolocation or vision to form an internal "image" or percept of complex shapes, with spatial information that is readily integrated across these senses. If a whale were able to update bathymetric maps from successive clicks, this might be integrated with vestibular input for acoustic orientation.

Use of Bottom Reverberation for Navigation

The last paragraph suggests that marine animals may be able to do more than just perceive the depth of the seafloor, but may also be able to perceive the bearings of echoes from targets that are not directly below. Figure 3 shows a

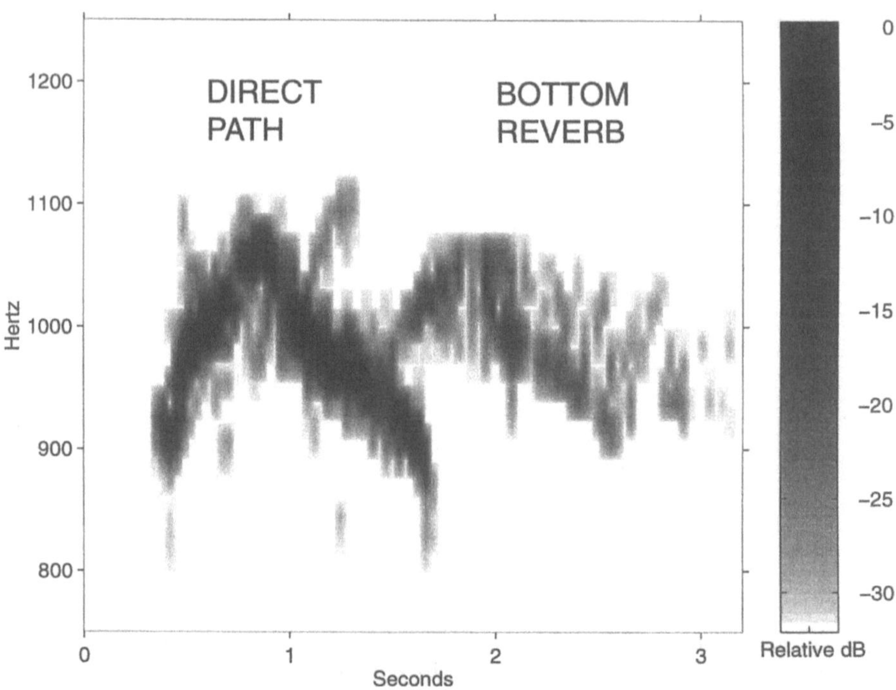

Fig. 3. Spectrogram of a unit from the song of a humpback whale, *Megaptera novaeangliae*, from 0.5 to 1.8 seconds and a bottom echo of the same sound from 1.6 to 3 seconds.

simple dominant echo from one sound of a humpback song. The whale recorded here was swimming in deep water a few kilometers offshore of an oceanic island. If the whale whose song is shown in Figure 3 could localize the bearing of the echo in the setting recorded here, these echoes would be likely to indicate the direction of rapidly shoaling water. Many recordings of humpback song in similar settings show such bottom reverberation. Tyack (1981) argues that humpback song appears to be a reproductive advertisement display with a primarily communicative function. During the breeding season, many humpbacks congregate on shallow banks where females have young and where there is a high density of singers. However, humpbacks sing far from the known concentrations of breeding females, and they sing during some phases of their annual migrations (Clapham & Mattila 1990). In these settings, bottom reverberation may provide them with useful information about their environment.

Several biologists have suggested that marine mammals might be able to sense echoes of low-frequency vocalizations from distant bathymetric features in order to orient or navigate (Norris 1967, 1969; Payne & Webb 1971; and Thompson et al. 1979). For example, if the 20-Hz signals of finback whales were designed to avoid overlap between outgoing pulse and returning echo, then the 1 sec duration of these signals coupled with the 10–30-sec interpulse interval would suggest an effective range of from 1 km to several tens of km. It has been suggested that these characteristics are not well suited for echo sounding in these whales, which are not thought to dive deeper than several hundreds of meters (McDonald et al. 1995). Payne and Webb (1971) evaluated the hypothesis that finbacks might use these pulses for a bathymetric sonar with ranges of tens of km. They concluded that signals with these frequencies and source levels could function for a long-range bathymetric sonar but they questioned the appropriateness of the very regular and relatively short repetition rate of the pulses for use as this kind of biosonar. In particular they question why animals slowly meandering in a restricted area would require such frequent updates of bathymetry. Payne and Webb (1971) and Watkins et al. (1987) both suggest that these signals appear primarily to have evolved for long-distance communication. These analyses all assume that whales must wait for the echo to return before they emit the next signal. However, blue whale calls are quite long, with both CF and FM segments, somewhat similar to the long CF/FM calls of bats. CF/FM bats can process an echo even while the signal is still being emitted. Beluga whales also are capable of emitting "packets" of clicks with interpulse intervals less than the round-trip travel time to the sonar target (Turl & Penner 1989). Whales may not have to process each pulse independently, waiting for the echo return before emitting the next pulse, but rather they may be able to process whole series of pulses in ways that improve their sonar performance. Clark (1993) has suggested that the low-frequency calls of finback or blue whales would produce easily identifiable echoes from seamounts hundreds of kilometers away, and he speculates that this would provide important cues for navigation during migration or transits across the deep ocean. Even if these signals have a primarily communicative function, these whales may also detect and respond to reverberation from bathymetric features.

Testing Sensory Mechanisms for Orientation in Cetaceans

My primary reason for bringing up the potential use of sounds for orientation or echo sounding is to direct our focus on the sensory mechanisms which marine mammals use to solve problems posed by their natural environment. By focusing on a particular problem of biological significance, we can explore the range of sensory mechanisms that an animal might use. Kinne (1975) emphasizes that animals of many species integrate information from a variety of senses in order to orient in space. Several different mechanisms have been suggested for long-range orientation by cetaceans, including detection of geomagnetic cues (Klinowska 1985; Walker et al. 1992; but also see Hui 1994), visual monitoring of celestial cues (e.g., Busnel & Dziedzic 1966), and landmarks (Pike 1962). The temperature, salinity, and movement of the surrounding water may also provide important cues for marine mammals (Norris 1974). Regarding sensory mechanisms for estimating depth, Kinne (1975) suggests that marine mammals may be able to detect increasing pressure with increasing depth. None of these mechanisms have been tested directly in orientation experiments.

Since marine mammals are very sensitive to a variety of sensory cues, it is unlikely that observational data alone will suffice to demonstrate which senses are necessary and sufficient for tasks such as orienting with respect to the seafloor. Experimentation with captive dolphins under controlled conditions has been critical for elucidating echolocation, but large whales are difficult to maintain in captivity. How might one experimentally test whether whales use echoes from their sounds in order to orient with respect to the seafloor? One opportunity is opened by a device called an echo repeater. These are used for testing man-made low-frequency sonar systems. The echo repeater constantly monitors sound with a hydrophone. When a potential sonar signal is detected, the repeater electronics modify the signal as if it had reflected from a specific kind of sonar target. An outgoing pulse is then generated using a sound projector in order to simulate a real echo from the target. Large targets such as the seafloor would be difficult to emulate, because a signal from one projector would not have the spatial complexity of reverberation from the large target. We would need more observation of tagged whales swimming near bathymetric features and potential obstacles in order to predict a response to a synthetic echo from an echo repeater. However, if, for example, sperm whales use echoes from their clicks in order to avoid collision with obstacles, then presumably they would slow in response to a synthetic echo indicating they were closing on an obstacle. If one could place such a repeater in the path of a whale whose movements were being monitored, one could evaluate how the whale responds to a synthetic echo indicating an apparent obstacle. One could use precisely the same outgoing signal as a control stimulus if its playback were not contingent with the whale having just produced its own sound. Most biosonars estimate range to a target by measuring the time delay between a vocalization and when the echo returns. If whales attend preferentially to echoes from their own sounds, they would be likely to respond more to the contingent echo playback. On the other hand, if whales simply habituate to such reverberation as noise, and if they respond to novel and unexpected stimuli, then they might be expected to show a stronger response to the noncontingent "echo."

Comparisons between different species may help elucidate different mechanisms for problems such as detecting the seafloor. Elephant seals, for example, have a dive pattern that can compete with sperm whales. When at sea, they spend 90% of their time underwater, and dive to depths of 1500 m or more. The average rate of descent of adult female elephant seals averages around 2 m/sec (Le Boeuf et al. 1988). Elephant seals are not known to vocalize underwater and have large eyes that appear to be specialized for vision at low light levels. We know little about the variety of sensory cues used by deep-diving mammals to orient at depth. While there is little downwelling light at the depths to which these animals can dive, especially at night, they might be able to use bioluminescence in order to detect prey or the seafloor. In order to resolve these questions, we must obtain detailed dive data along with the relevant sensory stimuli. Both elephant seals (Le Boeuf & Laws 1994; Fletcher et al. 1996) and sperm whales (Watkins et al. 1993) can be tagged with large enough packages to record or telemeter data such as depth of dive, acoustic data, and light levels. Deployment of such tags on these animals will help identify likely cues and detailed responses that will facilitate the development of experimental manipulation of sensory cues.

While discussions of long-range orientation among cetaceans typically focus on migrating baleen whales, most toothed whales are also highly mobile, ranging on the order of 100 kilometers/day, well in excess of terrestrial carnivores and primates (Wrangham et al. 1993). For example, killer whales are reported to swim at an average speed of 5 km/hr, and to have day ranges of 80–160 km/day (Kruse 1991). A young male pilot whale tracked in the North Atlantic covered 7588 km in a 95-day interval. He moved an average of 80 km/day with a maximum daily movement of 243 km (Mate 1989). Our developing ability to track the movements of these animals may be coupled with experimental manipulation of sensory cues in order to conduct field experiments on the sensory bases of orientation in marine mammals.

Reverberation from the Sea Surface

While most people are more familiar with the idea of sound reflecting off the seafloor, the interface between sea and air also reflects sound energy. Sound energy that reaches a sensor after scattering from the sea surface is called sea surface reverberation, and it can exert a significant influence on acoustic signals in the sea. Reverberation from the sea surface and the seafloor can greatly modify the original signal. However, these modifications do not necessarily just reflect degradation of the signal. The ways in which reverberation modifies the original signal can inform an animal about its environment. For example, diving mammals have a variety of ways to plan their dives, but it may often be useful for them to be able to determine how deep they are in order to estimate the time for ascent. Echoes from the sea surface could inform a diving animal of range to the surface.

Many marine mammals live in polar waters where the sea surface is covered with ice. For example, bowhead whales, *Balaena mysticetus*, migrate in the spring across the ice-covered Beaufort Sea, finding leads in the ice far offshore (Braham et al. 1980). Beluga whales have also been sighted many hundreds of kilometers north of the Beaufort Sea coast deep in Arctic pack ice. Finding breathing holes is a life-or-death problem for animals that dive under the ice. Downwelling light is likely to be an important stimulus during periods of daylight, but there are extended periods with low light in polar regions. Martin (1995) tracked beluga whales frequently diving to depths greater than 500 m under heavy ice. He suggests that they use these deep dives to scan for distant breathing holes, using either vision or acoustic cues. Seals have been shown to rely upon vibrissal sensation to find holes in ice when vision is limited, but this only functions at relatively short ranges (Watkins & Wartzok 1985). Ice makes such a good acoustic target that some of the first human sonar systems were developed to detect icebergs after the Titanic disaster (Clay & Medwin 1977; Hunt 1954). The problem of how cetaceans detect large ice obstacles or breathing holes in ice has been little studied. Clark (1989) and George et al. (1989) described how vocalizing bowhead whales, *Balaena mysticetus*, avoided a large floe of multiyear ice at ranges much farther than the limit of underwater visibility. Ellison et al. (1987) speculated that bowhead whales might be detecting the ice floe by listening for echoes from their own vocalizations.

There is a significant physical difference in how underwater sound reflects off an air interface compared to a denser surface. When sound in the dense aqueous medium reflects off of the air surface, there is a reversal in phase of the reflected wave (Horton 1959). This phase reversal is clearly visible when we analyze the details of the waveform of a signal. For example, Figure 4 shows on

→

Fig. 4. Left: waveform of the direct path of the click of a sperm whale at 150 to 200 msec and an apparent surface echo of the same click at 500 to 550 msec. Right: Cross-correlation of the direct-path click (from 150 to 200 msec) against the full waveform. The positive peak at 180 msec is the autocorrelation of the click against itself. The negative peak at 550 msec indicates that the surface echo has a phase that is reversed compared to the direct-path click (see text).

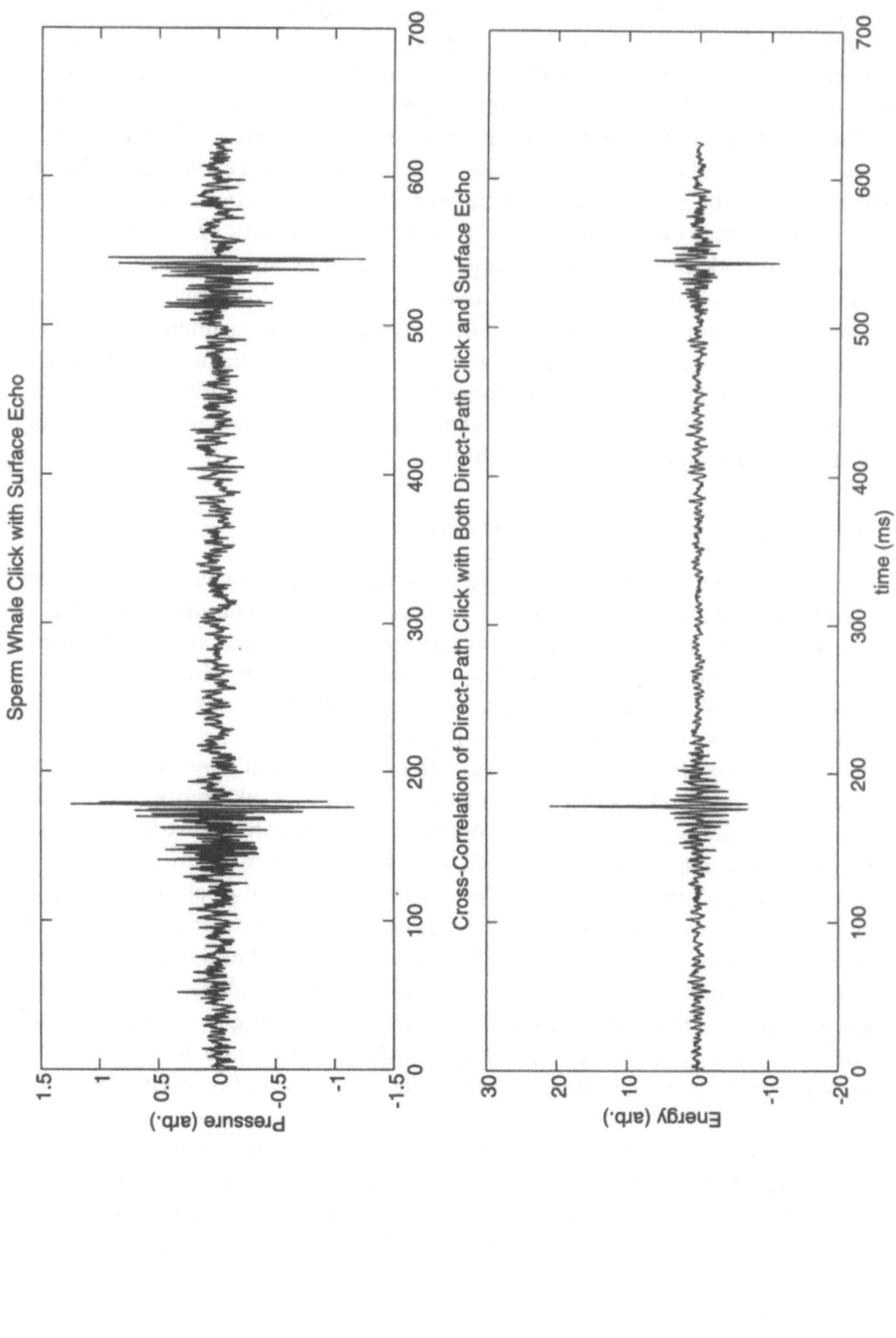
Sperm Whale Click with Surface Echo
Pressure (arb.)
Cross-Correlation of Direct-Path Click with Both Direct-Path Click and Surface Echo
Energy (arb.)
time (ms)

top the waveform of a sperm whale click. On the left is the direct path of the click, and on the right is what appears to be a later arrival of an echo from the sea surface. The phase reversal can be seen on the bottom of the figure, which shows the cross-correlation of the direct path version of the click with the rest of the waveform. The click shows a strong positive peak at 180 msec as it correlates with itself, and a strong negative peak where it correlates with the surface bounce at 550 msec. This negative peak indicates that positive pressures in the direct path correlate with negative pressures in the surface echo and vice versa. This is indicative of the phase reversal of a surface reflection. Many models of marine mammal hearing and echolocation treat the auditory system as one which simply integrates energy over a time interval (Au 1993). This differs from much work with bats, which suggests that some bats are sensitive to the differences between the phase of an outgoing signal and incoming echo (e.g., Menne et al. 1989). The peripheral auditory system of most mammals is capable of following waveforms at least at low frequencies. Given the potential biological relevance of the phase reversal from the sea surface, experiments on whether marine mammals can detect phase reversals may help us to determine whether marine mammals may have similar abilities.

Sensing Biological Targets in Volume Reverberation

When sound is scattered by objects or inhomogeneities within the water column, underwater acousticians call this volume reverberation. There are a large variety of potentially significant targets for marine mammals within the water column, including conspecifics, predators, and prey. Much of what we know about dolphin echolocation involves using high-frequency sound to detect echoes that reflect from solid objects in the water column. This mode of biosonar operation would function well to detect fish or similar small targets at ranges of up to 100 m or so. However, many of the fish or zooplankton prey of marine mammals are schooling species and occur in a highly patchy distribution both in time and space. Marine mammals that feed on them often face serious problems for finding aggregations of prey within large areas of ocean. Very little is known about how cetaceans find patches of prey on the order of 10 m in size at ranges on the order of kilometers. Physical inhomogeneities due to differences in the temperature or salinity of water may also scatter sound. These boundaries between different water masses often are correlated with high concentrations of planktonic prey (Parsons et al. 1984). If whales were able to detect these boundaries at large ranges, they might be able to use this to find high concentrations of prey.

A major difference in echolocation of marine vs aerial animals concerns the density of the medium versus the density of biologically relevant targets. Seawater is much denser than air. Many biologically significant targets are animals or plants, and these are often roughly the density of water. Biosonar targets of aerial echolocators tend to be much denser than the air medium, and they are efficient reflectors of airborne sound. However, since animate targets are roughly the same density as seawater, these excellent targets in air will reflect less sound energy when in an aqueous medium. One of the most important targets in the sea consists of air bubbles or gas-filled cavities (Clay & Medwin 1977). These are the biological targets that provide a difference in acoustic impedance as great as the fluid-filled targets that are so important in airborne biosonar. Bubbles also resonate at surprisingly low frequencies. Sound energy drives a bubble into a resonant oscillation at a wavelength much longer than that of the physical circumference of the bubble. For example, the resonant frequency, f, of a bubble with a 1 cm radius near the sea surface is 326 Hz (Urick 1983, p. 251). Yet the wavelength, λ, of this frequency in seawater with a sound speed c = 1500 m/sec is $\lambda = c/f = 1500/326 = 4.6$ m or 460 cm. An air-filled spherical bubble also has a very narrow frequency range for its maximum resonance (Stanton 1989).

Fish Swim Bladders and the Air-Filled Lungs of Mammals as Biosonar Targets

Bubbles are particularly relevant targets for our discussion because there are several targets of biological interest that are air-filled. The lungs of marine mammals are inflated during at least part of their dive, and this can produce a strong biosonar target (Love 1973). Many fish species that are prey for cetaceans have gas-filled swim bladders that are used for buoyancy regulation, as well as sound production and reception. Most swim bladder fish would have negative buoyancy if there were no gas in the bladder. These swim bladders are not spherical and they are constrained by the body of the fish, but their acoustic properties are roughly comparable to bubbles. The resonant frequencies of these swim bladders are surprisingly low in frequency given the size of the fish. For example, Batzler and Pickwell (1970) reported that a small anchovy at one atmosphere of pressure had a resonant frequency of 1275 Hz.

Many cetaceans feed on fish with swim bladders. Delphinids produce high-frequency echolocation clicks while feeding, but are also reported to produce a variety of pulsed sounds with energy below 5–10 kHz. These have typically been considered as communicative signals (e.g., Caldwell & Caldwell 1967), but Marten et al. (1988) suggest that they are also associated with feeding in bottlenose dolphins and killer whales. Many large whales also feed on small schooling fish such as anchovy, and many of these whales produce low-frequency sounds during the feeding season, but little is known about the function of these sounds. I believe that it would be well worth more detailed effort to compare the spectra of sounds produced by cetaceans against the target spectra of their prey. If the low-frequency resonance of fish with swim bladders were significant target characteristics, one would predict especially clear differences for species that feed on several different prey, some with gas-filled cavities and some with none. For example, individual humpback whales may feed on schooling fish at some times and on schooling crustaceans at others. If they use their sounds to detect swim bladders, then one would predict more vocalizations or more emphasis on frequencies near the resonance of the swim bladders when they feed on fish.

The deep scattering layer is one of the dominant targets for reverberation within the water column. This is a layer of organisms living in the open ocean, which shows a vertical migration on a diurnal cycle. Hersey and Backus (1962) review data on the frequency dependence of echoes from the deep scattering layer. They found that deep scattering layers tend to have a strong frequency peak in the range from 2.5 to 25 kHz. Many cetaceans feed on deep scattering layer organisms, particularly at night when the layer rises toward the surface. If they are feeding on organisms which reflect acoustic energy in the 2.5–25 kHz range, they might also benefit from echolocation signals containing energy in these lower frequencies in order to receive echoes backscattered from these targets. The optimal signal would depend not only upon the resonant characteristics of the target, which change with depth, but also upon absorption (which favors lower frequencies) and ambient noise (which usually favors higher frequencies). Echolocation signals designed for longer-range detection of aggregations of deep prey may be particularly useful as animals are making decisions about when and where to start a feeding dive.

When biologists think of animal biosonar, they usually think of an animal making a sound and then listening for echoes reflecting off of targets nearby. Many fish-finding sonars operate in a similar fashion, as have most of the sonars which biologists use to study organisms in the ocean. Sonar engineers call this a monostatic sonar, in which both the sound source and receiver are in the same place (top row Figure 5). In the biological case, this means that the animal receiving the echo also produced the original signal. Some man-made sonars are designed for a bistatic mode, in which the sound source is received by a distant receiver (e.g., Curran et al. 1994). Bistatic sonars can either detect sound backscattered from a target (middle row Figure 5), or if they are oriented in line with the source and target, on the opposite side of the target than the source (bottom row Figure 5), then they can detect attenuation of the source signal induced by passage through the target.

The acoustic properties of a target may differ depending upon whether it is measured by backscatter (the scattering target strength) or by forward propagation. The target strength measured by forward propagation can be called the extinction target strength, because what is measured is the reduction in sound energy after passage through the target. Clay and Medwin (1977) compare the amount of incident sound energy scattered from an air bubble to the total amount of sound extinction. The extinction target strength is greater than that measured by scattering in this theoretical analysis. This suggests a further advantage for

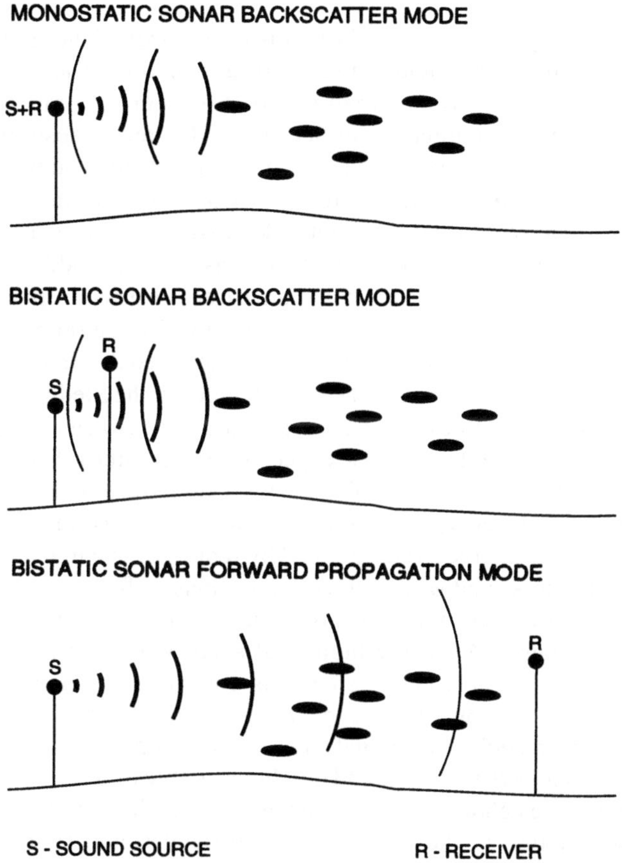

Fig. 5. Illustrations of three basic kinds of active sonar. Top: a monostatic sonar in the backscatter mode. The term monostatic implies that the source and receiver of the sonar are very close together (in biological terms, they are part of the same individual). A sonar operating in backscatter mode detects the echoes reflected back from a target. Middle: a bistatic sonar in the backscatter mode. A bistatic sonar has a receiver separated from the source (in biological terms, the receiver would be a different individual than the source). Bottom: a bistatic sonar in the forward propagation mode. A sonar operating in forward propagation mode does not detect echoes reflected from the target but rather detects modifications to the original signal that are caused by passage through the target.

forward propagation sonar to detect resonant gas-filled targets such as fish with swim bladders.

The limited data on detecting fish with forward propagation sonar suggest that fish with swim bladders may in some settings absorb significant sound energy in specific frequency ranges. Weston (1967) studied attenuation in signals received 137 km away from a 1-kHz source in the Bristol Channel. Just after sunset, he reports a drop of 30 dB in level of this 1-kHz noise. He associates this with the fish along the axis of the beam dispersing from daytime schools into a diffuse dispersion that would cause more attenuation. Diachok (1996) and Diachok and Ferla (1996) measured sound absorption at a range of 12 km in shallow water in the Mediterranean Sea. They also report absorption losses of up to 25 dB at sunrise from pelagic schooling fish. Hersey and Backus (1962) reported that the resonant frequency recorded from a deep scattering layer shifted upward in frequency as the layer migrated downward. As these fish swam deeper, the increasing pressure would compress the gas in their bladders, causing it to decrease in volume and buoyancy. Fish can use active gas transport mechanisms in order to maintain an appropriate bladder volume for neutral buoyancy as they change depth. Diachok and Ferla (1996) also associated this frequency shift with vertical migration of fish. These data were obtained in the Mediterranean in areas of high density of sardines and anchovies, and the resonant frequencies they observed were consistent with those expected from sardines. The data presented by Diachok (1996), Diachok and Ferla (1996), and Weston (1967) suggest that there would be quite easily detectable dropouts or attenuation of sound around 1 kHz for a source and receiver separated by ranges of one to tens of kilometers when the path intersected with high concentrations of fish.

This discussion of biosonar targets absorbing sound energy may sound unusual to biologists. A bistatic biological sonar would be one in which one animal can learn about a target after it is ensonified by a different animal. Dolphins have been shown to be capable of bistatic sonar in an experiment in which one dolphin clicked at a target and a second dolphin listened to the echoes (Xitco & Roitblat 1996). This second dolphin was able to select the correct target after hearing these echoes. Dolphins can also detect the presence or absence of a target by listening to the echoes of an artificial sound source (Scronce & Johnson 1976).

Differences in the Efficiency of Sonars for Backscatter and Forward Propagation

When a sonar receiver is removed from the source, this bistatic sonar is not limited to detecting echoes, but can also detect targets if they absorb part of the sonar signal's energy, so-called sound extinction. The bistatic forward propagation mode of sonar operation may offer significant advantages for detecting schools of fish over ranges of many kilometers. Figure 5 diagrams the operation of bistatic sonars where the receiver either detects echoes reflecting back (backscatter mode) or detects extinction in sounds passing through the targets (forward propagation mode). One major difference between these backscatter and forward propagation examples concerns the directionality of echo returns. When sound energy couples to an air bubble, the bubble resonates. This means that the bubble reradiates the sound energy impinging on it in all directions. There is no similar three-dimensional dissipation in the forward propagation case, with the exception of energy absorbed by the targets. This leads to a large difference in efficiency of the backscatter vs forward modes of sonar operation for these targets. For example, let us compare an omnidirectional sound source used to detect a bubble-like target 100 m away under conditions where sound energy propagates in all directions. As a sound propagates outward in all directions, the sound energy will be distributed over a sphere of radius $4\pi r^2$. If you measure energy at one point at a radius of r away, the sound energy will be diluted by a ratio of $1/(4\pi r^2)$ (inverse square law). This is called the spreading loss in underwater acoustics. In the monostatic backscatter case, the sound would dissipate in all directions for the 100 m from the source to the target for a loss of $1/(4\pi r^2) = 1/(4\pi\ 100^2)$. If the bubble then resonates the echo in all directions, there will be a similar loss of $1/(4\pi\ 100^2)$ on the return of the echo in the direction of the source. This leads to a total spreading loss of $1/(4\pi\ 100^2) \times 1/(4\pi\ 100^2) = 6 \times 10^{-11}$. Now let us consider a bistatic sonar using forward propagation in which the receiver is 100 m past the target. In both this and the monostatic backscatter case, the received sound must cover a path length of 200 m, but in the forward propagation mode, a bistatic sonar does not rely upon sound being reradiated from the target. Here the spreading loss is $1/(4\pi\ 200^2) = 2 \times 10^{-6}$, significantly less than in the equivalent monostatic example. The reduced spreading loss is not the only advantage for forward propagation sonar. Theoretical analyses conclude that the target strength of sound absorbed by a gas bubble is greater than the target strength of the backscattered sound (Clay & Medwin 1977).

Sensory Mechanisms Marine Mammals Might Use to Detect Patches of Prey at Ranges of > 1 km

The above discussion emphasizes the potential utility of low-frequency bistatic sonars for finding patches of fish at ranges of many kilometers. Is this of any relevance to cetaceans? The design features for sonars just discussed for this problem differ markedly from what we know of dolphin echolocation. Research on high-frequency echolocation in dolphins emphasizes its utility in detecting individual targets at ranges up to about 100 m. There is little evidence that it would function well at greater ranges. I am struck by how little attention has been paid to the problem of the sensory mechanisms by which marine mammals orient for navigation and find patches of prey that may be tens to hundreds of meters in size over the range of 1–100 km.

A major problem for foraging cetaceans involves locating evanescent patches of dense prey. Many cetaceans range over 100 kilometers or more per day (e.g., Kruse 1991; Mate 1989), and feed on prey that is highly patchy in time and space. Many toothed whales and dolphins that feed on schooling fish expend considerable energy on the search for prey. For example, dusky dolphins which are feeding primarily on schooling anchovy, swim 5 km/hr in shallow water where they seldom feed, but 16 km/hr in deeper water where they find schools of anchovy (Würsig & Würsig 1979). These dolphins often spread out in a line perpendicular to their travel and echolocate as they search for prey. If these animals spend much of their travel time searching for food, then any mechanisms to increase the effective range of their biosonar might yield substantial benefits, either in increasing their detections of prey or reducing the energetic cost of searching. Similarly, animals that feed on deep scattering layer organisms could benefit from a biosonar that could detect prey before an animal takes a deep dive.

However, the problem of finding evanescent patches of dense prey is even more acute for baleen whales. Unlike toothed whales which chase individual prey items, baleen whales feed by engulfing an entire mouthful of prey, a volume of hundreds of liters. The denser the concentration of prey in the mouthful, the higher the payoff. Baleen whales are not known to echolocate and nothing is known about the sensory mechanisms by which they find suitably dense patches of prey. There are a few reports of baleen whales establishing small home ranges during the feeding season (Dorsey 1983). However, the enormous scale over which most baleen whales range makes it unlikely that individual whales could find the most productive patches for feeding by local knowledge of a small area. Many baleen whales migrate thousands of kilometers from winter breeding grounds in the tropics in order to take advantage of the burst of productivity in polar waters during the summertime. Most baleen whales are adapted to meet most of their energetic needs for the entire year during this concentrated summer feeding season.

The feeding areas for individuals of some species are very large, comprising thousands of km^2. For example, in the northwestern Atlantic, a feeding area for humpback whales might comprise the southern Gulf of Maine, or waters from Iceland to Greenland. Within such an area, there are dramatic changes in prey distribution from year to year. Humpback whales feed on schools of prey, either fish such as herring, capelin, and sand lance, or invertebrates including euphausiids such as krill. A newborn humpback calf will follow a mother to her feeding area, and typically will adopt this as its own feeding area. The calf thus has an opportunity for learning general areas of high prey concentration while it is with the mother. However, the calf will wean within its first year of life, and show little later association with the mother. Productive areas for feeding can vary by hundreds of kilometers from year to year. For example, a calf might for its first five years spend most of the feeding season in Stellwagen Bank near Massachusetts, only to find insufficient prey there the sixth year, when it would have to find prey tens or hundreds of kilometers away. Watkins and Schevill (1979) suggest that distant whales may be attracted by the calls of feeding whales. The sensory mechanisms by which whales initially detect either plankton or fish remain little investigated, however. These whales clearly would benefit from mechanisms to detect patches of prey at ranges of tens of kilometers down to meters. Even a crude mechanism that slightly improved a random search could yield high benefits for whales competing for prey.

The optimal prey detector for baleen whales would yield a stronger signal for a larger and denser patch of prey. The data from Weston (1967) and Diachok and Ferla (1996) suggest that a bistatic sonar operating in forward propagation mode at the resonance frequency of swim bladders might detect more attenuation for dispersed prey than for dense schools. This would be suboptimal for detecting the densest schools at the greatest distance. There might be acoustic interactions between the swim bladders of fish in very dense schools, and these would be likely to broaden the bandwidth of the resonance and to lower the resonant frequency compared to single bladders (Clay & Medwin 1977). Whales might be able to assess the density of schools using acoustic features of these interactions between swim bladders. On the other hand, once a whale has detected dispersed prey, it might then be able to use behavioral mechanisms for concentrating dispersed prey. For example, humpback whales will often feed in social groups or emit large bubbles apparently in order to concentrate prey.

Can Cetaceans Use Biosonar in Bistatic or Forward Propagation Mode?

We have seen the potential utility of low-frequency sonar in a forward propagation mode for finding schools of fish at long ranges and we have also reviewed evidence that some cetaceans face significant problems in finding suitably dense patches of prey at ranges of hundreds or thousands of meters. In particular, the sensory mechanisms by which baleen whales find prey are completely unknown. I would now like to consider the question of whether whales might be able to exploit biosonar in either a forward propagation mode or a bistatic mode or both.

Monostatic Forward Propagation Sonar

It is possible for an individual cetacean to detect attenuation from fish in a monostatic mode. Recall the earlier discussions about bottom and surface reverberation. Suppose a whale were regularly listening to echoes from these kinds of reverberation. If the path taken by the appropriate sounds were to intersect large patches of fish with swim bladders, then the echoes might sound different than when such targets were not present. The whale might be able to use the attenuation of specific frequencies in the backscattered echo in order to detect pray in the intervening path. As an illustration of the phenomenon I am considering, note the sound from a humpback's song illustrated in Figure 3. This sound was produced during the breeding season away from the feeding grounds, but illustrates the acoustic characteristics of interest. The direct signal from this sound has both an upsweep and a downsweep that pass through the frequency band near 1 kHz at which many fish swim bladders have a resonant frequency. The downsweep has a long section of stable amplitude. If the signal reverberating from the bottom passed through many targets absorbing energy in a narrow frequency band, then the echo would be likely to have a dropout in that frequency. We do not know much about how large whales hear, but the small toothed whales whose hearing has been tested in captivity have excellent abilities to detect small intensity differences within narrow frequency bands. Bottlenose dolphins can detect frequency differences of 0.2%-0.4% in frequencies between 2 and 53 kHz and differences of 1.4% at 1 and 140 kHz (Jacobs 1972; Herman & Arbeit 1972; Thompson & Herman 1975). Popper (1980) reviews a variety of studies of intensity discrimination and suggests that bottlenose dolphins can probably detect changes in intensity as low as 1 dB. If larger cetaceans have anything like these abilities for intensity discrimination in narrow frequency bands, then they should be quite sensitive to attenuation from fish intervening in the path of their calls.

Bistatic Biosonar

For the rest of this section I would like to explore the question of whether some cetaceans might be able to use bistatic biosonar. Since marine mammals are highly social and live within the water column, it is possible that they might also use biosonar in a bistatic mode to detect bottom, surface, or volume reverberation. The basic idea behind biological bistatic sonar in a forward propagation mode is that whales might listen for changes in repeated vocalizations of other whales in order to detect targets. A key feature of a bistatic sonar is that the receiver must be able to compare the received signal to a stored representation of the undegraded signal. There are several different mechanisms known for marine mammals that could achieve this end. In order to study possible bistatic sonar, it is necessary to have a method that allows one to follow several individuals, simultaneously recording their vocalizations and behavior. This has only been studied on scales of kilometer separations in a very few situations such as humpback whales on their Hawaiian breeding grounds (Frankel et al. 1995; Tyack 1981) and bowhead whales migrating past Point Barrow, Alaska (Clark et al. 1986).

Bistatic sonar has already been discussed in one of these two cases: the context of bowhead whales sensing reverberation from sea ice. Ellison et al. (1987) present preliminary data suggesting that deep-keeled ice may produce strong echoes from the low-frequency calls of migrating bowhead whales. Clark (1989) and George et al. (1989) suggest that a migrating bowhead whale might use the echoes from the calls of other whales in order to detect an ice obstacle. Würsig and Clark (1993) also specifically discuss the possibility that bowhead whales might use a bistatic sonar in forward propagation mode. When migrating, a bowhead whale will often produce a series of calls with very similar acoustic features, called a signature call. Different migrating bowheads will often countercall with each individual producing a different signature call (Clark et al. 1986). As Würsig and Clark (1993) describe:

> In these counter-calling episodes, one whale calls with its signature call and other whales call with their signature calls within a matter of seconds.... these stereotypic calls may allow bowheads to monitor changes in the ice conditions throughout the group's area. In theory, this might be accomplished by comparing the amount of degradation in the stereotypic signals received from another animal. As ice conditions between and around the whales change, the characteristics of the received calls change. (pp. 189–190)

While detection of ice is an important problem for many polar marine mammals, I would like to also discuss the possibility that forward propagation bistatic sonar may have more general utility. For example, I have already discussed how the absorption of sound by swim bladders makes a potential signal for a bistatic biosonar. Let us examine the hypothesis whether cetaceans might listen for frequency-dependent attenuation in order to identify large concentrations of prey fish in the path between caller and receiver. In order for this to work as a bistatic biosonar, the animals would have to produce a series of stereotyped calls. The receiving whale could only identify the frequency-dependent attenuation by comparing the received signal to an expected signal stored in memory. This means that the receiving whale would have to have a clear expectation of what signal will be emitted by the source whale. The receiver could either compare the attenuation at the resonant frequency in successive calls or compare within a call the energy within and outside the frequencies absorbed by the prey. Animals able to make such a comparison, which also searched for food in dispersed groups, would then be able to follow quite simple decision rules in order to close in on a prey patch. For example, countercalling animals might simply swim in parallel when there was no attenuation correlated with prey, but swim toward a calling animal whenever they detected this attenuation.

There have been no studies of vocalizations and foraging behavior in marine mammals that are appropriate for testing the bistatic sonar hypothesis. However, many baleen whales, such as humpback whales, and toothed whales, such as most dolphins and many larger whales such as killer whales, produce calls with energy in the 1 kHz range and feed in groups on fish with swim bladders. I will select killer whales as an example to illustrate a context in which bistatic sonar might function. This species has been selected because we know more about vocalizations during foraging rather than because of any increased expectation they rely upon bistatic sonar.

Killer whales that feed on fish in the Pacific Northwest live in stable social groups, called pods, and they produce echolocation clicks, whistles, and pulsed calls. The pulsed calls separate into discrete and variable calls. The discrete calls can easily be categorized into stable call types while the variable calls cannot be sorted into such well-defined categories (Ford 1989, 1991). Most of the time when these killer whales forage, several pods forage together, suggesting an advantage to increasing group size for foraging. In general, these pods break up during foraging into small subgroups that disperse over areas of several square kilometers (Ford 1989). While dispersed, a foraging killer whale will often produce a series of the same discrete call, or two or more whales will exchange series of the same call. While all of the pulsed calls are thought to function primarily in social communication, the discrete calls are most associated with foraging and traveling; the more socializing occurring in a group, the higher the abundance of the more variable pulsed calls (Ford 1989).

The top row of Figure 6 shows spectrograms of two discrete calls of these fish-eating killer whales. Each of these calls, N7 and N9, is made up of two sections. The bottom row of Figure 6 shows spectra from the first section of seven different N7 calls on the left and of 12 different N9 calls on the right. These calls were recorded from the same pod but include calls recorded during different sessions. These may include repetitions from the same individual, but probably

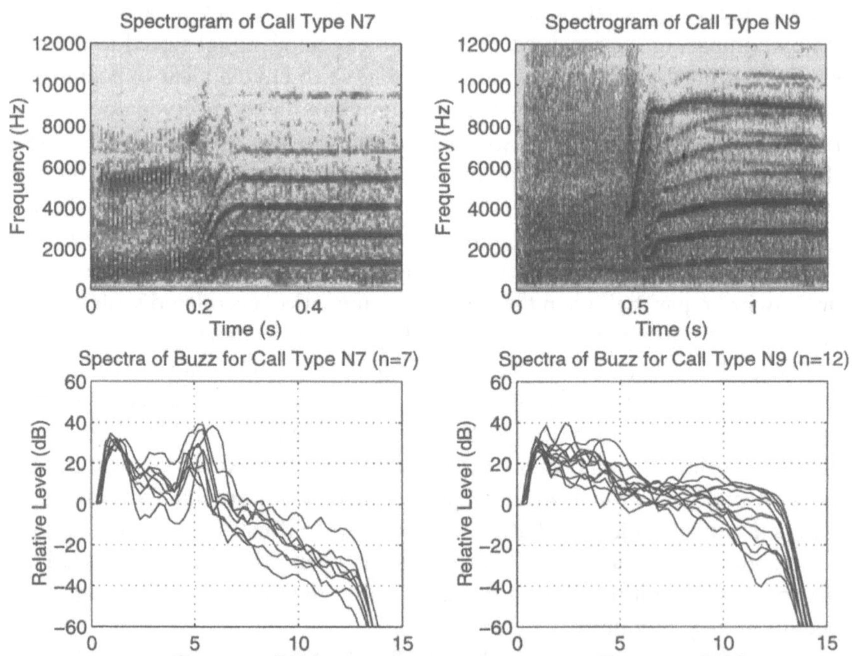

Fig. 6. Illustrations of two kinds of discrete calls made by fish-eating killer whales, *Orcinus orca*, in the Pacific Northwest. Top row: spectrograms of an N7 and an N9 call. Note how each of these calls is made up of two sections. The first section is called a buzz. Bottom row: spectra of the first section of each call. Each figure shows spectra from several calls (bottom left: 7 N7; bottom right: 12 N9) selected from several recording sessions from the A1 pod. Some of these calls may have come from the same individual, but some were likely to have been produced by different individuals from the same group. Each of these calls has relatively low variation in spectra, but the two calls differ in spectra; for example, N7 has a spectral peak near 5 kHz which is absent from N9.

include versions of the same call from different individuals. The spectra of each type of call vary by about ± 5 dB at any one frequency. Different discrete calls have different spectral maxima. For example the N7 calls illustrated here have a spectral maximum near 5 kHz, while the N9 calls do not.

Whether or not whales could detect fish in the path of the call of another whale depends upon a combination of how precisely they can predict the call structure and by the strength of frequency-dependent attenuation effect from the fish. Stanton (1989) reviews models showing that spherical bubbles have a sharp resonance 37 dB greater than frequencies well above resonance. The Weston (1967) data suggest a 30-dB change in attenuation from simple changes in fish distribution over a very long path length. Diachok and Ferla (1996) report absorption losses of up to 25 dB at the resonance frequencies of sardines over a 12-km path. In order to evaluate the feasibility of bistatic fish-finding biosonar, it will be critical to measure frequency-specific attenuation of sound as a source and receiver separated by several kilometers move from an area with few to many fish prey. If the attenuation effect is on the order of 10–20 dB, then there would be less need for the receiver to have a highly precise internal model of the source's signal. Such attenuation would be obvious given the variance in spectra illustrated in the killer whale calls in Figure 6. More evidence on the spectral stability both of calls within a series produced by one animal and of similar calls between different individuals will be required to evaluate whether stability of call production is consistent with use as a bistatic biosonar signal.

The data just reviewed from killer whales suggest to me that this species has a set of preadaptations for the evolution of bistatic sonar. They feed in groups which disperse when feeding on swim bladder fish, and they share a repertoire of discrete calls that they repeat and exchange while feeding. If their prey caused detectable changes in these signals, then it would be a small step to learn to use this degradation of the signal to learn about the density of prey between signaler and receiver. The fish-eating killer whales of the Pacific Northwest appear to have quite a stable set of prey species such as Pacific salmon (*Oncorhynchus* spp.), rockfish (*Sebastes* spp.) and herring (*Clupea harengus*). Some of these prey, such as rockfish, may not have distribution and behavior suitable for using forward propagation sonar to detect prey patches, while others, such as herring, may be more suitable. If animals capable of high-frequency monostatic sonar, such as killer whales, were to use lower-frequency sounds to detect patches of prey, then one might expect a transition from a low-frequency mode of searching for patches to using higher-frequency sonar for detecting and pursuing individual fish.

If low-frequency signals became adapted for prey detection, then one might expect them to have spectral peaks that match the resonant frequencies of their prey. In the case of killer whales, this might lead to the development of a repertoire of discrete calls that match the resonant frequencies of particular sizes of different prey species.

Killer whales meet the requirements of bistatic sonar by having stable groups in which individuals share stable repertoires of discrete calls. Most other cetacean species have much less stable groups. Long-term memory of stable calls would not be required if cetaceans started calling while together, and continued producing similar calls as they dispersed to search for food. The receiver would not even need to have a representation of the signaler's call if it could make an immediate and precise imitation of a call it heard, whether it was degraded or not. In this case, the original signaler could listen for changes in its original call induced by passage through the water between signaler and receiver and then by passage of the imitated call back from the original receiver to the original sender. This kind of immediate imitation of a call could function in a similar manner to the monostatic forward propagation mode using bottom reverberation discussed in the previous section.

Some of the cetaceans that may feed socially but live in fluid groups are capable of this kind of immediate imitation. Dolphins in captivity can imitate a novel synthetic sound within seconds of the first exposure, which implies a very rapid ability to form at least short-term auditory memories. Dolphins may imitate sounds spontaneously within a few seconds after the first exposure (Herman 1980), or after only a few exposures (Reiss & McCowan 1993).

The ability to learn vocalizations relatively rapidly would not only be useful for animals with fluid groups, but might also allow animals to adapt to rapidly changing prey or noise conditions. For example, humpback whales tend to feed in groups, but these do not tend to have a stable composition of individuals. Humpback whales also appear to be generalists, feeding on a wide variety of prey in many different habitats. We know very little about the vocalizations of humpback whales during the feeding season, but these whales have highly developed abilities to learn the sounds that make up the song during the breeding season. At any one time, the songs of different humpback whales within a population are remarkably similar, yet the song of each individual changes dramatically over time. Sounds may change in duration, frequency, and timbre on a monthly basis; they may disappear from the song entirely, and new sounds may appear in some other part of the song (Payne et al. 1983). Analysis of songs recorded off Bermuda for over two decades showed that once a particular song phrase disappeared, it never recurred (Payne & Payne 1985). The song of each individual is much more like different individuals recorded at the same time than it is to itself recorded, say, a year later (Guinee et al. 1983). This means that all singers change their songs more or less in synchrony over weeks, months, and years. This indicates a powerful selection for vocal convergence to track the long-term changes in the song. There are thus a diversity of mechanisms and time scales over which cetaceans could learn the precise acoustic characteristics of sounds produced by other whales for potential use in a bistatic biosonar.

Testing the Bistatic Biosonar Hypothesis

I have elaborated on speculations about bistatic biosonar in order to generate testable hypotheses about this potential form of biosonar. There are serious limits of purely descriptive studies of feeding in marine mammals. Feeding behavior often appears bewilderingly variable. This probably stems from our ignorance of the actual distribution of prey and of the sensory mechanisms and behavioral strategies employed by the animals in finding, capturing, and handling prey.

Understanding the opportunities afforded to acoustically specialized marine mammals by the physics of ocean acoustics will require close collaborations between underwater acousticians, acoustical oceanographers, and biologists. My own interest in the use of bistatic sonar to detect fish was recently stimulated by recent research of the underwater acoustician Orest Diachok (Diachok 1996; Diachok & Ferla 1996). This exemplifies the importance of close communication between underwater acousticians and biologists interested in how marine animals may use sound. It is difficult enough for humans to have a feel for the sensory world of terrestrial animals. Humans have such limited experience with the underwater world that it is even more difficult to develop an intuitive understanding of the sensory environment of marine animals. This makes it all the more important to rely upon an understanding of the physics of underwater acoustics.

All of the mechanisms for bistatic sonar discussed above would predict that animals would select the signals with spectra that match the absorption spectra of the prey on which they are working at any one time. Underwater acousticians will be able to test both reflected energy and attenuation produced by fish with swim bladders in areas where cetaceans are feeding on fish. If such studies were linked to mapping the distribution of fish, they could verify how useful bistatic biosonar in forward propagation mode might be for detecting fish at ranges that are useful for cetacean feeding. They would also be able to predict expected frequency ranges of optimal sonar signals and to measure the spectra of calls used by cetaceans as they feed on these prey. If descriptive studies were consistent with this biosonar hypothesis, it could be tested using passive acoustic targets. For example, Malme (1994) developed a simple air-filled target that emits no sound but provides a strong echo return from an active sonar. Artificial targets that mimicked the frequency-dependent attenuation of fish swim bladders could enable experimental testing of whether cetaceans do in fact respond to this kind of sonar signal. Whatever the fate of this particular hypothesis, I would argue that progress will come more rapidly with a tighter coupling between natural observations and generation and testing of specific hypotheses on the sensory mechanisms used by foraging cetaceans. Confronting some of the daunting problems solved by marine mammals may also free up our imagination to consider new applications of well-known sensory mechanisms.

Results from testing the bistatic biosonar hypothesis will be interesting whether they are positive or negative. If bistatic biosonar proves to be highly effective for detecting concentrations of fish, and if whales with the appropriate vocal abilities do not in fact use such a system, then this would suggest that abilities evolved for specific communicative purposes may not easily be coopted for the evolution of biosonar abilities. On the other hand, if any cetaceans are demonstrated to use bistatic biosonar in nature, then this would be a fundamentally new form of animal biosonar, one that combines elements of echolocation and of communication.

PARALLELS BETWEEN ACOUSTIC COMMUNICATION AND ECHOLOCATION

The marine mammal literature tends to assume a rigid dichotomy between echolocation and communication, with high-frequency clicks used exclusively for echolocation and all other sounds used exclusively for social communication. I have discussed how marine mammals may use for biosonar, low-frequency sounds that are typically thought of as serving a communicative function. For example, while humpback song appears to function as a reproductive advertisement display (Tyack 1981), whales may also learn about their environment from listening to echoes of bottom reverberation from sounds used in song. On the other side of this issue, there are also several dolphin species that appear to use "echolocation" clicks for social communication. Several of the species which specialize in high-frequency hearing, including the phocoenid porpoises and dolphins of the genus *Cephalorhynchus* are not known to produce any of the sounds typically associated with social communication in other dolphins (Amundin 1991; Dawson & Thorpe 1990). Both Amundin (1991) and Dawson (1991) suggest that the clicks used in echolocation may also function in social communication. Amundin (1991) associated relatively stereotyped patterns of repetition rate of "echolocation" clicks with specific social contexts. If predators of these species, such as killer whales, cannot hear the high frequencies of these clicks, then it may be advantageous to use them for communication as well as echolocation.

The suggestion that animals may use a bistatic biosonar, where the signaler and receiver are different individuals, further blurs the distinction between communication and echolocation. A phenomenon very similar to bistatic sonar has been discussed by Morton (1982, 1986) in relation to acoustic communication by songbirds. Male songbirds often use song as a territorial display to other males. If a male hears a song within his territory, he ought to respond more strongly than if the song is farther away. Morton has proposed that songbirds estimate the range of a singer by comparing the received song, which has been degraded by passage through the environment, to a remembered version of the undegraded song. It has been demonstrated that adult male Carolina wrens (*Thryothorus ludovicianus*) respond less to degraded than undegraded song (Richards 1981), and naive young wrens are more likely to learn undegraded than degraded songs (Morton et al. 1986). Great tits (*Parus major*) have also been shown to use this degradation to estimate the distance of a singer, and they are better at estimating the distance of familiar songs than unfamiliar songs (McGregor et al. 1983; McGregor & Krebs 1984). The McGregor and Krebs (1984) results suggest that songbirds need to compare degraded song to an internal representation of an undegraded song in order also to estimate the range of the song of a conspecific. These abilities in great tits are very similar to the ones that marine mammals would need in order to use a bistatic biosonar in forward propagation mode.

Morton (1986) elaborates on this ranging hypothesis by analyzing competitive communication strategies for singing birds and neighbors that use acoustic cues to estimate the range of the song. If territorial neighbors are in competition with one another, then Morton (1986) suggests that selection would favor singers with songs that sounded as close as possible in order to evoke a maximum response at maximum range. The listener on the other hand would be under selection to develop degradation assessment mechanisms that would allow the bird to detect all incursions but to ignore songs from outside of the territory that do not require a response. Morton suggests that large repertoires of complex songs may have evolved in part to interfere with distance assessment mechanisms. Morton also suggests that learning to produce a song may be the best way for a bird to develop a precise auditory representation of that song, and to develop the ability to compare the similarity of auditory input with this auditory memory (following motor theories of sound perception, e.g., Williams & Nottebohm 1985). Morton suggests that the evolution of vocal learning and the tendency of young birds to copy the songs of their neighbors may relate to this arms race between singer and listener concerning distance assessment mechanisms by enhancing the ability of listeners to learn the songs of neighbors.

Morton's evolutionary argument emphasizes the costs and benefits to a signaler for producing a communication signal as well as the costs and benefits to a receiver for ignoring or responding to the signal. This perspective has become common in analyses of animal communication (e.g., Dawkins & Krebs 1978). Myrberg (1981) also emphasizes that signals can be intercepted by unintended receivers, such as predators, to the detriment of the signaler. While interception is usually discussed for communication signals, it also can influence the evolution of echolocation. For example, many insects that are prey for bats have evolved countermeasures against bat biosonar (Miller 1983). Insects can detect the ultrasonic cries of bats and engage in a variety of behavioral strategies to avoid capture including flying away from a distant bat, and diving, falling, or unpredictable zigzag flight away from bats nearby. We know next to nothing about marine mammal equivalents of the bat-insect arms race, but there are interesting suggestions that interception may be important for marine mammals as well. For example, in the Puget Sound area of the Pacific Northwest, there are two populations of killer whales, *Orcinus orca*. One population feeds on marine mammals, a prey that is sensitive to the frequencies of killer whale clicks; the other population feeds on salmon, a prey that is likely to be much less sensitive. Barrett-Leonard et al. (1996) report that mammal-eating killer whales tend to produce clicks less often than do fish-eating killer whales. When mammal-eating killer whales do click, they vary the intensity, repetition rate, and spectral composition within click trains, apparently making these clicks more difficult for their acoustically sensitive prey to identify than the regular click series of fish-eating killer whales. These results suggest that echolocating whales may modify their echolocation strategy depending upon the auditory sensitivities of their prey. If individual whales can modify their echolocation in this way, this presents parallels with communicative processes in which a signaler modifies its signal depending upon its intended audience (e.g., Marler et al. 1990). Animals producing sounds used by others in a bistatic biosonar mode may also be influenced by the behavior of conspecific receivers.

There is even some suggestive evidence that some fish may be able to intercept high-frequency echolocation sounds of their predators. Some recent research suggests that fish such as alewives (*Alosa pseudoharengus*), herring (*Alosa aestivalis*), and cod (*Gadus morhua*) are able to detect intense sounds of frequencies much higher than is typical of their vocalizations (Astrup & Møhl 1993; Dunning et al. 1992; Nestler 1992). Some of these species are prey for echolocating odontocetes. The only known natural sources of sounds with the intensity and frequency of these ultrasonic stimuli are the clicks of echolocating toothed whales. It is unknown whether this sensitivity happens to be a by-product of an auditory system that evolved to operate at lower frequencies and intensities, or whether fish have evolved an adaptation to enable interception of predators (as was suggested for goldfish by Offut 1968). Just as the receiver may be a potent source of selection upon a communicative signaler (e.g., Endler 1992; Ryan 1994), so interception of echolocation signals may affect the evolution of hearing in the prey and of echolocation strategies in the echolocating animal.

Generalized Substrates for the Evolution of Specialized Echolocation Abilities

Pollak (1992) hypothesizes that all higher vertebrates have an ability to perceive echoes and to learn to use echoes for information about their environment. He bases this upon the independent origin of echolocation in animals as diverse as birds, bats, and cetaceans. However, we do not know how important the abilities to process echoes are for animals that have not evolved specialized high-frequency echolocation. For example, estimation of the elevation angle of a sound may require analysis of reverberation patterns for many animals. For some animals, this may involve reflection from the pinnae and other parts of the body (Kuhn 1987), but animals may also rely upon reflection or reverberation from the environment for estimating the azimuth or elevation of a sound. For example, Berkley (1987) reviews the importance of reverberation in rooms for sound localization by humans. Predictably reflective surfaces such as the sea surface might present similar opportunities for marine organisms. Sound localization may select for auditory processing skills that could then play a role in the evolution of more specialized forms of echolocation. For example, if animals rely upon interaural time delays to locate a sound, might these abilities for precise timing of sounds also be used for timing delays between signal and echo? Some of the initial stages in the evolution of specialized high-frequency biosonar may have evolved from the use of more generalized auditory abilities to solve simple biosonar tasks such as those I suggest for low-frequency signals. In this essay I have suggested factors that may have affected the evolution of echolocation abilities in marine mammals. There has been little examination of how much cetaceans were able to recruit general mammalian auditory abilities for echolocation vs requiring the evolution of more specialized audition, beyond their obvious abilities for more rapid processing of higher frequencies.

Psychologists have debated for decades about the relative importance of general vs specialized learning abilities. For example, some linguists have argued that perception of speech involves a highly specialized "module" that shares little information or central mechanisms with a module, say, for the perception of music (e.g., Fodor 1983). Some students of animal intelligence have argued in a similar vein that cognitive skills used by animals in social interactions may not generalize to nonsocial contexts (Cheney & Seyfarth 1990). Most students of biosonar and acoustic communication appear to assume implicitly that the use of sound to explore the inanimate environment involves a module completely independent from that used to communicate acoustically with conspecifics. I would like to suggest that this is an untested hypothesis and should be evaluated more explicitly.

Learning Mechanisms for Matching Biosonar or Communication Signals to Local Circumstances

Some species appear to have evolved mechanisms for vocal production that involve signals that cannot be modified by auditory input. However, many sonar and communication problems require modification of the outgoing signal depending upon the auditory environment. A simple example common to both sonar and communication involves switching output frequency to avoid jamming by another source operating at the same frequency. This ability to modify one's vocal output depending upon auditory input is called vocal learning. Vocal learning is critical for the development of human communication, and its benefits are so obvious that it is surprising to find that it is rare among animals.

Most animals appear to have mechanisms for patterning vocal signals that are remarkably resistant to modification. Even such drastic treatments as deafening at birth, which would derail vocal development in humans and many bird species, do not prevent terrestrial nonhuman mammals from developing normal vocalizations (e.g., Buchwald & Shipley 1985). Primates raised in isolation or with foster mothers of a different species still produce species-typical vocalizations even though these animals are constantly exposed to, and must learn to respond to, vocalizations that differ from the ones they produce themselves (Owren et al. 1993; Winter et al. 1973).

Kroodsma and Konishi (1991) suggest that vocal learning requires special neural connections between auditory input and vocal motor output, connections that are present in songbird species that learn their song, but that appear absent in bird species that do not learn their song. Discussions of the evolution of vocal learning in birds and mammals tend to emphasize the importance of communication. However, echolocation also requires the ability to modify one's vocal output depending upon what one hears. Animals with short echolocation pulses at the very least emit pulses so they do not overlap with incoming echoes. Bats that perform Doppler compensation modify their outgoing signal to maintain the appropriate frequency of the echo. Neural connections between auditory input and vocal motor output have also been reported for several bat species (Covey & Casseday 1995). This raises the question whether connections between auditory input and vocal output are limited to either a communicative or sonar function, or whether they may have more general functionality.

If these kinds of connections create the potential for general abilities of vocal learning, and if vocal learning plays important roles in either communication or echolocation, then the evolution of vocal learning for echolocation may preadapt animals to be able to use the abilities for communication, or vice versa. There are hints of evolutionary interactions between these two functions for vocal learning. Both echolocation and the ability to modify communication signals based upon auditory input are rare among animals, but both abilities tend to occur in the same taxa of animals. For example, the only nonhuman mammals for which there is evidence for vocal learning are bats (e.g., Jones & Ransome 1993) and marine mammals (e.g., Tyack & Sayigh 1997), and these are also our best examples of animal echolocation. Vocal learning in bats has been described both for sounds used in echolocation (Jones & Ransome 1993) and communication (Esser & Schmidt 1989). Vocal learning is well known for communication signals in whales and dolphins, and beluga whales are known to modify the frequency structure of their echolocation clicks depending upon ambient noise (Au et al. 1985).

Discussions of vocal learning tend to focus simply upon matching vocal output to auditory input. Most other forms of learning are linked to some more extrinsic reinforcement. West and King (1990) suggest that some birds may direct their vocal development by creating a diverse array of sounds and then selecting which sounds they repeat depending upon the social responses evoked by their sounds. For example, West and King (1990) studied starlings (*Sturnus vulgaris*) which are accomplished mimics. Starlings raised by humans imitate many sounds in their environment. These birds are particularly likely to produce excellent imitations of the sounds of humans with whom they interact. The only starlings to imitate human sounds were those that interacted with humans on a regular basis. West and King (1990) suggest that it is this social interaction that enables these birds to shape their vocalizations to produce such excellent imitations of human speech while ignoring socially irrelevant sounds such as the dishwasher. This vocal learning does not just involve matching vocal output to auditory input, but must also involve social reinforcement as well. These birds imitate speech so precisely that they must not only select preexisting sounds but to modify the sounds and slowly shape their vocal repertoire depending upon the responses which these modifications evoke.

West and King (1988) provide evidence that this kind of nonvocal behavioral feedback between conspecifics may also direct vocal development in more natural circumstances. Male cowbirds will modify the songs they sing depending upon how likely the sounds are to elicit distinctive wing movements from female cowbirds. West and King (1990) also suggest analogies between communicative and echolocation in describing the open system for vocal development involving social feedback:

> We propose that some birds use acoustic probes to test the contingent properties of their environment, an interpretation largely in keeping with concepts of communication as processes of social negotiation and manipulation. An analogy with the capacities of echo-locating animals may be appropriate. Like bats or dolphins emitting sounds to estimate distance, some birds may bounce sounds off the animate environment, using behavioral reverberations to gauge the effects of their vocal efforts. (p. 113)

Many marine mammals also have exceptional skills of vocal imitation. Captive bottlenose dolphins of both sexes are highly skilled at imitating synthetic pulsed sounds and whistles (Caldwell & Caldwell 1972; Evans 1967; Richards et al. 1984). A captive harbor seal, *Phoca vitulina*, was reported to imitate human speech with a regional accent (Ralls et al. 1985). Captive beluga whales, *Delphinapterus leucas*, are also reported to imitate human speech well enough for caretakers to "perceive these sounds as emphatic human conversation" (Ridgway et al. 1985) or even for words to be recognized (Eaton 1979). The songs of humpback whales, *Megaptera novaeangliae*, are very similar within a population but change progressively over time (Payne et al. 1983). The vocal convergence at any one time within a population of singing humpback whales coupled with the rapid changes in the song over time provide evidence for vocal matching in these animals. Both the evolutionary origins and current utility of these imitative skills in marine mammals remain obscure.

West and King (1990) focus upon the extraordinary flexibility of vocal development that proceeds by comparing production of an arbitrary sound with the communicative benefit of using that sound. The same general abilities for vocal learning and flexibility in vocal development might also enable animals to use sound to explore their inanimate environment. For example, even though humans do not have specialized echolocation abilities, both blind and sighted humans can detect, locate, and discriminate targets by listening to echoes (Rice 1969). Learning and practice can enhance these general abilities to use echoes. If humans are allowed to spend an hour or so trying out their own vocalizations to find one that yields good echo information, their performance with this signal is about as good as that with artificial signals designed for enhancing echoes.

Humans appear to have evolved vocal learning for the communication, but we can also use this skill to rapidly learn signals that can help us explore our environment acoustically. Animals with biosonar abilities as highly evolved as the communicative abilities of the starling may also derive benefit from an open system of vocal development in which the feedback for vocal learning derives from the utility of different sounds for exploring the environment. Once evolved for biosonar, an open system of vocal development might also facilitate the development of communicative skills. Most biologists tend to think of sonar and communication as involving distinctly separate central mechanisms. The open process of vocal development described by West and King (1988, 1990) suggests the possibility that a more generalized skill might function to hone either communicative or biosonar skills. Animals that use vocal learning to modify sounds used for echolocation and sounds used for communication, such as bats and marine mammals, are particularly interesting candidates for testing the interplay between highly specialized and canalized skills vs more open and general mechanisms in learning and development.

ACKNOWLEDGMENTS

I started thinking about this essay during a year at the Center for Advanced Study in the Behavioral Sciences in Stanford, CA. Thanks to the Center for an excellent environment to develop new ideas. Support was provided by a Mellon Independent Study Award from the Woods Hole Oceanographic Institution and from NSF grant #SBR-9022192 to the Center for Advanced Study in the Behavioral Sciences. I would like to thank Christopher Clark, Orest Diachok, Kurt Fristrup, Donald Griffin, Vincent Janik, Cynthia Moss, Donald Owings, and Nicholas Thompson for comments on an earlier version of this chapter. Thanks to Orest Diachok and David Farmer for informal discussions about the forward propagation mode of bistatic sonar, particularly with respect to fish with swim bladders as targets. I must take responsibility for any errors in my description of the underwater acoustics or in my extrapolations to bioacoustics, but I would like to credit Diachok with suggesting the utility of a bistatic mode of biosonar to detect fish over large ranges of kilometers or more. The recordings of sperm whales came from research cruises in the Caribbean with Dr. William A. Watkins. Thanks to Patrick Miller for providing the data used for Figure 6 and to Patrick Miller and Pamela Willis for help producing the figures. Don Griffin suggested the idea of a monostatic biosonar using echoes from bottom reverberation to operate in forward propagation mode. This is contribution number 9246 from the Woods Hole Oceanographic Institution.

REFERENCES

Amundin, M. (1991). Click repetition rate patterns in communicative sounds from the harbour porpoise, *Phocoena phocoena*. Chapter in Ph.D. thesis, Sound production in odontocetes with emphasis on the harbour porpoise, *Phocoena phocoena*, University of Stockholm

Astrup, J., Møhl, B. (1993). Detection of intense ultrasound by the cod *Gadus morhua*. *J Exp Biol 182:*71–80

Au, W. W. L. (1980). Echolocation signals of the Atlantic bottlenose dolphin (*Tursiops truncatus*) in open waters. In: Busnel R-G, Fish JF (eds) *Animal sonar systems*, New York: Plenum, pp 251–282

Au, W. W. L. (1993). *The sonar of dolphins*. New York: Springer

Au, W. W. L., Carder, D. A., Penner, R. H., Scronce, B. L. (1985). Demonstration of adaptation in beluga whale echolocation signals. *J Acoust Soc Am 77:*726–730

Au, W. W. L., Pawloski, J. L., Nachtigall, P. E., Blonz, M., Gisiner, R. C. (1995). Echolocation signals and transmission beam pattern of a false killer whale (*Pseudorca crassidens*). *J Acoust Soc Am 98:*51–59

Backus, R., Schevill, W. E. (1962). *Physeter* clicks. In: Norris KS (ed) *Whales, dolphins, and porpoises*. Berkeley: University of California Press, pp 510–528

Barrett-Lennard, L. G., Ford, J. F. B., Heise, K. A. (1996). The mixed blessing of echolocation: differences in sonar use by fish-eating and mammal-eating killer whales. *Anim Behav 51:*553–565

Batzler, W. E., Pickwell, G. V. (1970). Resonant acoustic scattering from gas-bladder fish. In: Farquhar GB (ed) *Proceedings of an international symposium on biological sound scattering in the ocean*. Washington DC: Govt Printing Office.

Berkley, D. A. (1987). Hearing in rooms. In: Yost WA, Gourevitch G (eds) *Directional hearing*. New York: Springer, pp 249–260

Bodenhamer, R. D., Pollak, G. D. (1983). Response characteristics of single units in the inferior colliculus of mustache bats to sinusoidally frequency modulated signals. *J Comp Physiol 153:*67–79

Braham, H. W., Fraker, M. A., Krogman, B. D. (1980). Spring migration of the Western Arctic population of bowhead whales. *Mar Fish Rev 42:*36–46

Brill, R. L., Pawloski, J. L., Helweg, D. A., Au, W. W., Moore, P. W. B. (1992). Target detection, shape discrimination, and signal characteristics of an echolocating false killer whale (*Pseudorca crassidens*). *J Acoust Soc Am 92:*1324–1330

Buchwald, J. S., Shipley, C. (1985). A comparative model of infant cry. In: Lester BM, Boukydis CFZ (eds) *Infant crying*. New York: Plenum, pp 279–305

Busnel, R.-G., Dziedzic, A. (1966). Acoustic signals of the pilot whale, *Globicephala melaena* and of the porpoises *Delphinus delphis* and *Phocoena phocoena*. In: Norris KS (ed) *Whales, dolphins, and porpoises.* Berkeley: University of California Press, pp 607–646

Caldwell, M. C., Caldwell, D. K. (1967). Intraspecific transfer of information via the pulsed sound in captive odontocete cetaceans. In: Busnel R-G (ed) Animal Sonar Systems. NATO Advanced Study Institute, Jouy-en-Josas: Laboratoire de physiologie acoustique, vol 2, pp 879–936

Caldwell, M. C., Caldwell, D. K. (1972). Vocal mimicry in the whistle mode by an Atlantic bottlenosed dolphin. *Cetology 9:* 1–8

Cheney, D. L., Seyfarth, R. M. (1990). *How monkeys see the world.* Chicago: University of Chicago Press

Clapham, P. J., Mattila, D. K. (1990). Humpback whale songs as indicators of migration routes. *Marine Mammal Science 6:* 155–160

Clark, C. W. (1989). The use of bowhead whale call tracks based on call characteristics as an independent means of determining tracking parameters. *Rep Int Whal Comm 39:* 111–113

Clark, C. W. (1993). Bioacoustics of baleen whales: from infrasonics to complex songs. *J Acoust Soc Am 94:* 1830, Abstract

Clark, C. W., Ellison, W. T., Beeman, K. (1986). Acoustic tracking of migrating bowhead whales. *Proceedings of the IEEE Oceans 86 Conference 86:*341–346

Clay, C. S., Medwin, H. (1977). *Acoustical oceanography.* New York: Wiley

Covey, E., Casseday, J. H. (1995). The lower brainstem auditory pathways. In A. N. Popper & R. R. Fay (Eds.), *Hearing by bats.* New York: Springer, pp 235–295

Curran, T. A. , Lemon, D., Ye, Z. (1994). The acoustic scintillation flowmeter: application for a new environmental tool. *Journal of the Canadian Hydrographic Association 49:*25–29.

Dawkins, R., Krebs, J. R. (1978). Animal signals: information or manipulation? In J. R. Krebs & N. B. Davies (Eds.), *Behavioural ecology.* Oxford: Blackwell, pp 282–309

Dawson, S., Thorpe, C. W. (1990). A quantitative analysis of the sounds of Hector's dolphin. *Ethology 86:* 131–145

Dawson, S. (1991). Clicks and communication: the behavioral and social contexts of Hector's dolphin vocalizations. *Ethology 88*: 265–276.

Diachok, O. (1996). Fish absorption spectroscopy. In J. Papadakis (Ed.), *Proceedings of the third European conference on underwater acoustics.* Luxembourg: EC Press

Diachok, O., Ferla, C. (1996). Measurement and simulation of the effects of absorptivity due to fish on transmission loss in shallow water. *Oceans 96 Conference Proceedings,* Piscataway NJ: IEEE Service Center

Dorsey, E. M. (1983). Exclusive adjoining ranges in individually identified minke whales (*Balaenoptera acutorostrata*) in Washington state. *Can J Zool 61:* 174–181

Dudok, van Heel, W. H. (1981). Investigations on cetacean sonar. III. A proposal for an ecological classification of cetaceans in relation to sonar. *Aquatic Mammals 8:*65–68

Dunning, D. J., Ross, Q. E., Geoghegan, P., Reichle, J. J., Menezes, J. K., Watson, J. K. (1992). Alewives avoid high-frequency sound. *North American Journal of Fisheries Management 12:*407–416

Eaton, R. L. (1979). A beluga whale imitates human speech. *Carnivore 2:*22–23

Ellison, W. T., Clark, C. W., Bishop, G. C. (1987). Potential use of surface reverberation by bowhead whales, *Balaena mysticetus*, in under-ice navigation. *Rep Int Whal Comm 37:*329–332

Endler, J. A. (1992). Signals, signal conditions, and the direction of evolution. *Am Nat 139S:*S125-S153

Esser, K.-H., Schmidt, U. (1989). Mother-infant communication in the lesser spear-nosed bat, *Phyllostomus discolor* (Chiroptera, Phyllostomatidae) — evidence for acoustic learning. *Ethology 82:* 156–168

Evans, W. E. (1967). Vocalization among marine mammals. In: Tavolga WN (ed) *Marine bioacoustics*. Oxford: Pergamon, vol 2, pp 159–186
Evans, W. E., Awbrey, F. T., Hackbarth, H. (1988). High frequency pulse produced by free ranging Commerson's dolphin *Cephalorhynchus commersonii* compared with those of phocoenids. *Rep Int Whal Comm Special Issue 9:*173–181
Fenton, M. B. (1995). Natural history and biosonar signals. In A. N. Popper & R. R. Fay (Eds.), *Hearing by bats.* New York: Springer, pp 37–86
Fletcher, S., Le Boeuf, B. J., Costa, D. P., Tyack, P. L., Blackwell, S. B. (1996). Onboard acoustic recording from diving northern elephant seals. *J Acoust Soc Am 100:*2531–2539
Fodor, J. A. (1983). *The modularity of mind.* Cambridge: MIT Press
Ford, J. K. B. (1989). Acoustic behavior of resident killer whales (*Orcinus orca*) off Vancouver Island, British Columbia. *Can J Zool 67:*727–745
Ford, J. K. B. (1991). Vocal traditions among resident killer whales (*Orcinus orca*) in coastal waters of British Columbia. *Can J Zool 69:*1454–1483
Frankel, A. S., Clark, C. W., Herman, L. M., Gabriele, C. M. (1995). Spatial distribution, habitat utilization, movements, and social interactions of humpback whales, *Megaptera novaeangliae*, off Hawaii using acoustic and visual techniques. *Can J Zool 73:*1134–1136
George, J. C., Clark, C., Carroll, G. M., Ellison, W. T. (1989). Observations on the ice-breaking and ice navigation behavior of migrating bowhead whales (*Balaena mysticetus*) near Point Barrow, Alaska, spring 1985. *Arctic 42:*24–30
Griffin, D. R. (1955). Hearing and acoustic orientation in marine animals. *Deep-Sea Research, supplement to 3:*406–417
Griffin, D. R. (1974). *Listening in the dark.* New York: Dover
Griffin, D. R. (1980). Early history of research on echolocation. In: Busnel R-G, Fish JF (eds) *Animal sonar systems.* New York: Plenum, pp 1–8
Grinnell, A. D. (1995). Hearing in bats: an overview. In: Popper AN, Fay RR (eds) *Hearing by bats.* New York: Springer, pp 1–36
Guinee, L., Chu, K., Dorsey, E. M. (1983). Changes over time in the songs of known individual humpback whales (*Megaptera novaeangliae*). In R. Payne (Ed.), *Communication and behavior of whales.* Boulder: Westview Press
Harvey, P. H., Pagel, M. D. (1991). *The comparative method in evolutionary biology.* Oxford: Oxford University Press
Hatakeyama, Y., Soeda, H. (1990). Studies on echolocation of porpoises taken in salmon gillnet fisheries. In J. A. Thomas & R. Kastelein (Eds.), *Sensory abilities of cetaceans.* New York: Plenum, pp 269–281
Herman, L. M. (1980). The communication systems of cetaceans. In L. M. Herman (Ed.), *Cetacean behavior: mechanisms and functions.* New York: Wiley-Interscience, pp 149–209
Herman, L. M., Arbeit, W. R. (1972). Frequency discrimination limens in the bottlenosed dolphin: 1–70 KC/S. *J Aud Res 2:*109–120
Hersey, J. B., Backus, R. H. (1962). Sound scattering by marine organisms. In M. N. Hill (Ed.), *The sea.* New York: Interscience Publishers, vol 1, pp 498–539
Horton, J. W. (1959). *Fundamentals of sonar.* Annapolis: United States Naval Institute
Hui, C. A. (1994). Lack of association between magnetic patterns and the distribution of free-ranging dolphins. *J Mammal 75:*399–405
Hunt, F. V. (1954). *Electroacoustics.* New York: Wiley and Harvard University Press
Jacobs, D. W. (1972). Auditory frequency discrimination in the Atlantic bottlenose dolphin *Tursiops truncatus* Montagu: a preliminary report. *J Acoust Soc Am 52:*696–698
Jones, G., Ransome, R. D. (1993). Echolocation calls of bats are influenced by maternal effects and change over a lifetime. *Proc Roy Soc Lond B 252:*125–128
Kamminga, C. (1988). Echolocation signal types of odontocetes. In P. E. Nachtigall & P. W. B. Moore (Eds.), *Animal sonar: processes and performance.* New York: Plenum, pp 9–22

Kamminga, C., Wiersma, H. (1981). Investigations on cetacean sonar II. Acoustical similarities and differences in odontocete sonar signals. *Aquatic Mammals 8:*41–62

Kamminga, C., Wiersma, H. (1982). Investigations on cetacean sonar V. The true nature of the sonar sound of *Cephalorhynchus commersonii. Aquatic Mammals 9:*95–104

Kellogg, W. N. (1961). *Porpoises and sonar.* Chicago: University of Chicago Press.

Ketten, D. R. (1994). Functional analyses of whale ears: adaptations for underwater hearing. *IEEE Proceedings in Underwater Acoustics 1:*264–270

Kick, S. A., Simmons, J. A. (1984). Automatic gain control in the bat's sonar receiver and the neuroethology of echolocation. *J Neurosci 4:*2725–2737

Kinne, O. (1975). *Marine ecology.* London: Wiley, volume 2, part 2

Klinowska, M. (1985). Cetacean live stranding dates relate to geomagnetic disturbances. *Aquatic Mammals 11:*109–119

Kroodsma, D., Konishi, M. (1991). A suboscine bird (eastern phoebe, *Sayornis phoebe*) develops song without auditory feedback. *Anim Behav 42:*477–487

Kruse, S. (1991). The interactions between killer whales and boats in Johnstone Strait, B.C. In K. Pryor & K. S. Norris (Eds.), *Dolphin societies.* Berkeley: University of California Press, pp 149–159

Kuhn, G. F. (1987). Physical acoustics and measurements pertaining to directional hearing. In W. A. Yost & G. Gourevitch (Eds.), *Directional hearing.* New York: Springer, pp 3–25

Le Boeuf, B. J., Costa, D. P., Huntley, A. C., Feldcamp, S. D. (1988). Continuous deep diving in female northern elephant seals, *Mirounga angustirostris. Can J Zool 66:*446–458

Le Boeuf, B. J., Laws, R. M. (1994). *Elephant seals.* Berkeley: University of California Press

Love, R. H. (1973). Target strengths of humpback whales *Megaptera novaeangliae. J Acoust Soc Am 54:*1312–1315

Malme, C. I. (1994). Development of a high target strength passive acoustic reflector for low-frequency sonar applications. *IEEE J Oceanic Eng 19:*438–448

Marler, P., Karakashian, S., Gyger, M. (1990). Do animals have the option of withholding signals when communication is inappropriate? In C. A. Ristau (Ed.), *Cognitive ethology: The minds of other animals* (essays in honor of Donald R Griffin). Hillsdale, NJ: Erlbaum

Marten, K., Norris, K. S., Moore, P. W. B., Englund, K. A. (1988). Loud impulse sounds in odontocete predation and social behavior. In P. E. Nachtigall & P. W. B. Moore (Eds.), *Animal sonar: processes and performance.* New York: Plenum, pp 567–579

Martin, A. R. (1995). How do whales find the next breathing site when travelling under heavy sea-ice? In: *Abstracts, Eleventh Biennial Conference on the Biology of Marine Mammals,* p 73

Mate, B. (1989). Satellite monitored radio tracking as a method for studying cetacean movements and behavior. *Rep Int Whal Comm 39:*389–391

McDonald, M. A., Hildebrand, J. A., Webb, S. C. (1995). Blue and fin whales observed on a seafloor array in the Northeast Pacific. *J Acoust Soc Am 98:*712–721

McGregor, P. K., Krebs, J. R. (1984). Sound degradation as a distance cue in great tit (*Parus major*) song. *Behav Ecol Sociobiol 16:*49–56

McGregor, P. K., Krebs, J. R., Ratcliffe, L. M. (1983). The response of great tits (*Parus major*) to the playback of degraded and undegraded songs: the effect of familiarity with the stimulus song type. *Auk 100:*898–906

Menne, D., Kaipf, I., Wagner, I., Ostwald, J., Schnitzler, H.-U. (1989). Range estimation by echolocation in the bat *Eptesicus fuscus*: trading of phase versus time cues. *J Acoust Soc Am 85:*2642–2650

Miller, L. A. (1983). How insects detect and avoid bats. In F. Huber & H. Markl (Eds.), *Neuroethology and behavioral physiology: roots and growing pains.* New York: Springer

Moore, P. W. B. (1980). Cetacean obstacle avoidance. In: Busnel R-G, Fish JF (eds) *Animal sonar systems.* New York: Plenum, pp 97–108

Moore, P. W. B., Pawloski, D. (1991). Investigation on the control of echolocation pulses in the dolphin (*Tursiops truncatus*). In J. Thomas & R. Kastelein (Eds.), *Sensory abilities of cetaceans*. New York: Plenum, pp 305–316

Morton, E. S. (1982). Grading, discreteness, redundancy, and motivation-structural rules. In D. E. Kroodsma & E. H. Miller (Eds.), *Acoustic communication in birds*. New York: Academic Press, vol 1, pp 183–212

Morton, E. S. (1986). Predictions from the ranging hypothesis for the evolution of long distance signals in birds. *Behaviour 99:*65–86

Morton, E. S., Gish, S. L., van der Voort, M. (1986). On the learning of degraded and undegraded songs in the Carolina wren. *Anim Behav 34:*815–820

Murchison, A. E. (1980). Detection range and range resolution of echolocating bottlenose porpoise (*Tursiops truncatus*). In R.-G. Busnel & J. F. Fish (Eds.), *Animal sonar systems*. New York: Plenum, pp 43–70

Myrberg, A. A., Jr. (1981). Sound communication and interception in fishes. In W. N. Tavolga, A. N. Popper & R. R. Fay (Eds.), *Hearing and sound communication in fishes*. New York: Springer

Nestler, J. M., Ploskey, G. R., Pickens, J., Menezes, J., Schilt, C. (1992). Responses of blueback herring to high-frequency sound and implications for reducing entrainment at hydropower dams. *North American Journal of Fisheries Management 12:*667–683

Neuweiler, G. (1990). Auditory adaptations for prey capture in echolocating bats. *Physiol Rev 70:*615–641

Neuweiler, G., Link, A., Marimuthu, G., Rübsamen, R. (1988). Detection of prey in echocluttering environments. In P. E. Nachtigall & P. W. B. Moore, (Eds.), *Animal sonar: processes and performance*. New York: Plenum, pp 613–618

Norris, K. S. (1967). Some observations on the migration and orientation of marine mammals. In R. M. Storm (Ed.), *Animal orientation and navigation*. Corvallis: Oregon State University Press

Norris, K. S. (1969). The echolocation of marine mammals. In H. T. Andersen (Ed.), *The biology of marine mammals*. New York: Academic Press

Norris, K. S. (1974). *The porpoise watcher*. New York: Norton

Norris, K. S., Møhl, B. (1983). Can odontocetes debilitate prey with sound? *Am Nat 122:*85–104

Offut, C. G. (1968). Auditory response in the goldfish. *J Aud Res 8:*391–400

Owren, M. J., Dieter, J. A., Seyfarth, R. M., Cheney, D. L. (1993). Vocalizations of rhesus (*Macaca mulatta*) and Japanese (*Macaca fuscata*) macaques cross-fostered between species show evidence of only limited modification. *Developmental Psychobiology 26:*389–406

Pack, A. A., Herman, L. M. (1995). Sensory integration in the bottlenosed dolphin: immediate recognition of complex shapes across the senses of echolocation and vision. *J Acoust Soc Am 98:*722–733

Papastavrou, V., Smith, S. C., Whitehead, H. (1989). Diving behaviour of the sperm whale, *Physeter macrocephalus*, off the Galapagos Islands. *Can J Zool 67:*839–846

Papi, F. (1992). *Animal homing*. New York: Chapman and Hall

Parsons, T. R., Takahashi, M., Hargrave, B. (1984). *Biological oceanographic processes*. Oxford: Pergamon

Payne, K., Payne, R. (1985). Large scale changes over 19 years in the songs of humpback whales in Bermuda. *Zeitschrift für Tierpsychologie 68:*89–114

Payne, K. B., Tyack, P., Payne, R. S. (1983). Progressive changes in the songs of humpback whales. In R. S. Payne (Ed.), *Communication and behavior of whales*. AAAS Selected Symposia Series. Boulder: Westview Press, pp 9–59

Payne, R. S., Webb, D. (1971). Orientation by means of long range acoustic signalling in baleen whales. *Ann NY Acad Sci 188:*110–141

Pike, G. (1962). Migration and feeding of the grey whale (*Eschrichtius gibbosus*). *J Fish Res Board Can 19:*815–838

Pollak, G. D. (1992). Adaptations of basic structures and mechanisms in the cochlea and central auditory pathway of the mustache bat. In A. N. Popper, R. R. Fay, & D. B. Webster (Eds.), *Evolutionary biology of hearing*. New York: Springer, pp 751–778

Popper, A. N. (1980). Sound emission and detection by delphinids. In L. M. Herman (Ed.), *Cetacean behavior: mechanisms and functions*. New York: Wiley-Interscience, pp 1–52

Ralls, K., Fiorelli, P., Gish, S. (1985). Vocalizations and vocal mimicry in captive harbor seals, *Phoca vitulina*. *Can J Zool 63:*1050–1056

Reiss, D., McCowan, B. (1993). Spontaneous vocal mimicry and production by bottlenose dolphins (*Tursiops truncatus*): evidence for vocal learning. *J Comp Psychol 107:*301–312

Renaud, D. L., Popper, A. N. (1975). Sound localization by the bottlenose porpoise, *Tursiops truncatus*. *J Exp Biol 63:*569–585

Rice, C. E. (1969). Perceptual enhancement in the early blind? *The Psychological Record 19:*1–14

Richards, D. G. (1981). Estimation of distance of singing conspecifics by the Carolina wren. *Auk 98:*127–133

Richards, D. G., Wolz, J. P., Herman, L. M. (1984). Vocal mimicry of computer-generated sounds and vocal labeling of objects by a bottlenosed dolphin, *Tursiops truncatus*. *J Comp Psychol 98:*10–28

Ridgway, S. H., Carder, D. A., Jeffries, M. M. (1985). Another "talking" male white whale. *Abstracts, Sixth Biennial Conference on the Biology of Marine Mammals*, p 67

Ryan, M. J. (1994). Mechanisms underlying sexual selection. In L. A. Real (Ed.), *Behavioral mechanisms in evolutionary ecology*. Chicago: University of Chicago Press, pp 190–215

Schmidt-Koenig, K. (1975). *Migration and homing in animals*. New York: Springer

Schnitzler, H.-U. (1968). Die Ultraschall-Ortungslaute der Hufeisen Fledermäuse (Chiroptera-Rhinolophidae) in verschiedenen Orientierungssituationen. *Z vergl Physiol 57:*376–408

Schnitzler, H.-U., Ostwald, J. (1983). Adaptation for the detection of fluttering insects by echolocation in horseshoe bats. In J. P. Ewert, R. R. Capranica, D. J. Ingle (Eds.), *Advances in vertebrate neuroethology*. New York: Plenum, pp 801–827

Schuller, G., Beuter, K., Schnitzler, H.-U. (1974). Responses to frequency shifted artificial echoes in the bat, *Rhinolophus ferrumequinum*. *J Comp Physiol A 89:*275–286

Schuller, G., Pollak, G. D. (1979). Disproportionate frequency representation in the inferior colliculus of Doppler-compensating greater horseshoe bats: evidence for an acoustic fovea. *J Comp Physiol A 132:*47–54

Schusterman, R. J. (1972). Visual acuity in pinnipeds. In H. E. Winn, & B. L. Olla (Eds.), *Behavior of marine animals*. New York: Plenum, vol 2, pp 469–492

Schusterman, R. J. (1981). Behavioral capabilities of seals and sea lions: a review of their hearing, visual learning and diving skills. *The Psychological Record 31:*125–143

Scronce, B. L., Johnson, C. S. (1976). Bistatic target detection by a bottlenosed porpoise. *J Acoust Soc Am 59:*1001–1002

Simmons, J. A., Ferragamo, M. J., Saillant, P. A., Haresign, T., Wotton, J. M., Dear, S. P., Lee, D. N. (1995). Auditory dimensions of acoustic images in echolocation. In A. N. Popper, R. R. Fay (Eds.), *Hearing by bats*. New York: Springer

Simmons, J. A., Kick, S. A. (1983). Interception of flying insects by bats. In F. Huber, H. Markl (Eds.), *Behavioral physiology and neuroethology: roots and growing points*. New York: Springer

Stanton, T. K. (1989). Simple approximate formulas for backscattering of sound by spherical and elongated objects. *J Acoust Soc Am 86:*1499–1510

Thompson, R. K. R., Herman, L. M. (1975). Underwater frequency discrimination in the bottlenosed dolphin (1–140 kHz) and human (1–8 kHz). *J Acoust Soc Am 57:*943–948

Thompson, T. J., Winn, H. E., Perkins, P. J. (1979). Mysticete sounds. In H. E. Winn, B. L. Olla (Eds.), *Behavior of marine animals*. New York: Plenum, vol 3, pp 403–431

Turl, C. W., Penner, R. H. (1989). Differences in echolocation click patterns of the beluga (*Delphinapterus leucas*) and the bottlenose dolphin (*Tursiops truncatus*). *J Acoust Soc Am 86:*497–502

Tyack, P. (1981). Interactions between singing Hawaiian humpback whales and conspecifics nearby. *Behav Ecol Sociobiol 8:*105–116

Tyack, P. L., Sayigh, L. S. (1997). Vocal learning in cetaceans. In C. Snowdon, M. Hausberger (Eds.), *Social influences on vocal development.* Cambridge: Cambridge University Press, pp 208–233

Urick, R. J. (1983). *Principals of underwater sound.* New York: McGraw-Hill

von der Emde, G., Menne, D. (1989). Discrimination of insect wingbeat-frequencies by the bat *Rhinolophus ferrumequinum. J Comp Physiol A 167:*423–430

Walker, M. M., Kirschvink, J. L., Ahmed, G., Dizon, A. E. (1992). Evidence that fin whales respond to the geomagnetic field during migration. *J Exp Biol 171:*67–78

Watkins, W. A., Daher, M. A., Fristrup, K. M., Howald, T. J., di Sciara, G. N. (1993). Sperm whales tagged with transponders and tracked underwater by sonar. *Marine Mammal Science 9:*55–67

Watkins, W. A., Schevill, W. E. (1979). Aerial observations of feeding behavior in four baleen whales: *Eubalaena glacialis, Balaenoptera borealis, Megaptera novaeangliae*, and *Balaenoptera physalus. J Mammal 60:*155–163

Watkins, W. A., Schevill, W. E., Best, P. B. (1977). Underwater sounds of *Cephalorhynchus heavisidii* (Mammalia: Cetacea). *J Mammal 58:*316–320

Watkins, W. A., Tyack, P., Moore, K. E., Bird, J. E. (1987). The 20-Hz signals of finback whales (*Balaenoptera physalus*). *J Acoust Soc Am 82:*1901–1912

Watkins, W. A., Wartzok, D. (1985). Sensory biophysics of marine mammals. *Marine Mammal Science 1:*219–260

Weilgart, L., Whitehead, H. (1988). Distinctive vocalizations from mature male sperm whales. *Can J Zool 66:*1931–1937

West, M. J., King, A. P. (1988). Female visual displays affect the development of male song in the cowbird. *Nature 334:*244–246

West, M. J., King, A. P. (1990). Mozart's starling. *Am Sci 78:*106–114

Weston, D. E. (1967). Sound propagation in the presence of bladder fish. In V. M. Albers (Ed.), *Underwater acoustics.* New York: Plenum, vol 2

Williams, H., Nottebohm, F. (1985). Auditory responses on avian vocal motor neurons: a motor theory for song perception in birds. *Science 229:*279–282

Winter, P., Handley, P., Ploog, D., Schott, D. (1973). Ontogeny of squirrel monkey calls under normal conditions and under acoustic isolation. *Behaviour 47:*230–239

Wrangham, R. W., Gittleman, J. L., Chapman, C. A. (1993). Constraints on group size in primates and carnivores: population density and day-range as assays of exploitation competition. *Behav Ecol Sociobiol 32:*199–209

Würsig, B., Clark, C. W. (1993). Behavior. In J. J. Burns, J. J. Montague, C. J. Cowles (Eds.), *The bowhead whale.* Lawrence, KS: The Society for Marine Mammalogy

Würsig, B., Würsig, M. (1979). Behavior and ecology of the dusky dolphin, *Lagenorhynchus obscurus*, in the south Atlantic. *Fish Bull 77:*871–890

Xitco, M. J., Jr., Roitblat, H. L. (1996). Object recognition through eavesdropping: passive echolocation in bottlenose dolphins. *Animal Learning & Behavior 24:*355–365

Chapter 9

AN AFFECT-CONDITIONING MODEL OF NONHUMAN PRIMATE VOCAL SIGNALING

Michael J. Owren[1]

Department of Psychology
Reed College

Drew Rendall

Department of Psychology
University of Pennsylvania

ABSTRACT

We outline a model of nonhuman primate vocal behavior, proposing that the function of calling is to influence the behavior of conspecific receivers and that a Pavlovian conditioning framework can account for important aspects of how such influence occurs. Callers are suggested to use vocalizations to elicit affective responses in others, thereby altering the behavior of these individuals. Responses can either be unconditioned, being produced directly by the signal itself, or conditioned, resulting from past interactions in which the sender both called and produced affective responses in the receiver through other means.

In this view, the social relationship between the sender and the receiver is an important determinant of what sorts of responses can be elicited and, hence, which calls are used. For instance, a sender that is subordinate to, or otherwise has little power over a given receiver also has little opportunity to use its calls

[1] Correspondence to: Michael J. Owren, Department of Psychology, 224 Uris Hall, Cornell University, Ithaca, New York 14853. Voice: (607) 255-3835; fax: (607) 255-8433; email: mjo9@cornell.edu.

Perspectives in Ethology, Volume 12
edited by Owings *et al.*, Plenum Press, New York, 1997

as predictors of negative affective responses. It therefore relies primarily on vocalizations that have unconditioned effects. We refer to such calls as squeaks, shrieks, and screams, and propose that sounds of this general type should occur in acoustically variable streams—thereby maximizing unconditioned affective responses in the receiver while minimizing habituation effects. If the sender is dominant to the receiver, in contrast, it has ample opportunity to pair threatening calls with negative outcomes and can routinely induce and subsequently elicit conditioned affective responses. Such responses result from experiences in which the sender has produced individually distinctive vocalizations prior to attacking or otherwise frightening the other animal. As a given receiver routinely hears many such calls, the identity of the sender is the most important predictor of upcoming events and this animal's individually distinctive acoustic cues play a primary role in mediating any conditioning that occurs. Vocalizations used as conditioned stimuli must therefore carry salient, discrete cues to individual identity. We argue that individually distinctive cues based on vocal-tract filtering are best suited to this role, and refer to such sounds as sonants and gruffs.

Sonant and gruff calls should also be used by both dominant and subordinate senders in order to elicit positive conditioned responses. Such calls might occur, for instance, when an animal approaches a subordinate individual for grooming and attempts to decrease its fear during the approach. A subordinate animal should pair such calls with grooming or other positive outcomes when interacting with a dominant, thereby being able to elicit positive conditioned responses in that individual on other occasions.

The affect-conditioning model suggests that nonhuman primate vocalizations need not have "meaning" in the sense of transmitting referential information from a sender to a receiver. This approach may thereby provide a unified conceptual framework in which a number of issues related to the structure of vocal repertoires, acoustic features of calls, repetition and variability in vocalizations, and the evolution of such signals can be understood. In emphasizing the ability of the sender to mold the affective state of the receiver through simple conditioning processes, the model underscores the inherent asymmetry of these two roles. This imbalance is suggested to be an important factor in the evolution of more sophisticated cognitive mechanisms, which allow receivers to modulate their own behavioral responses to calls by evaluating the significance of such signals in a flexible, context-dependent fashion.

INTRODUCTION

The vocal behavior of nonhuman primates (hereafter primates) has been examined from a variety of perspectives (e.g., Todt, Symmes, & Goedeking, 1988; Zimmermann, Newman, & Jürgens, 1995), producing progress on a range of theoretical issues. Within this diversity, however, a smaller number of themes have tended to recur. For instance, a number of studies have been geared to showing that primate calls can be referential—encoding information that is transmitted from sender to receiver. It has been of particular interest to demonstrate that the referents in question can be objects, events, or circumstances from the external environment (see Gouzoules, Gouzoules, & Ashley, 1995, and Hauser, 1996, for recent reviews). Such work has led to important advances in understanding animal cognition, helping to uncover both impressive capabilities and unexpected limitations in various species (e.g., Cheney & Seyfarth, 1990). However, the excitement over evidence of human-like symbolism in the communication of both primates and other animals has also distracted attention from other important aspects of signaling (e.g., Owings, 1994).

On the one hand, any approach that helps create a common framework for understanding psychological processes in humans and nonhumans is to be encouraged. As the prevailing paradigm for understanding human cognition is based on representation and processing of information, this concept represents a powerful tool that should be applied in animal work as well. On the other hand, the notion of information as a commodity that moves from sender to receiver is obviously metaphorical. Information does not literally reside in the energy of a signal, but represents an emergent property of the combined attributes of the individual producing the signal, the individual perceiving the signal, and the circumstances under which the signal is emitted (e.g., Smith, 1977; also see Owings & Morton, this volume). Approaches that capture other aspects of this complex interactive process are therefore also needed. We suggest, for example, that vocal communication must have originated in unspecialized responses occurring to unspecialized energy transmissions. If so, concepts like information and representation would not apply and it is not clear how information-based communication could evolve from such circumstançes. We therefore see a particular need for a framework that includes more fundamental principles—accounting for aspects of communication that preceded information-processing, came to be information processing, and arguably now coexist with information processing.

That framework is not presented here. However, we do outline a model that attempts to account for the basic design and function of primate vocalizations by treating the sounds as stimuli that senders use in order to elicit simple affective responses in receivers. The concepts and terminology involved are borrowed from learning theory, and are meant to capture aspects of the communication process that are not well-characterized by terms like cognition and representation. This approach is consistent with Mason's (1979) distinction between "wanting" and "knowing," and Owings' (1994) use of "conation" to designate the motivational and emotional processes that impel and guide behavior, but are not "cognition." Note that the contrast intended is not equivalent to the classic differentiation of *referential* from *motivational* signals. Whereas the former refers to encoding of relatively specific external designata, the latter analogously includes internal designata, for instance representing the signaler's internal states (see discussion by Marler, Evans, & Hauser, 1992). Both terms are therefore rooted in an informational perspective, which we, initially at least, wish to avoid.

Overview

Our model rests on a number of observations and proposals concerning the social lives of primates, a hypothesized distinction between vocalizations that function to elicit unconditioned and conditioned affective responses, respectively, and the implications of vocal-tract-based sound production for the forms of such signals. For clarity's sake, we first describe the model briefly. Some general issues in primate vocal behavior are then discussed, setting the stage for more detailed exploration of conditioning and acoustic cues to individual identity. We then describe the model, both conceptually and by example, and present some of its implications for the general issues raised in the early going.

Throughout the chapter, we broadly refer to the vocal behavior of "primates." Nonetheless, we recognize that significant variation exists in many aspects of the production, perception, and function of vocalizations among the more than 200 extant primate species. Furthermore, while our proposals have been inspired by a general pattern of social and vocal characteristics exemplified by macaques and baboons, important variation occurs among the species and subspecies of these groups as well. However, as presenting the model requires relatively detailed consideration of some important aspects of both learning theory and sound production, this generic sort of approach seems preferable. We have therefore also chosen to rely on selected and illustrative examples rather providing a more comprehensive data review, but go on to describe a number of specific, testable predictions.

Characteristics of the Primate Species of Interest

The model is being proposed for species with the following general characteristics. First, the animals live in an environment that affords close visual and auditory contact among individuals. Examples include open or lightly forested areas. Second, the animals live in large, complex, and long-lived social groups that include multiple adult males, adult females, subadults, and offspring. The social relationship of any two individuals is therefore shaped by repeated encounters that routinely occur in the course of everyday activity. Third, stable social hierarchies are present, such that these interactions are marked by dominance-related behavioral asymmetries. Fourth, animals are capable of recognizing group members (and others) both by their appearance and voice characteristics. Finally, the animals exhibit a complex vocal repertoire that includes tonal and noisy sounds, a range of typical production-amplitudes, significant variability in call acoustics within and among individuals, and variability in the calls used in any given circumstance.

Principles of the Affect-Conditioning Model

Vocalizations Elicit Affective Responses

We adopt the general position that the ultimate function of communication is to influence the behavior of another individual. While such influences can be effected in a variety of ways, vocalizations appear particularly well-suited for this purpose in that they can be used to elicit affective responses in receivers. These responses are hypothesized to benefit the sender by priming or biasing the receiver to behave in a way that is compatible with the caller's best interests. These affective effects occur as unconditioned responses, conditioned responses, or some combination of the two. One potentially important unconditioned response is the simple affective quality that a vocalization may have due to general characteristics of the mammalian auditory system. In other words, some calls are proposed to be inherently noxious or pleasant based on phylogenetically ancient auditory processes that are probably shared by many primates and other mammal species. In addition, unconditioned and broadly differentiated affective responses may occur to various calls due to more recent specializations that may be species- or genus-specific. For instance, calls used in affiliative contexts by a given species are likely to elicit positive unconditioned responses, while negative affect can be expected in response to calls typically produced in agonistic contexts.

Our strongest claim, though, is that conditioned affective responses play a central functional role in primate calling. Vocalizations are well-suited for this kind of learning as they are discrete and perceptually salient stimulus events that are controlled by the sender. The potential value of affective conditioning is proposed to depend on the nature of the interaction between the sender and receiver, as well as their respective positions in the social hierarchy. In the course of a social interaction involving two individuals, both the more dominant and the more subordinate animal can produce calls that elicit conditioned responses. However, the individual that is dominant in a given encounter inherently has greater control over the outcome of this interaction than does the other. As a result, the dominant animal can routinely pair its calls with other actions that elicit significant unconditioned affective responses in the subordinate. Such pairings produce conditioning, which the caller can thereafter use to elicit learned affective responses in this other animal in both affiliative and agonistic circumstances. The subordinate individual has less opportunity to shape the outcome of an interaction, especially in the case of agonistic encounters. It is therefore proposed to rely on unconditioned effects of calling in such circumstances, while exploiting both conditioned and unconditioned responses in affiliative situations.

Acoustic Cues to Individual Identity Mediate Conditioned Affective Responses

Based on learning theory, it is expected that acoustic cues to individual identity are important mediators of conditioned effects occurring in receivers. As these responses are shaped by the history of interactions occurring between any two animals, the identity of a caller is a crucial determinant of the significance of a vocalization for a given receiver. As a result, the acoustic cues to individual identity that occur in the context of one call-type or another are more predictive of upcoming events than is the occurrence of the call in and of itself. These cues therefore come to elicit conditioned affective responses.

Principles of acoustics and vocal production suggest that various primate vocalizations afford different opportunities for carrying individually distinctive cues to caller identity. The most stable, consistent cues are proposed to be the features related to vocal-tract filtering, meaning characteristic amplification and attenuation effects produced by cavities located above the larynx. In order for such cues to appear, however, a vocalization must include broadly distributed spectral energy that reveals these effects in detail. Both low-frequency, tonal calls with rich harmonic structure and noisy vocalizations of intermediate amplitude appear to be well-suited to this function. These sounds are referred to as *sonants* and *gruffs*, respectively. In contrast, high-frequency tonal calls and high-amplitude, noisy vocalizations appear to be poorly suited to showing vocal-tract filtering effects. These sounds are referred to as *squeaks*, *shrieks*, and *screams*.

Functional Constraints Have Shaped the Acoustic Design of Vocalizations

If vocalizations function to elicit unconditioned and conditioned responses, and conditioning is mediated by cues to individual identity, it follows that the acoustic design of a primate vocal repertoire reflects the constraints imposed by these factors. In eliciting unconditioned responses, animals can be expected to produce bouts of acoustically potent vocalizations, with both repetition and variability shown in the call stream. In contrast, sounds whose primary function is to elicit conditioned responses are expected to show design features that promote conditioning and take advantage of any such learning that has already been instilled. Inclusion of discrete, salient cues to individual identity is paramount, which we propose to be provided primarily by spectral-patterning effects related to vocal-tract filtering.

SOME ISSUES OF INTEREST CONCERNING PRIMATE VOCALIZATIONS

Setting the stage for more detailed examination of both the principles and implications of our approach, a number of questions concerning primate vocalizations will now be raised. Although many different elements of such signaling have been discussed over the years, only a few integrative theories have emerged in this area. As the affect-conditioning model purports to identify some general principles of primate calling, we need first to touch on and then later return to various aspects of this behavior.

Signal Function

A fundamental part of studying communication is to determine the function of the signals being used. In work on nonhumans, this topic has spawned fruitful, but sometimes contentious and polarizing debate (see Hauser, 1996, for a recent review of these issues). One important question was outlined above—how to characterize the content or meaning of signals. A second, related question has been whether signals are inherently cooperative, involving information sharing, or are selfish in nature, primarily benefiting the sender. While a variety of frameworks and terminologies have been proposed for understanding these two issues (e.g., Hauser, 1996; Owings, 1994), important commonalties occur among the various positions. For instance, both referential and motivational approaches to signal meaning implicitly involve an informational perspective, although neither need imply that signaling is inherently altruistic or even "honest" (e.g., Cheney & Seyfarth, 1990; Hinde, 1981). Efforts to distinguish between sharing and manipulation in communication have produced several convergent proposals, each emphasizing the selfish interests and active roles of both parties involved in a signaling event. For example, senders and receivers have been characterized as being "manipulators" and "mind readers" (Krebs & Dawkins, 1984) or "managers" and "assessors" (Owings, 1994; Owings and Morton, this volume), respectively.

We suggest that the diverse positions taken in these debates can be reconciled by starting from the fundamental assumption that in a well-established, species-typical communication system, signaling occurs because it has in the evolutionary past provided a net benefit to the fitness interests of the sender by influencing the immediate or later behavior of the receiver. From this perspective, it is inevitable that a number of strategies for achieving such influence will have evolved, with corresponding variety in details of signaling. For instance, a given repertoire is likely to include some calls that are primarily conative in function, as well as others that are better described as engaging cognitive systems. Depending on the circumstances, the influences that senders are able to exert may also be either beneficial or costly to receivers. To the extent that the effect of the sender's behavior is detrimental to the overall fitness of the receiver, however, natural selection will favor adaptations that decrease those costs. In this chapter, we focus primarily on the interests of the sender.

Acoustic Features of Calls and Repertoire Structure

A second important topic is why the vocal repertoires of various primates take the forms they do. A number of relevant principles have been identified in this regard, including constraints imposed by the transmission characteristics of the environment in which calls are used (e.g., Brown, Gomez, & Waser, 1995), possible relationships among acoustic gradedness, complexity of the information conveyed, and the degree to which signals in other modalities complement the acoustic event (e.g., Green & Marler, 1979; Marler, 1975), and a general relationship between the acoustics of calls and sender state (Morton, 1977, 1982). However, these and other observations, hypotheses, and principles (see for instance Fitch & Hauser, 1995) leave unaddressed many basic questions that might be asked by naïve observers of monkeys and apes. One such question is why, in some instances, calls with particular acoustic features seem to be used similarly by many primates, whereas in other cases acoustically similar calls may be used differently. Within a species, some calls seem to occur only in quite specific circumstances, while others are produced in a variety of situations. Furthermore, two acoustically dissimilar calls might be used in the same contexts, whereas other, more similar calls are used in differentiated contexts. Overall, there is no general framework that can account for the bewildering variety of acoustic form and repertoire design among primates.

Repetition and Acoustic Variability in Calling

A third, related issue concerns the use of multiple vocalizations. While calls sometimes occur singly, bouts of calling are very common. In these cases, the same call may occur a number of times. In other cases, the acoustic features of successive calls may change, either gradually or abruptly. When acoustic variation occurs, the animal may produce vocalizations that seem confined to a single identifiable call-type, jump back and forth between categories, or grade from one category to another. Complications that can arise are discussed by Green and Marler (1979), and are concretely illustrated in Green's (1975) description of the vocal repertoire of Japanese macaques (*Macaca fuscata*). However, broadly applicable principles that make sense of this overall puzzle are in short supply.

Cues to Individual Identity

A fourth question is what role is played by acoustic cues to individual identity in vocal communication processes. Evidence of either discrimination among animals or explicit recognition of individuals is available from several species (e.g., Snowdon, 1986; Rendall, Rodman, & Emond, 1996). While a critical review of the evidence does not appear to us to warrant a blanket assumption that individually distinctive acoustic cues occur in all calls of a given vocal repertoire, both discrimination and recognition of callers seem likely to occur for some calls in many, if not all species. It is not clear, however, what functional role is played by acoustic cues to individual identity, particularly among primates whose habitats allow group members to readily see one another in many of the typical contexts in which vocalizations are produced.

A common explanation for discrimination or recognition of individuals based on acoustic cues is that receivers can then respond to distress-related calling. For instance, primate mothers are proposed to be particularly attentive to distress vocalizations produced by their offspring, a claim that has been supported to varying degrees in several species. Adult primates, particularly females, are also suggested to make use of individually distinctive acoustic cues in calls in coming to the aid of related individuals or unrelated allies involved in agonistic social encounters. This proposal is consistent with the more general observation that a primate's behavior toward others is strongly shaped by kin relationships and interaction histories (see Smuts, Cheney, Seyfarth, Wrangham, & Struhsaker, 1987; Cheney & Seyfarth, 1990). As yet, however, relatively little quantitative evidence is actually available from even the best-studied species showing that adult animals that are out of visual contact with genetically related group members consistently do respond to such distress calls in the absence of other information.

A PROPOSED ROLE OF CONDITIONING IN ANIMAL SIGNALING

Principles of Conditioning

The most basic proposal of our model is that individual primates use vocalizations to produce affective responses in conspecific receivers, thereby influencing the subsequent behavior of those animals. This claim is elaborated using concepts from learning theory concerning elicited (involuntary) responses, and changes in those responses occurring through habituation and Pavlovian (classical) conditioning. Relevant principles and data can be reviewed in any good textbook on learning (e.g., Domjan, 1993; Schwartz, 1989). The vocabulary used here is the traditional one. However, contemporary theorists emphasize that this terminology more accurately describes procedural aspects of the learning process than its underlying mechanisms. Those mechanisms, in fact, appear to be indistinguishable from the sorts of processes implied by typical concepts from cognitive theory, particularly the notion of representation (discussed by Rescorla, 1988a, 1988b; see also Turkhan, 1989, and the commentaries therein). As the sorts of learning involved in habituation and Pavlovian conditioning are ubiquitous among animals and occur in even the simplest nervous systems, these principles appear to provide a promising starting point for eventually understanding "representation" and "processing" of information at all levels of neural organization. We believe that both the terminology and the perspective adopted here are inherently compatible with, rather than exclusive of, other learning-related or information-based formulations. Our specific hope is that by first differentiating between possible conative and cognitive roles of vocalizations, we may actually be contributing to the development of a larger framework in which these functions can be integrated.

Habituation

Habituation refers to the learned decrease in responsiveness that is typically shown by an organism when repeatedly exposed to a stimulus. The underlying process is explicitly distinguished from sensory adaptation and muscular fatigue, both of which produce response decrements but are not caused by changes in the central nervous system. In studying habituation, these alternative interpretations are therefore ruled out by showing that the decrease produced is specific to the response in question and that the behavior can immediately recover if a new stimulus is introduced. Following habituation to a particular stimulus, generalization of the effect to other, similar stimuli often occurs. Habituation is also known to involve separable short- and long-term effects on the elicited response.

Pavlovian Conditioning

Pavlovian conditioning takes place when the occurrence of a biologically significant event is reliably preceded by some other discernible stimulus. As a result of experiencing the relationship between these two events, organisms come to respond to the first, *conditioned stimulus* as if it predicts the occurrence of the more important, *unconditioned stimulus*. Through learning, then, the conditioned stimulus begins to elicit *conditioned responses* that occur before, and in the absence of the unconditioned stimulus. While these responses often resemble one or more of the *unconditioned responses* that are elicited by the unconditioned stimulus, they need not do so. Instead, these conditioned effects are best characterized as anticipatory reactions that allow the organism to respond more adaptively to the upcoming event.

Pavlovian conditioning has been demonstrated in many aspects of organismal function, and affective responses play a central role in such learning. For instance, conditioned motivational and emotional responses have been proposed to be important in guiding an organism's instrumental, or voluntary behaviors. In one frequently used laboratory paradigm, the *conditioned-emotional-response* procedure, hungry rats are first trained to repeatedly press a lever in order to obtain food rewards. A light or tone is then presented as a conditioned stimulus, followed by an aversive electric shock. After a few such trials, lever-pressing slows or ceases whenever the fear-inducing conditioned stimulus is presented, and this suppressive effect is used as a measure of the learning that has occurred. Another common testing situation involves hungry pigeons intermittently given access to food. In this *sign-tracking* procedure, food presentations are preceded by illumination of a small, circular key. Although not required to do so, the birds typically come to quickly approach and peck the key when it is lit, treating the conditioned stimulus as if it were the unconditioned stimulus. In both of these prototypical paradigms, then, associations formed between a predictive stimulus and a biologically significant event arguably elicit affective responses in the subjects, thereby shaping their voluntary behavior.

Depending on the particular stimuli and organism involved, the number of trials required for Pavlovian conditioning to occur can range significantly. However, contemporary studies tend to emphasize learning that can be demonstrated within a few, or a few dozen trials. In general, the *associative strength* said to accrue to a conditioned stimulus is directly related to the predictive value of this event vis-à-vis the unconditioned stimulus. Therefore, the strongest conditioning occurs when an unconditioned stimulus is always preceded by the conditioned stimulus, and neither stimulus occurs otherwise. Nonetheless, even modestly predictive relationships between the two stimuli produce some learning. In general, conditioning occurs more readily when the conditioned stimulus is discrete rather than diffuse, and when it is more, rather than less, perceptually salient to the organism in question. The *context* in which learning takes place can also be a critical factor, as associative strength is normally divided among all the predictive stimuli present. As a result, more associative strength accrues to a conditioned stimulus when it is uniquely predictive than when it is partially or wholly redundant with other cues. If the stimulus has predictive value across a number of different contexts, stronger conditioning occurs. Conversely, differentiated responses can readily develop to the same conditioned stimulus if that cue predicts different outcomes when appearing in two or more distinct contexts. Thus, the significance of a conditioned stimulus can be context-dependent, producing requisitely different responses.

A Role for Learning Theory in Naturally Occurring Animal Behavior

Concepts related to elicited responses and learned changes in those responses are applicable to many aspects of naturally occurring animal behavior.

The potential value of such applications was noted in Hinde's (1966) integrative approach to behavior, which explicitly attempted to bring laboratory-based learning principles to bear on questions and issues in ethology. An important subsequent development was the discovery by Garcia and others (e.g., Hinde & Stevenson-Hinde, 1973) that so-called "general" learning processes like Pavlovian conditioning do not operate uniformly across the conditioned and unconditioned stimuli experienced by a given species. Instead, it was found that conditioning can proceed either quickly or slowly (or not at all), depending on the particular stimuli used. Therefore, while conditioning-related phenomena occur widely among animals, it became clear that the mechanisms involved have also been strongly shaped by the particular learning needs of each species. Classic work followed in a number of areas (e.g., Marler & Terrace, 1984; Gould, 1986; Bolles & Beecher, 1988), demonstrating inherent connections between laboratory-based learning principles and naturally occurring behavior.

Overall, though, these principles have more often been used as methodological rather than as theoretical tools in investigating ethologically relevant capabilities and behaviors. For instance, conditioning-based preparations have been routinely used to probe sensory functions, perceptual processing of communication signals, and various aspects of cognition in animals (see reviews by Cynx & Clark, in press; Stebbins & Berkley, 1990; Roitblat & von Fersen, 1992; Wassermann, 1993). Recently, some field researchers have also taken advantage of habituation effects occurring to repeated stimulus presentations in testing hypotheses about processing occurring at other levels. One technique has been to examine transfer of habituation to novel calls or callers following repeated presentation of sounds in order to better understand signal meaning (e.g., Cheney & Seyfarth, 1990). Somewhat surprisingly, then, the role that habituation may play in the many situations in which animals themselves repeatedly produce calls has not been explored.

However, some work has directly demonstrated the potential importance of conditioning processes in natural behavior (reviewed by Domjan, 1992; Domjan & Hollis, 1988). Hollis (1984), for instance, found that male blue gourami fishes (*Trichogaster trichopterus*) housed in an aquarium were more successful in defending territories when the arrival of a conspecific intruder was signaled using a discrete, salient stimulus than in the absence of this predictive event. Similarly, Hollis, Cadieux, and Colbert (1989) found that the male was more likely to show courtship behavior if the arrival of a female was signaled than if the female appeared unexpectedly. In both cases, conditioned preparatory responses in the male evidently allowed it to respond more quickly and effectively to the new situation.

Pavlovian Conditioning in Stomatopod Behavior

Work by Caldwell and his colleagues (reviewed by Caldwell, 1986) provides a potential example of the processes we propose to be important in primate vocal signaling. Although conditioning concepts were not invoked in these studies, Caldwell's investigations of stomatopods (mantis shrimp) of the genus *Gonodactylus* demonstrate the role that Pavlovian conditioning may play in the ability of one animal to directly influence the behavior of another using individually distinctive signals. These stomatopods were shown to employ both threat displays (termed *meral spread*) and physical attack (blows delivered using a raptorial appendage) for territorial defense of crevices in preferred habitat. Further, Caldwell proposed that clumped distribution of suitable crevices and stability in territory ownership resulted in repeated, agonistic encounters among particular animals. Individual discrimination based on distinctive odor cues was demonstrated, and might also occur through visual cues presented during threat displays.

Caldwell found the meral spread to reliably precede physical attack when a resident stomatopod defended its territory against an intruder. As both the olfactory and visual cues that occur in this situation are evidently predictive of subsequent physical blows, we propose that conditioning takes place. Specifically, individually distinctive cues are paired with a biologically significant outcome, thereby potentially allowing the sender to elicit affect-like responses in the receiver during subsequent encounters between the two animals. In other words, in later interactions a sender becomes more likely to be able to use its odor cues and meral display alone to repel that particular intruder, having paired these stimuli with one or more physical blows during previous encounters.

A test of this interpretation is provided by molting-related behavior observed in these animals. During the molting stage, the stomatopod's raptorial appendage becomes ineffective as a weapon and the animal is therefore vulnerable to physical attack. During this phase, territory-holders can only use the meral display to repel intruders, and the latter become requisitely more successful in evicting residents from crevices. The key finding for our purposes is that in the days immediately preceding the molting phase, Caldwell found that residents increased the rate at which meral displays were followed by physical attack. The display was then used at elevated rates during the subsequent period of vulnerability, when it did not in fact signal impending attack. In these cases, the display was found to be more effective in fending off opponents if it had previously been paired with blows from the raptorial appendage.

Caldwell described these findings in terms of "bluffing" and "reputation." We suggest that these stomatopods are in essence conducting Pavlovian conditioning trials in which individuals pair their distinctive olfactory and visual cues with physical blows delivered to the opponent. Although Caldwell did not investigate specific pairing of odor cues with attack, it was reported that individual stomatopods in the premolting phase often pursued a fleeing opponent to deliver additional blows. As the contest had already been won in such an instance, the act of leaving the crevice served no evident immediate purpose for the resident and compromised its safety. However, such behavior arguably provides a powerful means of promoting further learning through additional pairings of the conditioned and unconditioned stimuli experienced by the intruder.

Extending Conditioning Principles to Communication

Unconditioned Responses

Two forms of signaling in which the signal itself acts as an unconditioned stimulus will be distinguished. In the first case, the signal elicits affective responses through relatively direct effects on the sensory system of the receiver. In the acoustic modality, then, we propose that due to general properties of the mammalian auditory system, a stimulus can elicit positive or negative reactions. We expect that negative responses are particularly common and are tied to acoustic characteristics like overall amplitude and noisiness. These hypothesized effects are illustrated by crying in human infants, some forms of which are reported to be extremely aversive to human observers (e.g., Zeskind & Lester, 1978; see also Halpern, Blake, & Hillenbrand, 1986). Other acoustic dimensions of signals may also have significant unconditioned effects. For example, exaggerated pitch contours typically found in the speech of human caretakers to young infants (e.g., Fernald, 1992) have been proposed to increase, decrease, or maintain arousal levels in these receivers, depending on the particular frequency-modulation pattern involved (Papousek, Papousek, & Symmes, 1991). Corroborating evidence has been found in laboratory-based learning studies using both natural and synthetic stimuli (e.g., Kaplan & Owren, 1994; Kaplan, Goldstein, Huckeby, Owren, & Panneton Cooper, 1995), and infants have also been found to explicitly prefer listening to voices that show arousing modulation patterns (e.g., Fernald & Kuhl, 1987; Werker & McLeod, 1989).

Signals are also arguably likely to elicit unconditioned responses as a result of species-specific selective histories. In another example from human behavior, laughter has been found to elicit characteristic affective responses. When repeatedly exposed to a brief laughter recording, for instance, listeners initially demonstrate positive affect, often laughing themselves. After a few presentations, however, the sound becomes an irritant, eliciting negative affect (e.g., Provine, 1996). The rapidity of this change suggests that neither of these affective reactions results from simple sensory responses per se. Among nonhumans, species-typical signals like alarm calls are found to reliably elicit alerting or arousal responses, even in the absence of previous experience with those signals. In primates, for instance, while well-organized responses to conspecific vocalizations typically emerge later in development than do adult-like production and usage (e.g., Seyfarth & Cheney, 1997), young animals react to various species-typical sounds well before emitting those calls themselves or responding appropriately to them. For instance, before acquiring the differentiated response patterns shown by older offspring and adults, a very young vervet (*Cercopithecus aethiops*) hearing an alarm call typically responds by looking or running toward its mother. While other factors like visual cues may be important here as well, early emergence of affectively toned unconditioned responses (like fearful startle to alarm calls) would provide an excellent foundation for subsequently acquiring more specific response patterns.

Conditioned Responses

The occurrence of a particular signal may also be predictive of a significant upcoming event, and therefore function as a conditioned stimulus. For instance, even in species that are not capable of individual discrimination or recognition, the occurrence of "anonymous" signals that are correlated with subsequent agonism or affiliation can arguably come to elicit learned responses. A similar rationale holds if species members can discriminate or recognize one another, but use signals in which individually distinctive cues are absent. Overall, however, conditioning should accrue most quickly to the individually distinctive features of a signal. As described, these are the cues that are likely to be the most predictive aspect of the signaling event for a particular receiver. We therefore expect that these features play an important role in the signaling systems of animals that are able to discriminate or recognize other individuals and experience repeated interactions with them.

Applications to Primates

Most primates exhibit exactly this lifestyle—individuals typically live among familiar conspecifics that are encountered numerous times each day. The majority of primate groups also exhibit social hierarchies, and any animal that is dominant to another can strongly affect that individual's everyday existence. In species like macaques and baboons, dominant individuals can routinely administer painful bites and blows to subordinates. Violent and intimidating behavior by dominant animals induces strong arousal and fear in others, and lower-ranking group members are frequently observed to avoid contact with higher-ranking ones. They cringe or move away when the dominant animals approach, often seem to avoid drawing attention to themselves, and are easily supplanted from food and water, grooming partners, or resting places. Subordinates also readily respond to both visual and vocal signals, which dominant animals can use to alter the behavior of lower-ranking group members with minimal energy expenditure. Although physical prowess and support from kin or other allies are important in maintaining social rank, energy-intensive physical contests occur much less frequently than do signals like stares, facial expressions, and calls.

We suggest that in addition to unconditioned affective responses induced by such signals, conditioned responses play an important, and perhaps primary role. These kinds of responses can be expected based on a history of interactions between the sender and receiver, or from instances in which the receiver has observed interactions between the sender and other animals. However, the significance of this sort of signal is also critically dependent on the relationship between the particular sender and receiver involved. For instance, while an individual routinely experiences interactions with higher-ranking animals in which threatening signals may be followed by pain and fear, signals produced by lower-ranking group members are not reliably predictive of such outcomes. Therefore, the individually distinctive cues that are embedded within a display or a call play a special role in determining their predictive value.

Vocalizations appear to be particularly well-suited for use as conditioned stimuli in this kind of learning process. First, they are salient, discrete events with clearly marked onsets and rapid energy fading. This form of signaling is therefore almost perfectly designed for the prototypical Pavlovian conditioning process in which a well-defined stimulus is associated with a biologically significant outcome. In fact, the same argument is applicable to short-lived visual signals like facial expressions. Laboratory-based studies suggest that as brief, discrete events, both kinds of signals should be inherently more effective as conditioned stimuli than more diffuse or longer-lived signals that have requisitely less predictive precision. Second, vocalizations are controlled by the sender and are difficult for receivers to avoid. While facial expressions are also controlled by the sender, the receiver must attend to these signals if they are to be effective. Among primates, subordinate animals often avoid visual signals given by higher-ranking group members by "studiously" looking in some other direction.

Overall, we believe that a conditioning-based framework like this one provides a useful alternative to an information-based perspective on the communication process. However, due to the proposed role of individually distinctive cues in eliciting conditioned responses in others, any constraints that species-typical production mechanisms place on the form of such cues are also likely to be very important in understanding a given signaling event. In the next section, vocal production mechanisms in primates are examined in detail.

ACOUSTIC CUES TO INDIVIDUAL IDENTITY IN PRIMATE VOCALIZATIONS

Calls produced by primates are often assumed to be individually distinctive, and there is significant empirical evidence to support this conclusion (e.g., Snowdon, 1986; Rendall et al., 1996). However, the data are also limited in scope, and we argue in this section that not all call-types provide equivalent opportunities for discriminable individual variation to occur. For the species targeted in this chapter, we propose that a basic distinction exists between cues related to vocal-tract resonances and other characteristics of calls. While a highly detailed discussion of the acoustical and sound-production principles underlying this claim is beyond the scope of the present work, more detailed reviews of these topics are readily available (e.g., Fitch & Hauser, 1995; Owren & Linker, 1995; Schön Ybarra, 1995).

Vocal Production Processes

Call production in primates (and in mammals generally) is shaped by physical characteristics of the vocal tract. As illustrated by the schematic, midsagittal drawing of a rhesus monkey (*M. mulatta*) head shown in Figure 1, two critical components can be distinguished. The *source energy* of a typical call is derived from vibrating the *vocal folds*, which are enclosed in the larynx. This energy excites the cavities located above the larynx, which comprise the *supralaryngeal vocal tract*. These cavities shape the spectral characteristics of the source energy in accordance with its input-output relation, or *transfer function*—the effect we have referred to as vocal-tract filtering. While recent work has demonstrated both similarities (Fitch & Hauser, 1995; Owren & Linker, 1995) and important differences (Schön Ybarra, 1995) between vocal production in humans and primates, this two-component, *source-filter* perspective is applicable in both cases.

The process involved in producing a complex tonal sound is illustrated in Figure 2. This figure is based on human speech production parameters, but is applicable to primate calling as well. In the particular case shown, the vocal folds are in regular, or *periodic* vibratory motion, and by opening and closing allow puffs of air to emanate from the *glottis* (the opening between the vocal folds). The glottal airflow waveform illustrated in Figure 2(a) reflects an opening and closing rate of 100 times per sec, which corresponds to a *fundamental frequency* (F_0 or *first harmonic*) of 100 Hz in the resulting sound. Vibratory movement in the vocal folds produces signals whose frequency spectra include energy not only at the F_0, but also at higher harmonics—spectral components occurring at integer multiples of the base rate of vibration. Figure 2(b) shows an idealized source-energy for a vowel sound. The harmonics are high in amplitude, but energy declines exponentially with increasing frequency.

The cavities and tissues of the supralaryngeal vocal tract strongly influence the glottal waveform through their resonance (amplifying) and antiresonance (damping) properties. Vocal-tract resonances, or *formants*, reinforce energy in specific frequency ranges. Figure 2(c) shows the transfer function of a relaxed human vocal tract, a "neutral" configuration whose filtering characteristics

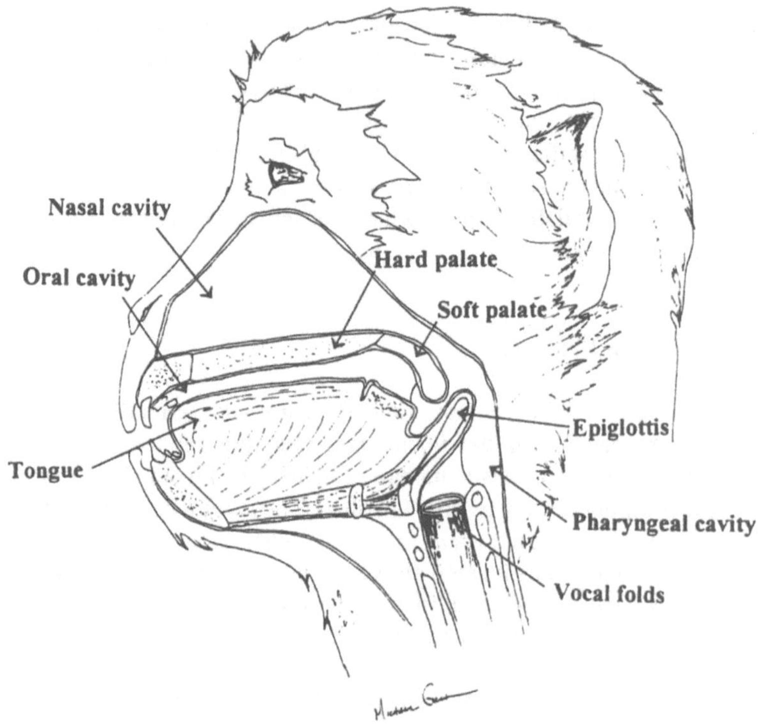

Fig. 1. A schematic, midsagittal view of a rhesus monkey vocal tract. Anatomical structures important to the basic sound-production process are labeled. Drawing by Michael Graham. After Rubin and Vatikiotis-Bateson (in press), used with permission.

closely resemble those evident in chacma baboon (*Papio cynocephalus ursinus*) *grunt* calls (Owren, Seyfarth, & Cheney, 1997). The characteristic frequency spectra of both vowels and these baboon grunts are marked by 4 to 5 prominent spectral peaks occurring below 5 kHz, each of which results from a vocal-tract resonance. The overall spectral pattern formed by these peaks plays a major role in determining the auditory quality of the sound. As illustrated in Figure 2(d), then, vocalization features reflect characteristics of both the source energy and vocal-tract filtering involved in the sound-production process. Important perceptual attributes like pitch, tonality, and timbre all result from the interaction of these source and filter components. Formants can also affect *noisy* sounds, whose underlying source waveforms lack periodic (cyclical) energy patterning. Such *aperiodic* source energies show a concomitant lack of orderly, patterned energy distribution in the frequency domain. Spectral patterns may nonetheless be imposed by the supralaryngeal vocal tract, as is illustrated by formant effects

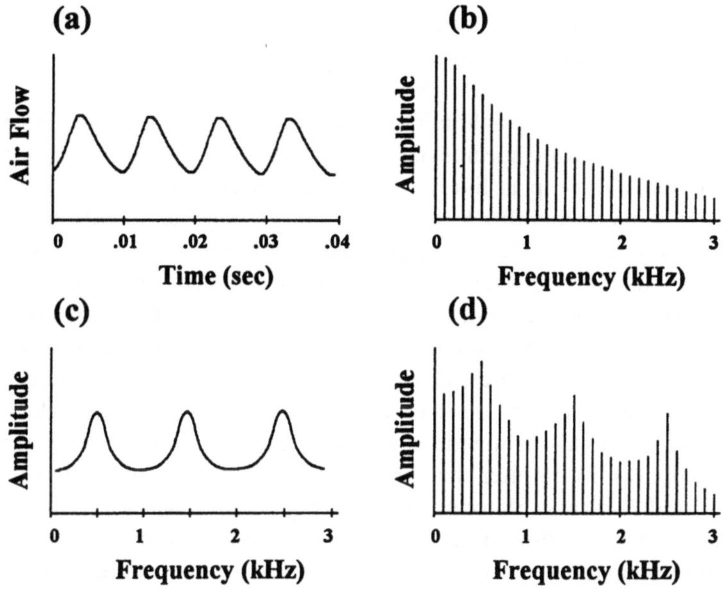

Fig. 2. An illustration of the vocal production process. (a) Periodic opening and closing of the vocal folds produces periodically varying glottal airflow. (b) The 100-Hz rate of vocal fold vibration in the glottal signal exhibits a harmonically structured spectrum with maximal energy at this fundamental frequency and its integer multiples. Energy at successive harmonics decreases exponentially. (c) A transfer function showing the resonances of a 17.5-cm vocal tract in neutral position, modeled as a straight tube closed at one end. (d) The resulting rich, tonal sound reflects both the harmonically related energy of the source and resonance characteristics of the subsequent filter. From Owren and Linker (1995), used with permission.

occurring in the spectra of whispered vowels. Humans produce these sounds using low-amplitude, broadband glottal turbulence as an energy source rather than periodic vibration.

Comparing Vocal Production in Humans and Primates

In general, only a few kinds of source energy are employed in speech production, and changes in vocal-fold vibration modes are not typically used as linguistic cues. Instead, important phonetic contrasts are produced by articulation—movements of the tongue, mandible, and lips that influence the shapes and concomitant resonance properties of the pharynx and oral cavity. These maneuvers produce differentiated filtering effects in the supralaryngeal vocal tract that shape the two source-energy types into a variety of perceptually distinctive sounds. In contrast, vocal production in primates typically reveals a wider range of source energies and less flexible modification of the filter component. These differences are evidently traceable to disparities in both vocal tract anatomy and neural control of sound-producing structures.

Laryngeal studies by Schön Ybarra (1995) and others, for example, have shown that many primate species have a rigid "lip" on the medial extremity of each vocal fold. This lip is likely to allow a number of vocal-fold vibration patterns, evidently accounting for the occurrence of various sounds that humans cannot easily emulate. These signals include high-amplitude, noise-based sounds, virtually pure-tone sinusoids, frequency sweeps that cover multiple octaves in a fraction of a second, and source energies that combine independently produced periodic and aperiodic components. Compared to humans, however, primates have thinner tongues, larynges positioned higher in the neck, and a relative lack of flexible soft tissues in the supralaryngeal vocal tract. Thus, while many species use tonal vocalizations with rich harmonic spectra, they appear to have less opportunity to alter the formant-related spectral patterning of those sounds by modifying articulator positioning (e.g., Lieberman, 1975). In addition, whereas speech production in humans requires significant involvement of neocortical brain structures as well as circuitry in brainstem, limbic, and midbrain areas, the neocortex appears to play a negligible role in primate vocal production (see reviews in Steklis & Raleigh, 1979, and Baer, Sasaki, & Harris, 1986). Several researchers have noted that articulatory maneuvers occurring in human vocal production can make use of neural connections that are absent in primates (e.g., Deacon, 1989). Thus, while some modification of vocal-tract resonances clearly does occur among various monkeys and apes (e.g., Hauser, 1996; Owren et al., 1997), one can generally conclude that these animals produce differentiated sounds primarily through changing laryngeal source-energy characteristics.

Implications for Individually Distinctive Acoustic Cues

Based on call acoustics and production processes, several different aspects of primate vocalizations can potentially provide cues to caller identity. Nonetheless, it can also be argued that some cues are inherently more revealing than others, and that the features of greatest importance for a particular call-type depend on the sound-production processes involved. The source-filter model can therefore be used as a general framework for understanding the origins of individual variation in the acoustics of vocalizations.

Cues Related to Source Energy

Individually distinctive acoustic cues might derive from several aspects of a call's source energy. For sounds based on periodic vocal-fold vibration, for instance, mean F_0 value might be used as a caller-specific attribute. However, we suggest that this feature is unlikely to reliably distinguish various members of a social group, especially where animals of all ages and both sexes are present and a variety of call-types are used. In humans, the length and mass of the vocal folds are known to be the primary determinants of vibration rate (Titze, 1994) and the same is likely to be true of primates. If so, however, any two callers with comparably sized vocal folds will show similar typical F_0 values. In general, such coincidences can be expected to occur regularly. Furthermore, various call-types in a repertoire often exhibit different average F_0 values or show significant frequency modulation, making vocal-fold vibration rate by itself unlikely to be a consistent source of unambiguous cues to caller identity. It has been found that F_0 measurements can play a role when vocalizations are sorted by individual caller, for example, in statistical testing of rhesus macaque coos (Rendall, Owren, & Rodman, in press) and baboon grunts (Owren et al., 1997). Predictably, however, the value of such measures decline as sample sizes are increased or other variables are entered in the classification equations used.

The detailed spectral characteristics of the source component of call production might also provide reliable cues to caller identity, as the shapes and tissue properties of an individual's vocal folds are reflected in its characteristic glottal waveform. In humans, physical characteristics of the vocal folds are known to contribute to voice quality, as is shown by aging and disease-related effects on these tissues (e.g., Baken, 1987; Titze, 1994). However, because normative speech essentially uses only two vocal-fold vibration modes, it is an excellent medium in which to reveal individual variation in vocal-fold characteristics. In primates, fine-grained, individually distinctive glottal-waveform differences are arguably less likely to be apparent due to the greater range of source energies used. Such cues are probably the most important in tonal sounds, where spectral components are arranged in a predictable, harmonically related series and idiosyncratic variation in this patterning is requisitely evident. As an example, we have observed that while very old female macaques often produce distinctive-sounding coos, the aging effects that are apparent in these sounds are much less evident in noisier vocalizations.

Noisy primate calls are in general much less likely to show individually distinctive spectral patterns due to the inherent randomness of energy distribution in these sounds. Such calls often consist of extended or pulsed energy bursts with broadband spectral energy (i.e., pure noise) or combinations of periodic and aperiodic components. Screams shown in Figure 3 illustrate some of the properties of noisy vocalizations produced by macaques and baboons. In light of both Schön Ybarra's (1995) anatomical observations and spectrographic evidence showing that noisiness in primate calls can grade in near-continuous fashion, we assume that the source energy in such calls is produced by aperiodic vocal-fold vibration. While larger-scale spectral patterning can occur in screams and other noisy calls, the energy of any particular frequency component varies randomly (or quasi-randomly) in these sorts of waveforms. Such fluctuations must tend to mask the more fine-grained, individually distinctive spectral differences that might be occurring.

Cues Related to the Filter

While the source-energy component of primate calling is therefore not likely to provide individually distinctive voice qualities across call-types, the filter component may produce identity cues that are both consistent and broadly applicable. Subtle differences in vocal-tract cavity shapes and sizes can potentially produce individually distinctive variation in spectral patterning in a variety of call-types, so long as the sounds reveal the filtering effects involved. Examples of such formant patterns are shown for tonal and noisy calls in Figures 4 and 5, respectively.

In each of the tonal call-types, the F_0 is relatively low and the sound has a dense, harmonically structured spectrum that readily reveals the amplification and attenuation effects of the vocal-tract filter. In discriminant-function analyses reported by Rendall et al. (in press) for rhesus coos and by Owren et al. (1997) for baboon grunts, spectral-peak characteristics related to vocal-tract filtering were found to be more important than other acoustic measures when sounds were sorted by individual caller. For the rhesus monkeys, playback experiments showed that listening animals could readily differentiate between coos produced by relatives and other, unrelated group members (Rendall et al., 1996). However, the effect of a vocal-tract resonance only becomes apparent in a call if energy occurs in the frequency ranges affected by that formant. Therefore, as F_0 values rise and the harmonics of a tonal sound become more widely spaced, fewer details of the vocal-tract filtering are represented in the call. Individually distinctive patterning therefore disappears if the F_0 is too high. This outcome is illustrated in comparing the coos and grunts to the tonal, but high-pitched shrieks in Figure 6.

Because the vocal-tract filter is inherently better "displayed" in a vocalization if the underlying source waveform has a broadband spectrum, noisy calls

Fig. 3. Noisy screams produced by (a) an adult female baboon, (b) a juvenile rhesus monkey, and (c) an infant baboon. The figure shows waveforms (top), wideband (300-Hz) spectrograms (second from top), and corresponding spectral "slices" computed over a 512-point segment centered on each sound's amplitude peak (bottom three panels). Each of the latter shows both fine-grained and smoothed versions of the segment's frequency-energy structure, based on Fourier transformation (jagged envelope) and linear predictive coding (smooth envelope), respectively. (Digital sampling rates were either 20 or 22 kHz for the sounds shown in this chapter, except where otherwise noted.) In these screams, energy at any particular frequency varies semirandomly from instant to instant. However, a simple overall spectral pattern is apparent in each case—reflecting regularities in the source waveform of the call, effects of resonances, or both.

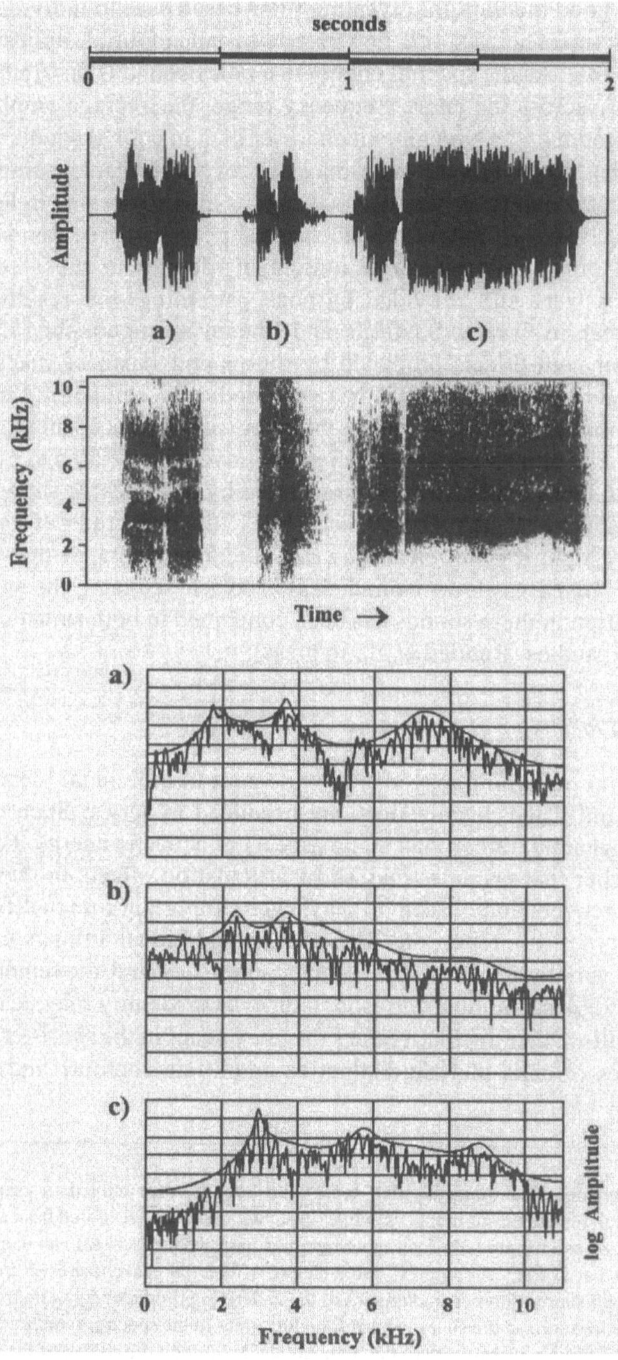
seconds
0
1
2
Amplitude
a)
b)
c)
Frequency (kHz)
10
8
6
4
2
0
Time →
a)
b)
c)
log Amplitude
0
2
4
6
8
10
Frequency (kHz)

can also be a good medium for revealing filter-based cues to individual identity. Complications arise here as well, however, with call amplitude appearing to play an important role. As the spectral energy in a noisy sound is distributed more or less uniformly across the entire frequency range, the average amplitude of an individual component is typically much lower than in a tonal sound. As a result, noisy calls that are relatively quiet may not reveal formant characteristics as clearly as louder sounds. While the noisy rhesus monkey *grunts* in Figure 5, for instance, clearly show formant-related spectral peak patterns, Rendall et al. (in press) found only modest cues to individual identity in these sounds. The strongest cues were still provided by peak patterning, but results were less compelling than in Owren, Seyfarth, and Cheney's (unpublished data) earlier tests of louder pant-threats produced by rhesus and Japanese macaques (also shown in Figure 5). Fitch (1997) has provided very compelling evidence of prominent formant effects in rhesus monkey pant-threats, showing that the frequencies of resulting peaks are predictably (inversely) related to overall vocal tract length in these animals. As demonstrated by the screams shown in Figure 3, however, spectral peaks tend to disappear if the amplitude of noisy sounds becomes very high. It is not clear why this outcome occurs, or that it is caused by amplitude increases alone. Nonetheless, as discussed later, the apparent lack of differentiation in these sounds has been confirmed in both statistical analyses and playback studies (Rendall et al., in press).

Cues Related to Temporal Patterning

Temporal patterning may also provide cues to individual identity, both in single calls and when vocalizations are produced in series. Such cues would result from dynamic, rather than static aspects of acoustic energy. For instance, we noted earlier that primates may show articulation effects in their calls, but that such maneuvers do not seem to play a very important role in differentiating various call-types in a repertoire. However, as individual animals undoubtedly show minor variation in the physical characteristics and movements of their tongues, mandibles, and lips, corresponding cues to identity may be available in articulated calls. Other distinctive cues that may occur in the short-term temporal characteristics of calls include distinctive amplitude contours and F_0 patterns

Fig. 4. Characteristics of harmonically rich, tonal calls. Sounds in (a) and (b) are coos produced by two different adult female rhesus monkeys, while (c) shows a grunt call recorded from an adult female baboon (organized as in Figure 3). Each spectrogram is narrowband (45-Hz) and the grunt call was digitally sampled at 11 kHz. A complex, stable pattern of spectral peaks is present in these sounds, reflecting the joint source-filter characteristics of the underlying production system. As is evident in comparing peak locations in the slices to corresponding areas in the spectrograms, each call displays a fine-grained frequency-energy pattern that remains stable through the course of the sound.

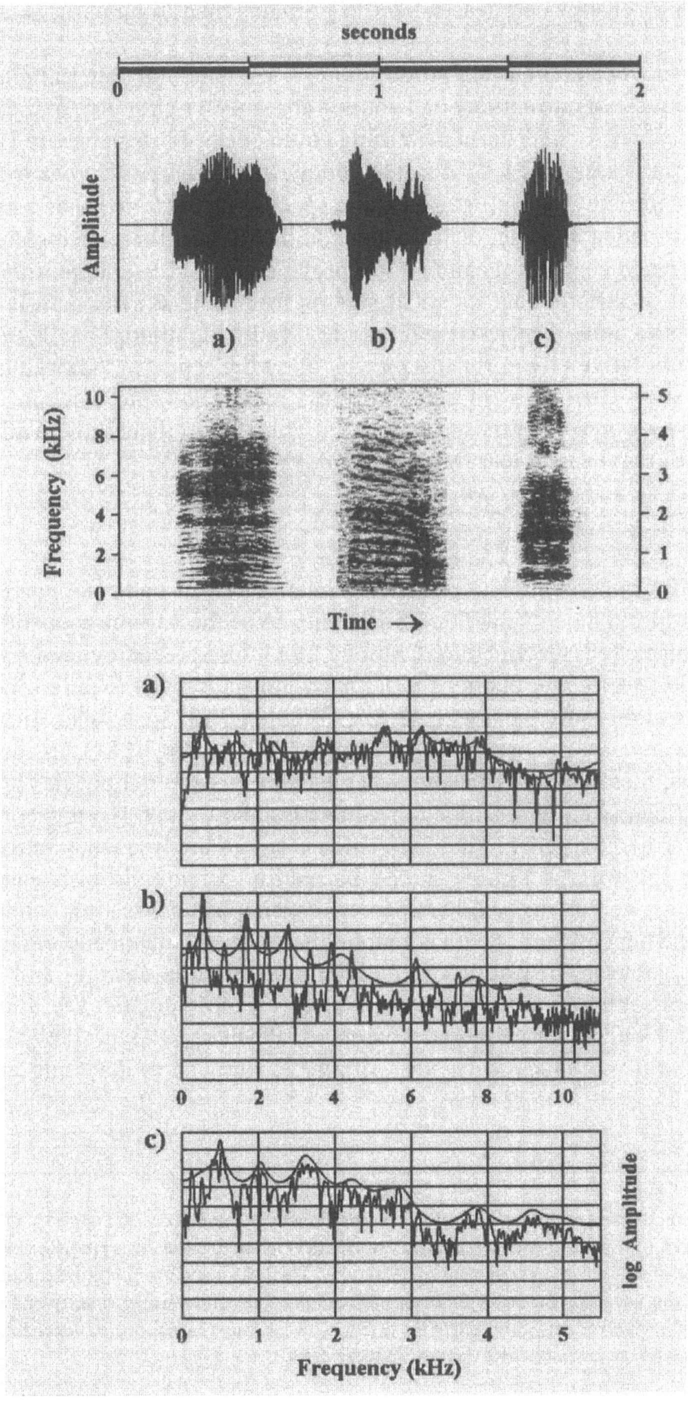
seconds
0
1
2
Amplitude
a)
b)
c)
Frequency (kHz)
10
8
6
4
2
0
5
4
3
2
1
0
Time →
a)
b)
0
2
4
6
8
10
c)
log Amplitude
0
1
2
3
4
5
Frequency (kHz)

(although F_0 changes are less salient to primates than to humans, as discussed below).

When vocalizations are produced in series, temporal patterning occurring over the course of one or more call bouts is arguably a rich source of individually distinctive cues. Longer-term patterning could occur, for instance, in F_0 characteristics, the arrangement of discrete energy bursts, amplitude changes over a long but continuous waveform, dynamic spectral characteristics of a harmonic-series or broadband noise, or from any combination of these elements. In each case, patterned changes in call features provide much greater opportunity for an individual to differentiate its vocal signals from others' calls than is possible using a static, unidimensional call feature. Producing multiple calls might also have a cumulative effect, for instance by allowing repeated "sampling" of one or more short-term cues. In either of these cases, however, listening animals would be slower to identify the calling individual than if cues are available in a single call, requiring at least a few seconds or more.

Evidence from Perceptual Studies

Auditory perception in primates is relatively well-understood, allowing at least an approximate evaluation of the salience of the various acoustic features we have discussed (see Stebbins & Moody, 1994, for a recent review). As a group, Old-World species are roughly similar in overall sensitivity to pure-tone stimuli across the audible frequency range (e.g., Stebbins, 1973; see also Hienz, Turkkan, & Harris, 1982; Owren, Hopp, Sinnott, & Petersen, 1988). In comparison to humans, these animals are somewhat less sensitive to frequencies below 500 Hz, significantly more sensitive to frequencies above 8000 Hz, and comparable in the intermediate range. Thus, acoustic energy in the frequency ranges represented in the various figures should be readily perceptible to macaques and baboons, as well as many other species. Psychophysical testing conducted by Sinnott and her colleagues with pure tones has shown that humans and macaques are comparable in their ability to discriminate duration changes and intensity increments (although intensity decrements were problematic; see Sinnott, Petersen, & Hopp, 1985; Sinnott, Owren, & Petersen, 1987a, 1987b). Moody (1994) found macaques to be very similar to humans in detecting amplitude

Fig. 5. Characteristics of other noisy calls in rhesus monkeys (organized as in Figure 3). These sounds are (a) a grunt, (c) a pant-threat, and (b) an intermediate version. Each call was produced by a different adult female and is shown in wideband spectrograms. The broadband energy of these sounds reveals features of the vocal-tract transfer function of the caller, but with varying fidelity and detail. The spectral peaks involved arguably reflect the effect of vocal-tract resonances, as vocal-fold vibration is unlikely to be the primary energy source in these calls.

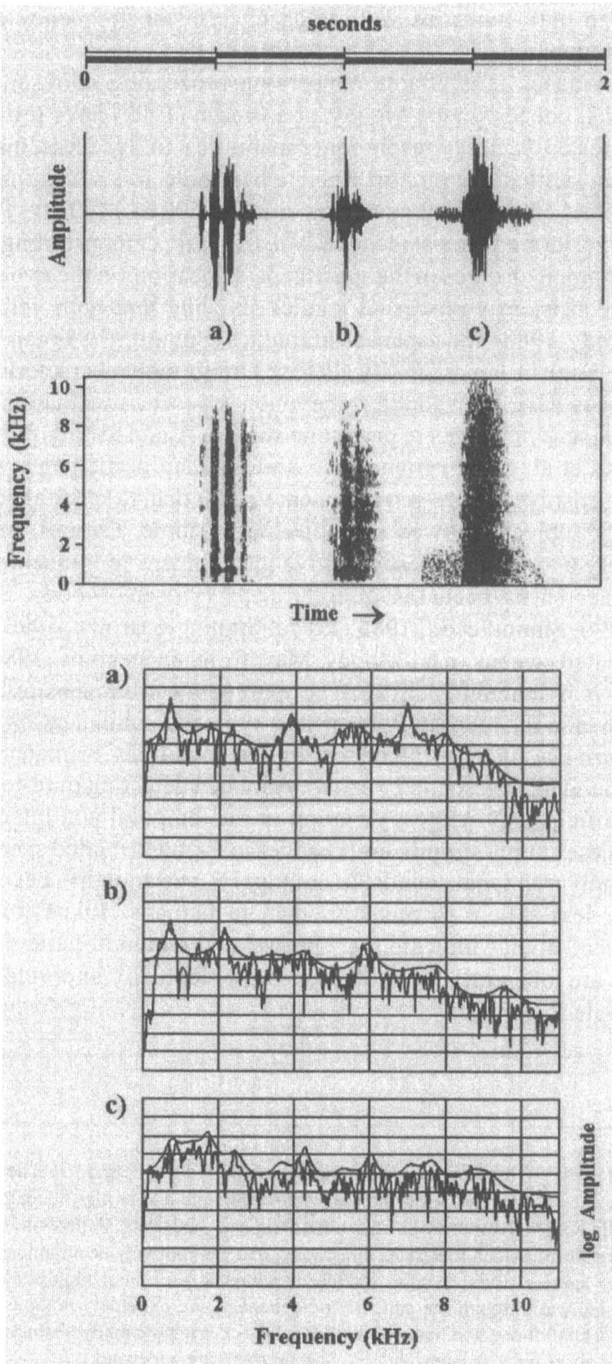
seconds
0
1
2
Amplitude
a)
b)
c)
10
8
6
4
2
0
Frequency (kHz)
Time →
a)
b)
c)
log Amplitude
0
2
4
6
8
10
Frequency (kHz)

modulation in pure tones, as well as changes in the frequency of repetitive amplitude modulations.

Less evidence is available concerning processing of resonance-related spectral peaks, but Sommers, Moody, and Prosen (1992) have provided telling data. They tested spectral resolution capabilities in Japanese macaques and humans using synthetic stimuli in which a harmonic series was shaped so as to mimic the effect of a formant occurring either at 500 or 1400 Hz. The monkeys and humans performed almost identically in this task, demonstrating comparable sensitivity to small changes in the spectral peak locations in the two sounds. This outcome is consistent with earlier studies showing that both yellow baboons (Hienz & Brady, 1988) and Japanese macaques (Sinnott, 1989) can be trained to discriminate English-language vowels based on formant characteristics. In contrast, macaques have been found to be much less sensitive than humans when detecting frequency changes in pure-tone stimuli. Again testing at 500 and 1400 Hz, Sommers et al. (1992) found that while human participants were significantly more sensitive to pure-tone frequency shifts than to spectral peak changes, their monkey subjects showed the opposite outcome. Overall, macaques are approximately 6 to 10 times less sensitive than humans to frequency changes in tones, whether shifts occur as discrete steps (Prosen, Moody, Sommers, & Stebbins, 1990; Sinnott et al., 1985; 1987a; Sinnott & Brown, 1993) or continuously modulated sweeps (e.g., Moody, May, Cole, & Stebbins, 1986).

While it is inherently difficult to extrapolate from sensitivity measurements obtained in the laboratory to species-typical communication processes, a notable pattern has emerged. Under controlled conditions, primates and humans have been found to be essentially equivalent when detecting pure-tone energy in intermediate frequency ranges, variation in the temporal and intensity characteristics of these simple stimuli, and changes in formant-related spectral features of harmonically rich tonal sounds. In contrast, F_0 variation has been found to be significantly less salient to monkeys than to humans. Taken together, these psychophysical studies indicate that spectral and temporal patterns in species-typical calls are inherently more likely to be perceptually important to primates than are simple F_0 changes.

Fig. 6. Characteristics of high-pitched, tonal shrieks (organized as in Figure 3). The first, longer call (a) was produced by a juvenile rhesus monkey, exhibiting both a very high fundamental frequency and a requisitely simple, harmonically structured frequency spectrum. A juvenile baboon gave the second, shorter call (b), which illustrates the grading that can routinely occur among sound types in these and other species. Here, a noise overlay is added to the initial high-pitched tonal output approximately halfway through the call. Both components are extremely noxious-sounding when produced at high amplitudes, and neither provides salient cues to potentially distinctive aspects of the calling animal's vocal-tract transfer function. Spectrograms are wideband.

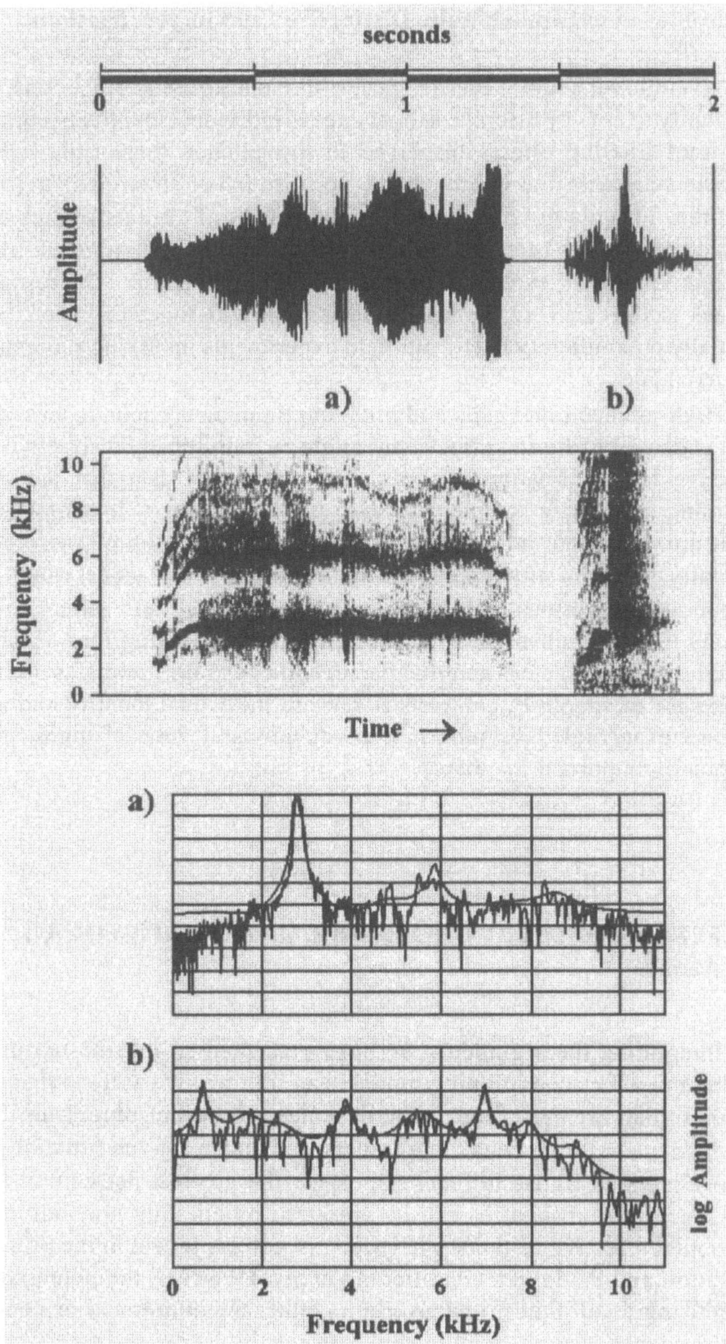
seconds
0
1
2
Amplitude
a)
b)
Frequency (kHz)
10
8
6
4
2
0
Time →
a)
b)
log Amplitude
0
2
4
6
8
10
Frequency (kHz)

Conclusions about Individually Distinctive Cues in Vocalizations

Overall, we suggest that in species like macaques and baboons, the best opportunity to convey discrete, salient cues to individual identity occurs through vocal-tract filtering effects displayed in low-pitched tonal calls with dense harmonic structures and in broadband, noisy sounds of intermediate (or lower) amplitude. This claim is supported by the results of various studies in which acoustic analysis has been used to sort vocalizations by individual caller, and by data from laboratory studies of primate auditory perception. The results of the playback studies cited so far are also consistent with this conclusion, but some potentially contradictory data from field experiments involving screams will be considered later.

High-pitched tonal calls and high-amplitude noisy sounds do not readily reveal vocal-tract filtering effects, but might be individually distinctive through distinctive temporal-patterning cues. The perceptual data that are available concerning sensitivity to intensity and duration changes in simple acoustic stimuli are consistent with this suggestion, although tests of more complex, temporally patterned stimulus sequences are needed. F0 changes, which are very common among various call-types, appear to be much less salient to primate receivers than to humans hearing the same sounds. Taking into account both production-related factors and this perceptual constraint, primates are unlikely to make use of simple F_0 contrasts as cues to individual identity and arguably must use exaggerated F_0 jumps or modulations if such changes are to be perceptually important to others.

AN AFFECT-CONDITIONING MODEL OF PRIMATE VOCAL SIGNALING

Integrating the arguments we have presented so far, the reasoning underlying the affect-conditioning model is as follows. We assume that communication signals are most fundamentally a means by which organisms influence the immediate or future behavior of others, and that in a given situation, senders use vocalizations whose influences provide the greatest net benefit to themselves. Important influences can be produced by eliciting unconditioned and conditioned affective responses in receivers, effects to which the principles of habituation and Pavlovian conditioning apply. However, the potential benefit of producing a call that primarily elicits either unconditioned or conditioned responses depends on the relationship between the sender and receiver. A caller that typically has little opportunity to control the outcome of its interactions with the receiver derives the greatest benefit from producing vocalizations that elicit unconditioned responses. In contrast, a caller that is able to control the outcomes of its interactions with the receiver is likely to benefit most from first inducing conditioned affective responses to its calls by pairing these sounds with unconditioned stimuli delivered directly to the receiver, then later taking advantage of the conditioning that has accrued by using the same calls as conditioned stimuli. Vocalizations are well-suited to either kind of function, as they are discrete, salient, controlled by the sender, and difficult for the receiver to avoid.

Calls can produce unconditioned affective effects both through their relatively generic effects on the auditory system or by tapping into more specific, evolved sensitivities in the species in question. Characteristics like high overall amplitude, noxious spectral qualities, abrupt transitions, high F_0, and pronounced F_0 modulation exemplify features that are likely to elicit unconditioned responses. We refer to these calls collectively as squeaks, shrieks, and screams. Such sounds are likely to be used repetitively or in bouts in order to maximize the unconditioned responses being elicited. As receivers inevitably habituate to repeated stimuli, the sender also varies the acoustic features of the sound stream. Conditioned affective responses occurring to vocalizations are specifically mediated by acoustic cues to the caller's identity. However, the primate vocal-production and auditory-perception systems shape the forms that such calls can take, depending on the particular call-type involved. Individually distinctive vocal-tract filtering effects can provide immediate and clearly perceptible cues to caller identity, both in low-pitched tonal vocalizations with dense harmonic structures and intermediate-amplitude, noisy sounds with broadband energy. These are the calls we have referred to as sonants and gruffs. Repetitive call production is not inherently as useful for such sounds as for calls whose function is to elicit unconditioned responses. High-pitched tonal calls and high-amplitude noisy sounds do not readily reveal vocal-tract filtering effects, but may provide identity cues through distinctive temporal patterning.

To illustrate the implications of these proposals, we now describe four scenarios in which two animals interact and one of them calls. In each scenario, the interests of the sender are evaluated based both on its relationship to the receiver and on the general function proposed for calling in that circumstance. In a given situation, the caller is either subordinate or dominant to the receiver, and the interaction is either agonistic or affiliative in nature. We then return to the general issues concerning primate vocal signaling that were raised earlier. The various topics are revisited in reverse order, and a variety of implications of the affect-conditioning model are considered. A number of predictions are also presented, both to flesh out the scope of the approach we are proposing and to make the model as specific and testable as possible.

Four Scenarios Illustrating the Use of Unconditioned and Conditioned Effects

Subordinate Caller in an Agonistic Interaction

The first scenario is the simplest. In this situation, an animal vocalizes during an agonistic encounter with a more dominant group member and calls function to discourage impending or ongoing aggression. Being subordinate, the caller's most effective tactic is to produce noxious-sounding vocalizations in a loud, repetitive fashion, eliciting aversive unconditioned responses in the dominant animal. High-amplitude, grating calls like shrieks and screams are best-suited to this purpose, and the deterrent effect of the subordinate's calling is proportional to the magnitude and noxiousness of the acoustic energy involved. Variation in the acoustic features of the vocalization series also occurs, in order to alleviate habituation in the unconditioned responses experienced by the receiver. The subordinate animal has relatively little opportunity to effect or take advantage of conditioning in the dominant individual, as it cannot exert much control in this kind of encounter. Therefore, the caller's best available vocal option is to use sheer magnitude and raw features of acoustic signals for inducing aversive unconditioned responses in opponents, even though this option is energetically expensive.

Dominant Caller in an Agonistic Interaction

In the second scenario, an animal vocalizes during an agonistic encounter with a subordinate individual. The function of calling is to elicit an affective response, such as fear, that makes the receiver more likely to depart or show submissive behavior. Here, the caller could arguably take advantage of both unconditioned and conditioned effects. For instance, high-amplitude, noxious sounds could be used to induce fear in the receiver as an unconditioned response. However, conditioning can be effected by producing "threat" vocalizations that include salient, discrete cues to individual identity and then engaging in violent behavior that traumatizes the subordinate. After a small number of such encounters, the dominant individual can effectively influence the behavior of the other animal by using its energetically inexpensive threat calls to elicit conditioned affective responses. The acoustic features of the calls reflect the need to embed the individually distinctive cues in a signal whose unconditioned effects add to, or are at least compatible with, the desired affective response. Such calls should therefore be from the sonant and gruff class.

Dominant Caller in an Affiliative Interaction

The third scenario is one in which an animal interacts affiliatively with a subordinate. Here, for example, calls might be produced as the dominant animal approaches and sits with the other, possibly grooming it or being groomed itself. As the approach or proximity of a dominant group member evidently induces anxiety in a subordinate individual, the calls used should encourage the receiver to participate in the interaction by decreasing its fearfulness or inducing positive affective responses. The calling animal has at least some control over the outcome of the interaction, and can therefore make use of conditioned effects. The dominant animal should again promote conditioning by using calls that embed salient, discrete cues to individual identity in energy that has compatible unconditioned effects and (usually) following those vocalizations with affiliative behaviors that lead to positive affective states in the receiver. These calls will thereafter be from the sonant and gruff class, but should be clearly differentiated from the calls used to elicit conditioned fear or other negative affective responses.

Subordinate Caller in an Affiliative Interaction

In the fourth and final scenario, an animal calls to a dominant group member in an affiliative context, for instance when approaching to groom that individual. Although it is subordinate in this situation, the caller has some inherent control and leverage—it can provide stimulation that evidently constitutes a strongly positive event for the other animal. The subordinate individual's interests are therefore analogous to those of a dominant animal that is acting affiliatively. Specifically, the subordinate benefits by associating individually distinctive cues in its calls with the pleasant outcome experienced by the other animal, and thereby promoting positive affective conditioning to those features. The caller gains some ability to use vocalizations to elicit a positive conditioned response in the dominant individual on other occasions, for instance making it more likely to participate in an affiliative interaction or less likely to be aggressive. These vocalizations will be from the sonant and gruff class, and should have unconditioned effects that are compatible with the positive conditioned affect being elicited.

Cues to Individual Identity

A variety of predictions concerning cues to individual identity can be derived from the affect-conditioning model, beginning simply by restating its assumptions as testable hypotheses. For instance, individually distinctive cues in calls labeled as sonants and gruffs are predicted to derive primarily from differentiated spectral patterning in these sounds. Squeaks, shrieks, and screams, in contrast, are proposed not to be readily identifiable based on the static features of individual calls. Dynamic features of calls may be identifiable, but we suspect that cues are more likely to emerge from temporal patterning occurring over a series of calls. As resonance frequencies and spacing are inversely related to vocal tract length (e.g., Fitch, 1997), calls of younger, smaller animals can be expected to show both higher and more widely spaced formants than the calls of older, larger individuals. As vocalizations of young animals also show higher F_0s, sonant and gruff calls in these individuals are predicted to be significantly less individually distinctive than comparable sounds produced by more mature conspecifics. In fact, very young primates may be unable to produce calls with salient spectral patterning cues, and hence might not be expected to use true sonants or gruffs. Evidence of all these predicted effects should be found in each important testing domain—acoustic analysis of calls, perceptual testing conducted in laboratory settings, and playback trials examining functional responses.

Cues to Individual Identity in Screams

As some of the evidence that is available from playback studies with screams may be inconsistent with our predictions, several relevant experiments will be reviewed. In primates, the best evidence of individual discrimination and recognition based on vocal cues has arguably been provided by Cheney and Seyfarth (1980). In their studies of wild vervet monkeys, these investigators used a hidden speaker to play screams of juveniles to their mothers and other adult females. A bout of calls was presented on each trial, averaging approximately 7 sec in length. When several females sitting together heard the screams, the caller's mother was the animal most likely to look in the direction of the speaker. In addition, the other females were found to be more likely to look at this mother than to look toward the source of the calls. Gouzoules, Gouzoules, and Marler (1986) have also tested the distinctiveness of screams, but did so by playing back single calls to free-ranging rhesus monkeys. An earlier study had distinguished 5 acoustic variants of screams, and two of these subtypes were tested (Gouzoules, Gouzoules, & Marler, 1984). For *noisy* screams, an example of which is shown in Figure 3, Gouzoules et al. (1986) reported that adult females were quicker to look, and looked longer when hearing calls of related rather than unrelated juveniles. No differences were found for arched screams.

Based on the acoustic features of these rhesus screams, we would not expect that the identity of the caller would be clearly revealed by a single exemplar of either type. Little spectral energy patterning is evident in noisy screams, for instance, while arched screams are piercing calls with very high F_0s. Consistent with this point of view, Rendall et al. (in press) found no evidence of kin-based discrimination when they played back noisy screams produced by adult females to other adults in the same rhesus groups that had been tested by Gouzoules et al. However, the response latencies and durations reported for the experimental groups in the two studies were actually almost identical. In both cases, responses were clearly slower and shorter than was the case when Rendall et al. (1996) tested adult females in these groups using coo calls. The difference between typical responses to the coos of female kin and nonkin was quite dramatic and also quite unlike response patterns reported in the two studies testing rhesus screams (see Rendall et al., in press, and Rendall, 1996, for further discussion).

The discrepancies among the various outcomes of these tests of vervets and macaques may have resulted from differences in the durations of the playback stimuli used. Gouzoules et al. (1984) presented individual screams because subjects that heard scream bouts in pilot trials approached or even charged the speaker. In other words, the most compelling evidence of individually distinctive acoustic cueing occurred when scream bouts were played—first in Cheney and Seyfarth's tests of vervets and later in Gouzoules et al.'s preliminary tests. When Rendall and his colleagues presented either single coos or single screams (by design, the methodologies were as similar as possible in the two cases), evidence of kin-based discrimination emerged only for coo calls. Results were different when screams were tested, showing no evidence of discrimination but matching the latencies and durations reported for a comparable condition in Gouzoules et al.'s (1986) experiment.

We interpret these outcomes as showing that cues to individual identity are present in individual coos, but are more likely to emerge over the course of a bout of screams than in the features of single calls. This proposed difference is supported by acoustic measurements and statistical classification results described by Rendall et al. (in press). In this work, coos were readily sorted by caller based both on acoustic measures related to vocal-tract filtering effects and other features. However, across calls, formant-related cues were primary, showing the least intraindividual variability and allowing the most accurate statistical classification. Screams, in contrast, were more homogeneous, did not show distinctive spectral patterning, and could not be successfully classified by caller.

Repetition and Acoustic Variability in Calling

Unconditioned and Conditioned Effects

A number of implications can be drawn from the proposal that the balance of power in a social relationship determines whether a caller should attempt to elicit unconditioned or conditioned affective responses. For instance, the argument that a subordinate individual under attack should elicit aversive unconditioned responses in the receiver is a general one, applicable to any situation in which an individual with little direct power over another individual attempts to influence that animal's behavior. By extrapolation, younger individuals and subordinate animals of all ages can be predicted to rely primarily on unconditioned effects of calling. A prototypical example is a young animal being weaned, who calls while unsuccessfully seeking caretaking behavior from its mother. Such sounds are predicted to act as unconditioned stimuli and should therefore be drawn from the squeak, shriek, and scream class, occur repetitively, and show significant acoustic variation. Anecdotally at least, when frustrated young animals call to their mothers or other caretakers, their vocalizations often occur in long bouts and are marked by features like exaggerated F_0 modulations and "melodramatic" plaintiveness. The calls are often also noxiously noisy or screechy, and are produced in seemingly endless streams that can be very annoying to human listeners.

Vocalizations that capitalize on conditioned effects should not be used in this fashion. For these calls, repetitive use might increase the immediate response, but simultaneously decrease the long-term value of the calls as predictive stimuli. Therefore, another implication of the model is that sonants and gruffs are less likely to be produced in long bouts than are squeaks, shrieks, and screams. However, animals producing sonant- and gruff-like calls may in fact be using them primarily to elicit unconditioned responses, or both kinds of responses. As noted, younger animals are expected to have less opportunity to produce calls with rich harmonic spectra and prominent filtering effects, and may therefore rely on repetition and variation. If so, features that are arguably related to unconditioned responses should be more prominent. If older, larger individuals use sonant-like calls in this way, otherwise individually distinctive spectral features should be made less prominent, for instance through increases in F_0.

Positive versus Negative Affect

While we consider positive affective responding to be an important component of the model, we also expect that primates do not have equivalent opportunities to elicit positive and negative affective responses in others. For instance, a dominant animal can induce negative affective states in another individual very quickly and effectively through directed actions like biting, hitting, kicking, scratching, chasing, or lunging. Multiple conditioning trials can occur in a given interaction, as the caller can repeatedly pair its vocalizations with these traumatic unconditioned stimuli. The situation is rather different for a dominant individual attempting to induce positive affective states. Such responses might be effected by grooming a subordinate, allowing it to groom, or simply tolerating its presence. However, each of these outcomes is rather diffuse and the dominant animal is inherently less able to control the subordinate's responses in such circumstances. A subordinate behaving in an affiliative fashion faces similar constraints. It might groom a dominant animal or allow this individual access to an infant, but cannot induce positive affective states in this individual in the direct, controlled manner afforded by using noxious sounds as negative unconditioned stimuli.

Using Calls as Conditioned Stimuli for Other Vocalizations

An additional tactic that a subordinate may use is to condition another animal by using its own calls as the unconditioned stimulus. For example, we have proposed that screams are aversive to receivers and can be used to shape their behavior. By predictably pairing calls from the sonant and gruff class with these biologically significant events, subordinate senders may be able to produce some conditioned effects in receivers. This tactic could be used both by adults facing attack from higher-ranking group members and by young animals seeking attention from caretakers. Bouts of vocalizations being used in this fashion are predicted to consist of some calls carrying prominent cues to individual identity interspersed with other vocalizations eliciting unconditioned effects. In other words, while both kinds of calls should appear, the vocalization sequences observed should reflect the differentiated roles being played by each type of call.

Discrete and Graded Signaling

Marler and others (reviewed by Green & Marler, 1979) have suggested that a distinction can be drawn between discrete and graded call-types, or discrete and graded vocal repertoires. A discrete call-type is one in which variation in acoustic features is clearly bounded, creating a category of sounds that is readily distinguished from other calls in the repertoire. A graded call-type, in contrast, is one whose acoustic features vary substantially, such that continuous gradation along one or more acoustic dimensions can bridge between this sound and other call categories. An entire vocal repertoire, then, might be labeled as being discrete or graded depending on whether the call-types involved are predominantly of one kind or the other. A general relationship has been noted between discreteness and gradedness and the degree to which the vocal signals in question are complemented by information that is simultaneously available in other modalities. Discrete repertoires (or call-types) are proposed to be of greatest value when vocalizations are used in the absence of other information, primarily when senders and receivers cannot see one another. Graded repertoires (or call-types) are more likely to involve sounds used at close quarters, where vocalizations are supplemented by other kinds of signals. Graded repertoires or vocalizations have also been suggested to encode more information than do their discrete counterparts, for instance through allowing acoustic variability to be partitioned into meaningful subcategories (e.g., Marler, 1975; Hauser, 1996). However, as the functional significance of acoustic variation also depends on perceptual processing in the species in question, it is often difficult to classify a call-type or a vocal repertoire unambiguously.

The affect-conditioning approach provides a different perspective. In this framework, acoustic variability is linked to minimizing habituation of elicited responding in receivers. Call-types in the squeak, shriek, and scream class are therefore predicted to always be subject to acoustic grading. Analogously, grading should routinely occur in sounds used by very young animals that cannot readily produce true sonant and gruff calls. In normative calling, different versions of similar calls may also be found, where one variant is used primarily to elicit unconditioned responses alone and another includes cues to individual identity and is used as a conditioned stimulus. Grading that occurs among call variants or separable call-types may reflect shifting or mixed tactics by a sender that is responding to the complexities of an unfolding interaction. From this viewpoint, no clear distinctions can be expected between discrete and graded vocal repertoires. Instead, all repertoires should show graded call-types, as the opportunity to elicit unconditioned responses using vocalizations can be expected in every primate species.

However, the general relationship between the discreteness or gradedness of signals and habitat characteristics can also be reinterpreted, by taking into account probable influences of vegetation density on the effectiveness of the vocalizations that function as unconditioned stimuli. When sender and receiver are separated by dense vegetation, sound energy traveling between them is subject to significant attenuation and degradation. While subordinates that are being physically tormented by dominant animals can still use noxious screams to good effect, calls that might have been used to elicit unconditioned responses in animals farther away in other sorts of interactions are arguably much less effective. In addition, senders are less likely to be able to approach or closely follow a given receiver at will in heavily vegetated habitats than in open environment. The effectiveness of conditioned responses, in contrast, should not be greatly affected by the relative density of vegetation. Conditioning trials can be conducted opportunistically whenever two animals are in close proximity, thereafter allowing senders to elicit responses based on learning rather than the sound energy per se. Salient cues to individual identity must be preserved in the calls used as conditioned stimuli, however, if this strategy is to be effective, thereby favoring greater stereotypy in these signals.

Overall, we suggest that the repertoires traditionally considered to be discrete are ones in which environmental constraints decrease the sender's ability to elicit unconditioned responses in receivers—except when the animals are close together. Under these circumstances, then, both habituation effects and the corresponding strategy of producing acoustically variable calls are significantly less important. Repertoires traditionally considered to be graded are ones in which callers can routinely elicit both conditioned and unconditioned effects. Calls used during physical attack or by young animals that are close to their mothers should not differ between repertoires otherwise considered to be either discrete or graded. The distinction should instead apply mainly to calls in the sonant and gruff class. In these cases, discrete call-types or repertoires are predicted to occur when acoustic variability cannot be used to elicit unconditioned responses in addition to conditioned effects based on individually distinctive spectral-patterning cues in sonants and gruffs. Graded call-types, in contrast, will occur when conditioned effects can be supplemented by unconditioned effects, or when sonant and gruff calls are modified so as to decrease cues to individual identity and increase their value as unconditioned stimuli. It is predicted that grading is much more likely to occur in acoustic dimensions related to unconditioned rather than conditioned effects.

Acoustic Features of Calls and Repertoire Structure

Functional Differentiation of Calls

As illustrated by a number of the preceding points, the functional distinction proposed between calls eliciting unconditioned and conditioned responses has implications for both the broad structure of a vocal repertoire and more detailed patterns of acoustic variation within each of the two broad call classes we have described. In contrast to other approaches, the affect-conditioning model does not necessarily segregate calls according to the social contexts in which they are produced, or on the basis of their acoustic features alone. Again, starting from the assumption that the fundamental function of vocalizations is to influence the behavior of others, we instead suggest that within the two call classes, natural selection pressures drive an ongoing process of acoustic differentiation. In other words, from relatively simple beginning stages, vocalizations of both general types are likely to diversify through continual emergence of acoustic variants that increase the ability of callers to elicit affective responses. Within the general constraints imposed by the differentiated functions of the two call classes, the process of differentiation can be expected to differ among species, in accordance with the much more specific constraints imposed by species-typical social organization, ecology, and habitat.

Motivation-Structural Rules

The affect-conditioning model also provides an alternative to Morton's (1977, 1982) proposed "motivation-structural" rules for linking internal motivational states of senders to the acoustic characteristics of their signals. Most importantly, Morton suggested that animals use harsh, low-frequency vocalizations with concomitant broadband spectra when in a hostile, aggressive state, and tonal, high-frequency sounds when frightened, appeasing, or affiliative. Mixed motivational states were proposed to be reflected in acoustic grading between these endpoints, for instance producing combinations like upward-moving noisy sounds or downward-moving tonal sounds. These patterns were linked to an overall relationship between a caller's body size and the pitch of its vocalizations. As larger animals are generally more threatening to others, an aggressive individual was described as seeking to make itself seem bigger by producing low-frequency vocalizations. Conversely, an affiliative or submissive animal emulates the higher-frequency sounds of smaller individuals.

These rules have been found to be consistent with some, but not all of the available data (see Hauser, 1996, for a recent review). While Hauser (1993) reported the predicted relationship between pitch and evident motivational state to hold in a variety of primate species, exceptions to the rules are also readily found. Owren, Dieter, Seyfarth, and Cheney (1992), for instance, noted that group-housed rhesus macaques produced noisy, broadband grunts in *play* and *infant* contexts that can both be described as affiliative situations involving contact-seeking. When a familiar human observer approached the large outdoor cage in which the group was housed, these animals also produced grunts, but with no indication of affiliative intent. In similarly housed Japanese macaques, tonal coos were recorded in each of these contexts. Furthermore, each of the two species could produce both coos and grunts, and sometimes used these very different-sounding calls interchangeably.

The motivation-structural rules are similar to the affect-conditioning model in emphasizing effects of signals on receivers, and in some cases the two approaches make similar predictions. However, the affect-conditioning perspective differentiates more specifically between unconditioned and conditioned effects, and suggests that the acoustic features of vocalizations reflect the calling strategy being employed rather than the caller's motivational states per se. For instance, according to the motivation-structural rules, an animal producing first high-pitched tonal sounds and noisy broadband screams in an agonistic situation is showing motivational states of fearful submission and aggression, respectively. A priori predictions about the calls that will be produced can be made only to the extent that the caller's internal states can be anticipated. The affect-conditioning model proposes that the animal is taking advantage of calls that have unconditioned effects on the receiver, and that calling should therefore show repetitive, but varying acoustic form. Vocalizations are predicted not to exhibit individually distinctive cues in spectral-patterning aspects of individual calls, unless such calls are being used in an identifiable conditioning process.

Motivation-structural rules predict that a dominant individual in an agonistic circumstance should give low-pitched, harsh calls, which is consistent with the affect-conditioning model. The latter, however, proposes that cues to individual identity play a primary functional role when primates call in this circumstance and that formant-based cuing is therefore more important than the tonality or noisiness of the sound. The occurrence of either tonal or noise-based threat calls may therefore evolve, depending on the particular species and vocal repertoire involved. In an affiliative circumstance, a subordinate animal can be expected to be fearful and therefore to produce high-frequency tonal calls according to the motivation-structural rules. In the affect-conditioning model, the form of the calls will depend on the degree of control the subordinate has over the other animal's subsequent affective state. Finally, the motivation-structural rules appear to make no clear prediction concerning a dominant animal in an affiliative situation, the affect-conditioning model predicts the occurrence of calls with cues to individual identity that are used to elicit conditioned effects.

Correlations between Calls and Behavior

In suggesting that vocalizations are most fundamentally related to influencing the behavior of a receiver, we have also shifted the emphasis away from functional explanations that propose that senders provide information about their internal states or upcoming behavior. Consistent with this approach, studies that have explicitly examined correlations between signals and subsequent sender behavior have shown that the relationship is typically more probabilistic than precise (e.g., Hauser, 1996). We predict instead that a stronger relationship should be found between the occurrence of signals of one form or another and the receiver's subsequent behavior. While natural selection will favor receivers that can resist influences that are detrimental to their overall fitness, the sender's control of signaling event inherently makes this role primary in an ongoing process of differentiation and innovation in the vocal repertoire. Therefore, the ability of senders to influence receivers should be at the leading edge of this process.

Signal Function: Evolution of Vocalizations

As discussed early in this chapter, we believe that the evolutionary origins of communication lie in conation-like rather than cognition-like functions, with the requisite caveat that the processes involved are fundamentally intertwined. An implication of the affect-conditioning framework is that the simplest and most ancient "communication" function would have been to influence the behavior of a receiver by producing an unconditioned stimulus that elicited an unconditioned response. The emergence of signals specialized for use as conditioned stimuli must have occurred later, but would have been an early development in the evolution of vocal repertoires. This bifurcation of function may have been particularly important in that it arguably laid the foundation for more sophisticated cognitive processing, particularly in receivers. While the learning process that takes place in Pavlovian conditioning procedure is basic and ubiquitous, it is nonetheless complex in that the association involved reflects predictive power and a relationship between two stimuli. As argued by Rescorla (1988a, 1988b), Pavlovian conditioning does not occur because stimuli are merely paired, with connections thereby being "stamped in." Instead, the underlying associations are inherently indistinguishable from simple cognitive structures and have identifiable representational properties. From this point of view, a satisfactory framework of either conditioning or cognition will be one that unifies the two disciplines, rather than separating them.

Continuing our evolutionary scenario, then, we suggest that the emergence of using calls as conditioned stimuli was an important factor in the evolution of cognition, in that selective pressure was created that then favored increasing powers of inference in receivers. While senders serve their own interests by signaling, receivers probably benefit by behaving as the sender desires in some circumstances, but not others. Whereas it is of benefit to senders to produce conditioned affect in receivers, such responses also constitute a form of knowledge—a representation of the signal, the sender, and past interactions. In cognitive terms, conditioned affective responses thereby encode information concerning the characteristics and behavior of other group members. Naturally, the receiver's best interests are served by making use of such information and being able to respond as flexibly as possible to the particular circumstances of each interaction. Selection pressure acting on receivers, then, may have contributed to the emergence of more sophisticated processing capabilities that built on the representational capacity inherent to affective responses. We suggest that this sort of information processing would be added to existing components, allowing modulation of the behavioral effects of activity occurring at affective levels.

CONCLUSIONS

A general theme of this chapter has been that the study of animal communication should not be restricted to information-based approaches. This suggestion is not new, having been anticipated by a variety of important empirical and theoretical developments, and similar recommendations from others. One important component, for instance, has been work highlighting the inherent difficulty of separating information potentially encoded in signals from inferences derived by receivers through active evaluation of both the signal and the context of signal emission (e.g., Smith, 1977; Marler et al., 1992; Leger, 1993). Another component has been proposals for how to enlarge the domain of study to include the role played by motivational and emotional processes in both sender and receiver (see Owings, 1994; Owings & Morton, this volume). The affect-conditioning model is inherently closer to the latter, emphasizing a conative rather than a cognitive perspective. We suggest that many primate calls do not involve "meaning," in the normal sense of this word, and that neither referential nor motivational information is necessarily encoded in such signals.

We also believe that many of the available data concerning primate vocal behavior are consistent with the spirit of our approach. This intuition is based on the evident functional importance of vocalizations in "coordinating" interindividual relationships and social behavior in primates, a theme that has emerged again and again in various forms over the history of acoustic primatology. Our specific contribution lies mainly in suggesting that conditioning-related constructs can be applied to understanding how such coordination might occur. In our view, this sort of function is probably more fundamental for the broad sweep of primate calls than representation of either external or internal designata, and has been largely unaddressed by information-based interpretations. Furthermore, only in a few cases has the purported information content or designata of primate vocalizations been specified, and the cognitive mechanisms implicitly thought to process such information are left as hypothetical constructs. Overall, while it can indisputably be useful to view communication processes in informational terms, the general inability to describe either signal content or the processing mechanisms involved points up the pressing need for additional approaches.

The approach we have outlined is meant to be compatible with findings like those of Bauers and de Waal (1991), who reported that female stumptailed macaques (*M. arctoides*) were more likely to engage in affiliative approach and contact after directing coo calls to one another. Similarly, Silk, Seyfarth, and Cheney (1996) found that in chacma baboons, adult females routinely produced grunt vocalizations when reestablishing affiliative contact with a subordinate animal that had recently been an opponent in an aggressive interaction. In fact, females were virtually never found to engage in nonaggressive interactions with former opponents during the postconflict period unless they grunted during their approach. Both in this case and in interactions that were not preceded by aggressive interactions (Cheney, Seyfarth, & Silk, 1996), grunts played a critical role in facilitating affiliative social encounters. Like Bauers and de Waal (1991), we see calls like these as "social tools" and suggest that over time, natural selection has favored those individual primates that were the most effective in using these tools to influence others. The predictable and observable outcome is that every species now exhibits a "toolbox" full of vocalizations.

The affect-conditioning model may be found to be broadly applicable—in accordance with the ubiquity of phenomena like elicited responses, habituation, and Pavlovian conditioning. Applied on a case-by-case basis, the principles we have proposed could be useful in understanding a variety of communication systems, from the simplest to the most complex. This flirtation with hyperbole is borne out by a growing recognition that even in human speech—the most complex communication system of all—individual variability plays a central and necessary role in normative, linguistic processes (e.g., Johnson & Mullenix, 1997). Pisoni and Lively (1995), for instance, review compelling evidence that voice characteristics of individual human talkers play an important role in "abstract" phonemic representations. While long thought to be a barrier to the speech-decoding process, individual variability has been discovered to be informative to listeners making linguistic judgments. After more than 40 years of largely unsuccessful effort to eliminate or dramatically reduce such effects in the acoustic descriptions of speech sounds (Miller, 1989), researchers in speech perception, word recognition, and computerized speech recognition have independently concluded that individual characteristics should instead be explicitly included in speech representations (Johnson, 1995). For both humans and primates, then, we therefore suggest that examining the simpler, fundamental components of a communication system is a necessary prerequisite for understanding the principles that govern its more complex functions.

ACKNOWLEDGMENTS

The authors benefited significantly from discussions with, and helpful manuscript comments from Jo-Anne Bachorowski, Dorothy Cheney, Tecumseh Fitch, Allen Neuringer, Don Owings, Peter Rodman, Robert Seyfarth, and Nicholas Thompson. Thanks go to Michael Graham for donating the drawing of a rhesus monkey head shown in Figure 1, to Robert Seyfarth and Dorothy Cheney, who recorded the baboon calls used here, and to Don Owings for pointing out the work of R. Caldwell and colleagues. Chapter preparation was partially supported by awards from the Dean's Development Fund at Reed College to M.J.O. and the NSERC of Canada to D. R.

REFERENCES

Baer, T., Sasaki, C., & Harris, K. S. (1986). *Laryngeal function in phonation and respiration*. Boston, MA: College-Hill Press.

Baken, R. J. (1987). *The clinical measurement of speech and voice*. Boston, MA: College-Hill Press.

Bauers, K. A., & de Waal, F. B. M. (1991). "Coo" vocalizations in stumptailed macaques: A controlled functional analysis. *Behaviour, 119,* 143–160.

Bolles, R. C., & Beecher, M. D. (1988). *Evolution and learning*. Hillsdale, NJ: Lawrence Erlbaum Associates.

Brown, C. H., Gomez, R., & Waser, P. M. (1995). Old World monkey vocalizations: Adaptation to the local habitat? *Animal Behaviour, 50,* 945–961.

Caldwell, R. (1986). The deceptive use of reputation by stomatopods. In R. W. Mitchell & N. S. Thompson (Eds.), *Deception: Perspectives on human and nonhuman deceit* (pp. 129–145). New York: State University of New York Press.

Cheney, D. L., & Seyfarth, R. M. (1980). Vocal recognition in free-ranging vervet monkeys. *Animal Behaviour, 28,* 362–367.

Cheney, D. L., & Seyfarth, R. M. (1990). *How monkeys see the world*. Chicago, IL: University of Chicago Press.

Cheney, D. L., Seyfarth, R. M., & Silk, J. B. (1995). The role of grunts in reconciling opponents and facilitating interactions among adult female baboons. *Animal Behaviour, 50,* 249–257.

Cynx, J., & Clark, S. J. (in press). The laboratory use of conditional and natural responses in the study of avian perception. In S. L. Hopp, M. J. Owren, & C. S. Evans (Eds.), *Animal acoustic communication: Sound analysis and research methods*. Heidelberg: Springer-Verlag.

Deacon, T. (1989). The neural circuitry underlying primate calls and human language. *Human Evolution, 4,* 367–401.

Domjan, M. (1992). Adult learning and mate choice: Possibilities and experimental evidence. *American Zoologist, 32,* 48–61.

Domjan, M. (1993). *Domjan and Burkhard's: The principles of learning and behavior,* (3rd ed.). Pacific Grove, CA: Brooks/Cole.

Domjan, M., & Hollis, K. L. (1988). Reproductive behavior: A potential model system for adaptive specializations in learning. In R. C. Bolles & M. D. Beecher (Eds.), *Evolution and learning* (pp. 213–327). Hillsdale, NJ: Lawrence Erlbaum Associates.

Fernald, A. (1992). Human maternal vocalizations to infants as biologically relevant signals: An evolutionary perspective. In J. Barkow, L. Cosmides, & J. Tooby (Eds.), *The adapted mind: Evolutionary psychology and the generation of culture* (pp. 391–428). New York: Oxford University Press.

Fernald, A., & Kuhl, P. (1987). Acoustic determinants of infant preference for motherese speech. *Infant Behavior and Development, 10,* 279–293.

Fitch, W. T. (1994). *Vocal tract length perception and the evolution of language.* Unpublished doctoral dissertation, Brown University, Providence, RI.

Fitch, W. T. (1997). Vocal tract length and formant frequency dispersion correlate with body size in rhesus monkeys. *Journal of the Acoustical Society of America.*

Fitch, W. T., & Hauser, M. D. (1995). Vocal production in nonhuman primates: Acoustics, physiology, and functional constraints on 'honest' advertisement. *American Journal of Primatology, 37,* 191–219.

Gould, J. L. (1986). The biology of learning. *Annual Review of Psychology, 37,* 163–192.

Gouzoules, H., Gouzoules, S., & Ashley, J. (1995). Representational signaling in non-human primates vocal communication. In E. Zimmermann, J. Newman , & U. Jürgens (Eds.), *Current topics in primate vocal communication* (pp. 235–252). New York: Plenum Press.

Gouzoules, H., Gouzoules, S., & Marler, P. (1984). Rhesus monkey (*Macaca mulatta*) screams: Representational signaling in the recruitment of agonistic aid. *Animal Behaviour, 32,* 182–193.

Gouzoules, H., Gouzoules, S., & Marler, P. (1986). Vocal communication: A vehicle for the study of social relationships. In R. G. Rawlins & M. J. Kessler (Eds.), *The Cayo Santiago macaques: History, behavior, and biology* (pp. 111–129). Albany: State University of New York Press.

Green, S. (1975). Variation of vocal pattern with social situation in the Japanese monkey (*Macaca fuscata*): A field study. In L. A. Rosenblum (Ed.), *Primate behavior, Vol. 4* (pp. 1–102). New York: Academic Press.

Green, S., & Marler, P. (1979). The analysis of animal communication. In P. Marler & J. G. Vandenbergh (Eds.), *Handbook of behavioral neurobiology, Vol. 3: Social behavior and communication* (pp. 73–158). New York: Plenum Press.

Halpern, D. L., Blake, R., & Hillenbrand, J. (1986). Psychoacoustics of a chilling sound. *Perception & Psychophysics, 39,* 77–80.

Hauser, M. D. (1992). Articulatory and social factors influence the acoustic structure of rhesus monkey vocalizations: A learned mode of production? *Journal of the Acoustical Society of America, 91,* 2175–2179.

Hauser, M. D. (1993). The evolution of nonhuman primate vocalizations: Effects of phylogeny, body weight and social context. *American Naturalist, 142,* 538–542.

Hauser, M. D. (1996). *The evolution of communication.* Cambridge, MA: MIT Press.

Hauser, M. D., Evans, C. S., & Marler, P. (1993). The role of articulation in the production of rhesus monkey, *Macaca mulatta*, vocalizations. *Animal Behaviour, 45,* 423–433.

Hienz, R. D., & Brady, J. V. (1988). The acquisition of vowel discriminations by nonhuman primates. *Journal of the Acoustical Society of America, 84,* 186–194.

Hienz, R. D., Turkkan, J. S., & Harris, A. H. (1982). Pure tone thresholds in the yellow baboon (*Papio cynocephalus*). *Hearing Research, 8,* 71–75.

Hinde, R. A. (1966). *Animal behaviour: A synthesis of ethology and comparative psychology*. New York: McGraw-Hill.

Hinde, R. A. (1981). Animal signals: Ethological and games-theory approaches are not incompatible. *Animal Behaviour, 29,* 535–542.

Hollis, K. L. (1984). The biological function of Pavlovian conditioning: The best defense is a good offense. *Journal of Experimental Psychology: Animal Learning and Behavior, 10,* 413–425.

Hollis, K. L., Cadieux, E. L., & Colbert, M. M. (1989). The biological function of Pavlovian conditioning: A mechanism for mating success in the blue gourami (*Trichogaster trichopterus*). *Journal of Comparative Psychology, 103,* 115–121.

Johnson, K. (1995). Talker variability in vowel perception. *Journal of the Acoustical Society of America, 98,* 2949–2950.

Johnson, K., & Mullenix, J. W. (Eds.). (1997). *Talker variability in speech processing*. San Diego: Academic Press.

Kaplan, P. S., Goldstein, M. H., Huckeby, E. R., Owren, M. J., & Panneton Cooper, R. (1995). Dishabituation of visual attention in infant- versus adult-directed speech: Effects of frequency modulation and spectral composition. *Infant Behavior and Development, 18,* 209–223.

Kaplan, P. S., & Owren, M. J. (1994). Dishabituation of infant visual attention in 4-month-olds by infant-directed frequency-modulated sweeps. *Infant Behavior and Development, 17,* 347–358.

Krebs, J. R., & Dawkins, R. (1984). Animal signals: Mind-reading and manipulation. In J. R. Krebs & N. B. Davies (Eds.), *Behavioural ecology: An evolutionary approach* (2nd ed., pp. 380–402). Sunderland, MA: Sinauer Associates.

Leger, D. W. (1993). Contextual sources of information and responses to animal communication signals. *Psychological Bulletin, 113,* 295–304.

Lieberman, P. (1975). *On the origins of language: An introduction to the evolution of human speech*. New York: Macmillan Publishing.

Marler, P. (1975). On the origin of speech from animal sounds. In J. F. Kavanaugh & J. E. Cutting (Eds.), *The role of speech in language* (pp. 11–37). Cambridge, MA: MIT Press.

Marler, P., Evans, C. S., & Hauser, M. D. (1992). Animal signals: Motivational, referential, or both? In H. Papousek, U. Jürgens, & M. Papousek (Eds.), *Nonverbal vocal communication: Comparative and developmental approaches* (pp. 66–86). New York: Cambridge University Press.

Marler, P., & Terrace, H. (1984). *The biology of learning*. Berlin: Springer-Verlag.

Mason, W. A. (1979). Wanting and knowing: A biological perspective on maternal deprivation. In E. Thoman, (Ed.), *Origins of the infant's social responsiveness* (pp. 225–249). Hillsdale, NJ: Lawrence Erlbaum Associates.

Miller, J. D. (1989). Auditory-perceptual interpretation of the vowel. *Journal of the Acoustical Society of America, 85,* 2114–2134.

Moody, D. B., May, B. J., Cole, D. M., & Stebbins, W. C. (1986). The role of frequency modulation in the perception of complex stimuli by primates. *Experimental Biology, 45,* 219–232.

Morton, E. S. (1977). On the occurrence and significance of motivation-structural rules in some bird and mammal sounds. *American Naturalist, 111,* 855–869.

Morton, E. S. (1982). Grading, discreteness, redundancy, and motivation-structural rules. In D. E. Kroodsma & E. H. Miller (Eds.), *Acoustic communication in birds* (pp. 183–212). New York: Academic Press.

Owings, D. H. (1994). How monkeys feel about the world: A review of "How monkeys see the world." *Language & Communication, 14,* 15–30.

Owren, M. J., Dieter, J. A., Seyfarth, R. M., & Cheney, D. L. (1992). Vocalizations of rhesus (*Macaca mulatta*) and Japanese (*M. fuscata*) macaques cross-fostered between species show evidence of only limited modification. *Developmental Psychobiology, 26,* 389–406.

Owren, M. J., Hopp, S. L., Sinnott, J. M., & Petersen, M. R. (1988). Absolute auditory thresholds in three Old World monkey species (*Cercopithecus aethiops, C. neglectus, Macaca fuscata*) and humans. *Journal of Comparative Psychology, 102,* 99–107.

Owren, M. J., & Linker, C. D. (1995). Some analysis techniques that may be useful to acoustic primatologists. In E. Zimmermann, J. Newman , & Ü. Jürgens (Eds.), *Current topics in primate vocal communication* (pp. 1–27). New York: Plenum Press.

Owren, M. J., Seyfarth, R. M., & Cheney, D. L. (1997). The acoustic features of vowel-like grunt calls in chacma baboons (*Papio cynocephalus ursinus*): Implications for production processes. *Journal of the Acoustical Society of America, 101*, 2951–2963.

Papousek, M., Papousek, H., & Symmes, D. (1991). The meanings and melodies in motherese in tone and stress languages. *Infant Behavior and Development, 14,* 415–440.

Pisoni, D. B., & Lively, S. E. (1995). Variability and invariance in speech perception: A new look at some old problems in perceptual learning. In W. Strange (Ed.), *Speech perception and linguistic experience* (pp. 433–459). New York: York Press.

Prosen, C. A., Moody, D. B., Sommers, M. S., & Stebbins, W. C. (1990). Frequency discrimination in the monkey. *Journal of the Acoustical Society of America, 88,* 2152–2158.

Provine, R. R. (1996). Laughter. *American Scientist, 84,* 38–45.

Rendall, D. (1996). *Social communication and vocal recognition in free-ranging rhesus monkeys (Macaca mulatta)*. Unpublished doctoral dissertation, University of California, Davis.

Rendall, D., Owren, M. J., & Rodman, P. S. (in press). Rhesus monkey vocalizations: Individuality, vocal recognition, and the determinants of acoustic design. *Journal of the Acoustical Society of America.*

Rendall, D., Rodman, P. S., & Emond, R. E. (1996). Vocal recognition of individuals and kin in free-ranging rhesus monkeys. *Animal Behaviour, 51,* 1007–1015.

Rescorla, R. A. (1988a). Behavioral studies of Pavlovian conditioning. *Annual Review of Neuroscience, 11,* 329–352.

Rescorla, R. A. (1988b). Pavlovian conditioning: It's not what you think it is. *American Psychologist, 43,* 151–160.

Roitblat, H. L., & von Fersen, L. (1992). Comparative cognition: Representations and processes in learning and memory. *Annual Review of Psychology, 43,* 671–710.

Rubin, P., & Vatikiotis-Bateson, E. (in press). Measuring and modeling speech production. In S.L. Hopp, M.J. Owren & C.S. Evans (Eds.) Animal acoustic communication: sound analysis and research methods. Heidelber: Springer-Verlag.

Schön Ybarra, M. (1995). A comparative approach to the nonhuman primate vocal tract: Implications for sound production. In E. Zimmermann, J. D. Newman, & U. Jürgens (Eds.), *Current topics in primate vocal communication* (pp. 185–198). New York: Plenum Press.

Schwartz, B. (1989). *Psychology of learning and behavior* (3rd ed.). New York: W. W. Norton & Company.

Seyfarth, R. M., & Cheney, D. L. (1997). Some general features of vocal development in nonhuman primates. In C. T. Snowdon & M. Hausberger (Eds.), *Social influences on vocal development,* (pp 249–273).New York: Cambridge University Press.

Silk, J. B., Cheney, D. L., & Seyfarth, R. M. (1996). The form and function of post-conflict interactions between female baboons. *Animal Behaviour, 52,* 259–268.

Sinnott, J. M. (1989). Detection and discrimination of synthetic English vowels by Old World monkeys (*Cercopithecus, Macaca*) and humans. *Journal of the Acoustical Society of America, 86,* 557–565.

Sinnott, J. M., & Brown, C. H. (1993). Effects of varying signal and noise levels on pure-tone frequency discrimination in humans and monkeys. *Journal of the Acoustical Society of America, 93,* 1535–1540.

Sinnott, J. M., Owren, M. J., & Petersen, M. R. (1987a). Auditory frequency discrimination in primates: Species differences (*Cercopithecus, Macaca, Homo*) and humans. *Journal of Comparative Psychology, 101,* 126–131.

Sinnott, J. M., Owren, M. J., & Petersen, M. R. (1987b). Auditory duration discrimination in Old World monkeys (*Macaca, Cercopithecus*) and humans. *Journal of the Acoustical Society of America, 82,* 465–470.

Sinnott, J. M., Petersen, M. R., & Hopp, S. L. (1985). Frequency and intensity discrimination in humans and monkeys. *Journal of the Acoustical Society of America, 78,* 1977–1985.

Smith, W. J. (1977). *The behavior of communicating.* Cambridge, MA: Harvard University Press.

Smuts, B. B., Cheney, D. L., Seyfarth, R. M., Wrangham, R. W., & Struhsaker, T. T. (1987). *Primate societies.* Chicago: University of Chicago Press.

Snowdon, C. T. (1986). Vocal communication. In G. Mitchell & J. Erwin (Eds.), *Comparative primate biology, Volume 2A: Behavior, conservation, and ecology* (pp. 495–530). New York: Liss.

Sommers, M. S., Moody, D. B., & Prosen, C. A. (1992). Formant frequency discrimination by Japanese macaques (Macaca fuscata). *Journal of the Acoustical Society of America, 91,* 3499–3510.

Stebbins, W. C. (1973). Hearing of Old World monkeys (Cercopithecinae). *American Journal of Physical Anthropology, 38,* 357–364.

Stebbins, W. C., & Berkley, M. A. (Eds.). (1990). *Comparative perception, Vol. II: Complex signals.* New York: John Wiley & Sons.

Stebbins, W. C., & Moody, D. B. (1994). How monkeys hear the world: Auditory perception in nonhuman primates. In R. R. Fay & A. N. Popper (Eds.), *Comparative hearing: Mammals.* Berlin: Springer-Verlag.

Steklis, H. D., & Raleigh, M. J. (Eds.). (1979). *Neurobiology of social communication in primates: An evolutionary perspective.* New York: Academic Press.

Titze, I. R. (1994). *Principles of voice production.* Englewood Cliffs, NJ: Prentice-Hall.

Todt, D., Goedeking, P., & Symmes, D. (Eds.). (1988). *Primate vocal communication.* Berlin: Springer-Verlag.

Turkhan, J. S. (1989). Pavlovian conditioning: The new hegemony. *Behavioral and Brain Sciences, 12,* 121–179.

Wassermann, E. A. (1993). Comparative cognition: Beginning the second century of the study of animals intelligence. *Psychological Bulletin, 113,* 211–228.

Werker, J., & McLeod, P. (1989). Infant preference for both male and female infant-directed talk: A developmental study of attentional and affective responsiveness. *Canadian Journal of Psychology, 43,* 230–246.

Zeskind, P. S., & Lester, B. (1978). Acoustic features and auditory perception of the cries of newborns with prenatal and perinatal complications. *Child Development, 49,* 580–589.

Zimmermann, E., Newman, J., & Jürgens, Ü. (Eds.). (1995). *Current topics in primate vocal communication.* New York: Plenum Press.

Chapter 10

SPEECH ACTS AND ANIMAL SIGNALS

Andrew G. Horn

Department of Biology
Dalhousie University
Halifax, N.S. B3H 4J1, Canada

ABSTRACT

In everyday speech, we often use words more to do things (e.g., greet, make bets, accuse, ask, marry, etc.) than to make statements of fact. In the philosophy of language, viewing utterances as these sorts of speech acts (Austin 1962), or as moves in a language game (Wittgenstein 1958), challenged the view of communication as an exchange of true or false propositions. For ethologists, applying similar concepts to animal signals may help keep concepts like information, manipulation, and honesty in their proper perspective. This essay shows the many parallels between speech acts and animal signals, and touches on their implications. According to this perspective, the main function of signals is not to state facts, although facts (e.g., about honesty, intentions, and external referents) are crucial for the evolutionary stability of signals. Just as the speech acts in a marriage ceremony resist translation into facts outside of the social system of which they are a part, most animal signals will likely resist translation into general classes of messages or functions. Nonetheless, if the rules that govern the use of signals are sufficiently understood, the parts played by manipulation and information in the evolution of those rules can also be understood.

Perspectives in Ethology, Volume 12
edited by Owings *et al.*, Plenum Press, New York, 1997

INTRODUCTION

The study of animal communication uses concepts that at first sight appear familiar and straightforward, but on reflection become academic and obtuse. Concepts like meaning, reference, manipulation, honesty, and cost have spawned complex debates even though they are not particularly technical terms (e.g., Hinde 1981, 1985, Smith 1981, 1986a). Part of this unexpected subtlety is the nature of the beast; communication involves two coevolving parties whose interests may or may not coincide. Most of the complexity, however, comes from our preconceptions about how communication works.

Our common sense tells us that the main function of animal signals is to state facts. Just as we might say, "The chair is here" or "I am hungry," roosters crow to say they are dominant and songbirds sing to say "I'm here." We assume that our everyday use of language consists of such propositions about the state of the world, and then go on to assume that the main task for studies of animal communication is to translate the propositions that each signal represents.

If we think beyond our first impressions, however, the commonsense view of signals as statements of facts (or falsehoods) does not match the everyday use of language after all. Many if not most of our utterances to each other do not function primarily to make statements so much as to do things. We make bets, christen ships, bequeath possessions, and get married with words. "I am sorry" may be a statement about my emotional state that may be true or false, but "I apologize" is a social act that, while it may not be as binding as other actions (like signing a legal document, or drawing one's own blood as security), is nonetheless accepted as a commitment, not just a statement of fact. If you say "I'm sorry," and you're not, your sentence was false; you are not sorry. However, if you say "I apologize" when you don't "really mean it," then you may have been insincere, but you apologized nevertheless.

The analysis of these utterances, known as performatives or, more loosely, speech acts (Austin 1962, Searle 1969, 1979), and the related, but independently derived (Furberg 1971), idea of language games (Wittgenstein 1958), triggered a small revolution in the philosophy of language (Levinson 1983). Seeing language as a form of social behavior, just like fighting and wooing, might not seem revolutionary to ethologists, who have always been concerned with the biological functions of signaling. At the time the idea of speech acts appeared, however, the philosophy of language was dominated by logical positivism, which views language as a series of true or false propositions. The imprecision of ordinary language use was seen as evidence of the meaningless of everyday speech, rather than insufficiencies in the theory itself (e.g., Ayer 1936). The theory of speech acts revised this dry and academic view of language by reexamining the mechanics of ordinary speech—in effect replacing what I called the commonsense view of language with one that made more common sense.

Although ethologists might easily accept that animal signals do things, they have had trouble squaring this hard-nosed functional stance with the commonsense notion that signals state facts. Some ethologists have tried, for example, to view animal signals as uninformative perceptual tools for manipulating receivers (Dawkins and Krebs 1978), but this approach does not satisfy our intuition that communication has *something* to do with conveying information (for detailed critiques, see Owings and Morton, this volume, Smith 1986a), an intuition borne out by recent models of signal evolution (Grafen 1990, Godfray 1991).

In keeping with the title of this series, what follows is a perspective I think will be useful to ethologists studying animal signals, rather than a broadside against this or that view of signaling. Seeing everyday language in terms of speech acts satisfies both our biological intuition that signals are meant to *do* something, and our commonsense intuition that signals work at least partly by conveying information. Rather than replacing previous views of animal communication, this view keeps previous approaches, and their potentially muddy or contentious concepts, like information and manipulation, in their proper perspective. I proceed by reviewing the basic ideas in the speech act theory as given in Austin's (1962) original lucid formulation, indicating parallels to conceptual issues in animal communication as they arise.

SPEECH ACTS AND LANGUAGE GAMES

We all recognize that words are tools for getting things done. We shop with words, we warn with words, we woo with words, so the idea that sentences have social functions is not revolutionary. However, the idea that those functions are more important in communication than the facts the words represent goes against our fundamental ideas about how communication works. We think of our words as stand-ins for facts, that then are interpreted, and only then serve their various functions.

Many things we say, however, do not need to be stand-ins for facts in order to perform their functions. Consider the following sentences:

I dub thee Sir Walter
I object
I sentence you to ten years of hard labor
I give you my word
I warn you that trespassers will be prosecuted

(modified from Austin 1962)

In a sense, all these sentences are true simply by their being said. I might warn you that trespassers will be prosecuted when in fact trespassers will be wined and dined, but I have warned you nevertheless. Similarly, if I promise to do something, my promise might be insincere or I might be unable to fulfill it, but I still promised.

Thus what such sentences do is separable from and more fundamental than what they state. By the mere act of saying them, one has performed a social act, often whether or not the facts one has implied (e.g., one's future behavior, in the case of a promise) are true (Austin 1962).

Many animal signals work in the same way. When a stag shakes its antlers at an opponent, it might be signaling various things—its size, its fighting ability, and so on. At a more fundamental level, however, it is threatening the other deer. Whether the various potential messages of the signal (size, prowess, etc.) are true or not, the deer still threatened its opponent. All these various messages might modulate the threat, or even render it ineffective, but the threat is still a threat.

If sentences and animal signals are simply social acts, it makes no more sense to call them true or false than to call attacks or matings true or false. A promise might be insincere or unfulfilled, but what does it mean to say that it is false? A threat is a social act, just as much as an attack is a social act. If we don't talk of attacks as true or false, why should we call threats true or false, rather than effective or ineffective? The same reasoning follows for many other pairs of analogous signal and nonsignal behaviors, like courtship and mating, or begging and foraging. They may or may not work, but they are not true or false.

At first glance, this distinction between facts and acts may seem to be the causal-functional distinction so familiar to ethologists (e.g., Tinbergen 1963): threats signal the fact that an attack may be imminent (causal mechanism), but ultimately they serve as social acts for driving away opponents (adaptive function). This interpretation misses the point, however. Speech acts do not need to be true or even to state anything to be effective. Truth and falsity do play a critical role in whether or not these signaling acts work, but their role is subtler than the cause-function dichotomy implies.

TRUTH AND CONSEQUENCES IN ANIMAL SIGNALS

Truth conditions play a role in the establishment and maintenance of effective speech acts, but not in their day-to-day functioning. For many speech acts to be effective, the speakers have to have particular thoughts, feelings, or intentions. For example, if I apologize, I should, on average, be truly sorry. Many speech acts also require consequent conduct as well. If I promise to do something, I should do it. I must really be a priest, justice of the peace, or other authorized officer to conduct a marriage, even though one doesn't say "I pronounce you man and wife" to say that one is a priest (Austin 1962).

Similarly, information plays a critical role in the evolution of effective signals. If nestlings do not beg in proportion to their need, then over evolutionary time parents devalue the begging signal (Godfray 1991). Threats must be followed by attack sufficiently often for them to be effective, and singers must be unpaired males on average for their songs to attract females. *Contra* the "common sense" view of signaling, these ultimate correlations do not mean that begging, for example, is a statement of need, but instead that information about fighting ability or need or pairing status *validates* the signal over evolutionary time. Begging, threatening, and courting are social acts whose effectiveness depends on their correlation (on average) with something relevant to receivers (Grafen 1990). Put another way, true and false information are some of the ultimate reasons for the signal's effectiveness as a tool for social interactions, but that does not mean that either the signal's main (evolutionary) function or its main (day-to-day) means of functioning is the conveyance of that information.

This interpretation flips the causal-functional distinction raised earlier. Now the information that signals convey is one of the ultimate reasons why signals are effective, rather than being one of the proximate causes for the signals' effectiveness. Yet putting information on the "ultimate" side of this traditional dichotomy would incorrectly imply that information only plays a role over evolutionary time. As Smith (1977) emphasized, however, signals also "evolve" within the life spans of individuals. Therefore what has been said of the "ultimate" role of information in the evolution of signals can equally be said of its "proximate" role in learning, at least in higher organisms in which such things are learned.

Specifically, given the dependence of signals on validity claims of truth and falsity, there will clearly be selection on individuals to "second-guess" evolution, and to evaluate the validity of signals on the spot. Although parent birds might respond blindly to their begging chicks, because *on average* begging and need are correlated, if nestlings often try to monopolize feedings, then parents might be selected to look for multiple cues of need. Parents might have a perceptual device, perhaps modified by learning, that compares the input of all these possible cues and computes an appropriate response.

The ultimate in this sort of short-circuiting of evolution is human language. We hear what other people say, and quickly compute what they presuppose, what they imply, what they are intending to accomplish, and so on, before we come up with a response. Animals, too, may "mind-read" (Krebs and Dawkins 1984), "assess" (Owings and Morton, this volume), or "generate predictive scenarios" (Smith 1990). Yet all this information processing primarily serves to evaluate an appropriate response to the social act the signal has performed, not to decode some sort of propositional content of the signal. All this thinking ultimately boils down to social action.

Where does this leave information? Whether we are talking in ultimate or proximate terms, information is what makes sense of social interactions. It is the properties of a situation that allow us to rationalize the animal's behavior in terms that make sense to us. In roosters, crowing and the frequency of each crow correlates with dominance, and roosters respond differently to the crows of dominants and subordinates (Leonard and Horn 1995), which we can rationalize by suggesting that roosters crow (at least partly) to state their dominance. Kicks, slaps, and pecks also enforce dominance, but to explain these behaviors we needn't appeal to information, because the efficacy of these attacks is based on everyday physical laws. Rationalizing why a physical blow is a deterrent seems unnecessary, whereas the effectiveness of a mere sound when the stakes are so high requires explanation.

MEANING IS USE

At this point one might still argue that my interpretation of sentences like "I promise" and "I apologize" does not apply to what sentences *mean*, but to how they are *used* (e.g., Smith 1986b). "I promise" might be an indirect statement of the set of propositions that are required for me to promise something, e.g., that I am capable of carrying out the promise and will actually carry through (Searle 1969, 1979). "I promise" would then be a statement of true or false facts, the most important being that I will do what I said I would do, i.e., that I have not abused the promise. If speech acts reduce to statements like this, then the commonsense view of how communication works, as a set of propositions, stands unchallenged.

Even statements of fact, however, are hard to translate without reference to how the statements are used. "This color is red" is impossible to translate without actually pointing out a red object. Even then, suppose one points to a red chair to demonstrate what red is. It isn't clear from that act alone that neither the shape of the chair, the number of chairs, nor the height of the chair is being referred to by the term "red." To define "red" or any other term, a whole series of actions are needed. The listener must be shown how the term is used before it can be understood (Wittgenstein 1958).

The main reason why we don't have to go through such elaborate acts every time we define a term is that we have other words to explain our meaning. "This color" specifies that it is the color of the chair, not its shape or height, that "red" refers to (Quine 1960). Still, ultimately those other words are definable in terms of actions. We are reminded of this fact by the development of language in children, who lack the rich vocabulary of adults, so that action-based learning is now known to be critical (Halliday 1975). In decoding the language of foreign cultures, the initial stage always consists of actions, before enough vocabulary has built up to translate the signals (Quine 1960).

Similarly, the most referential of signals, that seem to be black-and-white statements of fact—like food calls and predator alarm calls (Gyger et al. 1987,

Cheney and Seyfarth 1990)—are probably impossible to translate without spelling out the social context in which they are given, and the future behaviors that they commit the animal to (Smith 1990). Does the referential alarm call of a vervet monkey (*Cercopithecus aethiops*) mean "Leopard!" or "Head for the trees!" (Cheney and Seyfarth 1990, p. 309)? Even if it does mean "Leopard!", what is a leopard to a vervet monkey, but a collection of particular behavioral dispositions to respond to leopards?

This line of argument suggests that translating animal signals should consist of clarifying the rules that govern the use of those signals, rather than figuring out what facts they might stand for (see also Wiley 1983). That is why rationalizations of signaling behavior that invoke information usually involve information about behavior (e.g., Smith 1977). It is also why explaining animal communication can be so difficult.

HOW TO TRANSLATE ANIMAL SIGNALS

We can explain signals in at least two ways. First, signaling should make evolutionary sense; signals should correlate on average with some sort of state of affairs, or else receivers would ignore them. Second, signaling should make sense in the day-to-day functioning of social systems; a singing blackbird should not suddenly give up its territory to a silent intruder.

These two levels correspond to two ways we can translate any set of signals (Harris 1984). First, we can translate the signal into our own language—the method of external comparison. External comparison is in effect what an English-French dictionary accomplishes: an English word for each French word and vice versa. Second, we can translate the signal by contrasting it with the other signals in the repertoire. This is what a normal dictionary accomplishes; each word is defined in terms of other words in the lexicon. Put another way, one US dollar equals 1.33 Canadian dollars (external comparison) or 100 US cents (internal comparison).

Internal contrasts establish the rules of language use; the relationships between words, their uses, and their consequences. As the internal rules of the communication system are worked out, one can clarify the external relationships that justify, or validate, their use. This involves external comparisons between signals and facts, intentions, and conduct, i.e., the validity claims of speech acts. Note that the internal/external distinction made here applies equally to both senders and receivers. Internal contrasts make up the code of conduct that both parties follow, and the validation of this code may result from evolutionary pressure on senders from receivers (e.g., guards against being manipulated) or on receivers from senders (e.g., guards against being "mind-read"; Krebs and Dawkins 1984).

Austin (1962) made a parallel distinction for speech acts between abuses and misfires. Abuses violate external relationships between speech acts and truth conditions that should hold on average for the speech acts to be valid. Misfires violate internal rules that must be followed for listeners to recognize the acts as being appropriate. If the words in a marriage ceremony are to perform marriages, there must be something recognized as a marriage ceremony in the first place. This ceremony has worked partly because, over the course of history, people making the promises in the ceremony have on average followed them—i.e., because of a correlation between the ceremony and social behavior external to the ceremony. From time to time, someone might masquerade as the priest, but as long as this abuse isn't revealed, the ceremony is still accepted as a marriage by the participants. If a shop clerk in Tonawanda, New York, declares war on Russia, however, the declaration misfires, because no one recognizes an utterance outside of Congress as a declaration of war. The internal rules of the procedure itself, rather than its external background conditions, are being violated.

Currently, almost all research in animal communication uses the method of external comparison. The message of a signal is decoded by correlating the occurrence of a signal with various states or events, like dominance, affiliative behavior, or the appearance of a particular type of predator (Smith 1977). Ethologists with more functional interests might assign a function to each particular signal, associating each with a particular biological job that the animal must get done, like fighting, mating, or warning (Wilson 1975). Either way, the signal repertoire ends up being decoded by our translating the signals into our own terms, so that we have a catalogue of signals and their messages and/or functions that holds across a wide range of species.

This approach works when the rules of signal use and the types of information that validate them are general. In some mate attraction signals, for example, the signal may simply have to be heard over long distances to be effective. The method of external comparison may then lead us to quite reasonably translate the signal simply as advertising the signaller's presence or as functioning in mate attraction.

As soon as any layers of complexity are added, however, the situation requires thinking in terms of more proximate functions: subsidiary hurdles that must be overcome before lifetime reproductive success can be improved (Horan 1982). A bird defending a territory with song must not only be heard, but also (in its dual role as a sender-receiver) must regulate its song rate and the types of songs it sings in complex ways that in effect negotiate its territory boundary with neighbors. We may try to translate even these detailed acts by using external comparisons, and so may say that a bird singing in a particular way is advertising its readiness to interact aggressively or its indecisiveness, for example (Smith 1996). The bird *is* interacting, however; the songs it sings *replace* physical interaction. Like the speech acts in a marriage ceremony, the songs a bird sings are part of a functional system of social acts that must be performed so that broader functions like territory defense can be served. Explaining these acts requires working out the rules of the communication system in detail, i.e., translation by the method of internal contrasts (see also Beer 1975, 1982).

This argument is obviously an appeal for more piecemeal, analytical studies that involve extensive fieldwork to figure out the rules of interactions (e.g., Dabelsteen and McGregor 1996, Smith 1996). However, broad-brush, synthetic approaches are fine, too. Games theory models, for example, start with general goals and add specific constraints piecemeal to find the specifics of the signaling systems that emerge. The main point is that neither approach is likely to end up with a list of messages or a list of functions, but with some sort of theory of social interaction.

GENERAL TYPES OF VALIDITY

What would such a general theory look like? Wittgenstein (1958) feels the most we can hope in explaining human language is a set of family resemblances. Language is a game, and any general explanation of what words mean is likely to be as fuzzy as any general explanation of the similarities between chess and tennis. We can point out the most obvious features, but push any further and we become bogged down in specifics. Animals in particular play by their own rules. We can work out their rules by employing the method of internal contrasts, but their signals cannot be translated into any information we might understand. Attempts to generalize those rules or to translate language by comparison to any sorts of external standards are doomed to failure. If a lion could talk, we could not understand him.

Austin (1962) was more persistent. For all his emphasis on speech acts, he recognized that sentences still range from the more propositional ("The cat is on the mat") to the more ritualized ("I dub thee Sir Walter"). Much of his central work on speech acts, *How to Do Things with Words*, is devoted to an unsuccessful attempt to either distinguish or collapse all these gradations (see Levinson 1983 for other attempts).

In the end, Austin admits that *every* utterance makes a number of things happen, or, in his scheme, has three *forces*. First, saying something can have an effect on the listener that is not prescribed by any given convention. Yelling "Oh no!" in a crowded theater might alarm people, in a different way than yelling "Fire!" warns them. Saying "The cat is on the mat" might terrify someone with ailurophobia, even though it isn't a standard way to scare people. Second, saying something can do something by following a procedure that has evolved and is maintained in the way I showed earlier in this essay: by being a speech act.

Yelling "Fire!" is a standard way to warn people that a room is burning, so much so that children are reprimanded for yelling it as a joke, in a way that they are not reprimanded for just yelling "Oh no!" Third, saying something can state facts, like "The moon circles the earth" or "I have six brothers and three sisters."

In principle, most things we say have all three forces in varying proportions, so that it is neither easy nor perhaps desirable to distinguish these forces precisely. If I say "There's a bull in that field," am I trying to frighten you, to warn you, or just to inform you? The way in which the statement is to be taken will, of course, depend on the context. If you are a mammalogist on tour, then maybe I merely want to show you that there are bulls in Nova Scotia, and am merely stating a fact. If you are an ordinary person on a hike, I'm probably warning you. Similarly, whatever the force of an utterance, the generality of that force and hence our ability to translate the utterance beyond the particular context in which it is used will vary tremendously across different speech acts (a theme followed up by Austin's followers and critics; see Levinson 1983).

In the same way, animal signals likely have three forces in varying proportions. At one level, they are mere perceptual goads to action, manipulating the sensory systems of receivers to the sender's advantage (Dawkins and Krebs 1978). At another level, as argued throughout this essay, they are replacements for social behavior, that don't function to say anything as such. At still another level, they do imply particular information, without which their effectiveness as social acts would be selected against. The ease with which they can be translated into terms we understand will similarly vary depending on how specific each of these forces is to the situation in question.

Thus applying the theory of speech acts to animal communication might mediate debates over whether animal signals function to manipulate or inform. Animal signals both manipulate *and* inform, and each of these functions predominates for certain signals. But animal signals are first and foremost tools for social behavior, for which something less blunt than manipulation but not as finely honed as information is required. All signals have force, but only some convey facts.

ACKNOWLEDGMENTS

I thank the editors for their encouragement and helpful comments. Marty Leonard and Danny Weary also kindly read embarrassingly bad drafts. For helpful discussions and tips on relevant readings I also thank Tom Dickinson, Bruce Falls, Jerry Hogan, Henry E. Horn, and Rich Horn. D.H. Owings suggested the vervet monkey example.

REFERENCES

Austin, J. L. (1962). *How to Do Things with Words*. Oxford: Clarendon Press.

Ayer, A. J. (1936). *Language, Truth, and Logic*. London: Victor Gallancz.

Beer, C. G. (1975). Multiple functions and gull displays. In G. Baerends, C. Beer, and A. Manning (Eds.), *Function and Evolution of Behavior: Essays in Honor of Niko Tinbergen* (pp. 16–54). Oxford: Clarendon Press.

Beer, C. G. (1982). Conceptual issues in the study of communication. In D. E. Kroodsma and E. H. Miller (Eds.), *Acoustic Communication in Birds, Vol. 2*, (pp. 279–310). New York: Academic Press.

Cheney, D. L. and R. M. Seyfarth. (1990). *How Monkeys See the World*. Chicago, IL: University of Chicago Press.

Dabelsteen, T. and P. K. McGregor. (1996). Dynamic acoustic communication and interactive playback. In D. E. Kroodsma and E. H. Miller (Eds.), *Ecology and Evolution of Acoustic Communication in Birds* (pp. 398–408). Ithaca, NY: Cornell University Press.

Dawkins, R. and J. R. Krebs. (1978). Animal signals: information or manipulation? In J. R. Krebs and N.B. Davies (Eds.), *Behavioural Ecology: An Evolutionary Approach* (pp. 282–309). Sunderland, MA: Sinauer.

Furberg, M. (1971). *Saying and Meaning: A Main Theme in J. L. Austin's Philosophy*. Oxford: Blackwell.

Godfray, H. C. J. (1991). Signalling of need by offspring to their parents. *Nature, 352,* 328–330.

Grafen, A. (1990). Biological signals as handicaps. *Journal of Theoretical Biology, 144,* 517–546.

Gyger, M., S. J. Karakashian, and P. Marler. (1987). Semantics of an avian alarm call system: the male domestic fowl. *Behaviour, 102,* 15–40.

Halliday, M. A. K. (1975). *Learning How to Mean: Explorations in the Development of Language*. London: Arnold.

Harris, R. (1984). Must monkeys mean? In R. Harré and V. Reynolds (Eds.), *The Meaning of Primate Signals* (pp.116–137). Cambridge: Cambridge University Press.

Hinde, R. A. (1981). Animal signals: ethological and games theory approaches are not incompatible. *Anim. Behav., 29,* 535–542.

Hinde, R. A. (1985). Was "the expression of the emotions" a misleading phrase? *Anim. Behav., 33,* 985–992.

Horan, B. (1982). Functional explanation in sociobiology. *Biology and Philosophy, 4,* 131–158.

Krebs, J. R. and R. Dawkins. (1984). Animal signals: mind-reading and manipulation. In J. R. Krebs and N. B. Davies (Eds.), *Behavioural Ecology: An Evolutionary Approach,* (2nd ed., pp. 380–402). Sunderland, MA: Sinauer.

Leonard, M. L. and A. G. Horn. (1995). Crowing as a status signal in roosters. *Anim. Behav., 49,* 1283–1290.

Levinson, S. C. (1983). *Pragmatics*. Cambridge: Cambridge University Press.

Quine, W. V. O. (1960). *Word and Object*. Cambridge, MA: MIT Press.

Searle, J. R. (1969). *Speech Acts*. Cambridge: Cambridge University Press.

Searle, J. R. (1979). *Expression and Meaning*. Cambridge: Cambridge University Press.

Smith, W. J. (1977). *The Behavior of Communicating*. Cambridge, MA: Harvard University Press.

Smith, W. J. (1981). Referents of animal communication. *Anim. Behav., 29,* 1273–1275.

Smith, W. J. (1986a). An "informational" perspective on manipulation. In R. W. Mitchell and N. S. Thompson (Eds.), *Deception: Perspectives on Human and Nonhuman Deceit*. SUNY Press.

Smith, W. J. (1986b). Signaling behavior: Contributions of different repertoires. In R. J. Schusterman, J. A. Thomas, and F. G. Wood (Ed.s), *Dolphin Cognition and Behavior: A Comparative Approach* (pp. 315–330). Hillsdale, NJ: Lawrence Erlbaum Associates.

Smith, W. J. (1990). Animal communication and the study of cognition. In C. A. Ristau (Ed.), *Cognitive Ethology: The Minds of Other Animals* (*Essays In Honor Of Donald R. Griffin*) (pp. 209–230). Hillsdale, NJ: Lawrence Erlbaum Associates.

Smith, W. J. (1996). Using interactive playback to study how songs and singing contribute to communication and behavior. In D. E. Kroodsma and E. H. Miller (Ed.s), *Ecology and Evolution of Acoustic Communication in Birds*. (pp. 377–397). Ithaca, NY: Cornell University Press.

Tinbergen, N. (1963). On the aims and methods of ethology. *Z. Tierpsychol.*, *20*, 410–433.

Wiley, R. H. (1983). The evolution of communication: Information and manipulation. In T. R. Halliday and P. J. B. Slater (Eds.), *Animal Behaviour, Vol. 2: Communication* (pp. 82–113). San Francisco: Freeman Co.

Wilson, E. O. (1975). *Sociobiology: The New Synthesis*. Cambridge, MA: Harvard University Press.

Wittgenstein, L. (1958). *Philosophical Investigations*. Oxford: Blackwell.

Chapter 11

THE ROLE OF INFORMATION IN COMMUNICATION: AN ASSESSMENT/MANAGEMENT APPROACH

Donald H. Owings

Department of Psychology
University of California, Davis
Davis, California 95616-8686

Eugene S. Morton

National Zoological Park
Smithsonian Institution
Washington, D.C. 20008

ABSTRACT

The assumption that communication involves the transfer or withholding of information underlies most current research and interpretation of animal communication. This assumption implies that information plays a causal role in communication. Here we suggest that information, when accorded a central causal role, has limited our understanding of communication.

We offer another point of view, *assessment/management*. Assessment/management is founded upon the fundamental biological process of regulation, rather than the anthropocentric concept of information exchange. Management is synonymous with regulation. Organisms are regulatory systems whose effector outputs include signals. Signals are designed to regulate or manage the behavior of others in the manager's interest.

According to assessment/management, the process of assessment is more active than has been generally recognized, and is responsible for the "informa-

Perspectives in Ethology, Volume 12
edited by Owings *et al.*, Plenum Press, New York, 1997

tional" couplings between individuals. Signals are used to manage the behavior of others by operating on the couplings generated by others' assessment activities, not simply by sharing or withholding information.

The idea of feedback plays a powerful bridging role in an assessment/management approach. Feedback ties managing and assessing together because it is both central to regulation and a major source of assessment cues. Feedback also provides a link between proximate and ultimate frameworks because natural selection is an ultimate form of feedback.

An assessment/management point of view takes us beyond the current informational perspective by: (1) focusing our attention on the form of communicative behavior, at all levels, and its structural and pragmatic roots, (2) identifying the motivational aspects of communicative systems, in addition to their cognitive components, as sources of constraint and opportunity for management, (3) highlighting the costs of participating in communication, (4) considering age-specific as well as adult-focused interpretations of developmental changes, (5) offering a broader framework for thinking about signals with high "referential specificity," and (6) discouraging overuse of the concept of deception in communication.

DEVELOPMENT AND USE OF INFORMATION IN RESEARCH ON COMMUNICATION

Information Has Been Accorded a Causal Role

Information has appealed to contemporary biologists because of its connection to two bodies of theory. Engineering Theory (Shannon and Weaver, 1949) defined information as reduction of uncertainty and measured it in terms of binomial units of choice called bits (see also Young, 1954; Wiley, 1983). This definition gave "information" the technical precision of a mathematical construct, but it violated many intuitions about the idea. For instance, in engineering theory the concept of false or true information has no meaning, since both reduce uncertainty equally. Semiotic Theory defined information as a property that permits organisms to choose among courses of action. In its appeal to messages, rather than bits, as units of information, the semiotic treatment came closer than the engineering approach to accommodating such concepts as meaning or knowledge that are intuitively associated with information in ordinary language (Cherry, 1957; Smith, 1977). When biologists combined the two approaches, the intuitive appeal of a commonsense concept was united with the rigor of a mathematical construct under the heading of "the information perspective."

Development of the concept of information was part of a cognitive revolution in the behavioral and biological sciences in which workers in those disciplines began to think of organisms as information-processing systems (e.g.,

Dyer, 1994). As Oyama (1985) has pointed out, information was suddenly everywhere:

> In an increasingly technological, computerized world, information is a prime commodity, and when it is used in biological theorizing it is granted a kind of atomistic autonomy as it moves from place to place, is gathered, stored, imprinted and translated.... Information, the modern source of form, is seen to reside in molecules, cells, tissues, 'the environment,' often latent but causally potent, allowing these entities to recognize, select and instruct each other,... to regulate, control, induce, direct, and determine events of all kinds. When something marvelous happens...the question is always 'Where did the information come from?' (pp. 1–2)

As a part of this revolution, information became a key concept in the animal behavior literature, despite some variation in its definition (Smith, this volume), and recent controversy about the exact role it should play. All proponents of an informational perspective share the view that the impact of signals on the behavior of others is mediated by an exchange of information.

The "Manipulational" Critique of the Informational Perspective Left the Informational Perspective Intact as the Prevailing Proximate Framework

A proposed alternative to the informational perspective, communication as manipulation, provided a portion of a comprehensive framework by focusing attention on the adaptive value *to the signaler* of signal emission. Burghardt (1970) was the first to highlight beneficial consequences to the signaler, rather than mutualistic transfer of information, as central to our understanding of the process of communication. Apparently unaware of Burghardt's contribution, Dawkins and Krebs (1978) developed this viewpoint more completely, and chose the label "manipulation" for it, in order to criticize both the informational perspective and the idea that communicative systems were mutualistic. When Dawkins and Krebs applied the label "manipulation" to both direct physical action and signaling, their point was that the two share a pragmatic quality that does not require informational thinking. At the same time, the negative connotations of the term "manipulation" emphasized self-interested action to the potential detriment of other communicants.

Reactions to Dawkins and Krebs (1978) included both praise of their use of the logic of natural selection, and identification of a serious problem with their formulation. Myrberg (1981), Beer (1982), Wiley (1982, 1983) and Smith (1986a) insisted that, because "information" and "manipulation" deal with different time scales, they are complementary rather than alternative concepts. Information, several critics argued, has to do with the proximate coupling between signaling and other aspects of the signaling situation. The term "manipulation" was used metaphorically and is actually defined in ultimate terms, i.e., of the average impact of the signal on signaler and perceiver fitness. The proximate mechanism even for manipulation, several argued, is the transmission or withholding of information.

In an update of the manipulation approach, Krebs and Dawkins (1984) acknowledged the importance of the concept of information for an understanding of the proximate dynamics of animal communication. The question of whether signals inform in the semantic sense, they maintained, can be treated as a hypothesis for each signal in question, and evaluated on the basis of specific observations or experiments. This proximate empirical matter, they concluded, is distinct from the ultimate theoretical question of whether natural selection favors informative signals (see also Krebs and Davies, 1987).

Thus, the manipulation concept was effective in clarifying the implications of natural selection for ultimate explanations of animal communication. But, it did not offer a convincing proximate formulation that substituted for that provided by the informational perspective (see especially Smith, 1986a). As a result, the proximate part of the synthesis that emerged was still narrowly organized around the concept of information (e.g., see Guilford and Dawkins, 1991; Zuk, 1991), but involved greater interest in the possibility of exploitative activities like deception. Theoretical and empirical studies explored the conditions under which the fitness consequences of deceptive activities were negative and positive (e.g., Wiley, 1983). One conclusion from such work has been that withholding information should in many circumstances be favored over conveying false information, because the signaler is more likely to be caught and experience negative social consequences from the latter (see Cheney and Seyfarth, 1985; Hauser and Marler, 1993).

The Informational Perspective Is Implicitly Anthropomorphic

Central as the concept of information has become to the biological and behavioral sciences, there is some reason to believe that its use seriously misleads us. Despite their ubiquitous contemporary use, terms like information, sender, receiver, message, meaning, and deception have their origin and their intuitive base in discussions of human communication. To call "information transfer" what animals, organs, or even cells do when they interact is implicitly to use human communication as a model for all forms of interaction among organismic entities. Now, it is quite clear that human communication is a biological process that belongs to a much broader class of biological processes with which it shares many attributes. But it is equally clear that human communication is a uniquely specialized and complex member of the class and, therefore, totally unsuitable to serve as a type specimen for the class as a whole. Such usage has the same power to sow confusion as would referring to whales' flippers as hands or horses' hooves as fingernails. No self-respecting comparative anatomist would use the names of highly specialized or unusual structures and processes as the names of the general class of structures or processes of which they are a part. The time-honored biological practice is to use the least specific terms, in this case to refer to the whale's flipper as a forelimb and the horse's hoof as a digit. Nothing in such usage precludes the possibility that there are some animals—raccoons, prosimians, and many anthropoids—for which the term "hand" might be appropriate. Nevertheless, the comparative analysis and evolutionary explanation of animal interaction requires a terminology that can be uncontroversially applied to all members of the class without begging essential evolutionary questions.

For these reasons we sought a way of talking about interaction among organisms that was based on the broadest possible biological principles. We found this way of talking in Cannon's (1935) concept of homeostasis and in its modern derivative, the control system approaches of Bowlby (1969) and Powers (1973), and nonlinear dynamical systems theory (Fogel and Thelen, 1987; Cole, 1994).

Broadening Communication Research with Another Approach, Assessment/Management

Management. Viewed broadly, communicative behavior is part of the general regulatory processes of life. A regulatory system can be thought of as a coordinated collection of parts whose activity tends to maintain some variable at a relatively constant value. Cannon's (1935) concept of homeostasis illustrates this idea; he coined this term to refer to the mechanisms responsible for maintaining some condition of the body, such as its temperature, at a constant value. Cannon's view of biological mechanisms as regulatory systems foreshadowed the development of cybernetics and general systems theory, conceptual frameworks that treat regulation as a formal property of behavioral systems (Bowlby, 1969; Powers, 1973; Toates, 1980). As such, a regulatory view of behavior is descriptive, not explanatory (Thompson, this volume), and can be adopted as an alternative to more mechanistic descriptive formulations such as behaviorism (Pepper, 1948; Mitchell, 1986) in which behaviors are treated as elicited by stimuli rather than as the outputs of regulatory systems.

When an animal regulates its body temperature some aspects of this regulation are based on processes internal to the body; internally generated metabolic heat, for example, can aid in raising body temperatures that have dropped below normal. Other aspects of such regulation depend upon use of environmental resources; an animal might cool off by moving to shade, for example, or warm up by moving into sunlight. Attachment theory (Bowlby, 1969) has built upon this regulatory approach by highlighting the idea that youngsters regulate their own states by using parents and other conspecifics as environmental resources. In other words, infants might regulate such variables as their body temperatures by regulating, or managing, the behavior of their parents.

Regulation of egg temperature by birds provides a useful example. Incubating parents monitor egg temperature, and take corrective action when the temperature deviates from optimal. However, as the time of hatching approaches and passes, parents monitor offspring less closely and contribute less precisely to regulation of their offsprings' temperature (Evans, 1990a, 1990b). Embryonic and neonatal birds begin to take a more active role at this time. For example, white pelicans of this age begin to emit "squawk" calls, vocalizing at higher rates as their temperature deviates further from about 37°C, and calling at lower rates as the 37°C body temperature is restored (Evans, 1992). When the parents hear their young squawking, or even playbacks of squawking, they engage in brooding activities over their young which should result in adjustments in chick or egg temperature. Even embryos are feedback sensitive; they can regulate their own temperature by calling when placed in an experimental incubator with a voice-activated heating and cooling system. So, squawking is one of several effector outputs used by embryos and chicks to regulate their body temperature; squawking contributes to temperature regulation by managing the incubation and brooding behavior of parents.

Bowlby's attachment theory emphasizes that the give-and-take of such early social interactions helps to meet more than just the developing individual's *immediate* needs; social consequences also facilitate development of effective social management skills (e.g., Mason, 1979; West, King and Freeberg, 1994, 1997). Such developmental effects of interactions can be illustrated by West, King and Freeberg's work on the ontogeny of competence as a singer in male brown-headed cowbirds. Young males were reared with the opportunity to interact with older conspecific females, but not with older males. Feedback from females contributed to the ontogeny of songs that were exceptionally effective at inducing female copulatory solicitations in playback experiments. However, the absence of feedback from older males had a profound effect on the competence of these young males. Even though they had very potent songs, they did not deploy them effectively. Normal males sing as part of energetic efforts to court females, approaching them, maneuvering to get in front of them, fluffing their feathers as they get close, and mounting when the female invites it. In contrast, even though male-deprived males sang at females and even occasionally mounted them, they lacked the normal males' enthusiasm and focus, quickly lapsing into a state of disinterest, perching, and "vocalizing in a manner closer to hiccuping than interacting." Despite their more potent songs, deprived males achieved many fewer consortships than normal males did.

A regulatory framework is maximally compatible with the logic of natural selection. Proximate feedback is central to regulation, and natural selection is an ultimate feedback process. This idea parallels in many ways Thompson's natural design approach, which identifies a variety of proximate and ultimate sources of behavioral design (1986, this volume). Combining regulatory and evolutionary approaches further highlights the pragmatic nature of communicative activities, by making it clear that these activities are shaped by both their ultimate and proximate consequences. While the regulatory analogy is compatible with an informational perspective, the two are not logically linked. In contrast, regulation is fundamental to an assessment/management approach.

Assessment. Animals are active, not just in their self-interested management of the behavior of others, but also in their quest for the cues needed for making decisions that meet their own needs. As Gibson (1966) has noted:

> The input of the sensory nerve is not the basis of perception as we have been taught for centuries, but only half of it. It is only the basis for passive sense impressions... . The active senses cannot simply be the initiator of *signals* in nerve fibers or *messages* to the brain; instead they are analogous to *tentacles* and *feelers* [last two italics added]. And the function of the brain when looped with its perceptual organ is not to decode signals, not to interpret messages, nor to accept images. These old analogies no longer apply.... The perceptual systems, including the nerve centers at various levels up to the brain, are ways of seeking and extracting information about the environment from the flowing array of ambient energy. (p. 5)

Gibson's approach implies that all individuals live in a tangle of "feelers" maintained by others as channels for their assessment activities. Virtually any activity is likely to produce effects that "catch" in this tangle, incidentally "tugging" on some feelers and thereby generating consequences for the individual. For example, the odor of recently eaten food lingers on the breaths of Norway rats. Other rats, in active pursuit of cues about which novel foods are safe to eat, sniff the breaths of members of their social group and come to prefer foods that they have smelled there. The development of this preference does not depend simply on exposure to the smell of the food. The odor must be paired with a particular component of rat breath—carbon disulfide—a common constituent of exhaled air. The occurrence of this social learning does not demand active transmission of information. Indeed, the rat who exhales the food cues can be anesthetized, and the pairing of food odor and carbon disulfide will still affect the subsequent food preferences of the sniffing rat (Galef and Wigmore, 1983; Galef, Mason, Preti and Bean, 1988).

Because of active assessment, rats cannot even eat without influencing the behavior of other individuals. Inevitable social consequences of these sorts shape signaling behavior, to the extent that the consequences benefit cue producers. So, individuals do not need to "connect" themselves with others by sending information; they are already connected through many assessment channels. This modern approach contrasts with the longstanding assumption of "receiver" passivity, which has sustained continued reliance on the causal power of information transfer. This residual dependence disappears if assessment is given the evolutionary and proximate importance due to it, for now assessment systems get much more of the credit for the evolution and proximate functioning of communication systems (e.g., as in Endler, 1992; Krebs and Dawkins, 1984; Markl, 1985; Ryan, 1994; and Weldon, 1983). In turn, signals are characterized as having their impact, not by making information available, but by capitalizing on the couplings established by assessment processes. When we modernize our view of how assessment systems work, we must also modernize our view of the mechanism whereby signals influence the behavior of others.

These ideas about assessment are a natural extension of ongoing developments in the study of communication. Researchers have become more sensitive to the implications of active assessment in the past 20 years. For example, a common objection to Dawkins and Krebs' (1978) manipulation view of animal communication was that they gave management much of the power in communicative systems, and discounted the influence of assessment (Cheney and Seyfarth, 1985; Hennessy et al., 1981; Krebs and Dawkins, 1984; Markl, 1985; Morton, 1982; Owings and Hennessy, 1984; Smith, 1986a). Reiterating a theme developed by Zahavi (1977), these critics argued that the potential for signals to manipulate others is constrained by the self-interest of those others. That is, target assessment systems are not responsive to signals that have no links to important conditions. Concepts such as assessment (Parker, 1974), mind reading (Krebs and Dawkins, 1984), probing (Owings and Hennessy, 1984; Markl, 1985), eavesdropping or signal interception (Myrberg, 1981), skeptical receivers (Caryl, 1982; Moynihan, 1982), and receiver psychology (Guilford and Dawkins, 1991) illustrate the increasing sensitivity of researchers in animal communication to the powerful structuring influence of active assessment on communicative systems and behavior. The A/M approach adds the point that we also need to consider the *proximate* effects of assessment; virtually all previous treatments have emphasized *ultimate* effects.

Assessment and Management Mutually Imply One Another

Management implies assessment. That is, effective management of the behavior of others requires simultaneous involvement in assessment of such matters as the identity and state of the target, and of the effects of inputs to the target on its behavior. Indeed, feedback loops are central features of a regulatory approach to behavior. Conversely, assessment implies management. That is, the quest for cues about potential targets of management always occurs in the context of the would-be manager's goals. Gibson (1979) has expressed this idea with the concept of "affordance" in his ecological approach to perception. Affordances are what animals perceive: "The *affordances* of the environment are what it *offers* the animal, either for good or ill" (p. 127). A female mammal may afford her infant warmth, nourishment, refuge from danger, punishment, and so forth, and these are the functional frames of reference that structure the infant's perception of its mother. So, managing and assessing mutually imply one another, just as buying and selling do; they are two sides of the same process of communication. Management is the process of changing/maintaining conditions in one's own interests. Assessment is the process of changing behavior to adjust to current conditions.

The dynamic interplay between assessment and management provides insight into the question "If you are correct in saying that signals do not function simply by providing information, why do they seem so informative?" The A/M answer is that links between the situations and structure of signals are forged by situational constraints. Such an approach helps us to understand why so many bird and mammal species have evolved convergent low-pitched, broad-band growls as threat vocalizations. This convergence reflects the constraints involved in at least three steps. (1) Body size constrains both pitch of vocalizations and danger posed to others. (2) Assessment systems are constrained to use reliable cues, so they judge risk in part on the basis of body-size related factors, such as call pitch. (3) Management systems are constrained by this assessment rule of thumb to use pitch to maximize their apparent threat to others (Morton, 1977). Thus "informative" signals result from a process of attempts to manage and attempts to assess. An emphasis on the information made available by these vocalizations neglects or deemphasizes the processes that produce communication. These broader structural, functional, developmental, and evolutionary contexts account for much of the variation in animal communication. We feel that the informational perspective falls far short of encompassing these contexts. Furthermore, attempts to unite animal communication with other fields interested in animal signals will fail because of the procrustean focus of the informational perspective (e.g., Zuk, 1991).

Implications of A/M for the Study of Animal Communication

In the A/M approach, the central question about communicative behavior, from the perspective of management, is not "what does it stand for?," but "what does it serve to accomplish, and under what constraints?" By substituting the latter question for the former, we approach the management side of communication in a way that recognizes the distinct properties of assessment and management. By looking to the interface of assessment and management, we maximize our chances of understanding the complex causal processes underlying communication.

The Structural and Pragmatic Roots of Form

By emphasizing what signals "stand for" rather than how their structure covaries with circumstances, the informational perspective has deflected our attention from nonarbitrary sources of signal meaning. Identifying nonarbitrary sources of meaning can contribute to an understanding of the evolution *and* proximate functioning of signals. The sources of "informative" cues in the rattling of rattlesnakes illustrate this point. Rattling probably evolved as an aposematic signal to warn adversaries about the rattlesnake's venomousness (Klauber, 1972; Greene, 1988), but this sound leaks additional cues about the danger posed by the rattling snake to its adversaries (Rowe and Owings, 1990, in press). Larger snakes are more dangerous, and they produce rattling sounds of higher amplitude and lower dominant frequency. Similarly, these ectothermic predators are also more dangerous when they are warmer, and warmer individuals produce rattling sounds of audibly higher amplitude and click rate.

This "informativeness" of rattling requires little appeal to the process of specialization for communication to account for its origins. Cues about size and temperature are richly available not only in rattling, but also in the noncommunicative act of striking (Rowe and Owings, 1990). Rattling and striking are similar in having become potential sources of assessment cues through the physical and physiological factors that constrain and account for variation in their form (Hennessy et al., 1981; Owings and Hennessy, 1984). The body size and temperature cues available in rattling have their origins in relatively straightforward physical and physiological constraints. These include size-related differences in the resonant frequencies of rattles, and the greater force and speed of contraction of larger, warmer muscles.

Actively assessing California ground squirrels have discovered these incidental cues and put them to use in their own interest. When venom-resistant adults confront rattlesnakes in defense of their pups, they can be assertive enough to evoke rattling. Playback studies indicate that such feedback makes a difference; these squirrels behave more cautiously to the sounds of larger and warmer snakes (Swaisgood, 1994). Such acoustic assessment cues could be especially valuable in the darkness of nursery burrows, where many such encounters take place.

Size-related cues available in rattling do appear to have been modified somewhat by selection arising from use of these cues by adversaries (Rowe and Owings, 1996). If selection has favored enemies that can assess the vulnerability of rattlesnake adversaries, such appraisal would then select for rattlesnakes that sound as large and formidable as possible. Differences in the scaling of snake body size with rattling and striking are consistent with this hypothesis. Parameters of the noncommunicative act of striking increase linearly with rattlesnake body size. In contrast, parameters of the communicative act of rattling increase logarithmically, reaching an asymptote at moderate snake sizes. Such a pattern may reflect selection on rattlesnakes for the communicative ability to "sound big" at the earliest possible age. There is evidence of some pressure to outgrow "sounding small" as quickly as possible. Small rattlesnakes appear to be more vulnerable to their adversaries (Hersek, 1990); only two rattlesnakes have been mauled by California ground squirrels during observations of many natural encounters, and both of these were less than one year old.

The above observations suggest that rattling by rattlesnakes has become a "derived activity" (Tinbergen, 1952). Rattling originated as an aposematic signal, but subsequently acquired the additional function of making the rattler sound as large as possible. The very useful concept of derived activities has been applied only rarely in the recent communication literature (but see Krebs and Dawkins, 1984; Fogel and Thelen, 1987; and Dawkins and Guilford, this volume). This concept accounts for the structure of signals in part by appealing to the nonsignal and signal acts from which they have been derived evolutionarily and/or developmentally (e.g., see Kruijt, 1964; Tinbergen, 1952). In this approach signal structure is not at all arbitrary, and current signal structure and functions are linked to prior structure and functions as well as to evolutionary success in previous generations. The links between signal structure and important circumstances, which are the bases for accurate assessment, often reflect the constraints that were originally responsible for covariation between "candidate" signals and their situations. (Baring the teeth to threaten someone is another example. Retracting the lips from the teeth is a necessary preparation for biting; through its association with biting, lip retraction has the potential to become a threat [Darwin, 1872; Andrew, 1972; Krebs and Dawkins, 1984].)

According to A/M, one of the major sources of constraint on signal structure is the assessment systems of signal targets (Hennessy et al., 1981; Owings and Hennessy, 1984). It is the biases, cues and concerns of assessment systems that signals are designed (sensu Thompson, this volume) to exploit. Similarly, Guilford and Dawkins (1991; Dawkins and Guilford, this volume) have identified the psychological mechanisms of targets of signals as major sources of constraint on signal structure. Specifically, they have argued that a significant portion of selection on signal structure is not for the information to be conveyed, but for *efficacy* in such activities as exploiting the properties of the perceptual and memory systems of targets. For example, they propose that aposematic signals have high contrast because contrast is memorable, facilitating acquisition by predators of aversions to the pattern.

Dawkins and Guilford have taken a step in the right direction, away from informational thinking, but they have not gone far enough. They retain an informational perspective in the concept of the *strategic or content design* of signals. "Strategic design is concerned with how a signal is constructed by natural selection to provide the information necessary to make a receiver respond" (1991, p. 1). We suggest that this separate category of signal design may be unnecessary. Signals may be effective for managing the behavior of others for a variety of reasons. Signals may effectively capitalize on perceptual systems, by having a highly stimulating structure. Signals may be memorable because they capitalize on the contrast sensitivity of perceptual systems. And, signals may be salient because they deploy and perhaps even exaggerate some cue critical to others for assessment. For example, the structure of close-contact signals is based on physical constraints in the production of sound frequencies; smaller individuals are more limited than larger ones in their capacity to produce low-pitched sounds. Individuals assess the size of others in part by attending to these pitch-based auditory cues. A "motivation/structural rules" (M-S) model has been based on this historical relationship (Morton, 1977, 1982). Management systems capitalize on the use in assessment of this size/sound-pitch relationship; individuals "sound bigger" when more aggressively motivated and "sound smaller" when less aggressively motivated, as when approaching mates. This hypothesis does not demand the addition of a second, strategic category of design; efficacy is as applicable here as to the other cases.

Higher levels of the form of managerial activity. Effective regulation of the behavior of others depends critically on the availability of dimensions for variation in managerial action. The more dimensions available, the more precisely an individual can adjust its managerial efforts to maintain or move conditions toward preferred values. In early views of animal communication, this capacity for adjustment was thought to be quite limited, because display repertoires are small. Nowadays, however, it is recognized that switching to another display is only one of several means of adjusting managerial action (Owings and Hennessy, 1984). Animals also have additional repertoires, including ways to (1) modulate display structure, and (2) combine displays (Smith, 1986b). These additional repertoires greatly enrich an individual's managerial potential. For example, combining displays allows parallel management of the behavior of more than one target. British songbirds called stonechats emit both *whit* and *chack* calls when a predator approaches the nest in which they are rearing their young (Greig-Smith, 1980). *Whitting* serves to reduce the conspicuousness of the nestlings by suppressing their food begging vocalizations. *Chacking*, on the other hand, is varied independently of *whitting*, and is part of an effort to lead the predator away from the vicinity of the nest.

The identification of multiple repertoires of signaling has significantly extended our understanding of animal communication. Among other contributions, this formulation preserves the very useful concept of displays, while providing a way of thinking about the extensive and meaningful variation in display structure uncovered in the past 15–20 years (Smith, 1986b).

Although adopting an A/M approach increases the likelihood of such discoveries, it is not an absolute prerequisite for insights of this sort. For example, John Smith has proposed the idea of multiple repertoires, and has long been the clearest and most consistent articulator of an informational approach to communication (e.g., see his chapter in this volume). However, we suggest that Smith is limited by his continued emphasis of informational thinking about these multiple repertoires in his proximate work on communication. He thinks of signals, and rules for modulating and combining signals, and even formalized interactions as specializations for providing information (1986b, this volume). His proximate inquiries into the information provided by these repertoires, to the relative exclusion of thinking about proximate pursuit of self-interest, limit his insights into the roots of the causal power of all of these repertoires. An A/M approach goes beyond information, looking at the details of each participant's maneuvering in terms of how this represents proximate pursuit of self-interest under situational constraints. In particular, A/M steers our attention to the use and impact of immediate feedback, in addition to developmental and ultimate feedback, as a source of insight into the proximate functional organization and causes of communicative behavior.

The case of *whitting* by stonechats, described above, provides an example of how attention to immediate feedback can deepen our insights into communication. The repetitive emission of *whits* appears to be a case of tonic communication (Schleidt, 1973). Tonic signaling does not elicit a discrete reaction by targets with each emission; it maintains or fosters a longer-term effect with repeated inputs, and often performs that function on more distant targets (Smith, 1991). Tonic signaling often involves rhythmic repetition, a pattern that may serve to manage the behavior of targets through processes analogous to the entrainment of circadian rhythms (Owings, 1994). However, rhythmic signal emission may have an additional source—the absence of the perturbing effects of immediate feedback. Over smaller temporal and spatial scales, which are less conducive to tonic communication, the greater immediacy of feedback could have a larger disruptive effect on such rhythms.

The patterning of signaling in tonic time frames can be understood in part in terms of the temporal aspects of situational constraints. For example, California ground squirrels use tail flagging only to deal with the problem of snake predation (Hennessy et al., 1981), but most episodes of tail flagging (90 percent) occur outside the immediate presence of snakes (Hersek and Owings, 1993). Without the concept of tonic communication, these many episodes of "spontaneous" tail flagging might be labeled as false alarms. But, these episodes are linked to the threat of snake predation, and reflect the unusually tonic threat posed by rattlesnakes. Unlike the raptors and mammalian predators that also prey on these squirrels, rattlesnakes may move into a local area for many days (Hersek, 1990). If a ground squirrel has seen a rattlesnake once, chances are that it will run across it again, and again. The tonic patterning of tail flagging may be designed to deal with this particularly tonic source of predatory threat.

Convergence of communicative systems. If the structure of signals reflects the constraints of their typical situations, then unrelated species experiencing similar constraints might be expected to converge in signal structure, and therefore in signal meaning to perceivers. The best known example of such convergence is the structural differentiation of "seet" and mobbing calls shared by central European passerines of several families (Marler, 1955, 1959). The constraints of predatory situations apparently are consistent enough that even some mammalian species have converged upon similar patterns of call differentiation, e.g., in ground squirrels and marmosets (Melchior, 1971; Owings and Virginia, 1978; see also Vencl, 1977).

Marler's original explanation was that auditory specializations for sound localization by potential perceivers in these predatory situations have shaped the differential structure of these calls. "Seets" are ventriloquial sounds used when callers should minimize their locatability, whereas mobbing calls are structured for maximum localizability, e.g., to facilitate attraction of "help" in mobbing. Some experimental results have been consistent with this hypothesis (Brown, 1982), and others have not (Shalter and Schleidt, 1977; Shalter, 1978). A study of a specific predator-prey system also indicated that seets minimize the caller's conspicuousness to the predator, but do so via a different route (Klump, Kretzschmar and Curio, 1986). The frequency of great tit seet calls is high enough to be much more audible to conspecifics than to European sparrow hawks, whose audiograms roll off at lower frequencies than that of the great tit. A third nonexclusive alternative is that this structural difference between seet and mobbing calls reflects the constraints summarized by the motivation/structural rules hypothesis (Morton 1977). Many signals used in mobbing are chevron-shaped "barks" whose rise then fall in frequency reflects indecision about attack or retreat. Bark structure vacillates between the high tonal and low harsh endpoints, where retreat and attack, respectively, are relatively likely to occur. This signal is often motivated by a stimulus of great salience to the signaler and is salient to others perceiving only the bark and, therefore, attractive to them. In contrast, the high tonal sounds associated with evasion of dangerous aerial predators reflect the signaler's fear endpoint, which causes listening birds to become immobile.

All three of the above hypotheses are consistent with the view that signal structure is shaped by situational constraints, of which the assessment systems of others are an important source. These constraints account for what others can infer from these signals. Studies dealing with the consequences, contexts, and constraints of communicative behavior, rather than focusing narrowly on information transfer, are most likely to discover these adaptive patterns underlying convergent communicative systems (e.g., Marler, 1959; Maier, Rasa and Scheich, 1983; Dubois and Martens, 1984; Ryan, 1985).

Multiple signals with equivalent "meanings." The limits of informational thinking have been especially evident in studies of singing by songbirds, perhaps the most intensively studied type of vocalizing in the animal communication literature. For example, the evolution of large song repertoires has been the target of a variety of hypotheses quite compatible with A/M, including the dishabituation hypothesis (variation in song structure minimizes the habituation of targets to the singer's signals; Kroodsma, 1978), the female stimulation hypotheses (multiple song types are more stimulating to females; Searcy, 1984), and the Beau Geste hypothesis (multiple song types deter intruders by simulating the presence of more than one resident male; Krebs, 1977).

However, at least one feature of bird song has not seemed compatible with the A/M expectation that signal structure is nonarbitrarily constrained by the situation. Learning is necessary for normal song development, and different songs often appear to be functionally equivalent (i.e., appear to "mean" the same thing). So, song structure seems arbitrary. Indeed, these features of song have led to the use of bird song as an animal model of cultural evolution (Lynch et al., 1989). The focus of A/M on the constraints arising from the assessment systems of targets provides a way of thinking about these phenomena. The structure of songs is only arbitrary in the same sense that sexually selected features of signals are arbitrary. In both cases, signal structure is constrained by biases in the assessment systems of targets, whatever the proximate or ultimate sources of these biases. The ranging hypothesis illustrates this way of thinking; it offers a functional biological explanation for both the ability to learn songs and the adaptive choice of different song types during social interactions (Morton, 1986). According to this model, avian perceivers of conspecific songs estimate their distance from the singer using the amount of sound degradation as a cue. Furthermore, perceivers are able to make this estimation only if they have previously memorized the undegraded structure of the same song type (or elements of song types). This method of ranging singer distance provides the biological foundation for the evolution of song type differentiation and geographic distribution. Memorizing the song types of immediate neighbors, and thus being able to judge the distance of singers, permits the listening individual to maintain its own energetic schedule. Likewise, when a trespass is likely, singing the same song type as sung by the intruder takes advantage of the intruder's ability to range a song in his memory. The defender's song threat is more effective as a repellent since its close or declining distance is perceived by the intruder.

Multiple agonistic displays provide another example of multiple signals with equivalent "meanings" whose function and evolution are best explained by the A/M approach. Enquist et al. (1985) and Enquist (1985), working on fulmar displays, found that the cost and effectiveness of different displays are positively correlated—high-cost displays are used when the value of the contested resources is higher. This exemplifies our insistence that constraints on the signaler, rather than information, best account for variation in signal structure. In contrast, Paton and Caryl's work on the eleven agonistic displays of skuas seems much more resistant to this explanation (Paton, 1986; Paton and Caryl, 1986). There was little evidence of differential social impact of these displays, and little support for Enquist's or other functional hypotheses. They concluded that it was inappropriate to label even structurally quite distinct social acts as different displays if they did not differ in their impact on targets (i.e., if they did not have different "meanings"). However, from the perspective of management there is more to functionally distinct displays than elicitation of different reactions. The foundation of the A/M approach, regulation to reduce the difference between preferred and actual conditions, provides a functional explanation: multiple displays that evoke similar responses may serve to achieve the same outcome but be structurally adapted to do so from the constraints of different starting points. The following quote hints at this idea: "Perhaps the birds simply use a range of postures in contests because they are attacked or start an attack from different physical positions" (Krebs and Davies, 1987, p. 339; see also Tinbergen, 1952, pp. 18–19). In a nonagonistic context, structurally distinct "trill" variants of pygmy marmosets do not evoke different reactions in playback experiments. These trills are used to maintain contact at different distances, and maintenance of contact at greater distances constrains trills to be more localizable, i.e., to contain more frequency and amplitude modulation (Snowdon and Hodun, 1981).

The Pervasive Role of Motivation in Communication

Emphasis on the exchange of information is at least implicitly cognitive, in the sense that animals are treated primarily as information-processing systems (e.g., Dyer, 1994). Such an emphasis focuses our attention on the cognitive mechanisms underlying communication, a theme that has proven to be fruitful (Cheney and Seyfarth, 1990). Nevertheless, cognitive questions have been overemphasized, to the detriment of the investigation of motivational factors. As a result, little attention has been paid to the motivational consequences of perceiving signals, even though the effects of motivational states on production of signals have long been a subject of interest (Darwin, 1872; Gouzoules, Gouzoules and Marler, 1985; Hinde, 1985; Marler, 1984; but see Morton, 1977; Jurgens, 1979; de Waal, 1988; Owings, 1994).

When the effects of motivational/emotional signals have been considered, the treatment has typically been more cognitive than motivational. That is, discussions have typically centered around the information transmitted to the target about the signaler's motivational/emotional state, and the cognitive modifications that result. One exception was Dawkins and Krebs' (1978) influential "manipulation" paper, in which the motivational impact of signals was a significant theme:

> There may be little difference between regarding bird song as music and regarding it as hypnosis. 'Hypnotic' rhythm and 'haunting' melody are clichés in the description of human music. The drug-like effect of the nightingale's song on the poet's nervous system ('a drowsy numbness pains my sens, as though of hemlock I had drunk) might be at least as influential on the nervous system of another nightingale. (p. 307)

However, even this paper offered little discussion of the actual motivational mechanisms involved.

Exploration of the cognitive foundations of communication relies on motivational mechanisms, even when we do not explicitly acknowledge them. Studies of antipredator signaling systems, for example, have played a key role in the communication literature in part because of the high priority placed on avoiding predation. The immediate responses usually evoked by such signals make them ideal for playback studies. For a more complete proximate account of communication, we need to explore not only cognitive systems, but also the motivational mechanisms that enable an individual's cognitive, perceptual and managerial activities to be appropriately prioritized for its current needs and situation.

Motivational systems ensure that an individual's effort is distributed over all of its important goals in ways that meet the distinctive properties of each need (Simon, 1994). Some needs must be satisfied immediately when particular occasions arise; for example, one rarely defers action on a predator in order to finish a meal. Feeding, on the other hand, can often be set aside temporarily, because significant amounts of energy can be stored by the body. But, the need to eat becomes more urgent as energy stores are depleted. So, natural selection should favor motivational mechanisms that specify current urgencies of an animal's many needs, redirection mechanisms to defer currently active goals for more urgent or beneficial ones, and attentional mechanisms to focus an animal on sources relevant to currently activated goals.

The effects of motivationally mediated attentional shifts can be felt at very early levels of sensory systems. Neurons called mitral cells in the olfactory bulb of sheep, for example, are only one synapse removed from the receptor cells of the olfactory system. Nevertheless, the mitral cells of adult females undergo a major shift in responsivity as a result of the vaginal and cervical stimulation that accompanies parturition. Cells that were sensitive primarily to food odors prior to parturition become selectively responsive to lamb odors after the female gives birth (Kendrick, Levy and Keverne, 1992). Such attentional shifts play a key role in the establishment of a bond between mother and offspring that makes it possible for the mother to confine her parenting to her own young.

The prioritizing effects of signals have been demonstrated in the work by Wingfield and his colleagues on the role of testosterone in aggression by songbirds (e.g., Wingfield et al., 1987; Wingfield and Hahn, 1994). Peaks in plasma testosterone levels are confined to the period of intense aggression while breeding is going on, at least in migratory populations of such species as song and white-crowned sparrows. Simulated intrusion experiments with caged birds demonstrate that these testosterone peaks are produced by sustained aggressive interactions, and that playback of song is also required to produce this effect. This hormonal change sustains aggressive behavior. Indeed, if testosterone peaks are experimentally prolonged with testosterone implants, these normally monogamous males persist in aggression rather than making the shift to parenting. As a result, they often succeed in acquiring a second mate and clutch, but experience a reduction in reproductive success because their offspring experience greater mortality, apparently from the absence of paternal feeding (Hegner and Wingfield, 1987).

Changes in motivational state are common results of social stimulation, and are often hormonally mediated. "Information transfer" may not be the most useful way to characterize such communicative effects. Dewsbury (1988), for example, argued compellingly that the nonejaculatory intromissions used by many male rodents are communicative rather than copulatory. From an A/M perspective, we would characterize his argument in the following way. When the vaginal stimulation of copulation occurs, female assessment mechanisms initiate reductions in receptivity to subsequent males, and hormonal preparations for pregnancy. Nonejaculatory intromissions by males capitalize on this assessment mechanism. By applying more vaginal stimulation than required for copulation, males expedite the female's reduction in sexual receptivity, and the neuroendocrine changes that initiate pregnancy. Such hormonal changes involve more than just changes in knowledge; they mediate a reordering of priorities (e.g., Hegner and Wingfield, 1987). Although males capitalize on assessment cues used by females, little is added to our understanding by thinking of males as transferring information to females.

The Costs of Participating in Communication

Communicative behavior incurs savings, or costs, only in relation to the costs of other behaviors. For example, bird song might evolve because it partially replaces energetically more costly activities like patrolling territories or fighting (Dawkins and Krebs, 1978; Morton, 1986). Presumably, animals more efficient in energy use will leave more descendants than those less efficient. Similarly, predation risk to an individual may increase with exposure while signaling. When vocalizing is known to be energetically costly and risky the concept of information may be little used [e.g., communication in the Tungara frog (Ryan, 1985, 1994; Ryan et al., 1982); see also Enquist et al., 1985]. Instead, the costs and benefits of vocal behavior are combined with a knowledge of the mechanisms involved to develop an explanation for the evolution and functioning of complex signaling. Note that we include the behavior of participating in communication *and* the signal itself in this explanation.

Cost varies not only with signal structure but also with different patterns of repeated signal emission. Female songbirds, for example, base their assessment of a male's territory on both his song types and his persistence in singing (Morton, 1986; Reid, 1987). This discriminative response to singing suggests that females are interested in the amount of time a male's singing takes away from his feeding. The more singing time he can endure, the more food-rich his territory. Males appear to have "discovered" the females' use of this cue and are now capitalizing on it; several studies have shown an immediate increase in singing rate by males coincident with removal of the mate (e.g., Johnson, 1983; Krebs et al., 1981), and with experimental supplementation of food on territories (Reid, 1987). Zahavi's handicap principle has long recognized the role of signal costs in signal evolution (Zahavi, 1977); perceivers are selectively responsive to costly signals, he argues, because they are difficult to fake and therefore reliable information sources. Dawkins and Guilford (1991) have recently pointed out that costs are not confined to the signaler side of communication; assessment can also have negative consequences, and to the extent that it does, the potential persists for "dishonest" signaling. Such reciprocal interplay between assessment and management is fundamental to the A/M approach. We would add that this interaction between processes also transpires in proximate time scales, in addition to the ultimate time frames emphasized by these individuals. Recognition of these proximate processes reduces the need for the strong focus on information still characteristic of these individuals' proximate statements.

Ontogeny, Adult Focus, and Age Specificity

The A/M view suggests at least two changes to the traditional informational approach to the development of communicative behavior. First, consider how proximate feedback from the social consequences of signal emission might influence signal structure or use (e.g., see Johnston, 1988). Second, when age differences are discovered in how a particular signal is used, do not assume that younger animals are simply incompletely developed, and do not judge the behavior of young only by adult standards. The A/M view leads us to assume that apparent mistakes by young may very well be adaptive in their own right; that is, youngsters may be complete but different regulatory systems. One should explore how the youngsters' different signal usage might be appropriate for their developmental stage. The possibility of such interpretations has been acknowledged (Beer, 1979; Smith, 1985; Marler, Evans and Hauser, 1992), but such considerations have played little role in empirical studies of communication (see Harper, 1972; Roush and Snowdon, 1994; and Mateo, 1996, for exceptions).

Research on the ontogeny of song structure is consistent with the first suggestion above. Young white-crowned sparrow males tutored by recordings of songs selectively learned only conspecific song. When interacting with a live bird instead, they learned that bird's song even when it was heterospecific (e.g., Baptista and Petrinovich 1984). Similarly, the song structure of young brown-headed cowbirds is shaped by the responses of conspecifics (West and King 1985). Pepperberg's (1988) application of social modeling theory to animal communication has yielded a wealth of hypotheses regarding the impact of social context and feedback on the ontogeny of communicative abilities.

Song learning is generally conceived as an intermediary stage in the development of "crystallized" or adult singing. This view does not explain the insertion of appeasement sounds in subsong by young male Carolina wrens (Helgeson, 1980). The appeasement may be a form of "father management" that reduces adult male aggression during the important postfledgling period, a time of increasing parent-young conflict, of the young male's life history. Furthermore, after dispersal from the natal territory, young wrens will switch from full song to subsong when confronted by a territorial adult male (Morton, 1982). This implies that research into subsong function has a rich potential for important discoveries far beyond the traditional view of subsong as simply a steppingstone to crystallized song.

Age-specific considerations have already proven useful in a reinterpretation of Cheney and Seyfarth's (1990) work on the ontogeny of vervet monkey use of alarm calls (Owings, 1994). So, we will deal with it only briefly here. Infant vervets behave as though they are more ignorant than adults, for example, by calling in reaction to both predators and nonpredators. The authors' approach to this developmental problem clearly implies that the infants are "incomplete," and emphasizes the question of how infants achieve adult levels of competence.

However, youngsters may be recruiting adult aid and guidance rather than making mistakes, for example, when they bark at warthogs rather than leopards. Indeed, Cheney and Seyfarth's own data indicate that infants cope effectively with predators through their parents, as predicted by an age-specific view, rather than functioning simply like incomplete adults.

Research on antisnake behavior by black-tailed prairie dogs has yielded results consistent with this age-specific view of signal use. These prairie dogs use two general types of antipredator calls—barks and jump-yips. Barks are associated with higher levels of vulnerability than jump-yips are. Adults usually bark while avoiding predators to which they are more vulnerable, such as coyotes and hawks (King, 1955; Smith et al., 1977), but jump-yip while confronting predators such as snakes to which they are less vulnerable (Owings and Owings, 1979). In a field study involving experimental exposure of prairie dogs to snakes, adults emitted only jump-yips, the typical snake-evoked call, whereas pups both jump-yipped and barked (Owings and Loughry, 1985). Barking seemed appropriate for pups, given their higher age-specific vulnerability. Two other outcomes were consistent with such an interpretation. First, pup jump-yips were significantly more bark-like in structure and pattern of emission than adult jump-yips. Second, preliminary evidence suggested that pups and adults were equally able to distinguish vocally among species and size classes of snakes, but that they made different distinctions in ways appropriate for their differing age-specific vulnerabilities. Similar age-differences have been discovered in the use of tail flagging by California ground squirrels (Hersek and Owings, 1994). Tail flagging is a visual signal used by these squirrels to deal with the problem of snake predation; it is used both "phasically," to deal with the immediate presence of snakes, and tonically, to deal with the prospect of snakes even when they are not immediately present. Pups are more vulnerable than adults to snake predation; consistent with an age-specific approach, pups engage in significantly more tonic tail flagging than adults do. Pups also distinguish more clearly than adults do among various contexts of tail flagging. Pup tail flagging covaries with conditions more clearly than adult tail flagging does, (1) while dealing directly with snakes, as compared with tail flagging outside the immediate presence of snakes, and (2) while tail flagging on days when snakes had been observed on the site, as compared to days when no snakes were observed.

Referential Specificity in Animal Signals

A signal is deemed referentially specific when it conveys "sufficient information about an event for receivers to select appropriate responses" without the aid of contextual information (Macedonia and Evans, 1993). Perhaps the most visible examples come from Cheney and Seyfarth's (1990) studies of the antipredator calls of vervet monkeys, which appear to serve as labels for the predators that threaten this species, such as leopards, eagles, and snakes. Vervets emit audibly distinct calls to deal with these different types of predators, and respond to playbacks of different calls in ways appropriate for the associated type of predator. These results illustrate the minimum of two criteria that need to be met before a signaling system can be judged referential. Emission of the signal must be confined to situations in which the putative referent is present. And, perceivers of the signal must respond to the playback of the signal as though the referent is present.

Referential information has been contrasted with "motivational" information, that is, information about the signaler's internal state rather than about external objects or events (Marler, 1984). Such variation is not thought to be categorical, though. Motivational and referential are, instead, two ends of a continuum, with intermediate points representing signals that mix the two forms of information in varying proportions (Marler et al., 1992).

How is referential specificity treated in an A/M approach? First, the contrast between the processes of assessment and management in communication is fundamental. Even though communication involves an interplay between these two processes, the specific mechanisms involved are not necessarily intricately coordinated or coevolved. This point follows logically from the idea that assessment systems are specialized for many purposes other than communication, and so may bring preexisting biases to a communicative context that shape the features of signals through feedback processes (e.g., Ryan, 1994). Nevertheless, the literature on referential specificity has, so far, assumed symmetry of the relevant mechanisms. That is, a referentially specific signal has been assumed to involve referential specificity in both the management system and associated assessment mechanisms of typical targets. Little consideration has been given to the possibility that a management system might meet the criteria for referential specificity, but not the associated assessment system, or vice versa (Owings, 1994). But this is probably a common mix. For example, we know from Cheney and Seyfarth's (1990) work on vervet monkeys that manager and assessor systems can function at different levels even within the same individual. The reactions of six- to seven-month-old vervets to call playbacks are indistinguishable from the responses of adults, but the emission of calls in adult-like fashion, confining each to a particular type of predator, does not develop fully for another 18 months. So, there is A/M symmetry when these juveniles respond to adult calls, but not in the equally likely case of adults responding to juvenile calls. Such asymmetry may be common in communication not only between different age classes, but also between different species.

Two additional points are important regarding referential specificity from an A/M approach. First, the term *referential* specificity would be modified, to the less interpretive one of *situational* specificity. From the perspective of management, signals do not *refer* to anything; they are pragmatic acts emitted to produce an effect of variable specificity. Second, the issue of referential specificity would also be treated as important primarily as a subset of the broader topic of the relationship between signal structure and situations of use. Remember that, according to A/M, relations between signals and situations arise from the shaping effects of situational constraints. A minimum requirement for situationally specific signals of the "referentially specific" sort might be the possession by signal targets of the cognitive ability to deal with quite specific categories of entities, such as "leopard," "snake," and "eagle." Such a cognitive ability would set the stage for managing the behavior of others through the use of signals with that level of situational specificity. Thus, the cognitive abilities of targets of management comprise one class of constraints on the specificity of the relationship between signal structure and situation. This class of constraints is the one most compatible with an informational view of signaling, i.e., "if my conspecifics can understand about leopards, then I can signal to them about leopards."

The above cognitive source of situational specificity offers little guidance regarding the expected structure of signals. Macedonia and Evans (1993) have discussed sources of selection for contrasts in signal structure in an essay on the meaning of mammalian antipredator vocalizations. They begin with two simplifying steps, to make their task more tractable. First, they set aside the difficult question of underlying cognitive mechanisms, and explore "functionally referential" signals, i.e., signals that are confined to specific predators and evoke predator-specific responses, whatever the associated cognitive mechanism. Second, they argue that the judgment of functional referentiality is least ambiguous when dealing with qualitatively different signals, rather than with different variants of the same signal type. (A signal that exhibits graded variation might be more likely to convey information about graded changes in motivational states.) Such qualitative differences in the structure of antipredator signals, they propose, might be favored by selection where the escape responses required to avoid the different classes of predators are incompatible with each other (e.g., vervet monkeys climb trees to evade leopards, but leave trees to deal with eagles). The rationale for this hypothesis is not entirely clear. Perhaps they mean that telling another individual to engage in conflicting activities might most effectively be accomplished with signals of conflicting structures; the contrast in structure would then supplement the contrast in meaning.

An A/M approach would also lead us to seek more pragmatic constraints. Acredolo and Goodwyn's (1990) work on communicative signing in normal human infants provides an example of this sort of constraint. Prior to developing vocal proficiency, human infants develop and deploy communicative gestures in their interactions with parents. The form of gestures for objects typically arise either within interactive routines involving the parent (e.g., mimicking the mother's blowing on a fish mobile as a general symbol for fish), or through the child's own interaction with the object (e.g., a throwing motion to symbolize ball). Thus, the forms of these signals are not arbitrary; they emerge from the interactions that the infant typically has in association with the objects. This proximate process of creating signals is analogous to the ultimate process of evolving a signal from an intention movement (Tinbergen, 1952). Regarding the origins of form differences in antipredator calls, these results would lead us to explore how, for example, different evasive maneuvers might differentially distort the vocal cavity during calling; that is, how call structure emerges from interactive routines with the predator. Ohala (1984) has made an analogous "by-product" argument regarding the origins of the appeasing "grin" and aggressive "o-face" in primates. These visual displays, he hypothesized, were originally simply the facial adjustments needed to raise voice pitch in appeasement (shortening the vocal tract with the grin); and to lower voice pitch in a threatening growl (lengthening the vocal tract via lip protrusion in the o-face). These facial expressions subsequently became emancipated from sound production, serving as signals in their own right.

Finally, A/M's focus on the broad issue of constraints can direct our attention to sources of situational specificity that would not appropriately be called referential even in an informational perspective. For example, Macedonia and Evans' hypothesis, described above, would lead to the prediction that California ground squirrels should have what they call functionally referential antipredator signals. Incompatible antipredator responses are required for two types of predator; the burrows in which these squirrels live are sources of refuge from most mammalian predators that threaten them (not counting mustelids), but are sources of danger from rattlesnakes. Consistent with this hypothesis, the squirrels use very different signals for the different types of predators, chatter calling for mammalian predators, but waving their fluffed tail back and forth (tail flagging) for snakes (Owings, Hennessy, Leger and Gladney, 1986). By their operational definition, then, this is a functionally referential signaling system. However, the term seems inappropriate, for the following reason. The most likely reason for the structural difference in signals is that the predators themselves are targets of these signals (Hennessy and Owings, 1988; Hersek and Owings, 1993). The switch to visual signals with snakes apparently has been favored because, unlike mammalian predators, the snakes apparently cannot hear the calls of these squirrels. The specificity of signal structure to predator type arises from the properties of the signal target, not what is typically called the referent. Extending Macedonia and Evans' argument, in our terminology, we might say that qualitative differences in constraint on signal structure arise from a variety of sources, which can, for example, include the sensory specializations of signal targets.

DECEPTION

As mentioned above, strong interest in the study of deception has been stimulated by the logic of natural selection, especially when combined with a focus on information. The following quote from Dawkins and Krebs' (1978) "manipulation" paper provides a statement on the utility of the concept of deception that is also appropriate from an A/M perspective.

> There can be no doubt that an informational view of animal communication is helped by a consideration of assessment and perhaps also deception. But we prefer to avoid the very idea of information, whether true or false. Wilson remarks that 'If a zoologist were required to select just one word that characterizes animal communication systems, he might well settle on 'redundancy'. Animal displays as they occur in nature tend to be very repetitious, in extreme cases approaching the point of what seems like inanity to the human observer' (Wilson 1975). But it is only redundant and inane if you think the animals are trying to convey *information*. Substitute terms like manipulation, propaganda, persuasion, or advertising, and the 'redundancy' starts to make sense. (p. 304)

Consistent with the above quote, an A/M approach does not challenge the evidence in general that animals engage in communicative behavior that is exploitative; it questions excessive application of the anthropocentric term "deception" to the wide array of forms of exploitative behavior. In particular, an A/M perspective finds little utility in treating exploitative communication in terms of information exchange. There is little doubt that a modest subset of exploitative behavior can be labeled as deceptive, i.e., as withholding information or purveying false information (Hauser, in press), but we need descriptive labels that cover a wider range of cases. Indeed, the idea of withholding information begs for some alternative treatment such as that provided by A/M. It makes sense to treat withholding information as exploitative only in the context of an expectancy that the information will be available. That expectancy must be held by the target of exploitation, and is most appropriately treated as an aspect of the target's assessment system. Thus, the idea of "withholding information" is just one example of a broader concept—exploiting the assessment systems of others—the means whereby management is said to work.

How else can an information-based treatment lead us astray? The egg-mimicking spots on the anal fins of mouth-brooding cichlids provide one example; they were originally explained as working through a misperception by females that they were actually eggs (Wickler, 1962). More recent evidence, however, indicates that these spots function through their salience rather than their deceptive properties. Placement of these egg spots may draw the female's attention to a feature indicative of body size, thereby facilitating the male's advertisement of his size to females, and promoting his success in attracting mates (Hert, 1989; Guilford and Dawkins, 1991). The widespread mimicry of eyes, for example in the tails of peacocks, illustrates the same point. It is not surprising that eye-like patterns have acquired salience to assessment systems since an animal's eyes are most consistently visible to another when it is attending to the other. Mimicry of eyes may capitalize on this salience to catch the attention of targets (Guilford and Dawkins, 1991), and perhaps just as important, may also capitalize on the arousing effects of eye-like schema (Coss,

1979) to foster the long-term effects of signal input by facilitating the target's memory of the communicative occasion (Coss, personal communication; Guilford and Dawkins, 1991; see Brown and Kulik, 1977, for an example of the effects of arousal on memory). Analogously, television and magazine advertisements attempt to sell products by depicting them in association with very attractive women or men. Few viewers expect to become that attractive by purchasing those products; but, the product may become more appealing to potential purchasers through its association with attractive individuals.

CONCLUSION

When a parent wants a child to come to the dinner table, he can carry the child there or summon her to "come eat." There is a fundamental similarity between these two means; both are pragmatic ways for the parent to manage the child in order to accomplish his parental goals. But, there is also a fundamental difference in the features of the child that are used to get her to the table. When the parent carries his child, his actions take advantage of the physical properties of the child, e.g., lifting beneath the arms, and supporting at "folding" points in the body such as the back of the knees. When a child is summoned, the parent's actions take advantage of the psychological properties of the child, i.e., the assessment systems that translate input into action. Just as the carrying parent must grab and support the child by the physical handles, the summoning parent must grab and guide the child by the psychological appendages extended by the assessing child into its environment. Since assessment systems are complex and hierarchically structured, the managerial actions that capitalize on them must be similarly complex. Major insights into the structure and patterning of communicative behavior can be provided by examining the properties of the assessment systems that it must work through, as well as the goals and ultimate functions of that behavior.

The assessment/management approach illustrated by the above passage contrasts with an informational perspective. This informational perspective is founded upon the appealing amalgamation of semantic and information theoretic definitions of information, coupled with an underemphasis of the process of assessment in communication. As a result, only signaling has consistently been given active status in communication, with the consequence that information exchange has been needed to mediate the impact of signals on other animals. However, when assessment by others is credited with establishing the couplings between individuals, there is no utility in the concept of information in accounting for the activities of signalers. The door is then opened to a more pragmatic view of signaling and one that yields a proximate formulation that is easily synthesized with ultimate approaches to the study of animal communication.

Assessment/management does just that, treating signals as acts shaped by both their proximate and ultimate consequences. This formulation only eliminates information-like ideas from communication *when discussing the process of managing*. From the perspective of assessing, the term may continue to be useful shorthand for a description of what the process of assessment yields. Communication is not caused by information transfer; it emerges when the assessment and management activities of different individuals mutually constrain one another, and is both a product of and constraint on those individual activities.

ACKNOWLEDGMENTS

Various incarnations of this manuscript have benefited from the critical and constructive commentary of the following individuals: Mike Beecher, Diane Boellstorff, Gordon Burghardt, Dorothy Cheney, Dick Coss, Kim Derrickson, John Eisenberg, Karen Ericksen, Stevan Harnad, Marc Hauser, Tim Johnston, Dale Lott, Jim Loughry, Vickie McDonald, Bill Mason, Sally Mendoza, Fernando Nottebohm, Susan Oyama, Irene Pepperberg, Lewis Petrinovich, Matt Rowe, Robert Seyfarth, W. John Smith, David Spector, Bridget Stutchbury, Ron Swaisgood, Nick Thompson, Steve Towers, Meredith West, Amotz Zahavi, and a number of anonymous reviewers. Nick Thompson's guidance was especially critical in bringing this essay to its final form. This does not imply, of course, that any of the above endorse what we have to say.

REFERENCES

Acredolo, L. P. and Goodwyn, S. W. (1990) Sign language in babies: The significance of symbolic gesturing for understanding language development. In Vasta, R. (ed) *Ann. Child Devel., 7,* 1–42, Jessica Kingsley Publishers.

Andrew, R. J. (1972) The information potentially available in mammalian displays. In Hinde, R. A. (ed) *Non-verbal Communication* (pp. 179–204). Cambridge: Cambridge University Press.

Baptista, L. F. and Petrinovich, L. (1984) Social interaction, sensitive phases and the song template hypothesis in the white-crowned sparrow. *Anim. Behav., 32,* 172–181.

Beer, C. G. (1979) Vocal communication between laughing gull parents and chicks. *Behaviour, 70,* 118–146.

Beer, C. G. (1982) Conceptual issues in the study of communication. In Kroodsma, D. E. and Miller, E. H. (eds) *Acoustic Communication in Birds: Vol. 2* (pp. 279–310). New York: Academic Press.

Bowlby, J. (1969) *Attachment.* New York: Basic Books.

Brown, C. H. (1982) Ventriloquial and locatable vocalizations in birds. *Z. Tierpsychol., 59,* 338–350.

Brown, R. and Kulik, J. (1977) Flashbulb memories. *Cognition, 5,* 73–99.

Burghardt, G. M. (1970) Defining "communication." In Johnston, J. W., Jr., Moulton, D. G. and Turk, A. (eds) *Advances in Chemoreception. Vol. I: Communication by Chemical Signals* (pp. 5–18). New York: Appleton-Century-Crofts.

Cannon, W. B. (1935) Stresses and strains of homeostasis. *Am. J. Med. Sci., 189,* 1–14.

Caryl, P. G. (1982) Telling the truth about intentions. *J. Theoret. Biol., 97,* 679–689.

Cheney, D. L. and Seyfarth, R. M. (1985) Vervet monkey alarm calls: manipulation through shared information? *Behaviour, 94,* 150–166.

Cheney, D. L. and Seyfarth, R. M. (1990) *How Monkeys See the World: Inside the Mind of Another Species.* Chicago, IL: University of Chicago Press.

Cherry C. (1957) *On Human Communication: A Review, A Survey, and A Criticism.* Cambridge, MA: Technology Press of MIT.

Cole, B. J. (1994) Chaos and behavior: the perspective of nonlinear dynamics. In Real, L. A. (ed) *Behavioral Mechanisms in Evolutionary Ecology* (pp. 423–443). Chicago, IL: University of Chicago Press.

Coss, R. G. (1979) Perceptual determinants of gaze aversion by normal and psychotic children: The role of two facing eyes. *Behaviour, 69,* 228–254.

Darwin, C. (1872) *The Expression of the Emotions in Man and Animals.* New York: Appleton.

Dawkins, M. S. and Guilford, T. (1991) The corruption of honest signalling. *Anim. Behav., 41,* 865–873.

Dawkins, R. and Krebs, J. R. (1978) Animal signals: information or manipulation? In Krebs, J. R. and Davies, N. B. (ed) *Behavioural Ecology: An Evolutionary Approach* (pp. 282–309). Sunderland, MA: Sinauer.

De Waal, F. (1988) Emotional control. *Behav. Brain Sci., 11,* 254.

Dewsbury, D. A. (1988) Copulatory behavior as courtship communication. *Ethology, 79,* 218–234.

Dubois, A. and Martens, J. (1984) A case of probable vocal convergence between frogs and a bird in Himalayan torrents. *J. Ornithol., 125,* 455–463.

Dyer, F. C. (1994) Spatial cognition and navigation in insects. In Real, L A (ed) *Behavioral Mechanisms in Evolutionary Ecology* (pp. 66–98). Chicago, IL: University of Chicago Press.

Endler, J. A. (1992) Signals, signal conditions, and the direction of evolution. *Am. Nat., 139,* S125-S153.

Enquist, M. (1985) Communication during aggressive interactions with particular reference to variation in choice of behaviour. *Anim. Behav., 33,* 1152–1161.

Enquist, M., Plane, E., and Roed, J. (1985) Aggressive communication in fulmars (*Fulmaris glacialis*) competing for food. *Anim. Behav., 33,* 1107–1120.

Evans, R. M. (1990a) Embryonic fine tuning of pipped egg temperature in the American white pelican. *Anim. Behav., 40,* 963–968.

Evans, R. M. (1990b) Vocal regulation of temperature by avian embryos: a laboratory study with pipped eggs of the American white pelican. *Anim. Behav., 40,* 969–979.

Evans, R. M. (1992) Embryonic and neonatal vocal elicitation of parental brooding and feeding responses in American white pelicans. *Anim. Behav., 44,* 667–675.

Fogel, A. and Thelen, E. (1987) Development of early expressive and communicative action: Reinterpreting the evidence from a dynamic systems perspective. *Dev. Psychol., 23,* 747–761.

Galef, B. G., Mason, J. R., Preti, G., and Bean, N. J. (1988) Carbon disulfide: A semiochemical mediating socially-induced diet choice in rats. *Physiol. Behav., 42,* 119–124.

Galef, B. G. and Wigmore, S. W. (1983) Transfer of information concerning distant foods: A laboratory investigation of the "information-centre" hypothesis. *Anim. Behav., 31,* 748–758.

Gibson, J. J. (1966) *The Senses Considered as Perceptual Systems.* Boston: Houghton Mifflin.

Gibson, J. J. (1979) *The Ecological Approach to Visual Perception.* Hillsdale, NJ: Lawrence Erlbaum Associates.

Gouzoules, H., Gouzoules, S. and Marler, P. (1985) External reference and affective signaling in mammalian vocal communication. In Zivin, G. (ed) *The Development of Expressive Behavior: Biology-Environment Interactions* (pp. 77–101). New York: Academic Press.

Greene, H. W. (1988) Antipredator mechanisms in reptiles. In Gans, C. and Huey, R. B. (eds) *Biology of the Reptilia 16, Ecology B: Defense and Life History* (pp. 1–152). New York: Alan R. Liss.

Greig-Smith, P. W. (1980) Parental investment in nest defence by stonechats (*Saxicola torquata*). *Anim. Behav., 28,* 604–619.

Guilford, T. and Dawkins, M. S. (1991) Receiver psychology and the evolution of animal signals. *Anim. Behav., 42,* 1–14.

Harper, L. V. (1972) The transition from filial to reproductive function of "coitus-related" responses in young guinea pigs. *Dev. Psychobiol., 5,* 21–34.

Hauser, M. D. (1997) Minding the behavior of deception. In Whiten, A. and Byrne, R. W. (eds) *Machiavellian Intelligence II.* pp 112–143 Cambridge: Cambridge University Press.

Hauser, M. D. and Marler, P. (1993) Food-associated calls in rhesus macaques (*Macaca mulatta*): II. Costs and benefits of call production and suppression. *Behav. Ecol., 4,* 206–212.

Hegner, R. E. and Wingfield, J. C. (1987) Effects of experimental manipulation of testosterone levels on parental investment and breeding success in male house sparrows. *Auk, 104,* 462–469.

Helgeson, N. (1980) *Development of Song in the Carolina Wren.* University of Maryland, unpublished M.S. thesis.

Hennessy, D. F. and Owings, D. H. (1988) Rattlesnakes create a context for localizing their search for potential prey. *Ethology, 77,* 317–329.

Hennessy, D. F., Owings, D. H., Rowe, M. P., Coss, R. G. and Leger, D. W. (1981) The information afforded by a variable signal: constraints on snake-elicited tail flagging by California ground squirrels. *Behaviour, 78,* 188–226.

Hersek, M. J. (1990) *Behavior of Predator and Prey in a Highly Coevolved System: Northern Pacific Rattlesnakes and California Ground Squirrels.* Ph. D. dissertation, University of California, Davis.

Hersek, M. J. and Owings, D. H. (1993) Tail flagging by adult California ground squirrels: A tonic signal that serves different functions for males and females. *Anim. Behav., 46,* 129–138.

Hersek, M. J. and Owings, D. H. (1994) Tail flagging by young California ground squirrels, *Spermophilus beecheyi*: Age-specific participation in a tonic communicative system. *Anim. Behav., 48,* 803–811.

Hert, E. (1989) The function of egg-spots in an African mouth-brooding cichlid fish. *Anim. Behav., 37,* 726–732.

Hinde, R. A. (1985) Was 'the expression of the emotions' a misleading phrase? *Anim. Behav., 33,* 985–992.

Johnson, L. S. (1983) Effect of mate loss on song performance in the plain titmouse. *Condor, 85,* 378–380.

Johnston, T. D. (1988) Developmental explanation and the ontogeny of birdsong: Nature/nurture redux. *Behav. Brain Sci., 11,* 617–663.

Jurgens, U. (1979) Vocalization as an emotional indicator, a neuroethological study in the squirrel monkey. *Behaviour, 69,* 88–117.

Kendrick, K. M., Levy, F., and Keverne, E. B. (1992) Changes in sensory processing of olfactory signals induced by birth in sheep. *Science, 256,* 833–836.

King, J. A. (1955) Social behavior, social organization, and population dynamics in a black-tailed prairie dog town in the Black Hills of South Dakota. *Contrib. Lab. Vert. Biol., Univ. Michigan, 67,* 1–123.

Klauber, L. M. (1972) *Rattlesnakes: Their Habits, Life Histories and Influence on Mankind, Vols. 1, 2.* Berkeley, CA: University of California Press.

Klump, G. M., Kretzschmar, E. and Curio, E. (1986) The hearing of an avian predator and its avian prey. *Behav. Ecol. Sociobiol., 18,* 317–323.

Krebs, J. R. (1977) The significance of song repertoires: the Beau Geste hypothesis. *Anim. Behav., 25,* 475–478.

Krebs, J. R., Avery, M., and Cowie, R. J. (1981) Effect of removal of mate on singing behaviour of great tits. *Anim. Behav., 29,* 635–637.

Krebs, J. R. and Davies, N.B. (1987) *An Introduction to Behavioral Ecology,* (2nd ed.). Oxford: Blackwell Scientific Publications.

Krebs, J. R. and Dawkins, R. (1984) Animal signals: mind-reading and manipulation. In Krebs, J. R. and Davies, N. B. (eds) *Behavioural Ecology: An Evolutionary Approach,* (2nd ed., pp. 380–402). Sunderland, MA: Sinauer.

Kroodsma, D. E. (1978) Continuity and versatility in bird song: support for the monotony-threshold hypothesis. *Nature, 274,* 681–683.

Kruijt, J. P. (1964) Ontogeny of social behaviour in Burmese red junglefowl (*Gallus gallus spadiceus*) Bonnaterre. *Behaviour Suppl., 12,* 1–201.

Lynch, A., Baker, G. M., Jenkins, A. J., and Plunkett, P. F. (1989) A model of cultural evolution of chaffinch song derived with the meme concept. *Am. Nat., 133,* 634–653.

Macedonia, J. M. and Evans, C. E. (1993) Variation among mammalian alarm call systems and the problem of meaning in animal signals. *Ethology, 93,* 177–197.

Maier, Y., Rasa, O. A. E., and Scheich, H. (1983) Call-system similarity in a ground-living social bird and a mammal in the brush habitat. *Behav. Ecol. Sociobiol., 12,* 5–10.

Markl, H. (1985) Manipulation, modulation, information, cognition: some of the riddles of communication. In Holldobler, B. and Lindauer, M (eds) *Experimental Behavioral Ecology and Sociobiology* (pp. 163–194). Sunderland, MA: Sinauer.

Marler, P. (1955) Characteristics of some animal calls. *Nature, 176,* 6–8.

Marler, P. (1959) Developments in the study of animal communication. In Bell, P. R. (ed) *Darwin's Biological Work* (pp. 150–206). New York: John Wiley.

Marler, P. (1984) Animal communication: affect or cognition? In Scherer, K. and Ekman, P. (eds) *Approaches to Emotion* (pp. 345–365). Hillsdale, NJ: Lawrence Erlbaum Associates.

Marler, P., Evans, C. S., and Hauser, M D (1992) Animal signals: Motivational, referential, or both? In Papousek, H., Jurgens, U., and Papousek, M. (eds) *Nonverbal Vocal Communication: Comparative and Developmental Approaches* (pp. 66–86). Cambridge: Cambridge University Press.

Mason, W. A. (1979) Wanting and knowing: A biological perspective on maternal deprivation. In Thoman, E. (ed) *Origins of the Infant's Social Responsiveness* (pp. 225–249). Hillsdale, NJ: Lawrence Erlbaum Associates.

Mateo, J. M. (1996) The development of alarm-call response behaviour in free-living Belding's ground squirrels. *Anim. Behav. 52,* 489–505.

Melchior, H. R. (1971) Characteristics of Arctic ground squirrel alarm calls. *Oecologia, 7,* 184–190.

Mitchell, R. W. (1986) A framework for discussing deception. In Mitchell, R., and Thompson, N. S. (eds) *Deception: Perspectives on Human and Nonhuman Deceit* (pp. 3–40). Albany, NY: SUNY Press.

Morton, E. S. (1977) On the occurrence and significance of motivation-structural rules in some bird and mammal sounds. *Am. Nat., 111,* 855–869.

Morton, E. S. (1982) Grading, discreteness, redundancy, and motivation-structural rules. In Kroodsma, D. E., and Miller, E. H. (eds) *Acoustic Communication in Birds* (pp. 183–212). New York: Academic Press.

Morton, E. S. (1986) Predictions from the ranging hypothesis for the evolution of long distance signals in birds. *Behaviour, 99,* 65–86.

Moynihan, M. (1982) Why is lying about intentions rare during some kinds of contests? *J. Theoret. Biol., 97,* 7–12.

Myrberg, A. A., Jr. (1981) Sound communication and interception in fishes. In Tavolga, W. N., Popper, A. N. and Fay, R. R. (eds) *Hearing and Sound Communication in Fishes* (pp. 395–426). New York: Springer-Verlag.

Ohala, J. J. (1984) An ethological perspective on common cross-language utilization of Fo of voice. *Phonetica, 41,* 1–16.

Owings, D. H. (1994) How monkeys feel about the world. A review of *How Monkeys See the World. Lang. Commun., 14,* 15–30.

Owings, D. H. and Hennessy, D. F. (1984) The importance of variation in sciurid visual and vocal communication. In Murie, J. O. and Michener, G. R. (eds) *The Biology of Ground-Dwelling Squirrels* (pp. 169–200). Lincoln, NE: University of Nebraska Press.

Owings, D. H., Hennessy, D. F., Leger, D. W., and Gladney, A. B. (1986) Different functions of "alarm" calling for different time scales: A preliminary report on ground squirrels. *Behaviour, 99,* 101–116.

Owings, D. H. and Loughry, W. J. (1985) Variation in snake-elicited jump-yipping by black-tailed prairie dogs: Ontogeny and snake specificity. *Z. Tierpsychol., 70,* 177–200.

Owings, D. H. and Owings, S. C. (1979) Snake-directed behavior by black-tailed prairie dogs. *Z. Tierpsychol., 49,* 35–54.

Owings, D. H. and Virginia, R. A. (1978) Alarm calls of California ground squirrels (*Spermophilus beecheyi*). *Z. Tierpsychol., 46,* 58–78.

Oyama, S. (1985) *The Ontogeny of Information: Developmental Systems and Evolution.* Cambridge: Cambridge University Press.

Parker, G. A. (1974) Assessment strategy and the evolution of fighting behavior. *J. Theoret. Biol., 47,* 223–243.

Paton, D. (1986) Communication by agonistic displays: II. Perceived information and the definition of agonistic displays. *Behaviour, 99,* 157–175.

Paton, D. and Caryl, P. G. (1986) Communication by agonistic displays: I. Variation in information content between samples. *Behaviour, 98,* 213–239.

Pepper, S. C. (1948) *World hypotheses: A study in evidence.* Berkeley, CA: University of California Press.

Pepperberg, I. M. (1988) The importance of social interaction and observation in the acquisition of communicative competence: Possible parallels between avian and human learning. In Zentall, T. R. and Galef, B. G., Jr. (eds) *Social Learning: Psychological and Biological Perspectives* (pp. 279–299). Hillsdale, NJ: Lawrence Erlbaum Associates.

Powers, W. T. (1973) Feedback: Beyond behaviorism. *Science, 179,* 351–356.

Reid, M. L. (1987) Costliness and reliability in the singing vigour of Ipswich sparrows. *Anim. Behav., 35,* 1735–1743.

Roush, R. S. and Snowdon, C. T. (1994) Ontogeny of food-associated calls in cotton-top tamarins. *Anim. Behav., 47,* 263–273.

Rowe, M. P. and Owings, D. H. (1990) Probing, assessment, and management during interactions between ground squirrels and rattlesnakes. Part 1. Risks related to rattlesnake size and body temperature. *Ethology, 86,* 237–249.

Rowe, M. P. and Owings, D. H. (1996) Probing, assessment, and management during interactions between ground squirrels (Rodentia: Sciuridae) and rattlesnakes (Squamata: Viperidae). Part 2. Cues afforded by rattlesnake rattling. *Ethology. 102,* 856–874.

Ryan, M. J. (1985) *The Tungara Frog: A Study in Sexual Selection and Communication.* Chicago, IL: University of Chicago Press.

Ryan, M. J. (1994) Mechanisms underlying sexual selection. In Real, L. .A (ed) *Behavioral mechanisms in evolutionary ecology* (pp. 190–215). Chicago, IL: University of Chicago Press.

Ryan, M., Rand, M. D., and Tuttle, A. S. (1982) Bat predation and sexual advertisement in a neotropical frog. *Am. Nat., 119,* 136–139.

Schleidt, W. M. (1973) Tonic communication: continual effects of discrete signs in animal communication systems. *J. Theoret. Biol., 42,* 359–386.

Searcy, W. A. (1984) Song repertoire size and female preferences in song sparrows. *Behav. Ecol. Sociobiol., 14,* 281–286.

Shalter, M. D. (1978) Location of passerine seeet and mobbing calls by goshawks and pygmy owls. *Z. Tierpsychol., 46,* 260–267.

Shalter, M. D. and Schleidt, W. M. (1977) The ability of barn owls to discriminate and localize avian alarm calls. *Ibis, 119,* 22–27.

Shannon, C. E. and Weaver, W. (1949) *The Mathematical Theory of Communication.* Urbana, IL: University of Illinois Press.

Simon, H. A. (1994) The bottleneck of attention: connecting thought with motivation. In Spaulding, W. D. (ed) *Integrative Views of Motivation, Cognition, and Emotion. Nebraska Symposium on Motivation, 41,* 1–21.

Smith, W. J. (1977) *The Behavior of Communicating.* Cambridge, MA: Harvard University Press.

Smith, W. J. (1985) Comparative study of the ontogeny of communication. In Gollin, E. S. (ed) *The Comparative Development of Adaptive Skills: Evolutionary Implications,* Hillsdale, NJ: Lawrence Erlbaum Associates.

Smith, W. J. (1986a) An "informational" perspective on manipulation. In Mitchell, R. W. and Thompson, N. S. (eds) *Deception: Perspectives on human and nonhuman deceit* pp 71–86. Albany, New York: SUNY Press.

Smith, W. J. (1986b) Signaling behavior: Contributions of different repertoires. In Schusterman, R. J., Thomas, J. A., and Wood, F. G. (eds) *Dolphin cognition and behavior: A comparative approach* (pp. 315–330). Hillsdale, NJ: Lawrence Erlbaum Associates.

Smith, W. J. (1991) Singing is based on two markedly different kinds of signaling. *Journal of Theoretical Biology, 152,* 241–253.

Smith, W. J., Smith, S. L., Oppenheimer, E. C. and Devilla, J. G. (1977) Vocalizations of the black-tailed prairie dog, *Cynomys ludovicianus. Animal Behaviour, 25,* 152–164.

Snowdon, C. T. and Hodun, A. (1981) Acoustic adaptations in pygmy marmoset contact calls: locational cues vary with distances between conspecifics. *Behavioral Ecology and Sociobiology, 9,* 295–300.

Swaisgood, R. R. (1994) *Assessment of Rattlesnake Dangerousness by California Ground Squirrels.* Ph.D. dissertation. Davis, CA: University of California.

Thompson, N. S. (1986) Ethology and the birth of comparative teleonomy. In Campan, R. and Dayan, R. *Relevance of models and theories in ethology,* ed. Privat, International Ethological Conference.

Tinbergen, N. (1952) "Derived" activities; their causation, biological significance, origin and emancipation during evolution. *Quarterly Review of Biology, 27,* 1–32.

Toates, F. (1980) *Animal behaviour: A systems approach.* New York: John Wiley.

Vencl, F. (1977) A case of convergence in vocal signals between marmosets and birds. *American Naturalist, 111,* 777–782.

Weldon, P. J. (1983) The evolution of alarm pheromones. In Muller-Schwarze, D. and Silverstein, R. M. (eds) *Chemical signals in vertebrates, Vol. 3.* pp 309–312 New York: Plenum Press.

West, M. J. and King, A. P. (1985) Learning by performing: An ecological theme for the study of vocal learning. In Johnston, T. D. and Pietrewicz, A. T. (eds) *Issues in the ecological study of learning* (pp. 245–272). Hillsdale, NJ: Lawrence Erlbaum Associates.

West, M. J., King, A. P., and Freeberg, T. M. (1994) The nature and nurture of neo-phenotypes: A case history. In Real, L. A. (ed) *Behavioral mechanisms in evolutionary ecology* (pp. 238–257). Chicago, IL: University of Chicago Press.

West, M. J., King, A. P., and Freeberg, T. M. (in press) Building a social agenda for the study of bird song. In Snowdon, C. T. and Hausberger, M. (eds) *Social influences on vocal development.* (1997) Cambridge: Cambridge University Press.

Wickler, W. (1962) 'Egg-dummies' as natural releasers in mouth-breeding cichlids. *Nature, London, 194,* 1092–1094.

Wiley, R. H. (1982) Adaptations for acoustic communication in birds: Sound transmission and signal detection. In Kroodsma, D. E. and Miller, E. H. (eds) *Acoustic communication in birds, Vol. 1* (pp. 131–170). New York: Academic Press.

Wiley, R. H. (1983) The evolution of communication: information and manipulation. In Halliday, T. R. and Slater, P. J. B. (eds) *Animal Behaviour, Vol. 2: Communication* (pp. 156–189). Oxford: Blackwell.

Wilson, E. O. (1975) *Sociobiology: The New Synthesis*. Cambridge, MA: Harvard University Press.

Wingfield, J. C., Ball, G. F., Dufty, A. M., Jr., Hegner, R. E., and Ramenofsky, M. (1987) Testosterone and aggression in birds. *Am. Sci., 75,* 602–608.

Wingfield, J. C. and Hahn, T. P. (1994) Testosterone and territorial behaviour in sedentary and migratory sparrows. *Anim. Behav., 47,* 77–89.

Young, J. Z. (1954) Memory, heredity, and information. In Huxley, J., Ford, A. C., and Hardy, E. B. (eds) *Evolution as a Process* (pp. 281–299). London: Allen and Unwin.

Zahavi, A. (1977) Reliability in communication systems and the evolution of altruism. In Stonehouse, B. and Perrins, C. M. (eds) *Evolutionary Ecology* (pp. 253–260). London: Macmillan.

Zuk, M. (1991) Sexual ornaments as animal signals. *TREE, 6,* 228–231.

Chapter 12

COMMUNICATION AND NATURAL DESIGN

Nicholas S. Thompson

Departments of Biology and Psychology
Clark University
Worcester, Massachusetts 01610-1477

ABSTRACT

The basic value of the natural design perspective is that it provides a rich language for the description of social interactions that is neither covertly mentalistic nor barrenly behavioristic. A natural design analysis begins by clearly defining and identifying those properties of behavior that require special explanatory principles. Communication is design to mediate a design. Behavior between two animals is communicative only if one of the two animals has a behavioral design, *and* the other animal displays structures or behaviors that mediate this behavioral design *and* these structures or behaviors are evidently designed for their mediating role. Many phenomena in animal behavior are communications by this definition. Two classic cases are analyzed from the natural design perspective, the alarm calls of the vervet monkey and the song of the white-throated sparrow. This analysis makes clear that these two cases differ not in the referentiality of the communication, as is often supposed, but in the kinds of designs that are mediated.

INTRODUCTION

Confusions between description and explanation often seem to occur in conversations concerning communication. If we say that a bird's song or a baby's

Perspectives in Ethology, Volume 12
edited by Owings *et al.*, Plenum Press, New York, 1997

cry is a communication, are we saying how the behavior came to be? Or are we merely providing a way to classify the behavior, leaving for later the work of explaining it. The view presented in this paper is that communication is a descriptive category and that our first job in doing research on communication must be to identify the formal properties that define it.

For the last two decades I have been developing an approach to the study of animal behavior that I call the natural design perspective. The natural design perspective holds that before we engage in evolutionary, physiological, or developmental *explanation* we must first engage in teleonomic *description*: we must first identify those properties of behavior that demand special explanatory principles. If you drop a dead bird out of a tenth story window, simple physics will suffice to describe the corpse's trajectory. But if you drop a living bird out of the same window, what happens will move you to evolutionary, motivational, physiological, and developmental explanations. Providing a means to describe the special properties of organisms—*a priori to the explanation of these properties*—is what the natural design perspective seeks to do (Sommerhoff, 1950).

My basic strategy in developing this perspective has been to turn it loose on problems in the field of animal behavior and show how it helps to rationalize our practices as ethologists, sociobiologists and comparative psychologists. Often in ethology a thread of reasonableness connects our practices but the reasons we give for these practices are incoherent. Looking at the world from a natural design perspective helps to make intelligible many anomalies in the study of animal behavior and evolution. It makes clear why evolutionary explanations of behavior seem so often to be circular (Thompson, 1981) and how these explanations, while rarely strictly speaking circular, are often abused (Lipton and Thompson, 1988). It explains why the idea of a teleonomic science of biology went awry (Thompson, 1987) and it rationalizes the central concerns of ethology (Thompson, 1986) and comparative psychology (Thompson, 1987) as well as peculiarities in their historical development. It points to a defect in how Hempelian explanations are applied to behavior (Derr and Thompson, 1992), even by the master himself (Hempel and Oppenheim, 1948), and shows a way around Brentano's irreducibility thesis in the study of intentionality (Thompson and Derr, 1993). It elucidates what is uniquely puzzling about the natural phenomenon called deception and assists in developing an account of animal/human play (Mitchell and Thompson, 1986). It helps to demonstrate what is and what is not useful about "manipulation" as an evolutionary concept (Thompson and Derr, in press).

Broadly speaking, each of these situations is characterized by a confusion between description and explanation. Many of the key concepts of ethology are ambiguous: do they identify a phenomenon that needs to be explained or do they identify the causes of a phenomenon that has been already identified? For instance, when we say that a structure is "adapted" or a behavior "motivated," are we saying something about its *form* or are we saying something about its *causal history*? Knowing the difference is crucial because if we think we have explained—when we have in fact only described—we are left thinking that we know more than we actually know about the phenomenon in question. One of the advantages of the natural design perspective is that it clearly separates as descriptive such concepts as motivation and adaptation and separates them from their associated explanatory concepts, reinforcement and natural selection. Thus, the perspective should be useful in clarifying any theoretical area of ethology where descriptive and explanatory concepts tend to be confounded.

A BRIEF REVIEW OF THE NATURAL DESIGN PERSPECTIVE

In its most general form, *design* is an association between two arrays, an array of structures and an array of uses. The most easily communicated examples are structures of material in space. For instance, consider the design of the tools in a mechanic's toolbox. The box contains specialized instruments, each tool appropriate to one of the circumstances the mechanic encounters, each fitted to the form of a particular type of bolt, screw, socket or fitting. When the mechanic wishes to turn a number 8 Phillips head screw, he reaches for a Phillips head screwdriver of the appropriate size; when he wishes to loosen a 15 mm bolt, he reaches for a 15 mm wrench, and so forth. Watching the mechanic at his work, one might say that a particular wrench is well designed for turning a particular bolt. As design is here understood, attribution of design in this situation means more than saying that the wrench turns the bolt: one can, after all, turn the bolt with pliers. It means that there is in general a relationship between the form of the tools in the box and the circumstances in which they are employed and that the wrench is exemplary of that relationship.

Although the most common uses of the concept of design are to describe a property of human artifacts, not all designed objects are made by humans. Organisms, their behavior, and their artifacts are also designed. *Natural design* is an association between an array of forms of organisms, behaviors, or artifacts and the array of their circumstances. Natural design comes in at least four forms, forms that I have called *adaptation*, *development*, *motivation*, and now, *communication*.

Adaptation is correlation between the form of organisms and their circumstances. If we, like Darwin and Wallace, travel from place to place, studying the form of organisms in relation to their circumstances, we will find correlations between the structure and behavior of organisms and the circumstances under which they live. We speak of an organism as being "adapted," when it is an example of such a correlation. So, for instance, we speak of a polar bear's coat being adapted to Arctic life not just because the polar bear is white and the Arctic is white, but because in general there is a correlation among mammals between

```
CIRCUMSTANCES                          TRAIT

Snow on ground most of year->          White fur year round

Snow of ground part of year->          varying fur color
                                               |
                                               |
                                               |
                                       +-----------------------+
                                       | winter ------> white  |
                                       | summer-------> brown  |
                                       +-----------------------+

Snow rarely on ground------->          Brown fur year round
```

Fig. 1. The adaptation of fur color to environment color.

coat color and environment color, and the polar bear's coat exemplifies that correlation. This correlation we can plainly observe without counting the number of cubs in the polar bear's den, and it is, I insist, what we actually MEAN by adaptation.

Development is correlation between age-specific behaviors and age-specific circumstances in which those behaviors typically occur. Before we say that development is occuring in any individual, we must first have observed in the species of which the individual is a member a typical sequence of behavior traits that are deployed by a growing organism as well as a typical sequence of contexts in which the organisms grow. Development consists in a meshing between these sequences such that a particular behavior is regularly deployed at the time of life when a particular environment is present.

For instance, as a kitten increases in size, it typically deploys different clusters of behaviors: first rooting and groping, then crawling, then walking, then straying short distance from the nest, then approaching strange objects, then wandering farther and farther from the mother and finally hunting entirely on its own. These behavior patterns occur in the face of a gradually altering environment provided by the mother. Early on, she frequently nurses and licks the kittens, assists them in elimination and restrains their attempts at independent locomotion. Gradually, however, she attends to them less frequently, approaches and restrains them less. Ultimately, she begins actually to fend them off when they try to nurse and leads them on extended sorties away from the nest. Each stage in the kitten's behavior is fitted to a corresponding stage in the mother's behavior. What makes the kittens' growth an instance of development is not only that the kittens display stereotyped sequence of behaviors as they get larger or that the environment displays a regular sequence of changes, but that there is a correlation between events in behavior and events in environment such that a particular cluster of behaviors is regularly deployed in each particular environmental circumstance. Notice that as we distinguished between adaptation and its consequences above we here distinguish between development and its consequences. Effectiveness of the behavior in the circumstance is not part of the definition. Young animals often deploy behaviors such as sexual behaviors in circumstances in which they are not obviously effective: they do not bring about their adult consequences. Still, if these behaviors typically occur as part of an age-related sequence of behavior changes that are related to changes in the young animal's environment, they are constituents of development, by definition.

Motivation consists in two nested levels of design. If we study the individual animal in its natural habitat as it conducts its daily and yearly round, we can develop its "ethogram," the catalogue of its behaviors and the circumstances under which those behaviors are deployed. We will discover, in general, at least two levels of correlation in an ethogram. At the first level, is the correlation between the animal's motives and its circumstances. In the prolonged absence of water, the animal seeks water; in the prolonged absence of food, it seeks food; during the appropriate seasonal conditions, it seeks a sexual partner. At the next level of design, each of these motives consists of correlations between the animal's immediate circumstances and the behavior it deploys, each combination of circumstance and behavior increasing the likelihood of the achievement of the animal's purpose. So, in an animal that is seeking food, there is a variety of circumstances, food present, the odor of food present, the presence of stimuli which have been related to food in the past, to which the animal will respond in characteristic ways, ways that all share the property of increasing the likelihood that food will be consumed.

In this usage, the term "motivated" is synonymous with the word "purposive." In earlier publications, I have avoided the term "motive" because it invites circular reasoning. As the term is often used, the term appears to refer indifferently to the purposive organization of behavior and to the physiological and neural mechanisms that underlie that organization. But it cannot refer to both, because if it does, then the statement that motivation organizes behavior is empirically empty. Thus I have often used the word "purpose" to refer to the formal property of goal directedness that I wish to discuss. Unfortunately, for some, the term "purpose" carries its own unfortunate baggage of referring to conscious intent. Since I have terminological trouble whichever word I choose, I have decided to lay claim to "motive" for my own purposes; i.e., *to refer ONLY to the goal-directed organization of behavior, and NOT to the causes of that goal directedness, whatever they may be.*

My conception of motivation is based on Sommerhoff's analogy to a self-aiming gun (Sommerhoff, 1950). Imagine a gun mounted on the prow of a ship. Further imagine that the gun is *designed* to shoot at the target: i.e., there is an array of target positions and an array of gun positions and a function between these two arrays such that the projectile from the gun is likely to strike the target.

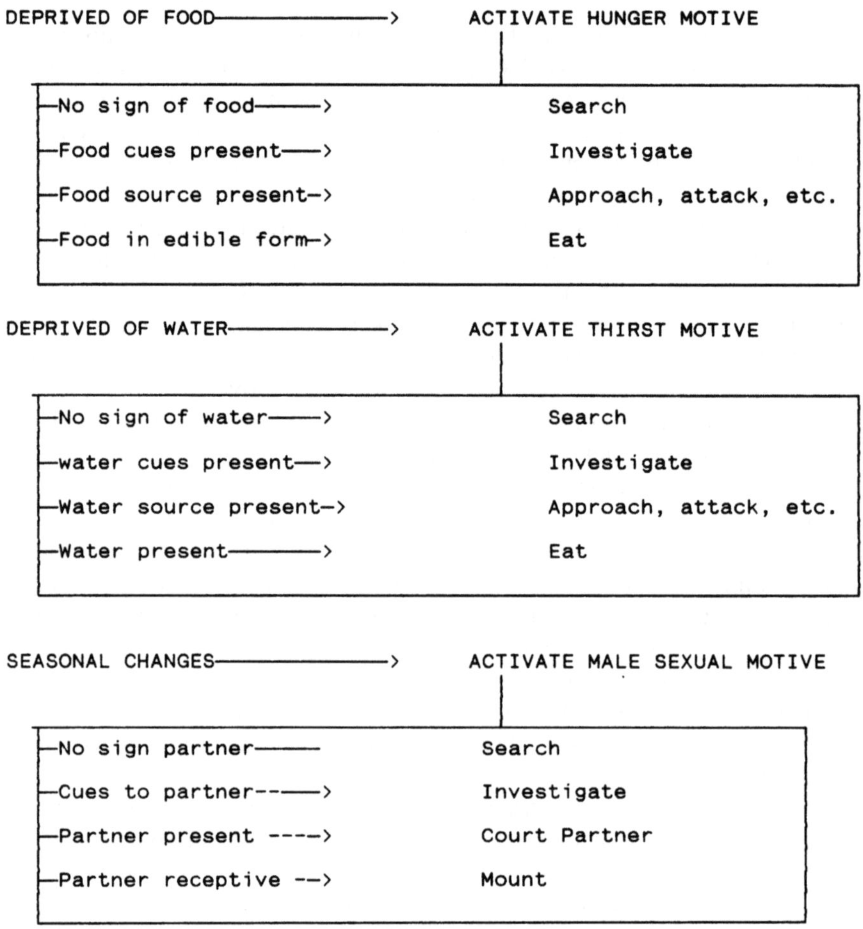

Fig. 2. MOTIVATION: A complex two-level design.

Second, imagine a gun that is designed to shoot at surface targets. Once again there is an array of target positions and an array of gun positions and a correlation between them. But the function that connects the two arrays will be different in the two kinds of gun because a surface target is capable of different sorts of motion from an aerial target. For one thing, the trajectory of an aerial object is much more ballistic than that of a surface object. A surface object is much more able to twist and turn because of the greater mass of the material it is attached to.

Third, imagine a self-aiming gun that shoots at subsurface targets, such as torpedoes. As before, there is an array of target position and an array of gun positions and a function that relates them, and as before, this function must be different because of the medium in which the target is moving. In this case, the function must not only take account of the greater possibilities for evasive maneuver due to the target's contact with the water, but it must also take account of the water surface's capacity to refract light and its capacity to deflect a projectile that approaches at too shallow an angle.

Finally imagine that a military engineer has the bright idea of mounting a single gun on the prow of the ship that can shoot at all three kinds of targets. Guided by a complex computer program, this single weapon could behave in accordance with three different designs: one for aerial targets, one for surface targets, and one for subsurface targets. Each design would relate particular trajectories of the target to particular firings of the gun; and each of these designs would, in turn be related to the type of target. So, the gun would display two levels of design, the lower level that relates trajectory to firing and the higher level that relates the lower level design to target type.

The motivational structure of behavior is analogous to such a multifunction gun. To say that an animal is thirsty is to say that the animal "shoots at water": there is an array of circumstances relevant to water, an array of behaviors relevant to water-seeking, and a function that connects the two arrays such that the animal has a high probability of finding water. To say that an animal is hungry is to say that the animal "shoots at food": there is an array of circumstances relevant to food, an array of behaviors relevant to food-seeking, and a function that connects the two arrays such that the animal has a high probability of finding food. To say that an animal is sexually motivated is to say that the animal "shoots at sex": there is an array of circumstances relevant to sex, an array of behaviors relevant to sex-seeking, and a function that connects the two arrays such that the animal has a high probability of finding a sexual partner.

Of the three, hunger, thirst, and sex, sex is the only motive that is essentially social. Social motives characteristically involve a meshing of the design arrays of the participants: i.e., the behaviors of one participant's design arrays become the circumstances for the others'.

Like the multifunctioned gun, the animal displays a second, higher level of design. At this higher level, the design of the purposive system relates circumstances to the deployment of the lower level designs. Each of its designs, its design for water-seeking, its design for food-seeking and its design for seeking a partner of the opposite sex, is itself matched to a set of circumstances, the absence of water, the absence of food, and the seasonal circumstances appropriate to mating.

This conception of multiple hierarchical layers of design is a useful way to *describe* many of the phenomena that ethologists and sociobiologists are required to explain.

WHAT SORT OF A DESIGN IS COMMUNICATION?

Communication is a form of design to mediate a design.[1] Thus, for a given act to be a communication between two animals, at least three conditions must be met: (1) One of the animals must have a behavioral design; (2) the second animal must have a structure or behavior that mediates that design; and (3) the mediating structure or behavior must show evidence of design for that role. Cries for help are examples of human communications. When a drowning man cries for help he mediates a design between his circumstances and the behavior of his rescuers. Lifeguards are people whose behavior is designed for the protection of swimmers: that is, they have an array of behaviors that are matched to circumstances in such a way as to protect from drowning the swimmers under their responsibility. They scan the water constantly, they intervene and prevent horseplay and other activities that carry a risk of drowning, they call swimmers from the water who seem to be beyond their depth or endurance, and, if necessary, they enter the water to rescue a swimmer who seems likely to drown. The function of a successful cry for help is to switch the behavior of the lifeguard to behavior appropriate to a drowning emergency, even though the lifeguard herself may not be able to see that such an emergency is in progress. The effect of the call is to invoke the design of her behavior so that her behavior becomes appropriate to the actual circumstances of the swimmer, even if those circumstances are not otherwise perceived. Even though the lifeguard may not be able to see the caller, she still begins to behave as a person in the presence of a drowning emergency.

Because the cry for help mediates a design, it meets the first two criteria for communication. It also meets the third: it is designed to mediate. Crying "Help!" is part of an array of behaviors that a swimmer might deploy upon finding himself in trouble. He might stop thrashing and try to float, he might look around for some sort of floating debris, he might pace himself and try to swim slowly toward a nearby shore. Or he might cry for help. Each technique will be deployed depending on the circumstances of the swimmer. If shore is near, he might swim for it; if a floating buoy is nearby, he might cling to it, or if a lifeguard is nearby, he might call for help. Thus, evidence in the behavior of drowning men supports the conclusion that crying help is behavior that is designed to mediate lifesaving.

By definition, communications always promote the appropriateness of behavior to the situation that obtains. The effect of the cry for help is to correct

[1] By mediate is not meant "negotiate" but something closer to what physiological psychologists invoke when they speak of "mediating mechanisms." An example of a mediating mechanism in this sense is the transmission of an automobile, which stands between the engine of the car and the wheels and mediates the relationship between engine speed and wheel speed.

a mismatch in the array of the lifeguard. Before the cry for help, the lifeguard was deploying general scanning behaviors, behaviors that were appropriate for a nonemergency situation. But, in fact, an emergency was in progress. Unbeknownst to the lifeguard, a man was drowning. To that extent, the lifeguard's behavior was momentarily ill-designed. After the cry for help, the lifeguard deploys lifesaving behaviors, behaviors which are appropriate for the actual state of affairs.

But not all uses of "Help!" are communications; some are deceptions. A deception is a design to defeat a design (Thompson, 1986).[2] Imagine that the hollering swimmer is a distant admirer of the lifeguard who wishes to meet her. What if he was only pretending to be drowning in the hope of getting her to swim out to him? In this case, the swimmer's design array would be quite different, an array of techniques for making new acquaintances of which crying "help" in a crowded swimming pool might be one. But the lifeguard's design array would be the same. The only difference for her would be that—prior to the cry for help—her behavior would be in keeping with the actual situation in the water before her. Thus, the effect of his cry for help would be to make her behavior incompatible with the actual situation before her. In this sense, his deceptive behavior is behavior designed to defeat the water-safety design of her behavior, not to mediate it.

Communications can be complex or simple depending on the complexity of the designs they mediate. The example of the swimmer is simple because it involves only two designs: first, the matching between the array of water-safety techniques and the array of water situations that constitutes the lifeguard's design for lifesaving; second, the matching between the array of swimmer's self-preservative techniques and the array of emergency situations that constitutes the swimmer's design for drowning-avoidance. One of the lifeguard's techniques is to respond to cries for help and one of the swimmer's techniques is to call for help. But imagine that our swimmer, instead of just crying out "Help" had called out, "I may be in a little bit of trouble, please keep an eye on me." This example reveals that nested within the technique of calling for help is a possible range of vocalizations, each related to the situation of the swimmer: "Throw me a line" might be appropriate for a swimmer who was experiencing leg cramps but whose arms were OK. "I'm going down," might be the thing to say to elicit an immediate rescue attempt. Choice among these alternative vocalizations has the effect of eliciting a second level of design from the lifeguard: selecting the appropriate rescue technique from among the array of rescue techniques in her repertoire.

[2] Defeat is here used in the sense of "subvert": "I defeated the design of the lawn mower safety power cutoff by tying the dead-man release bar to the handle of the mower with a bit of twine."

DO ANIMALS COMMUNICATE?

From a natural design perspective, animals communicate if their behavior manifests design to mediate design. Their behavior displays design to mediate design if:

- One of the two animals has a behavioral design; and
- The other animal manifests structures or behaviors that mediate this behavioral design; and
- These structures or behaviors are evidently designed for their mediation role.

Many phenomena in animal behavior are communications by this definition. Designs to mediate designs can come in the form of development, adaptation, or motivation. For instance, an example of adaptations that mediate designs are species-specific colorations or behaviors that mediate courtship behaviors in conspecifics. They are adaptations in the sense that each species displays structures or behaviors that are matched to the sensitivities of courtship partners of the same species. They are mediators in the sense that they facilitate a matching between the behavior of the partner and the species identity of the "communicator." They are designed for mediation in the sense that they are suited to the perceptual capacities of the receiver.

An example of developments that mediate designs are age-specific structures or behaviors that facilitate designs of full-grown conspecifics. For instance, many young primates have pelage colorations that change dramatically at the end of infancy. Adults in the troop are designed to behave discriminatively to young animals when they are infants and less so thereafter. The coat change is a developmental communication in the sense that it facilitates a matching between the age of the young primate and the behavior of its fellow group members. It is designed for mediation in that the infant color coat sharply contrasts the infant with the coat colors of the adults around it.

The most interesting and controversial examples of designs to mediate design are those that mediate motivational designs. With the rest of this essay, I would like to consider two sorts of mediated motivational designs: those where the mediated design is to events external to the mediator and those where the mediated design is itself a design of the mediator.

Designs Mediated to Events in the Environment of the Mediator ("Referential Communication")

One broad case of mediated motivational designs are those that make the receiver's behavior well-designed for some state of affairs in the world outside

CIRCUMSTANCES	TRAIT
Leopard--------------------->	Run for trees
Eagle----------------------->	Look up, run for brush
Snake----------------------->	Stand tall, look around feet.

Fig. 3. Vervet monkey's direct predator aversion design.

the sender or the receiver. This kind of design mediation is best illustrated by the alarm calls of the vervet monkey (Cheney and Seyfarth, 1990). The vervet monkey makes three acoustically distinct alarm calls, one to large predatory mammals such as leopards, one to eagles, and one to snakes. And each call is responded to in a manner appropriate to the predator. Vervets seeing a snake or hearing a snake call, stand tall and look in the grass about their feet; vervets seeing a leopard or hearing a leopard call, run to trees, and vervets seeing an eagle or hearing an eagle call look up and/or run into brush.

There are three designs immediately evident in this situation: between the evasive behavior of the caller and the type of predator (Figure 3), between the communicative behavior of the caller and the type of predator (Figure 4), and finally between the communicative behavior of the caller and the behavior of the hearer (Figure 5). None of these designs seem to constitute communication per se.

The effect of these three designs is to mediate a fourth. Together they assure that the call hearer's *behavior* is well-designed for the caller's *circumstances*. (Please see Figure 6.) When the caller sees a snake, the hearer deploys snake-appropriate behavior; when the caller sees an eagle, the hearer deploys eagle-appropriate behavior; and when the caller sees a leopard, the hearer deploys leopard-appropriate behavior. The caller's vocalizations link the hearer's responses with the caller's circumstances. In effect, they "stand in" for the caller's circumstances in producing an appropriate response on the part of the hearer.

CIRCUMSTANCES	TRAIT
Leopard in sight------------>	Give "Leopard" call
Eagle in sight-------------->	Give "Eagle" call
Snake in sight-------------->	Give "Snake" call

Fig. 4. Vervet monkey's prey signaling design.

CIRCUMSTANCES	TRAIT
"Leopard" call ------------->	Run for trees.
"Eagle" call --------------->	Look up, run for brush
"Snake" call --------------->	Stand tall, look around feet.

Fig. 5. Vervet monkey's call response design.

Thus, the vervet's system is a particularly clear case of "design that mediates design."

But is it design *to* mediate design? Observations on vervets provide two kinds of evidence that its calls are designed for their effects. The first is evidence that "design mediation" is one of a variety of techniques that the vervet monkey uses under various circumstances that have a common outcome. As a social species the vervet has many behaviors that are beneficial to fellow group members and induce them to behavior cooperatively (Figure 7). For instance, vervets not only warn one another, they groom one another, and they collectively mob predators when group members are attacked. That warning behaviors are part of a higher-order design reassures us that they are not simply behaviors that mediate design but are behaviors that are designed to mediate design.

The second kind of evidence that vervets' calls are designed for their effects is evidence that the signalling behavior itself belongs to a class of behaviors that are pitched to the receptor- and response-capacities of their hearers. In other words, we should find that design mediators as a class share properties and are different, as a class, from behaviors designed for their effects on rocks, water, and other inanimate objects. Such differentiation is evident in the calls of the vervet and not in the direct evasive actions of the vervet (Figure 8). The characteristics of displays in general and vocal displays in particular are predictable from the receptor characteristics and perceptual sensitivities of the receiving organism. Because alarm calls share this "hearer friendly" property with other vervet vocal displays, we are led to say that the vervet's calls not only

CIRCUMSTANCES	TRAIT
Leopard present, not visible>	Run for trees
Eagle present, not visible-->	Look up, run for brush
Snake present, not visible-->	Stand tall, look around feet

Fig. 6. Vervet's *mediated* predator aversion design.

```
CIRCUMSTANCES                        TRAIT

parasites on skin----------->        groom

predator at distance-------->        deploy design to mediate
                                     design (predator calls)

        Leopard in sight ---------> Give "Leopard" call
        Eagle in sight -----------> Give "Eagle" call
        Snake in sight -----------> Give "Snake" call

predator attacks------------>        mob
```

Fig. 7. Vervet monkey's co-group member protection design.

have the effect of mediating a design between the behavior of the hearer and the circumstances of the caller but that they are designed to do so. This second level of design is essential. A deaf monkey might avoid predators appropriately by using the evasive actions of other vervets as cues, but we would not say that the evading monkeys were communicating unless they also provide evidence of design for mediation in their behavior.

This discussion of the alarm calls of the vervet monkey illustrates the basic approach of the natural design perspective to communication. Communication is described in terms of the design arrays of the participants and the manner in which communicative behavior serves to place those design arrays in harmony each other and with the environment.

Designs that Mediate Designs between Social Partners

Many instances of "communication" are behaviors that mediate designs between the behavioral designs of two or more social partners. A simple and familiar example is the song of the white-throated sparrow. The white-throated

```
CIRCUMSTANCES                        TRAIT

Design mediators------------>        "receiver friendly": showy bright,
                                     loud, tuned.

Other behaviors------------->        Not showy, not bright, not loud.
```

Fig. 8. Design properties of designs to mediate design.

sparrow is said to communicate three sorts of information with its simple song: species identity, individual identity, and motivation. The simplicity and tonality of the song are related to the species identity of the singer, the particular arrangement of tones to the individual identity of the singer, and the length of the song to the motivation to defend the territory (Falls, 1969).

The details of white-throated sparrow territorial defense were unknown to me at the time of writing, so what follows must be understood as a plausible hypothetical. If white-throated sparrows are similar to many other territorial songbirds, they deploy two behavioral designs with respect to territories: one is the design to defend territory and the other is the design to avoid territorial conflict. The design to defend territory consists of a graded series of defensive behaviors each matched to the behavior of the intruder (Figures 9 and 10). The design to avoid territorial conflict consists of a graded series of evasive behaviors, each matched to the behavior of the territory holder. They are matched to clues of the location of the territory holder, including those provided by song. They permit a nonterritorial bird to survive and feed in the interstices of established territories or territorial birds to trespass on the territories of neighbors in search of food, extra pair copulations or other resources without attracting the attention of territory holders. How an encounter goes between a territory holder and an intruder will depend on the designs of the two birds when they encounter one another. If one bird's behavior is designed for territorial defense and the other for territorial conflict avoidance, then the territory holder will chase off the intruder and the interaction will be of short duration. If, on the other hand, both birds' behavior is designed for territorial defense, then there will be a fight and one of the birds will switch over into territorial conflict avoidance design and be

CIRCUMSTANCES	TRAIT
No sign of intruder	Deploy song: i.e., design to mediate territorial conflict coordinate design.
	attacked -----> avoid approached ---> retreat attacked -----> flee hear song ----> avoid no song ------> forage
Intruder seen near boundary -->	Approach, sing more
Intruder crosses boundary -->	Attack, sing more
Intruder reciprocates attack -->	Escalate attack

Fig. 9. Territorial defense design.

```
CIRCUMSTANCES                                  TRAIT

attacked --------------------->                flee

approached ------------------->                retreat

hear song -------------------->                avoid

no song ---------------------->                forage, explore
```

Fig. 10. Design to avoid territorial conflict.

chased off. Thus, the effect of the fight between the two animals will be to alter the design of one of the two from territorial defense to conflict avoidance design. Imagine a territorial intruder feeding undiscovered on a territory of a resident bird. Unless the intruder is challenging the owner for the territory, his behavior will be designed for avoiding confrontation. The song of the territory holder mediates that avoidance. In the absence of song, the territory intruder might inadvertently create a mismatch between his conflict avoidance design and his actual circumstances by blundering into the territory holder in the thick evergreen forests that the sparrows frequent. Song facilitates a correction of the mismatch by making it possible for the intruder to relate his behavior to the position of the territory holder even when he cannot see him.

RELATION OF THE NATURAL DESIGN PERSPECTIVE TO SOME OTHER CONTEMPORARY PERSPECTIVES ON ANIMAL COMMUNICATION

This redescription of two instances of animal "communication" is meant to demonstrate the potential of the natural design perspective to provide constructive ways of describing complex animal interactions that are not illicitly explanatory. What special advantages can be claimed for the natural design perspective? Here is not the place to provide a thoroughgoing comparison and contrast of the natural design perspective on communication in relation to its alternatives. But before closing I would like to provide a brief sketch of the direction that such an analysis might take.

The natural design perspective will have advantage over its alternatives just to the extent that it helps solve problems in the field of animal communication. What are those problems? They seem largely to be conceptual. Research on animal communication proceeds apace, but the researchers seem unable to agree on how to characterize the phenomenon they are studying. There are three prevalent characterizations, the information (Smith, 1977), manipulation (Dawkins, 1982), and management views.[3] Each has intuitive appeals and each has been heuristic in the sense that it has helped one or more researchers in the pursuit of a vigorous and productive research program.

But all three rely fatally on causal mentalism. The reliance on causal mentalism is not always as obvious as the kind some animal behaviorists have promoted (Griffin, 1976) and others would abjure (Davis, in press). I am not necessarily accusing animal communication theorists of thinking that animals are aware of their own intentions in the way that vernacular psychology suggests that humans are. But I do think all three views invoke a communicator's privileged knowledge of his own mental states as a crucial explanatory link, and that notion is faulted in two fundamental ways (Thompson, 1995).

The first way in which causal mentalism is faulted is that mental states are not the sort of entities that can cause behavior. In principle, causes must be distinguishable from the events they cause. If mental states are to be the causes of behavior then they must be specifiable independent of behavior, and there is no way to specify mental states independently of the relationships between circumstances and behavior that causal mentalists would have them explain. Thus, behaviors are constituents of mental states, not results of mental states, and a state cannot logically be said to cause its constituents (Derr and Thompson, 1992).

The second way in which causal mentalism is faulted is that it requires some sort of inherent privilege in the communicator's knowledge of "own" mental states, and there is no way to be sure that such a privilege exists (Thompson, 1995). The difficulties with the idea of privileged access to own mental states become clear when one thinks about what would be involved in any information-processing system knowing itself. Whatever system one invented to answer questions like, "What am I like?", the system could not answer

[3] One peril of writing an essay while interacting with such an illustrious company as the authors of this volume is that one may change one's views faster than one can write them down. One case in point is my view of W. John Smith's informational perspective on communication. While I am confident that I have heard many colleagues use the concept of information in the way I describe here, I am less confident than I was two months ago that W. John Smith is one of them. On the one hand, he has specifically and vehemently rejected the mentalistic and commodity transfer implications that I attribute to those who make use of an information account. On the other hand, in my own reading of his work, I cannot myself make sense of it without supplying those implications. Since deadlines made it impossible for Professor Smith and I to meet to sort this issue out before publication of this essay, it is my deep and abiding hope that some philanthropist will invite us (along with all the authors of this volume and any of our devoted readers who would care to come along) to some gracious Italian lakeside, some stark Austrian eyrie, or some luxurious Caribbean villa so that we may have appropriate surroundings in which to take on the painful labor of coming to an agreement on the many issues raised by this volume.

questions strictly speaking about itself. It could answer questions about the larger system of which it was part, but these answers would not be infallible.

Reversing the traditional sociobiological flow, I tend to think of individual humans as analogous to organizations, such as, say, The Administration In Washington. Everyday questions like, "What does the Administration think about the situation in the Middle East?" are asked and answered, but such exchanges are shot through with category errors. It is never the "Administration" that actually answers questions about itself but a specialist who is delegated to answer such questions. He reports his own view based on the partial information made available to him. He is not reporting about himself, nor does he have infallible information about the subject of his report. His answer is privileged only in that he is practiced at answering such questions and in that he may—or may not—have better sources of information than those "outside" the administration. Similarly, when you ask an individual to speak for him- or herself, it cannot logically be the individual that you hear from but a specialist subsystem that is designed to speak for the individual. That subsystem does not report about itself nor does it have infallible information about the subject of its report.

The three alternative views on communication vary in the degree to which they take as their point of departure the commonsense view of communication that is embedded in our ordinary language use of the term. The vernacular view of communication holds that the message is something apart from the communication of the message. In some sense we know the message before we communicate it. So, of the vervet monkey, a vernacular psychologist would say, the monkey sees the danger and communicates it with the appropriate call apart from his own reaction to it. Or the white-throated sparrow knows its own territorial motivation and consciously tries to communicate it with the appropriate number of triplets tagged onto the end of its song.

Closest to the vernacular view of communication is the informational view. It holds that the communicating animal is selected or reinforced for transferring information to the receiver about itself. According to this view, the vervet's bark reveals her disposition to flee ("There is a fair probability that I am about to take strong snake evasive action") and the sparrow's song his disposition to attack ("If you fly one more wing beat toward my territory I will come over there and beat the feathers off you"). What distinguishes the informational from the vernacular view is that its adherents can disclaim any implication of conscious intent in communication. They can say that the sender has been reinforced or selected for making this information available, but that it has no conscious intent to do so.

But the implication of privileged access still remains. For information-transfer views of communication to work, the communicator has to know his own dispositions apart from his enactment of them. While information-transfer theorists need not share the vernacular psychologist's view that there is conscious awareness of the information communicated by vervets and white-throated sparrows, they do imply that the communicator has accurate knowledge of self that may be communicated. This is essentially the doctrine of privileged access.

Dawkins' manipulation view starts out as an attack on the information-transfer view of communication. It takes as its model a dung beetle's pushing around of its bit of dung (Krebs and Davies, 1978). When animals interact, said Krebs and Dawkins, they are no more "communicating" with one another than a dung beetle is communicating with a dungball it rolls into its burrow. Just as the dung beetle uses the properties of the dungball to its own advantage, so communicating animals use each other's properties to their own advantage. The only difference is that in the case of the dung beetle the properties used are physical properties whereas in the case of the communicating animals, the properties used are properties of the nervous system.

But the dung beetle image is one of those aggressively reductive images that fails to stop the argument because it fails to address the issue the argument is about. When we are talking about communication we are trying to distinguish those properties of the behavior we are calling communicative that distinguish that behavior from rolling dungballs. The fact is that the rules of rolling a dungball and manipulating a nervous system are different and no account of the behavior we call communicative that does not characterize those differences will be very useful. Characterizing those difficulties requires considering how organisms differ from dungballs and such a consideration leads directly to the issues of design raised in this paper. One way out of this quandary is to assert that the crucial difference that a nervous system makes between a dungball and an organism is just that nervous systems are information processing systems and, therefore, if one would manipulate a nervous system, one would best do it by providing INFORMATION. But this "way out" leads directly to the information-transfer view that I have just discussed and that the manipulation theorists currently under discussion have ridiculed.

Dawkins and Krebs became aware of this problem and extended (1984) their notion of manipulation to include a special kind of manipulation relevant to the communication situation—manipulation by failing to disclose information. They treat the signaler as analogous to a poker player deciding to deploy a poker face (p. 397). But to make this idea meaningful we have to imagine that the animal has privileged access to his own intentions. And as I argued above, and elsewhere there is no particular reason to believe that the sender has information about what he is disposed to do that is *in principle* better than information possessed by his antagonist.

Of the three views of communication, the assessment/management perspective (Owings and Hennessey, 1984; Owings, Swaisgood, and Rowe, 1994; Rowe and Owings, 1990; Owings and Morton, this volume) is the natural design perspective's most formidable competitor because it comes closest to avoiding explanatory mentalism. It does so by employing an extension of the idea of homeostasis. Both participants in an interaction are viewed as creatures attempting to maintain critical variables within acceptable limits. The feedback loops that accomplish these ends may be narrower or wider. They may extend only to the reserves of the creature, or they may extend out into physical world around the creature, or they may extend still further into the social world of the creature. So, for instance, a hungry infant animal may manage its blood sugar account by metabolizing some starches, or by eating some foodstuffs that are to hand or by crying out until an adult animal brings it food. The signals of the crying infant work only because adults in its vicinity are constantly assessing its degree of need. Because the effectiveness of a communication presupposes another individual who is attempting to assess variables revealed by the communication, the view is called the management/assessment view. Of the three competing models, I am most attracted to the assessment/management perspective, perhaps because it is the only one of the three that is rooted primarily in physical imagery rather than in psychological imagery. But its weakness, from my point of view, also arises directly from this strength. Organisms differ from physical systems precisely in their possession of higher-order properties that I have called design. While many physical buffering processes serve to control variables in nature within limits, these physical processes aren't designed to do so: they just do. In the example of the baby above, the baby's cry does not simply control its parents, it is designed to do so; and the parents do not simply assess the baby's food state, they are designed to do so. And no account of the relation between them will be complete until this mutual design is described (Watson, 1995).

Moreover, the proponents of the M/A view have flirted with explanatory mentalism in the development of their perspective. Owings and Morton (present volume) reintroduce the concept of information in the course of their characterization of assessment because they feel that gathering information about states is the essence of assessment. But this extra grammatical level—"about states"—is entirely unnecessary. These authors would not feel that they had to introduce that extra step into an account of the monitoring of blood acidity by the carotid sinus: they would not feel that they needed to say that the carotid sinus gathers information "about" blood acidity; they would say only that the carotid sinus assesses blood acidity. So why do they feel they need to say, for instance, that the displayee monkey monitors information "about" the displaying monkey's probability of attack? Why can't they say that one monkey assesses another's probability of attack?

In response to these criticisms, Don Owings (personal communication) has argued that the natural design perspective on communication has its own weakness. In a commentary on an earlier draft of this manuscript, Owings suggested that there are many instances of designs to mediate design that I would have to recognize are *not* communications and therefore that my criterion is inadequate to pick out accurately only instances of what I would call communication. The example that he offered was of a parent carrying a child to the dinner table. The child is presumably designed to respond to the sight and smell of food by eating and the parent's act is designed to create the conditions to realize that design. The problem is not limited to one guilefully contrived human example. I can add many ethological examples, such as a mother stroking her baby's cheek with her nipple in order to induce nursing or a lamb's butting his mother's teats in order to induce milk letdown.

The objection is troubling, but I think it can be dealt with by insisting on a precise understanding of what it is for a process to mediate between two other processes. The word "mediate" is often wrongly used to mean "facilitate" or even simply "elicit." Its precise meaning is "to set up a relation between." The parent who moves the highchair or the mother who strokes the baby's cheek or the kid that butts his mother's flanks are all examples of behaviors that elicit a design, but except as these acts may be considered to be communications, they do not *mediate* designs. The intuition that all three examples capitalize on is that it is possible to elicit designed behaviors from another creature without communicating with it. But there is an important difference between eliciting designed behaviors and mediating a design. To be a mediator, a process must "stand in" for another process, in the way that the vervet's snake vocalization "stands in" for the snake in producing snake avoidant responses in hearers of the "snake call."

WHY THE NATURAL DESIGN PERSPECTIVE IS PREFERABLE

The basic value of the natural design perspective is that it provides a rich language for the description of social interactions that will substitute for and make unnecessary the use of intentional idioms in the study of animal behavior. Intentional idioms are expressions that contain a proposition about the world as the object of a mental verb such as wanting, knowing, believing, thinking and informing. Intentional idioms are an essential feature of causal mentalism since the mental terms that govern these idioms appear as causes in mentalistic accounts of behavior. The difficulty with intentional idioms is that the propositions they contain display two odd properties: existential inexistence and referential opacity (Thompson and Derr, 1993). If we say that Jones is human, we are prepared to commit ourselves both to the proposition that Jones exists and to the proposition that Jones has all the properties of a human being. But if we say that Smith thinks that Jones is human, we are not prepared to make either of those commitments because Smith may have imagined Jones and/or Smith may not know all the properties of a human being. In short, the line between intentional and nonintentional idioms is important because the possibilities of inference that arise from the propositions that appear within intentional idioms are very different from the possibilities of inference that arise from propositions made outside such idioms.

The differences in the possibilities of inference are crucial because science can only work with propositions that make possible substitution with truth. To take a clear contemporary example of scientific progress, one way of understanding the progress that has been made in our understanding of AIDS in the last decade is that we have been able to substitute more accurate descriptions of the agent that causes the disease. In the first place, when the disease first thrust itself upon our consciousness, we weren't sure what sort of an agent it was: environmental, bacterial, or viral. Once we knew it was a virus, we had to learn what sort of a virus it was. Nowadays, we are able to describe it as a very specific virus with a unique reproductive strategy. All the way through this latter process of specification, we have been able to make substitutions of the new information into our causal account while preserving the truth that HIV causes AIDS. How much more difficult to do science around the proposition that the American public believes that HIV causes AIDS! While we might hope that the American public would change its behavior as scientific knowledge about the virus changes, we can never predict that outcome in the scientific sense because we cannot confidently substitute into the statement "the American public believes that HIV causes AIDS" all the new descriptions of the AIDS virus that science has achieved. We cannot, for instance, infer that the American public knows that AIDS is caused "by an RNA virus with a conserved reverse transcriptase enzyme" because we have no way of inferring how much the American public knows about the new discoveries. Thus, as they are usually construed, intentional idioms such as "beliefs that," "knowledge that," or "information that" have no place in the scientific explanation of behavior.

These problems are difficulties not only for contrived human examples but also for describing the behavior of real animals in real situations. For instance, if a vervet gives an "eagle bark" and the other vervets leap for the bushes, then in the absence of the natural design perspective, we would have to choose between two undesirable descriptions. Avoiding intentional idioms, we could say that "eagle barks" are sounds that occur with high relative frequency in situations where a vervet begins to orient toward an eagle and are followed by a high frequency of eagle escape behaviors both in the caller and in animals near enough to hear the bark. Or adopting intentional idioms, we could say that "eagle barks" are given by vervets to inform other vervets that there is an eagle present. The first description fails to communicate the organization of what is taking place and the second opens us up to all the perils of explanatory mentalism. The whole trick of the communication debate seems to be to discover if there is anything of scientific value in the second account that is not in the first, or, perhaps more specifically, which of the things that the second account adds are scientifically valid.

Two things that the second account clearly adds are referential opacity and intentional inexistence: Nothing in the vervet's eagle bark assures either that an eagle exists or that the calling vervet is trying to inform other vervets that a member of the species *Stephanoaetus coranatus* is approaching. After all, the caller may be a juvenile misapplying the call or an adult abusing it and, in any case, we cannot expect that the caller has read up on his avian taxonomy.

Consider another example. If a macaque gives an open mouth threat, other macaques in the troop may attack or retreat depending on their relative dominance and who else is within reach that might help or oppose them in the altercation. A nonintentional account of these facts is that open mouth displays in macaque monkeys are followed with high relative frequency by attacks or retreats from monkeys toward which the threat was directed depending on their relative dominance. An informational account might be that the displaying monkey is informing the displayee that if the displayee takes another step forward the displayer has a high probability of beating the tar out of the displayee. The informational account once again adds the perils of intentionality: nothing guarantees that any such probability exists and nothing guarantees that the caller will beat the tar out of an animal that is the most recent grooming partner of the alpha male in the troop because the displayer may be wrong about his own probability of attack or so ignorant of the recent grooming history of the displayee as to know that that animal is the "most recent grooming partner of the alpha male of the troop."

The effect of introducing intentional language is to introduce a whole set of pseudoproblems about a commodity called "information" that is moved around when animals interact. People who talk about information transfer in animal interactions are not the first people to be accused of commoditizing interactions. In the late 1970s, the anthropologist Roy Wagner, *The Invention of Culture*, ridiculed the concept of culture as a commoditization of human interactions. He argued that culture was invented when 19th-century Western anthropologists encountered preliterate peoples and attempted to describe what they saw. Because of the societies from which they came, these anthropologists fell back on the imagery of the industrial life. They saw the preindustrial people as engaged in developing, manufacturing, and ultimately trading a product. That product was "culture." But of course, that product was an invention of the anthropologists. The peoples they studied, Wagner argued, were not engaged in any way in the project of developing, manufacturing, or trading their culture. The culture was simply a way of viewing the patterns of interaction among them. And the focus of anthropologists on culture caused them to neglect the description of human interaction for arguments concerning the origins of the structures that underlie the culture.

Similarly, the concept of interaction transfer can be seen as a commoditization of animal interactions. When we ethologists encounter nonhuman animals and attempt to explain what we see, we fall back on the imagery of economic life. We see the animals as acquiring, transmitting, and receiving a product. That product is information. But, of course, that product is an invention of ethologists. The creatures we study are not engaged in acquiring, transmitting, and receiving information. Communication is simply a way of characterizing the patterns of interaction among them. And the focus of ethologists on communication as a commodity transfer has caused them to neglect the description of animal interaction for arguments concerning the nature of the information that is transmitted in communication, where that information is before it is communicated, and other forms of silliness.

Given the downside to the information account, why do people use it? What are the benefits of introducing intentional language? Why is the correlational account inadequate? One inadequacy is that the correlation account leaves out a level of organization in the behavior. It's not simply that the calling vervet is responding in the presence of the eagle and the hearing vervet is responding to the calling vervet; the whole system seems to be organized to assure that the hearing vervet avoids eagles that the calling vervet sees. Similarly, it's not just that the displaying monkey responds to the displayee's proximity by giving the open mouth display and the displayee responds to the display by retreating or advancing, depending on the dominance situation; the whole system seems to be organized to assure that the displayee avoids an attack that the displayer's behavior foreshadows.

The informational account captures this higher level of organization by representing the caller vervet or the displaying macaque as attempting to inform and the hearing vervet or the displayee monkey attempting to gather information. But it pays, as I have argued, a terrible price of introducing a whole clutter of phantom issues.

So, the challenge for the future of the study of animal communication is to provide an account that captures that higher level of organization in animal interactions without the disabilities of the informational account. That the natural design perspective can accomplish this trick is what makes it preferable to other contemporary perspectives. Two heuristic consequences will flow from thinking of communication as a form of natural design: i.e., as a formal property of behavior, not as an event that explains behavior. First and foremost, it will turn attention away from phantom causes and commodities—away from what the animal is thinking or what information is being conveyed—and toward describing the patterns in the behavior that we can actually see. Second, it will postpone for the time being questions of selection. For most ethological purposes, selection on communication is a largely a hypothetical entity that is revealed only through its proposed effects on the patterns of behavior we observe in nature. Most of what we know about selection on communicative behavior is inferred from how that behavior is designed. Until the designs we call communicative are well described, invocation of selection to explain them is largely premature. How can we hope to bring natural selection to bear to explain a phenomenon before we have thoroughly described and identified the phenomenon we are explaining?

If the natural design perspective were to be widely and self-consciously adopted then all research on animal communication would have to be grounded in comparative study of ethograms—the sort of work that Niko Tinbergen did, for instance. Laboratory work would have to be closely coordinated with observation in the natural habitat. And less effort would be expended on models (they can, after all, always be made to work) and in the search for the crucial experiment or anecdotes (all experiments and anecdotes can, after all, be reinterpreted). Our understanding of animal behavior would arise only from the broadest possible description of the varieties of behavior that animals deploy and the circumstances under which they are deployed, and the relationships between them—i.e., in the study of the varieties of behavioral design.

ACKNOWLEDGMENTS

Patrick Bateson, Michael Beecher Patrick Derr, Chris Evans, Don Owings, W.J. Smith, Caleb Thompson, and many others all contributed valuable commentary during the development of this chapter. The chapter was begun while I was an occasional visitor in Peter Marler's Animal Communication Laboratory and in Don Owing's Animal Behavior Seminar at the University of California at Davis in 1992–93. I am deeply obligated to them for their hospitality.

REFERENCES

Cheney, D. L., & Seyfarth, R. M. (1990). *How monkeys see the world.* Chicago, IL: University of Chicago Press.

Davis, H. (in press). In R. Mitchell, L. Miles, & N. Thompson (Eds.), *Anthropomorphism and anecdotalism.* Albany, NY: State University of New York Press.

Dawkins, R. D. (1982). *The extended phenotype.* San Francisco, CA: Freeman.

Derr, P., & Thompson, N. S. (1992). Reconstruing Hempelian motivational explanation. *Behavior and Philosophy, 20,* 1.

Falls, J. B. (1969). Functions of territorial song in the white-throated sparrow. In R. A. Hinde (Ed.), *Bird vocalizations* (pp. 207–232). Cambridge, United Kingdom: Cambridge University Press.

Griffin, D. (1976). *The question of animal awareness.* New York: Rockefeller University Press.

Hempel, C. G., & Oppenheim, P. (1948). Studies in the logic of explanation. *Philosophy of Science, 15,* 135–175.

Krebs, J. R., & Davies, N. B. (Eds.). (1978). *Behavioural ethology: An evolutionary approach.* Oxford, United Kingdom: Blackwell.

Lipton, P., & Thompson, N. S. (1988). Comparative psychology and the recursive structure of filter explanations. *International Journal of Comparative Psychology, 1,* (4).

Mitchell, R. W., & Thompson, N. S. (1986). Deception in play between dogs and people. In R. W. Mitchell & N. S. Thompson (Eds.), *Deception: Perspectives on human and nonhuman deceit.* New York: State University of New York Press.

Owings, D. H., & Hennessey, D. F. (1984). The importance of variation in sciurid visual and vocal communication. In J. O. Murie & G. R. Michener (Eds.), *The biology of ground-dwelling squirrels* (pp. 169–200). Lincoln, NB: University of Nebraska Press.

Owings, D. H., Swaisgood, R. R., & Rowe, M. P. (1994). *Context and animal behavior IV: The contextual nature of a management/assessment approach to animal communication.* Preprint supplied by the first author.

Rowe, M. P., & Owings, D. H. (1990). Probing, assessment, and management during interactions between ground squirrels and rattlesnakes, Part 1: Risks related to rattlesnake size and body temperature. *Ethology, 86,* 237–249.

Smith, W. J. (1977). *The behavior of communicating.* Cambridge, MA: Harvard University Press.

Sommerhoff, G. (1950). *Analytical biology.* Oxford, United Kingdom: Oxford University Press.

Thompson, N. S. (1981). Toward a falsifiable theory of evolution. In P. P. G. Bateson & P. H. Klopfer (Eds.), *Perspectives in ethology, 4.* New York: Plenum Press.

Thompson, N. S. (1986). Ethology and the birth of comparative teleonomy. In R. Campan & R. Dayan (Eds.), *Relevance of models and theories in ethology* (pp. 13–23). Toulouse, France: Privat, International Ethological Conference.

Thompson, N. S. (1986). Deception and the concept of natural design. In R. W. Mitchell & N. S. Thompson (Eds.), *Deception: Perspectives on human and nonhuman deceit.* New York: State University of New York Press.

Thompson, N. S. (1987). The misappropriation of teleonomy. In P. P. G. Bateson & P. H. Klopfer (Eds.), *Perspectives in ethology, 6.* New York: Plenum Press.

Thompson, N. S. (1987). Natural design and the future of comparative psychology. *International Journal of Comparative Psychology, 101* (3), 282–286.

Thompson, N. S. (1994). The many perils of ejective anthropomorphism. *Behavior and Philosophy,* pp. 59–70.

Thompson, N. S., & Derr, P. (1993). The intentionality of some ethological terms. *Behavior and Philosophy.* Double issue, *20* (2) & *21* (1).

Thompson, N. S., & Derr, P. G. (in press). On the use of mental terms in behavioral ecology and sociobiology.

Wagner, R. (1975). *The invention of culture.* Englewood Cliffs, NJ: Prentice-Hall.

Watson, J. (1995). Mother-infant interaction dispositional properties and mutual designs. In N. S. Thompson (Ed.), *Perspectives in ethology, 11.* New York: Plenum Press.

INDEX

When indicated by *f, t*, or *n*, the subject may be found in a figure, table, or footnote, respectively.

MIX
Papier aus verantwortungsvollen Quellen
Paper from responsible sources
FSC® C105338

If you have any concerns about our products,
you can contact us on
ProductSafety@springernature.com

In case Publisher is established outside the EU,
the EU authorized representative is:
Springer Nature Customer Service Center GmbH
Europaplatz 3, 69115 Heidelberg, Germany

Printed by Libri Plureos GmbH
in Hamburg, Germany